INTRODUCTION TO ENZYMOLOGY

INTRODUCTION TO
ENZYMOLOGY

ALAN H. MEHLER

National Institutes of Health, Bethesda, Maryland

1957

ACADEMIC PRESS INC • PUBLISHERS • NEW YORK

PREFACE

Enzymology is a term that covers a broad area of biology and chemistry. The information accumulating in this area is of interest and importance to workers and students with diverse backgrounds and objectives. This book is designed to develop in the reader sufficient familiarity with enzymes that he is encouraged to continue his contact with the rapidly expanding body of knowledge about biological catalysis and is prepared to include information about enzymes in his scientific thinking and to apply this information to his work. The organization of this book has evolved from a course in enzyme chemistry taught by the author at the National Institutes of Health, where the classes consist of investigators primarily interested in clinical medicine, organic chemistry, bacteriology, biophysics, plant physiology, and other subjects, in addition to students in biochemistry.

Enzyme chemistry is not a neat package that can be presented without loose ends. It is possible to develop a theoretical generalized representation of enzymes and illustrate the theory with examples. I believe it is more descriptive of the status of the field and more meaningful for the student to describe individual enzymes and enzyme systems in some detail and to use the information at hand to develop concepts. Accordingly, topics have been selected to indicate the large variety of chemical reactions known to be catalyzed by enzymes.

For the most part these topics have been arranged as components of metabolic sequences in order to emphasize the interrelations of enzyme activities, the chemical mechanisms employed by biological systems, and the multiple factors to be considered in interpreting biological phenomena. A large part of biochemistry has been included, but practical considerations force the elimination of many topics that have intrinsic interest and illustrative value equal to those included.

In each section certain aspects have been stressed to illustrate points of general significance. Since many such points can be established with several different systems, the treatment given to some topics may appear exaggerated, while that of others seems unduly perfunctory. I regret particularly that this uneven treatment eliminates many references to work of great value in the advancement of the field. The references included, however, are not intended to serve as complete documentation for the statements in the text; they are selected only to assist the reader

v

in making his acquaintance with the original literature and to help him gain some historical perspective.

The rapid growth of modern enzymology has not yet permitted complete descriptions of biochemical processes at the enzyme level. Some of the gaps and limits of our knowledge are indicated in this text. The reader will undoubtedly find that some of the voids have been filled during the time when this book was being prepared. The chemistry of the enzymes themselves may also be expected to be revealed in elaborate detail in the near future. It is my hope that the interpretations I have included in the following pages will serve as a logically sound basis for supplementation, rather than contradiction, by the research of the future.

In the preparation of this text I have been greatly aided by many of my colleagues at the National Institutes of Health and other institutions. I am happy to acknowledge the contributions of Drs. Bruce N. Ames, G. Gilbert Ashwell, Simon Black, Jules A. Gladner, Joseph L. Glenn, Ronald C. Greene, Osamu Hayaishi, Leon A. Heppel, William B. Jakoby, Leonard Laster, Bruce Levenberg, Elizabeth S. Maxwell, R. Carl Millican, Jesse C. Rabinowitz, Earl R. Stadtman, DeWitt Stetten, Jr., Celia W. Tabor, Herbert Tabor, Gordon M. Tomkins, Edith C. Wolff, Barbara E. Wright and Charles I. Wright, who have improved the text through scientific and literary suggestions. I am particularly grateful to Dr. Sidney P. Colowick for his criticisms on the entire manuscript, and to my wife, Anne, for her assistance in the technical aspects of producing this book and for her careful nurturing of manuscript and author.

ALAN H. MEHLER

Bethesda, Maryland
September, 1957

CONTENTS

CHAPTER I

INTRODUCTION

In recent years the study of enzymes has expanded to such proportions that not only has it become a specialized field, but it has become an important element in studies in many other disciplines. The present body of knowledge about enzymes has been derived from workers who entered the field from many directions, and consequently is composed of fragments which at first glance appear to have little relation to each other. Indeed, a systematic organization of sciences has no need for a category "Enzyme Chemistry." Nevertheless, certain elements of physical chemistry, organic chemistry, physiology, microbiology, and other well-defined areas have been applied to an area of special interest, and from an expanding study of biochemistry, enzymology has grown until its special problems and interests warrant special texts. The true justification for this arbitrary selection of information is the interest of the investigators who have focused their attention on enzymes as the active, and therefore interesting, element of biological systems.

Reactions in which biological material participates, now known to be enzyme catalyzed, were studied throughout the development of modern chemistry. The term catalysis was coined by Berzelius (1836) to describe the effect of substances that cause reactions to occur in their presence. In addition to inorganic reactions, the fermentation of sugar was included among the original examples of catalysis. The existence of substances that digest meat had been known since the experiments of Spallanzani (1783) and de Reaumur (1752). The conversion of starch to sugars was shown by Kirchhoff (1814) to be caused by an extract of wheat. In 1830 Robiquet and Boutron demonstrated hydrolysis of the glycoside amygdalin by a factor in bitter almonds.

The term catalysis was applied to the biological phenomena listed above and to several other reactions discovered in the next several years. The active component of bitter almonds was named *emulsin* in 1837. Other carbohydrate-splitting activities (ptyalin in saliva, amylase in malt) had already been described. Pepsin and trypsin, protein-digesting agents from the stomach and pancreas, were also discovered during this period. The activities of these materials were contrasted with the materials responsible for fermentation. Early theories of Willis and of Stahl to explain fermentation as a disruption caused by violent motion of

1

particles of the fermenting substance were adapted by Liebig to form a chemical theory of fermentation. Several investigators, notably Schwann, Cagniard-Latour, and Kützing, were meanwhile determining the dependence of fermentation on living cells. Their publications led to a controversy that raged for over twenty years, until Pasteur (1850) provided convincing evidence that fermentation was associated with living, growing organisms. The catalytic agents active only as components of living cells were designated *organized ferments*, while the cell-free activities were called *unorganized ferments*. The latter were named *enzymes* by Kühne.

Modern enzyme chemistry may be dated from the successful studies of Buchner, who in 1897 obtained cell-free preparations of yeast that were capable of fermenting sugar. Before Buchner's work there had been numerous intensive attempts that had failed either because of loss of activity or because of the difficulty in removing all viable cells. Indeed it is possible that Pasteur himself, who argued uncompromisingly for the association of fermentation with living cells, failed to anticipate Buchner only because he did not happen to use an appropriate strain of yeast. Buchner successfully defended his conclusions against all attacks, and as his work was confirmed, the enzyme theory became accepted. Today the word *enzyme* is used to designate all biological catalysts; the infrequently used *ferment* has the same connotation.

SOME GENERALIZATIONS ABOUT ENZYMES AS CATALYSTS

Catalysts accelerate the rates of chemical reactions. In the case of many enzyme-catalyzed reactions there is no detectable reaction in the absence of the catalyst. It should be noted, however, that enzymes do not add energy to a reacting system; their function is only to influence the rates, not the extents of reactions. Enzymes are remarkable among catalysts for their specificities and for their efficiencies of catalysis at low temperatures. A given enzyme generally affects only one type of chemical bond, usually in a restricted group of compounds. Absolute specificity, the ability to react with only one compound of all similar compounds tested, is frequently observed. All catalysis is positive, in that the catalyst increases a rate of reaction. So-called negative catalysis does not exist except as an indirect result of a positive reaction. If one reaction modifies a system by changing one or more of the reagents of a second reaction, the second reaction may appear to be inhibited by the catalyst of the first. There are, however, no mysterious forces by which a negative catalyst can prevent the interaction of other molecules.

Enzymes are the organic catalysts elaborated by all organisms. As catalysts, they are defined by their activities; primarily, all enzymes must be defined in terms of rates of chemical reactions. The rate of an enzyme-catalyzed reaction is proportional to the concentration of enzyme when all other conditions are constant (Fig. 1). Ultimately enzymes may be described as particular molecules, and measured in terms of some constituent, weight, etc., but the measurement of any property other than activity may fail to give a true determination of the amount of enzyme in a preparation because subtle alterations in the enzyme molecule may either increase or decrease the activity. At the present time there is no complete chemical description of even the simplest enzyme. One of the objectives of current research is to learn the relation between structure and function of enzymes. Until this goal is reached, the kinetic definition of each enzyme is the only one that describes the enzyme.

The kinetic description of an enzyme requires the measurement of a rate of a reaction. Although examples are known of enzyme reactions at

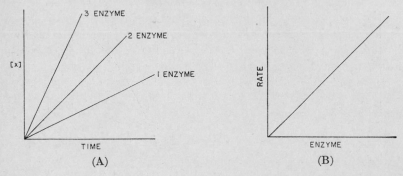

FIG. 1. (A) represents the increase in the concentration of x, the product of a reaction, with time in the presence of 1, 2, and 3 increments of enzyme. The slopes of the lines of (A) may be used as rates to give the plot of (B), which shows proportionality of rate to enzyme concentration.

interfaces, in general enzymes act in aqueous solutions. The compound or compounds initially present in solution which are modified through the agency of an enzyme are called *substrates*. The substrates are converted to *products*. To measure the rate of a reaction, it is necessary to measure the concentration of either a substrate or a product. The rate of change in concentration with time is a measure of the enzyme.

Reaction Rates. There are many embarrassing examples in the enzyme literature of errors in the assay method. These are generally avoided if it is ascertained that a rate of reaction is measured, and *that this is proportional to* enzyme concentration. The rate of reaction may be linear with time, independent of the concentrations of substrates or products. Such reactions are termed *zero order*. If the reaction rate is proportional to the concentration of one substrate it is called *first order* (Fig. 2). These are the simple types of reaction that are convenient to use in kinetic studies. The rate of a zero-order reaction is given by the equation $x = kt$, and is evaluated as the slope of the line, x/t. Any points on the line thus suffice for the rate measurement. A first-order reaction rate changes continually with time, as the rate-limiting substrate is consumed. A simple relation between the concentration of substrate $[S]$ and the rate v is $v = k[S]$ when k is a constant. From this equation another may be derived that relates k to the amount of substrate present at given times. In this equa-

$$k = \frac{1}{t} \ln \frac{a}{a - x}$$

tion a is the concentration of substrate at the beginning of a time period, t, and x is the concentration at the end of the period. The constant k may be evaluated by substituting experimental values for a, x, and t.

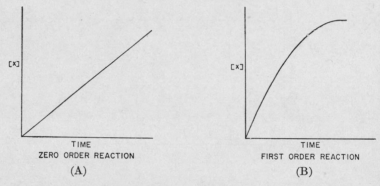

$[x]$ TIME $[x]$ TIME
ZERO ORDER REACTION   FIRST ORDER REACTION
(A)   (B)

Fig. 2. These curves illustrate the rate of formation of reaction product, x, when the reaction rate is (A) zero order or (B) first order.

It may be evaluated from any portion of the curve, and is proportional to the amount of enzyme. Graphically, a plot of log substrate concentration versus time gives a straight line with first-order reactions, and the negative slope of the line is proportional to the reaction constant. Occasionally it is not possible to use conditions that produce zero- or first-order rates. The rate may be a function of more than one substrate (second or higher order) or it may be influenced by the accumulation of products. The products may act as inhibitors or they may react back to yield the original substrate or other products. An obvious complication is the existence of additional reactions which may involve either substrates or products. When higher order reactions or other complications obscure rate measurements, various empirical devices are employed to establish an assay. The most familiar of these are measurement of initial rates that may tend to approach zero-order kinetics and therefore be proportional to enzyme, and the establishment of a reference curve of apparent activity versus enzyme concentration. The latter must be recalibrated whenever circumstances, such as state of enzyme purity, are changed. When amounts of enzyme are selected to give rates within the range used for calibration, empirical assay methods are often valid and extremely useful.

Valid Assays. A valid assay of an enzyme must measure the rate at which particular substrates are converted to specified products. Therefore, analysis of a single component of a system may give misleading results. Most importantly, a valid assay must measure a rate. Two values do not establish a rate because they do not describe the course of the

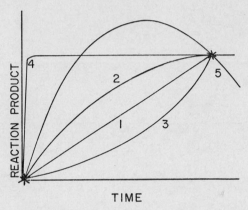

FIG. 3. Some representative routes to the same point. * = Measured values. Curve 1, zero-order reaction. Curve 2, first-order reaction. Curve 3, autocatalytic reaction. Curve 4, reaction gone to completion. Curve 5, secondary reaction.

reaction between the times of the analyses. One of the most common analytical errors is the measurement of a single value after a period of incubation of enzyme with substrate. It cannot be over-emphasized that single-point assays are not valid until the reaction has been studied thoroughly, and it is known that the points measured fall on a meaningful part of a curve (Fig. 3).

Finding that the amount of product formed or substrate removed is proportional to the amount of enzyme added is essential for a meaningful assay, but it is not sufficient to establish that the assay is valid, since a contribution of the enzyme preparation may be a substrate that determines the extent of a reaction.

Enzyme Units. Amounts of enzyme are conventionally expressed in *units*. A unit is an arbitrary designation that depends on the assay used. Under specified conditions a rate constant or any measure of substrate utilized or product formed in a given time can define a unit. Some conventional units are: micromoles O_2 consumed per hour, Δ optical density at a specified wavelength per minute, micrograms inorganic phosphorus (in phosphate) formed in 10 minutes, etc.

It is often desirable to express the degree of purification of an enzyme. When the chemical nature of enzymes was not known it was conventional to express the *specific activity* as units of enzyme per milligram dry weight. With the identification of enzymes with proteins, specific activity has been expressed as units per milligram nitrogen or per milligram protein as determined for example by the biuret reaction, by ultraviolet light absorption, by tyrosine (phenol) content, or by turbidity produced by protein precipitants. None of these methods is free from objections. Any of them may be of value for specific purposes and none may claim to give absolute values.

As mentioned above, units are defined for particular assay conditions. There are many conditions which must be controlled, including temperature, pH, ionic strength, specific ion concentrations, substrate concentrations, presence of activators, stabilizers, and inhibitors. The role of these factors will be illustrated for various enzymes in later sections. At this point it will merely be mentioned that there are no generalities that describe the effect of varying any of these conditions. Each enzyme must be studied as an individual case; some are indifferent to conditions that effect others profoundly, and some environmental changes, as temperature, influence competing phenomena, as rate of catalyzed reaction and rate of enzyme destruction.

Enzymes and Energy. The mechanisms by which enzymes act are only beginning to be discerned. Evidence to be discussed later for specific enzymes allows the generalization that enzymes participate directly in chemical reactions. In mediating a difficult over-all reaction, one with a large energy of activation, the enzyme does not supply energy. Instead, it circumvents the difficult reaction by substituting a sequence of reactions in which the enzyme participates, and which includes steps that proceed easily at ordinary temperatures. If the reaction $A + B \rightarrow C + D$ has a sufficiently high energy of activation, thermal agitation of the molecules of A and B at ordinary body temperatures may not be sufficient for the reaction to proceed measurably. If, however, the reaction $A + X \rightarrow AX$ proceeds easily, and $AX + B \rightarrow C + D + X$, a mechanism is provided for obtaining a net reaction in an indirect manner. The rate of reaction is increased according to the difference in activation energy of the uncatalyzed reaction and the rate-limiting step of the enzymatic pathway. In ways that are as yet not completely understood for any reaction, the activation energies for the partial reactions in an enzyme-catalyzed sequence are lower than the activation energy of the uncatalyzed reaction. The participation of an enzyme in chemical reactions that restore the enzyme to its original state has been described as a "cyclic" or "shuttle" process. Such processes may theoretically be repeated indefinitely, but in practice enzymes "wear out" through poorly defined reactions that may or may not be related to the reaction catalyzed.

Substrate Concentration and Rate of Reaction. In ordinary chemical reactions the rate of reaction increases with increase in concentration of reactants. The kinetics of enzyme-catalyzed reactions differ from those of uncatalyzed reactions because the enzyme must "adsorb" or "bind" substrates prior to conversion to products. The nature of the interaction between enzyme and substrate will be considered in some individual cases; in many systems the enzyme reacts efficiently with low concentrations of substrate (10^{-3} M or less), while in others the rate of reaction increases with substrate concentration up to very high concentrations. In general a maximum rate can be measured or calculated from the curve of rate versus substrate concentration. This maximum rate occurs when the concentration of substrate is sufficient to *saturate* the enzyme.

Michaelis Constant. A measure of the affinity of an enzyme for its substrate is the Michaelis constant, K_m. This constant is an equilibrium constant of the dissociation of an enzyme–substrate combination. It might be considered desirable to express enzyme–substrate affinity in terms of the amount of substrate needed to saturate the enzyme, but it is often difficult or technically impossible to measure the maximum rate directly, and it is obviously difficult to select the substrate concentration that just permits the maximum rate. These difficulties are avoided by determining instead the concentration of substrate required for permitting half the maximum rate. As described below, this value can be obtained by measuring reaction velocities in the presence of various rate-limiting amounts of substrate, without direct experimental determination of the maximum rate. The original formulation[1] for the determination of the constant now known as K_m assumes that the rate of equilibration between enzyme and substrate is fast compared with the subsequent reaction leading to products; that is, in the conversion of a compound A to B, the velocity constants k_1 and k_2 are large compared

$$A + \text{Enzyme} \underset{k_2}{\overset{k_1}{\rightleftharpoons}} A \cdot \text{Enzyme} \overset{k_3}{\rightarrow} \text{Enzyme} + B$$

with k_3. In this case the equilibrium constant,

$$K_{eq} = [A] \cdot [\text{Enzyme}]/[A \cdot \text{Enzyme}]$$

[1] L. Michaelis and M. L. Menten, *Biochem. Z.* **49**, 339 (1913).

determines the concentration of $A \cdot$ Enzyme at a given substrate concentration. The over-all rate of reaction is proportional to $A \cdot$ Enzyme concentration, and therefore is used to measure the amount of enzyme–substrate combination present at any time. Ordinarily the concentration of enzyme is very small compared with substrate, so it is assumed that A is not diminished through formation of $A \cdot$ Enzyme, and the constant is evaluated by determining the substrate concentration that permits a half-maximal rate. At this substrate concentration the concentrations of free and complexed enzyme are equal, and $K_{eq} = [A]$. This K_{eq} is K_m.

The derivation above contains assumptions that may not be justified. A more complete derivation shows that what is actually measured is $k_2 + k_3/k_1$, not k_2/k_1. If k_3 is *not* small compared with k_2, the measured K_m will not be a true dissociation constant.

In the first derivation of the Michaelis constant given, Enzyme was used to represent free enzyme, not that part bound to substrate. An alternate expression of the same concepts is obtained by letting $E = $ total enzyme, $ES = $ the concentration of complex, and $S = $ substrate whose concentration is essentially constant. Then

$$K_m = [E - ES]S/ES$$

and
$$ES = \frac{[E][S]}{K_m + S}$$

The velocity of the over-all reaction, v, is equal to a constant, k_3, times the concentration of ES. Therefore,

$$v = k_3[ES] = \frac{k_3[E][S]}{K_m + S}$$

When all of the enzyme is present as enzyme–substrate complex, the velocity is maximum, so for $k_3 E$ the term V_{max} may be substituted. Then

$$v = \frac{V_{max}[S]}{K_m + S} \tag{1}$$

or, rearranging,

$$K_m = \frac{V_{max}}{v} - 1 \tag{2}$$

Several methods have been used to evaluate the constants K_m and V_{max}. The one most often used is that of Lineweaver and Burk,[2] who rearranged equation (1) to obtain the form

$$\frac{1}{v} = \frac{K_m}{V_{max}} \times \frac{1}{S} + \frac{1}{V_{max}}$$

[2] H. Lineweaver and D. Burk, *J. Am. Chem. Soc.* **56**, 658 (1934).

When $1/v$ is plotted against $1/S$, a straight line is obtained with a slope of K_m/V_{max} and an intercept of $1/V_{max}$ (Fig. 4). Other useful formulations are

$$\frac{S}{v} = \frac{1}{V_{max}} \times S + \frac{K_m}{V_{max}} \qquad \text{(Lineweaver and Burk)}$$

$$\frac{v}{S} \times K_m = V_{max} - v \qquad \text{(Augustinsson)[3]}$$

Each substrate for a given enzyme for both forward and reverse reactions has a characteristic K_m under specified reaction conditions. It is possible to determine K_m values for all of the substrates involved in reversible reactions and to determine V_{max} in each direction. The reaction rates in the two directions may be so different that the amount of enzyme

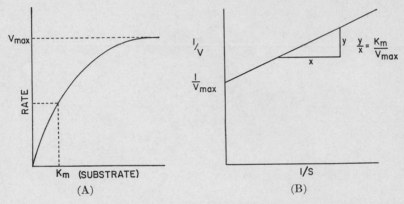

FIG. 4. (A) indicates the increase in reaction rate with increase in substrate concentration and constant amounts of enzyme. The maximum rate is obtained graphically, and K_m is determined from the curve as the value on the abscissa corresponding to ½ the maximum rate. (B) shows a typical Lineweaver-Burk plot.

required for measuring a rate in one direction may be many times the amount needed for the reverse reaction, but as long as the enzyme concentration is small compared with the substrate concentration, K_m values are not influenced by enzyme concentration. V_{max} values are proportional to the amount of enzyme used.

Inhibition. Graphic analysis of kinetic data according to the method of Lineweaver and Burk has a useful application in the study of inhibitors. Many substances limit the activity of enzymes by reacting with the protein or some other component in such a way as to destroy or decrease the catalytic ability. Other materials inhibit by forming the same sort of complex that a substrate does. In this latter case, the two materials

[3] K. B. Augustinsson, *Acta Physiol. Scand.* **15**, *Suppl.* 52 (1948).

compete for the enzyme, and the net rate of reaction is a function of the concentrations of substrate and inhibitor. A dissociation constant for the inhibitor, K_I, corresponds to the Michaelis constant of the substrate, K_m or K_s.

Lineweaver-Burk plots can be used to show that an inhibitor is a competitor with respect to substrate and to evaluate the affinity of the enzyme for the inhibitor. When plots are made for experiments carried out with and without inhibitor, two lines are obtained that extrapolate to the same intercept, V_{max}. The slopes of the two lines will be different; the presence of inhibitor causes the slope to be greater (Fig. 5).

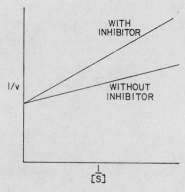

Fig. 5. Lineweaver-Burk plot of a reaction, with the rate measured at various substrated concentrations in the presence and absence of a competitive inhibitor.

The difference is in the apparent K_m; the apparent K_m is equal to

$$K_m + \frac{K_m \,(\text{Inhibitor})}{K_I}$$

It is often convenient to evaluate K_I by varying the inhibitor concentration and maintaining a constant substrate concentration. A formulation recently devised by Adams[4] shows that

$$\frac{1}{v} - \frac{1}{V_{max}} = \frac{K_s}{V_{max}S} + \frac{K_s I}{V_{max}SK_I}$$

When $1/v - 1/V_{max}$ is plotted against I, the intercept is $K_s/V_{max}S$ and the slope is $K_s/V_{max}SK_I$, so the ratio of intercept to slope is K_I.

Enzymes may be inhibited by materials that react with sites other than those that interact directly with a substrate and by reagents that bind sites irreversibly. When these types of inhibition occur, Lineweaver-Burk plots of the inhibited reaction do not have the same intercept as

[4] E. Adams, J. Biol. Chem. 217, 325 (1955).

found for the uninhibited system; V_{max} is decreased by the inhibitor. K_m may or may not be affected by a noncompetitive inhibitor, so the slope may or may not be altered. The difference in intercept is the usual criterion for distinguishing between competitive and noncompetitive inhibitions. A noncompetitive inhibitor may also compete with substrate so a qualitative observation of partial reversal of an inhibition by excess substrate is not adequate to describe the action of an inhibitor.[5]

Determinations of K_m and K_I have become routine in enzyme chemistry because of the theoretical and practical information derived from these values. From the theoretical point of view, the affinity of an enzyme for a substrate is important in determining the nature of the bonds between them, especially when the relative affinities for a variety of related structures (both substrates and inhibitors) can be determined. Variations in K_m with changes in the physical environment give additional information about the nature of the binding forces. Practically, the K_m permits decisions to be made about the concentrations required to obtain desired rates of reaction, the relative success of different enzymes that use the same substrate, the feasibility of using an enzyme assay for analysis of substrate at given concentrations, the ability of an enzyme to "pull" another reaction by removing a product, the relative rates of reaction when two substrates are present, and the relative effects of competitive inhibitors.

The Haldane Relationship. Another of the properties of enzyme systems frequently measured is the equilibrium constant of the over-all reaction. This is the means for determining a fundamental thermodynamic property, the free energy (F) of a reaction. Free energy will be discussed later. At this point a relation between enzyme kinetics and equilibrium is of interest. The equilibrium constant for a reaction

$$A \underset{k_2}{\overset{k_1}{\rightleftharpoons}} B$$

is $[B]/[A]$ and is also equal to k_1/k_2. With nonenzymatic reactions the velocities in the two directions are simply the products of reagent concentration and rate constant. In enzyme-catalyzed reactions, the velocities do not increase without limit as substrate concentration increases. The equilibrium constant can be evaluated as a function of reaction velocities, however, in the case of enzyme reactions also. The relationship between enzymatic reaction rates and chemical equilibrium was derived first by Haldane, and bears his name. Simple derivations were devised by Horecker and Kornberg and by Colowick, whose reasoning is

[5] A theoretical analysis of enzyme–inhibitor interaction and experimental applications to an enzyme system are described by A. Goldstein in *J. Gen. Physiol.* **27,** 529 (1944).

given. For the over-all reaction $A \rightleftharpoons B$, a Michaelis constant exists for each substrate:

$$E + A \rightleftharpoons EA \xrightarrow{k_{3_A}} E + B \qquad E + B \rightleftharpoons EB \xrightarrow{k_{3_B}} E + A$$

$$K_A = \frac{[E][A]}{[EA]} \qquad K_B = \frac{[E][B]}{[EB]} \tag{1}$$

$$\text{dividing:} \frac{K_A}{K_B} = \frac{[A][EB]}{[B][EA]} \tag{2}$$

$$\text{Rearranging} \frac{[B]}{[A]} = \frac{[EB]K_B}{[EA]K_A} \tag{3}$$

At equilibrium the rates of reaction in the two directions are equal. The maximum rates in each direction, V_A and V_B, are obtained when all of the enzyme is bound to one substrate. The rate of reaction in each direction at equilibrium is that fraction of V_{max} given by the actual enzyme–substrate complex divided by total enzyme.

$$V_A = E_{tot}k_{3_A} \qquad V_B = E_{tot}k_{3_B}$$

$$v = \frac{V_A[EA]}{[E_{tot}]} = \frac{V_B[EB]}{[E_{tot}]} \tag{4}$$

$$\frac{V_A}{V_B} = \frac{[EB]}{[EA]} \tag{5}$$

Substituting the terms of (5) into (3)

$$\frac{[B]}{[A]} = K_{eq} = \frac{V_A}{V_B} \times \frac{K_B}{K_A}$$

(The Haldane relationship between the equilibrium constant, maximum velocities, and Michaelis constants.)

For more complex reactions, the Michaelis constants for all components of the system must be included. When the substrates and products have similar affinities for the enzyme, the equilibrium constant approaches the ratio of the maximum rates, which can often be measured as initial rates. Examples will be shown later of reactions in which the correction of the ratio of rates by the ratio of Michaelis constants is more than an order of magnitude.

GENERAL REFERENCES

Alberty, R. A. (1956). *Advances in Enzymol.* **17**, 1.

Haldane, J. B. S. (1930). "Enzymes." Longmans, Green & Co., New York.

Huennekens, F. M. (1953). *In* "Technique of Organic Chemistry" (S. L. Friess and A. Weissberger, eds.), Vol. 8, p. 535. Interscience, New York.

Massart, L. (1950). *In* "The Enzymes" (J. B. Sumner and K. Myrbäck, eds.), Vol. I, Part 1, p. 307. Academic Press, New York.

A very thorough, analytical historical review of the embryonic and neonatal development of enzyme chemistry has been recorded elegantly by A. Harden in his monograph "Alcoholic Fermentation," Longmans, Green & Co., London, 1923.

Chapter II

HYDROLYSIS OF PEPTIDES AND PROTEINS

Enzymes that hydrolyze proteins and other compounds composed of amino acids were among the first biological catalysts to be discovered, and they have continued to be prominent in studies of enzyme structure, kinetics, activation, and mechanism of action. The crystallization of the enzyme urease by Sumner was followed by the crystallization of various proteolytic enzymes in the laboratory of Northrop. These studies established that catalytic activity is associated with what appear to be pure proteins; all well-defined enzymes isolated subsequently have also proved to be proteins, although many contain additional components. The study of enzymatic reactions involving proteins as substrate, therefore, gives insight into the chemical nature of enzymes as well as the mechanisms by which they act.

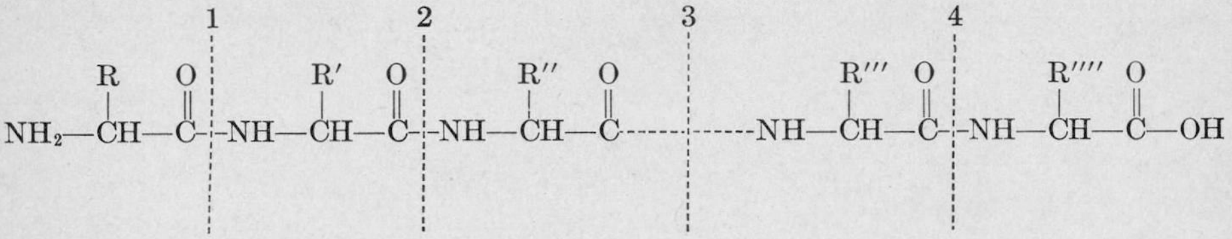

$$\begin{array}{ccccc}
1 & 2 & 3 & & 4 \\
\text{R} \quad \text{O} & \text{R}' \quad \text{O} & \text{R}'' \quad \text{O} & \text{R}''' \quad \text{O} & \text{R}'''' \quad \text{O}
\end{array}$$

NH$_2$—CH—C—NH—CH—C—NH—CH—C ------------ NH—CH—C—NH—CH—C—OH

Fig. 6. Peptide chain showing N-terminal linkage (1) and C-terminal linkage (4), which are hydrolyzed by appropriate exopeptidases, and additional peptide bonds (2) and (3), which are hydrolyzed only by endopeptidases.

Originally proteolytic enzymes (proteases) were studied by their ability to digest crude substrates; later highly purified proteins were used. Proteins are polymers of 20 different α-amino acids in which the amino acid residues form substituted amides. Amides of amino acids are called peptides. In addition to the proteases, peptidases, enzymes capable of hydrolyzing small synthetic peptides, were found by Fisher.[1] Considerable progress was made when Bergmann and his collaborators discovered that small synthetic substrates could also serve as substrates for proteolytic enzymes.[2]

The distinction between enzymes capable of attacking large protein molecules and those that attack only small peptides is not as complete as once thought. Two large groups of peptide-splitting enzymes have

[1] E. Fisher and P. Bergell, *Ber.* **36**, 2592 (1903).
[2] M. Bergmann, *Advances in Enzymol.* **2**, 49 (1942).

been designated endo- and exopeptidases. Endopeptidases characteristically split bonds adjacent to specific amino acid residues wherever they occur in a protein. Exopeptidases appear to be capable of splitting only terminal bonds in a peptide chain; some of these attack the terminal residues of intact proteins, while others react only with specific small molecules (see Fig. 6).

EXOPEPTIDASES

The hydrolysis of a peptide bond

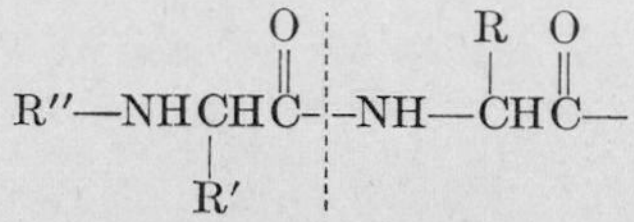

$$R''-NHCHC(=O)-NH-CHC(=O)-$$

(with R' on the first CH, R on the second CH)

can be followed in several ways. The products of hydrolysis are

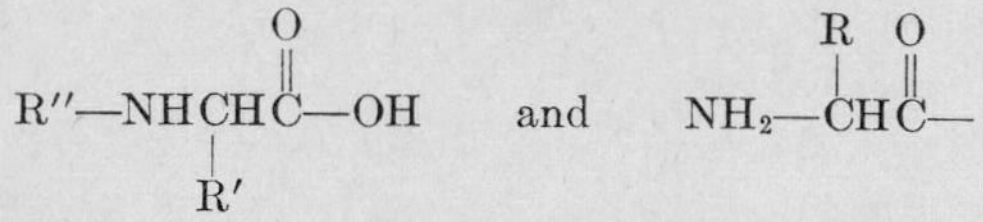

$$R''-NHCHC(=O)-OH \quad \text{and} \quad NH_2-CHC(=O)-$$

(with R' on the first CH, R on the second CH)

The appearance of the amino group can be measured by its reaction with nitrous acid, reaction with ninhydrin (colorimetrically or gasometrically), titration in acetone, and titration in the presence of formaldehyde. The appearance of a carboxyl group can be measured by titration in ethanol. Specific methods can be used for the determination of certain amino acids when these are liberated by hydrolysis.

Early work established that several different catalysts exist which differ in their abilities to attack various substrates. Glycylglycine and acylated glycylglycine were attacked by different preparations; dipeptides and tripeptides also differed in sensitivity to various enzyme preparations. As more enzymes were purified and more substrates became available and were tested, a number of peptidases were distinguished. The following descriptions are of some representative enzymes.

Carboxypeptidase. Carboxypeptidase occurs in extracts of pancreas.[3a] The pancreas contains an inactive precursor of the enzyme, a *zymogen*. In the case of carboxypeptidase the precursor has been partially purified. The accumulation of zymogens seems to be characteristic for those proteolytic enzymes that are secreted. In neutral solutions procarboxypeptidase is converted to active carboxypeptidase by the action of other enzymes present in the crude extracts. The mechanism of this activation is not completely known, although it appears that the reaction involves splitting of peptide bonds by other proteolytic enzymes.

[3a] M. L. Anson, *J. Gen. Physiol.* **20**, 663 (1937).

Activated carboxypeptidase is readily crystallized and recrystallized from dilute solution by proper pH adjustments. The recrystallized material contains only protein. This protein is composed of the usual amino acids, and to date there is nothing in the chemical analysis of the enzyme to indicate which structures confer the specific catalytic activity. The protein has a molecular weight of about 34,300 and has an isoelectric point at pH 6. Carboxypeptidase is inhibited by sulfide, cyanide, citrate, and other polyvalent anions. Inhibitions by these compounds usually indicate the presence of a metal ion cofactor, but such evidence is not conclusive. Corroborative evidence in cases in which metal is tightly bound to the protein is often difficult to obtain. Many times recrystallized carboxypeptidase was found on ashing to contain significant amounts of magnesium, but recent evidence has indicated that the activity of this enzyme depends on its content of tightly bound zinc.[3b]

The name carboxypeptidase is intended to indicate the requirement of the enzyme for an unsubstituted carboxyl group. A representative substrate is an acylated amino acid such as chloroacetyltyrosine. This substrate demonstrates the minimal requirements of carboxypeptidase:

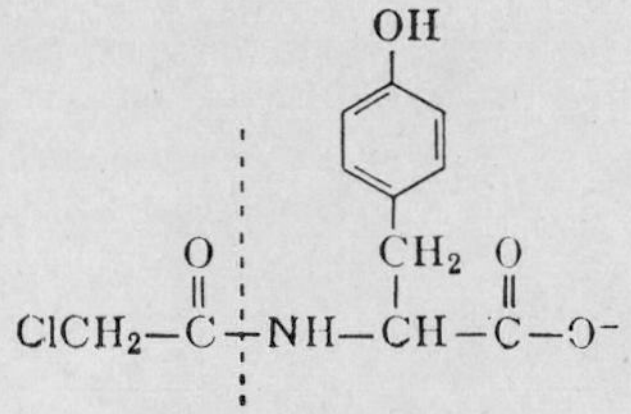

Chloroacetyltyrosine

a hydrolyzable bond involving a substituent on the carbon α to a free carboxyl group. If the carboxyl group is bound, as in an amide, the resulting molecule is not attacked by this enzyme. The hydrolyzable bond in most substrates studied is an amide, but esters, such as acetylphenyllactic acid, are also hydrolyzed.[3c] It seems probable that peptides are the natural substrates, since many susceptible molecules, including proteins, are found in the intestinal contents. The susceptibility of other types of bonds is of interest in determining the nature of the catalytic reaction. As yet other potential substrates, such as thioesters and appropriate ketones, have not been tested.

A large number of synthetic peptides have been used in studying the properties of carboxypeptidase (see Table 1). Modifications of the structure of the substrate have been found to influence both the affinity for the enzyme, K_m, and the over-all rate of reaction, k_3. The susceptibility

[3b] B. L. Vallee and H. Neurath, J. Am. Chem. Soc. **76**, 5006 (1954).
[3c] J. E. Snoke and H. Neurath, J. Biol. Chem. **181**, 789 (1949).

to hydrolysis of many substrates has been expressed as a *proteolytic coefficient, C*.[4] This is a form of specific activity which is defined as the first-order rate constant per milligram protein nitrogen under arbitrary conditions. Comparisons of C values have been of great value in obtaining a qualitative picture of the interaction of enzyme and substrate, but it has been suggested that K_m and k_3 are more meaningful values for describing the binding of substrate and the rate-limiting step in the catalytic process. The most important part of the structure in determining the rate of hydrolysis by carboxypeptidase is the side chain of the terminal acid, R. Maximum rates are found with compounds in which the C-terminal group is phenylalanine, and decreasing rates are obtained with tyrosine, tryptophan, leucine, methionine, and isoleucine. Peptides with C-terminal imino acids, cysteine, cystine, and basic amino acids are not hydrolyzed.

TABLE 1

SPECIFICITY OF CARBOXYPEPTIDASE[3d]

Substrate	C'
Carbobenzoxyglycyl-L-phenylalanine	13.
Carbobenzoxyglycyl-L-tyrosine	6.2
Carbobenzoxyglycyl-L-leucine	2.6
Carbobenzoxyglycyl-L-isoleucine	0.54
Carbobenzoxyglycyl-L-alanine	0.038
Carbobenzoxyglycylaminoisobutyric acid	0.013
Carbobenzoxyglycylglycine	0.0024

While simple acyl amino acids serve as substrates for carboxypeptidase, more rapid hydrolysis occurs with more highly substituted molecules. Derivatives of α-amino acids are good substrates, but the presence of a free amino group on the residue contributing the carboxyl of the bond to be split completely eliminates enzymatic activity. Free amino groups on more distant residues have very little if any effect. The position of the secondary peptide is critical; derivatives of glycine are split over 1000 times as rapidly as the corresponding derivatives of β-alanine. Interaction of R′ with the enzyme is indicated by the decreased rate obtained when glutamic acid is substituted for glycine, and the increased rate with analogous tryptophan compounds. The nature of the secondary peptides (R″) has some influence on the rate; sulfonamides in this position form substrates that are split more slowly than the corresponding

[3d] From E. L. Smith. See General References.

[4] M. Bergmann and J. S. Fruton, *J. Biol. Chem.* **145**, 247 (1942). These investigators suggested that enzymes bind their substrates at several points, and that different rates indicate altered affinity at individual sites.

carboxylic amides. The secondary peptide as well as the primary (hydrolyzable) bond must be in the L-configuration. These influences of substrate structure on affinity and rate of reaction have been used in extensions of the polyaffinity theory of Bergmann and Fruton.[4] Together with similar information about inhibitors, these data allow inferences to be drawn about the chemical nature of groups on the enzyme that would interact with specific parts of the substrate. When more complete studies on the amino acid sequences of carboxypeptidase are available, this information should be useful in constructing a model of the "active surface."

Carboxypeptidase activity is not restricted to small molecules. It hydrolyzes large synthetic chains, such as polyglutamic acid, and, more important, it hydrolyzes C-terminal groups from intact proteins. This property is currently being put to use in the analysis of protein structures.

There is some discrepancy in the literature about the sensitivity to carboxypeptidase of peptides with certain C-terminal residues. Some differences were explained recently, when a very similar enzyme was found to occur in some carboxypeptidase preparations.[5] This is also a carboxypeptidase, but its substrates have the C-terminal basic amino acids arginine or lysine. Its activity, thus, complements that of the classical carboxypeptidase.

Leucine Aminopeptidase. Leucine aminopeptidase was named at a time when only a few authentic peptides were available for study. The ability of intestinal preparations to split peptides containing N-terminal leucine was shown by Linderstrøm-Lang[6] in 1929, and the enzyme responsible was partially purified by Johnson *et al.*[7] in 1936. The purified enzyme requires a divalent cation, Mg^{++} or Mn^{++}, for activity. The properties of this enzyme have been studied extensively by Smith and his collaborators.[8] They have found that the corresponding activity in hog kidney can be purified more highly than the best preparations obtained from intestinal mucosa. The properties described below are for the hog kidney enzyme, but the corresponding enzymes from other sources appear to have similar properties.

Extracts of acetone-dried kidney were purified about 1600 times by ammonium sulfate and acetone fractionation and by paper electrophoresis. The purest preparations reveal only single components in both electrophoresis and ultracentrifuge studies. The purified enzyme appears to contain only amino acids and has a molecular weight near 300,000.

[5] J. E. Folk, *J. Am. Chem. Soc.* **78**, 3541 (1956).
[6] K. Linderstrøm-Lang, *Z. physiol. Chem.* **182**, 151 (1929).
[7] M. J. Johnson, G. H. Johnson, and W. H. Peterson, *J. Biol. Chem.* **116**, 515 (1936).
[8] D. H. Spackman, E. L. Smith, and D. M. Brown, *J. Biol. Chem.* **212**, 255 (1955);
E. L. Smith and D. H. Spackman, *ibid.* **212**, 271 (1955).

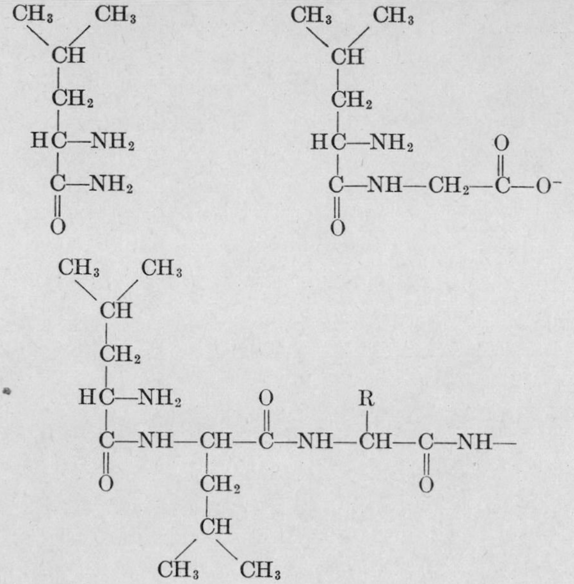

$$
\begin{array}{cc}
\text{CH}_3 \quad \text{CH}_3 & \text{CH}_3 \quad \text{CH}_3 \\
\diagdown \quad \diagup & \diagdown \quad \diagup \\
\text{CH} & \text{CH} \\
| & | \\
\text{CH}_2 & \text{CH}_2 \\
| & | \\
\text{H}\overset{|}{\text{C}}\text{—NH}_2 & \text{H}\overset{|}{\text{C}}\text{—NH}_2 \qquad \text{O} \\
\overset{|}{\text{C}}\text{—NH}_2 & \overset{|}{\text{C}}\text{—NH—CH}_2\text{—}\overset{||}{\text{C}}\text{—O}^- \\
\overset{||}{\text{O}} & \overset{||}{\text{O}}
\end{array}
$$

$$
\begin{array}{c}
\text{CH}_3 \quad \text{CH}_3 \\
\diagdown \quad \diagup \\
\text{CH} \\
| \\
\text{CH}_2 \\
| \\
\text{H}\overset{|}{\text{C}}\text{—NH}_2 \qquad \text{O} \qquad \text{R} \\
\overset{|}{\text{C}}\text{—NH—CH—}\overset{||}{\text{C}}\text{—NH—CH—}\overset{|}{\text{C}}\text{—NH—} \\
\overset{||}{\text{O}} \qquad \overset{|}{\text{C}}\text{H}_2 \qquad \overset{||}{\text{O}} \\
| \\
\text{CH} \\
\diagup \quad \diagdown \\
\text{CH}_3 \quad \text{CH}_3
\end{array}
$$

Representative Substrates of Leucine Aminopeptidase

The enzyme requires Mg^{++} or Mn^{++}; Ca^{++} and Co^{++} have no effect; Ni^{++}, Zn^{++}, and Fe^{++} inhibit somewhat and Cd^{++}, Cu^{++}, Hg^{++}, and Pb^{++} inhibit strongly.

Leucine aminopeptidase reacts slowly with Mg^{++} or Mn^{++} to form an active enzyme. The rate and extent of activation are functions of metal concentration. The combination with Mn^{++} is faster than that with Mg^{++}, and results in greater ultimate activity. Nevertheless, the fact that citrate inactivates crude preparations of the enzyme indicates that Mg^{++}, not Mn^{++}, is associated with the enzyme in nature. Citrate inhibits the Mg^{++}-activated, but not the Mn^{++}-activated enzyme.[9] The formation of active enzyme is apparently a simple reversible reaction in which one atom of metal combines with one molecule of protein. A dissociation constant could be calculated for the enzyme–Mn combination by measuring the rate of hydrolysis of L-leucinamide as a function of Mn^{++} concentration. The resulting data fit the equation

$$
K = \frac{(\text{Total Enzyme} - \text{Active Enzyme}) \cdot \text{Mn}}{\text{Active Enzyme}}
$$

At half-maximal activity $K = (Mn^{++}) = 4 \times 10^{-4} M$. The pH optimum for leucine aminopeptidase lies between pH 8 and pH 9, but the precise shape of the pH–activity curve depends upon both the metal cofactor and the substrate selected.

[9] E. L. Smith, *J. Biol. Chem.* **173**, 553 (1948).

Studies with analogous substrates show that the enzyme splits only amides, not esters. The N-terminal amino acid must be L, and the L-configuration is greatly preferred in the adjacent amino acid. Any of a large number of amino acids in the N-terminal position may be attacked, but all substrates must have free amino groups. An exception to the general mode of action is the hydrolysis of glycyl-L-leucinamide. Instead of attacking the glycyl–leucine bond, the enzyme first splits the amide bond, then much more slowly liberates free glycine and leucine (I). Apparently the small glycyl residue can be accepted by the enzyme as a modified amino group, and the reaction proceeds as with leucinamide.

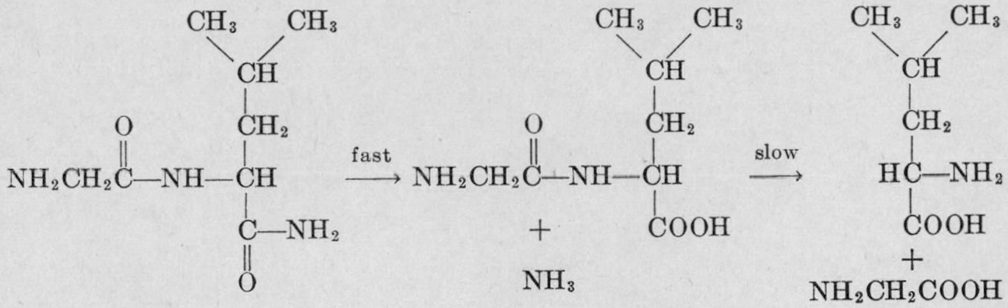

(I)

The proteolytic coefficient is largest for compounds with N-terminal leucine and norleucine, but several other amino acids may be attacked at sufficient rates to make this enzyme a generally useful reagent. Its activity is not restricted to small peptides, and it has been used in the stepwise degradation of proteins from the amino end, in the same manner that carboxypeptidase has been used in the analysis of the carboxyl ends of protein chains.

Prolidase. Prolidase was found by Bergmann and Fruton in intestinal mucosa in 1937. This enzyme is widely distributed in animal tissues, and

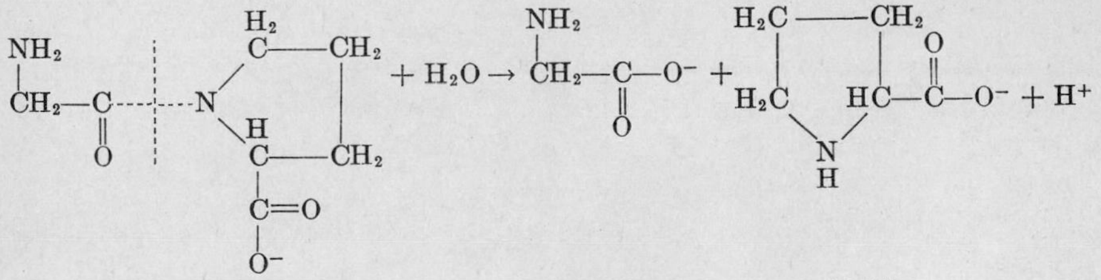

Glycylproline   Glycine   Proline

(II)

has unique specificities in that it is the only peptidase that splits bonds involving the imino N of proline, and this is the only type of bond prolidase attacks (II). The enzyme has been purified 12,000 times from pig kidney by Davis and Smith, who obtained a simple protein with a molecular weight around 100,000 that appeared to be 90 per cent pure by physical criteria.[10] To achieve this high degree of purification both Mn^{++} and sulfhydryl compounds were required for stabilization of the enzyme.

Prolidase has an absolute requirement for Mn^{++}, with a dissociation constant near 5×10^{-5}. The reaction of enzyme with metal is slow, and is accelerated by glutathione. There will be many examples described later of enzymes that are activated by glutathione and other sulfhydryl compounds. In some cases the sulfhydryl material appears to act as a coenzyme, but in many cases its function seems to be to maintain the sulfur of cysteine residues of the enzyme in the —SH form. Enzymes that require intact —SH groups for their activity may be inhibited by several types of compounds that react with sulfhydryl groups. Two commonly used "sulfhydryl reagents" are iodoacetic acid (IAA),[11] often used as its amide, and p-chloromercuribenzoate (PCMB).[12] While these compounds conceivably may react with other groups of proteins, in almost all cases they react only with sulfhydryl groups. IAA reacts in an essentially irreversible manner to form a thioether. PCMB and other mercurials form highly undissociated mercuric sulfides; the dissociation constants are sufficiently low that the reaction is essentially stoichiometric and can be used to titrate sulfhydryl groups. These compounds do dissociate sufficiently to permit the inhibitor to be removed by the addition of large amounts of other mercury binders, usually sulfhydryl compounds.

In general PCMB is a more powerful reagent than iodoacetamide, and will bind sulfhydryl groups that the latter reagent cannot affect. Both reagents inactivate metal-free prolidase, but only PCMB is effective after Mn^{++} has combined with the enzyme. It may be deduced from these observations that the binding of Mn^{++} masks an essential sulfhydryl group of the enzyme. The most obvious mechanism would involve binding of the metal to the protein through a sulfur linkage, but an indirect mechanism, in which the metal maintains the polypeptide chain in a configuration that buries the sulfur atom within the molecule, may be involved.

This enzyme is strictly a dipeptidase, requiring both a free carboxyl and free amino group. Some slight deviations in structure are permitted,

[10] N. C. Davis and E. L. Smith, *Federation Proc.* **12**, 193 (1953).

[11] F. Dickens, *Biochem. J.* **27**, 1141 (1933); L. Rapkine, *Compt. rend. soc. biol.* **112**, 790, 1294 (1933).

[12] L. Hellerman, F. P. Chinard, and V. R. Dietz, *J. Biol. Chem.* **147**, 443 (1943). This type of reagent had been studied previously by L. Hellerman, M. E. Perkins, and W. M. Clark, *Proc. Natl. Acad. Sci. U.S.* **19**, 855 (1933).

but they decrease the rate markedly. β-Alanylproline is slowly hydrolyzed, and glycylsarcosine is split at 10 per cent the rate of glycylproline.

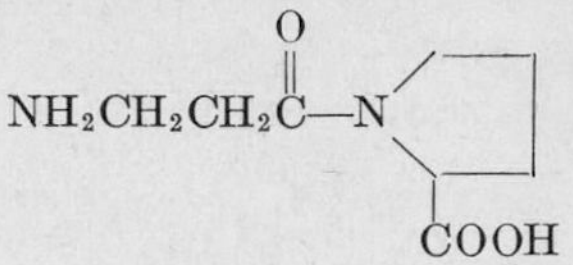

$$NH_2CH_2CH_2\overset{\overset{\displaystyle O}{\|}}{C}-N \quad\quad\quad NH_2CH_2\overset{\overset{\displaystyle O}{\|}}{C}-NCH_2COOH$$

COOH   CH_3

β-Alanylproline 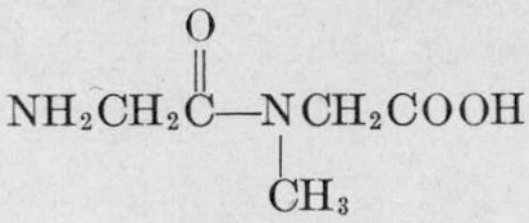   Glycylsarcosine

Substitutions on the ring also reduce the rate. Steric effects are of importance in determining the degree of inhibition; allohydroxyproline is split from a peptide more rapidly than the corresponding hydroxyproline compound.[13]

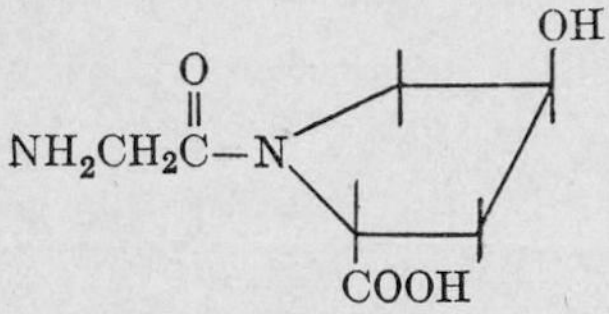

Glycylhydroxyproline 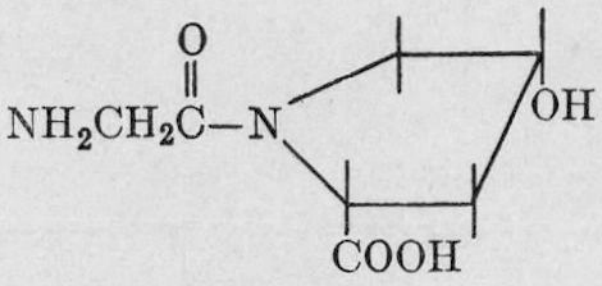   Glycylallohydroxyproline

Glycylglycine Dipeptidase. The hydrolysis of glycylglycine is catalyzed by an enzyme that attacks no other naturally occurring peptides. The only other known substrates are sarcosylglycine and glycyl-β-alanine, and both of these are hydrolyzed at slower rates. The enzyme obtained from muscle requires Co^{++}. These compounds, which react well with the enzyme, form complexes with Co^{++} that absorb light more strongly and at higher wavelengths than do other glycine compounds.[14] The enzyme requires substrates with both free carboxyl and free amino groups. It has therefore been suggested that the actual substrate for this peptidase is a chelate compound with Co^{++} bound to the substrate at three points: the free amino group, the amide, and the carboxyl group.

Glycyl-L-Leucine Dipeptidase. Glycyl-L-leucine dipeptidase has been found in various mammalian tissues. It is named for the only sensitive simple peptide; the only other known substrates are the N-methyl derivative, sarcosyl-L-leucine, and β-alanyl-L-leucine which are split slowly. This activity is of interest because of the differences in metal activation shown by enzymes from different sources. Enzymes from rabbit muscle and hog intestinal mucosa were found to be activated by Mn^{++}. Similar activities from human uterus and rat muscle were activated instead by Zn^{++}. Both types of enzyme are inactivated by Ca^{++}, which can be removed by phosphate buffers to leave active enzymes.[15]

[13] E. Adams, N. C. Davis, and E. L. Smith, *J. Biol. Chem.* **208**, 573 (1954).
[14] E. L. Smith, *J. Biol. Chem.* **173**, 571 (1948).
[15] E. L. Smith, *J. Biol. Chem.* **176**, 9 (1948).

Carnosinase. Carnosinase is an enzyme that splits carnosine, a naturally occurring peptide, β-alanyl-L-histidine. A preparation that is

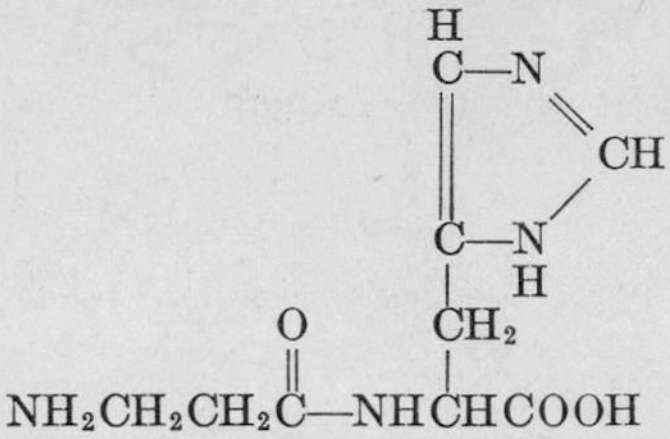

Carnosine

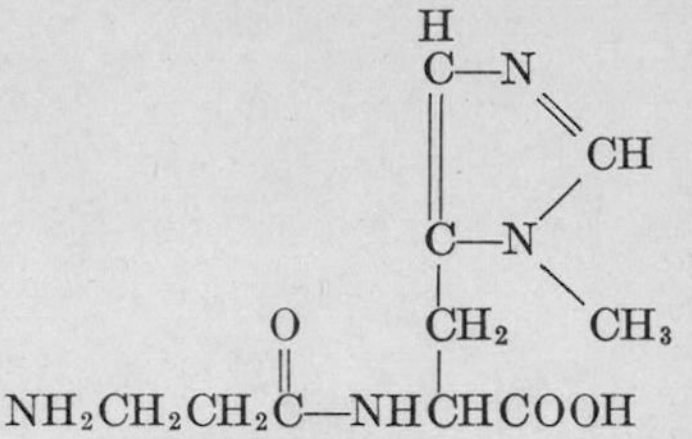

Anserine

activated by Zn^{++} or Mn^{++} was purified from hog kidney.[16] A Zn^{++} activated enzyme prepared from fish muscle splits the closely related anserine, β-alanyl-L-1-methylhistidine.[17] Kidney carnosinase has been shown to be a dipeptidase with the ability to attack peptides with glycine, D- and L-alanine and other simple amino acids in place of β-alanine. Simple derivatives of β-alanine are not hydrolyzed. The ability of carnosinase and anserinase to cross-react with the substrates of the other has not been reported, but it is to be expected that the methyl group on the ring will be found to have little influence on either enzyme.

An unusual specificity is shown by a *tripeptidase* detected in several animal tissues. This enzyme, aminotripeptidase, hydrolyzes the N-terminal residue from a tripeptide with both a free amino group and a free carboxyl group.[18] All of the amides must be of α-amino acids, and there is essentially no activity if D-amino acids occupy the N-terminal or adjacent positions. A D-amino acid in the C-terminal position reduces the rate of hydrolysis markedly.

Dehydropeptidases. A widely distributed type of activity has been found to attack a group of substrates that are not known to occur naturally, the dehydropeptides.[19] At least two enzymes with different specificities exist in animal tissues.[20] The net reaction requires two equivalents

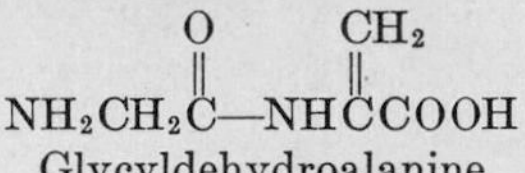

Glycyldehydroalanine
(a substrate for
dehydropeptidase I)

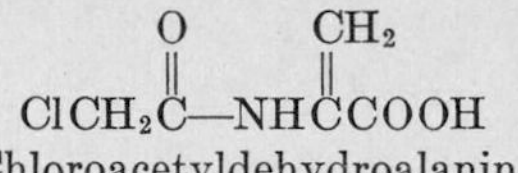

Chloroacetyldehydroalanine
(a substrate for
dehydropeptidase II)

of water, and results in the formation of one (the N-terminal) amino acid, an α-keto acid, and ammonia. The unsaturated amino acid or the

[16] H. T. Hansen and E. L. Smith, *J. Biol. Chem.* **179,** 789 (1949).
[17] N. R. Jones, *Biochem. J.* **60,** 81 (1955).
[18] J. S. Fruton, V. A. Smith, and P. E. Driscoll, *J. Biol. Chem.* **173,** 457 (1948).
[19] M. Bergmann and H. Schleich, *Z. physiol. Chem.* **205,** 65 (1932).
[20] J. P. Greenstein and F. M. Leuthardt, *J. Natl. Cancer Inst.* **6,** 197 (1946).

tautomeric imino acid might be postulated as a first product, but such compounds are not known in biological systems. It cannot be decided on the basis of available evidence whether the second hydrolysis occurs on the enzyme or in solution after liberation of an imino acid.

In addition to the enzyme described, many other exopeptidases are known. These include enzymes specific for peptides of D-amino acids,[21] prolyl peptides,[22] glutathione (γ-glutamylcysteinylglycine),[23] and alanyl peptides.[22] It is of interest that many of the peptidases occur in tissues that are not concerned with digestion. It is therefore possible that they may be involved in specific syntheses as well as in degradations.

MECHANISM OF EXOPEPTIDASE ACTION

Information of the type presented briefly for various of the exopeptidases has led to attempts to formulate a mechanism of action for peptide hydrolysis. There are two conflicting opinions about the function of the metal ions. One view, advocated strongly by Smith,[24] is that the metal forms a chelate complex with the peptide and also is bound to the protein. The protein also interacts with the substrate through ionic or van der Waals forces. When the substrate is properly complexed, a shift of electrons makes the carbon–nitrogen bond sensitive to hydroxyl ion attack.

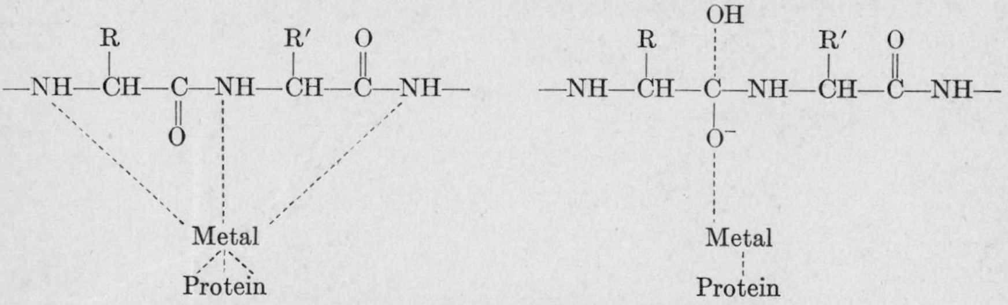

FIG. 7. Hypothetical combinations of cations with enzyme and substrate.

Another explanation[25] for the role of cations deliberately omits chelate compounds, and proposes instead that the metal forms a bridge between the protein and the oxygen of the peptide bond, and that this facilitates hydroxyl ion attack on the carbon atom (Fig. 7). The cation is thought also to increase the local OH⁻ concentration.

[21] J. Berger and M. Johnson, *J. Biol. Chem.* **133**, 639 (1940).
[22] E. L. Smith and M. Bergmann, *J. Biol. Chem.* **153**, 627 (1944).
[23] F. Binkley and K. Nakamura, *J. Biol. Chem.* **173**, 411 (1948).
[24] E. L. Smith, N. C. Davis, E. Adams, and D. H. Spackman *in* "Mechanisms of Enzyme Action" (W. D. McElroy and B. Glass, eds.), p. 291. Johns Hopkins Press, Baltimore, 1954.
[25] I. Klotz *in* "Mechanisms of Enzyme Action" (W. D. McElroy and B. Glass, eds.), p. 257. Johns Hopkins Press, Baltimore, 1954.

While the function of metals remains unclear, it is obvious that in some cases (as leucine aminopeptidase) the combination of enzyme and substrate is profoundly influenced by the metal. Other factors appear to be of even greater importance in determining enzyme–substrate affinity. The great variety of specific requirements displayed by various peptidases, all composed only of amino acids, indicates that adsorption sites are composed of very specific groupings with affinities for particular side chains and spaced to permit only appropriate peptides to approach. Those enzymes that require free carboxyl, free amino groups, or both probably contain arginine or lysine, glutamic or aspartic acids, or both in complementary positions. Affinities for uncharged groups in general are related to the affinities of similar structures for each other. In the case of prolidase, it appears that the hydroxyl group of hydroxyproline interferes to a limited extent in the approach of the ring, whereas the hydroxyl group on the other side of the ring in allohydroxyproline interferes much less. Another force available for binding enzyme to substrate is the hydrogen bond. It is apparent that the sum of several small forces strategically located suffices to give the enzymes great power for attracting, binding, and reacting with their specific substrates.

The specificities of enzymes may be derived from the characteristics of the "binding sites." In addition, certain restrictions may be placed on the substrate by steric factors; that is, there may be no specific requirements for certain groups, such as a side chain, but there may be space available only for a group in one of two possible optical isomers. One approach to describing the enzyme's "active surface" is to deduce properties of binding sites by measuring the properties of modified substrates. This cannot be done with many enzymes that exhibit very rigid requirements, but a large amount of information has been obtained in this area with peptidases and proteolytic enzymes.

The attachment of larger substrates, such as those of carboxypeptidase, leaves room for considering many sites of attachment, the purpose of all of which is to place one peptide bond in such a position that it is sensitive to attack. The primary attack may be made by some element of water (H^+, OH^-, H_2O) but may also be by a group of the enzyme. In the latter case, there must be a subsequent reaction with water.

ENDOPEPTIDASES

A large part of our fundamental knowledge about enzymes as proteins and of their behavior as chemical entities was derived from studies on proteolytic enzymes. Classical studies by Northrop and his collaborators[26]

[26] The accomplishments of this group were responsible for changing the study of enzymes from an obscure art concerned with biological mysteries to a branch of science investigating chemical and physical properties of catalysts of biologic origin.

resulted in the crystallization of pepsin, trypsin, chymotrypsin, and their zymogens, and for a quarter of a century this group has published extensively on the physical, chemical, and catalytic properties of these and other enzymes. It is no reflection on the pioneer efforts of these investigators to note that the materials with which they worked have certain properties that facilitated their studies, and that other enzymes often present much more difficult problems and the recommendations of this group often cannot be followed. For example, it is recommended that "several grams, or, better, several hundred grams, of each step in the process of purification" should be at hand before proceeding to the next step. In the case of enzymes that comprise several per cent of the starting material this recommendation is not difficult to follow, but it is obvious that several hundred different enzymes cannot each make up several per cent of the soluble protein of a cell. Conditions of relatively great stability were rather easily found for the first enzymes to be crystallized. Instability of other enzymes has often required very rapid work-ups to obtain material sufficient for certain studies. The need for speed imposes limitations on the methods that can be used.

A concept of catalytic efficiency was proposed to evaluate enzymes. This expresses the "efficiency" in terms of relative rates; the ratio of catalyzed to uncatalyzed rates is the measure of efficiency rather than absolute rates. This concept cannot be applied usefully to the large numbers of reactions that do not proceed measurably in the absence of enzymes, which thus have infinite efficiencies. A more useful function is the *turnover number*, which defines an absolute rate of reaction for the enzyme. The turnover number is the number of molecules of substrate converted to product by one molecule of enzyme in one minute under specified conditions.

Proteolytic enzymes were originally studied as extracellular digestive enzymes, and the first sources studied were the stomach and pancreas. Later intracellular proteolytic enzymes from nondigestive organs were studied. The intracellular proteolytic enzymes from all cells are called *cathepsins*. The isolation of cathepsins has not been so successful as the isolation of digestive enzymes because of their lower concentrations and greater labilities. In addition to animal sources, both plants and microorganisms have been studied as sources of proteolytic enzymes. Enzymes from these various sources show many different properties. Some of these will be discussed in connection with individual enzymes.

Pepsin. Pepsin is derived from a zymogen, pepsinogen, which occurs in gastric mucosa.[27] Pepsinogen has been obtained from various animals and crystallized from swine.[28] The various preparations are activated by

[27] J. N. Langley, *J. Physiol.* **3**, 246, 269 (1882).
[28] R. M. Herriott, *J. Gen. Physiol.* **21**, 501 (1938).

preformed pepsin to yield an active enzyme characteristic of the source, not the activator. In the activation process, a peptide with a molecular weight around 5000 is removed from the 42,000 molecular weight protein. After denaturing the pepsin with alkali, Herriott succeeded in crystallizing the peptide, pepsin inhibitor.[29] This peptide combines with pepsin at pH values above 5.4 and forms an inactive complex, but at lower pH values the inhibitor is destroyed by pepsin.

Pepsin has been crystallized from preparations of pepsinogen, from gastric juice, and, most conveniently, from commercial pepsin, which is approximately 20 per cent pure enzyme.[30] The separation of the pure enzyme is achieved with difficulty, as at least three other proteolytic enzymes may be present: rennin (the milk-coagulating enzyme), a gelatinase, and a cathepsin. Since the hydrolytic action of pepsin also causes casein to form an insoluble material, there was doubt about the distinction of rennin from pepsin. The doubts were resolved by the isolation of rennin and prorennin,[31] then crystalline rennin,[32a] which has very high milk-coagulating activity and only feeble proteolytic activity. Moreover, preparations of pepsin have been obtained which appear to be free from other enzymes, and show only a single component in electrophoresis measurements, and give the sharp break in solubility curves characteristic of pure compounds.[32b]

Pure pepsin contains one atom of phosphorus per mole.[30] Except for one phosphate group, the molecule is composed only of amino acid residues. The phosphate group has been removed from both pepsin and pepsinogen by certain phosphatases to yield proteins that retain proteolytic activity. Therefore, the phosphate group that gives pepsin its

[29] R. M. Herriott, *J. Gen. Physiol.* **29,** 325 (1941).

[30] J. H. Northrop, *J. Gen. Physiol.* **13,** 739 (1930); **16,** 615 (1933).

[31] H. Tauber and I. S. Kleiner, *J. Biol. Chem.* **96,** 745 (1932); I. S. Kleiner and H. Tauber, *ibid.* **106,** 501 (1939).

[32a] N. J. Berridge, *Nature* **151,** 473 (1943); C. L. Hankinson, *J. Dairy Sci.* **26,** 53 (1943).

[32b] This so-called phase-rule test is based on the fact that a pure compound dissolves in a solvent until the solution is saturated. At this point further addition of solute does not affect the composition of the solution. If the solute is composed of two or more compounds, these may dissolve independently or may interact to influence the distribution between solid and solution, as by forming "solid solutions" of one protein in another. In these cases the solubility curve bends gradually to the final value or contains one or more breaks before becoming horizontal. As with all other physical criteria of purity, the solubility test only indicates the possibility of homogeneity if a single sharp break in the curve is found. If experimental results give such a curve, it is desirable to repeat the measurements with other conditions and to apply other criteria of purity before concluding that a preparation consists of only a single molecular species.

unusual isoelectric point of less than 1 is not essential for its proteolytic activity.[33]

Careful acetylation of pepsin with ketene has yielded preparations with no free amino groups, but with full activity.[34] More extensive acetylation inactivates the enzyme. Tyrosine side chains appear to be somewhat more important to the enzyme, but some activity remains after partial iodination, and treatment with nitrous acid leaves 50 per cent of the activity.

Pepsin hydrolyzes proteins to products that are not precipitated by the conventional protein precipitants, such as trichloroacetic acid. Very little of the protein is converted to free amino acids, as many peptide bonds are not attacked by pepsin. Not all of the structural requirements for pepsin substrates have yet been determined. Studies with synthetic peptides have shown that the reaction rate is increased when an aromatic amino acid contributes the N of the amide,[35] and is fastest when both amino acids of the peptide to be split are aromatic.[36] In the latter case the pH optimum is near 2, as it is when many protein substrates are used. Many of the other synthetic substrates are split more rapidly near pH 4.

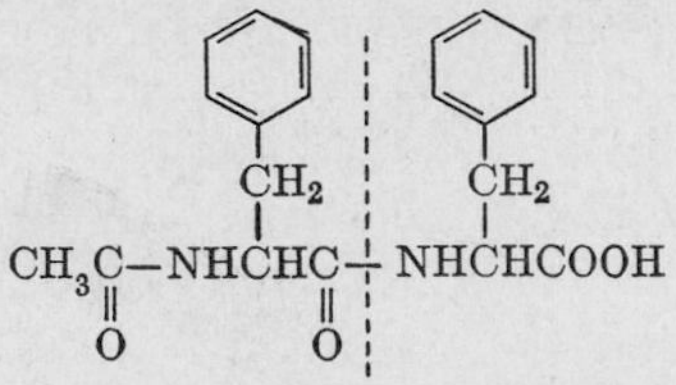

$$\text{CH}_3\text{C}-\text{NHCHC}-\text{NHCHCOOH}$$

Acetyl-L-phenylalanyl-L-phenylalanine
(a substrate for pepsin)

Peptides in which either component has the D-configuration or is a dehydro amino acid are not attacked. Differences in sensitivity of bonds in protein molecules and in small synthetic substrates indicate that adjacent structures exert great influence on the bonds that may be attacked. For example, one leucine–valine bond in insulin (whose amino acid sequence is known) is hydrolyzed rapidly, whereas this sequence in another part of the insulin molecule and in synthetic substrates resists hydrolysis.[37]

[33] G. E. Perlmann, *J. Am. Chem. Soc.* **74**, 6308 (1952).
[34] R. M. Herriott and J. H. Northrop, *J. Gen. Physiol.* **18**, 35 (1934).
[35] C. R. Harrington and R. V. Pitt-Rivers, *Biochem. J.* **38**, 417 (1944).
[36] L. E. Baker, *J. Biol. Chem.* **193**, 809 (1951).
[37] F. Sanger and H. Tuppy, *Biochem. J.* **49**, 481 (1951); F. Sanger and E. O. P. Thompson, *ibid.* **53**, 366 (1953).

Chymotrypsin. Chymotrypsin is a name for a group of closely related enzymes[38] all derived from the zymogen chymotrypsinogen.[39] Chymotrypsinogen has been crystallized in a very high state of purity. Hydrolysis by trypsin activates the zymogen. Activation has also been initiated by the bacterial protease, subtilisin. Chymotrypsinogen (M.W., 25,000) contains only one free α-amino group, contributed by an N-terminal cystine.[40] It has been speculated that the carboxyl end of the peptide chain is condensed with some other group in the molecule to form a cyclic compound with no C-terminal amino acid, since none has been found. During activation the first defined alteration in the structure of chymotrypsinogen is the splitting of an arginine–isoleucine bond.[41] This hydrolysis is the only alteration known to occur in the transformation of the zymogen into an active enzyme, but it seems reasonable to postulate concomitant changes in the physical structure; these may involve a new folding to create a catalytic site, or may expose a previously buried part of the protein. The first active enzyme to be formed is designated π-chymotrypsin.[42] This form is unstable because of autolytic changes. Removal of a dipeptide, serylarginine, from the chain exposed by tryptic action results in the formation of δ-chymotrypsin.[41] Other linkages in both π- and δ-chymotrypsin may be broken to yield additional forms of the enzyme. Three of these, α-, β-, and γ-chymotrypsin, have been obtained as crystalline compounds.[38] α-Chymotrypsin contains N-terminal isoleucine and alanine and C-terminal leucine and tyrosine.[43] β- and γ-chymotrypsins also exhibit these end groups together with variable amounts of other amino acids. It seems likely that all of the more highly degraded chymotrypsins, including α-, β-, and γ-, are mixtures of partially degraded enzyme; they exhibit slightly lower specific activity than π- or δ-chymotrypsin.

The chymotrypsins have been studied very extensively with synthetic substrates. The enzymes all have very similar substrate requirements. The most rapidly split substrates are those in which an aromatic amino acid contributes the carboxyl of the sensitive bond. The hydrolyzable bond may be an amide, hydroxamide, hydrazide, peptide, ester, thioester, or carbonyl carbon.[44] As in the case of carboxypeptidase, the nature of the "secondary peptide" is important in determining the rate of reac-

[38] M. Kunitz, *J. Gen. Physiol.* **32,** 207 (1938).
[39] M. Kunitz and J. H. Northrop, *J. Am. Physiol.* **18,** 433 (1935).
[40] F. R. Bettelheim, *J. Biol. Chem.* **212,** 235 (1955).
[41] W. J. Dreyer and H. Neurath, *J. Biol. Chem.* **217,** 527 (1955).
[42] C. F. Jacobsen, *Compt. rend. trav. lab. Carlsberg, Sér. chim.* **25,** 325 (1947).
[43] J. A. Gladner and H. Neurath, *J. Biol. Chem.* **206,** 911 (1954).
[44] D. G. Doherty and L. Thomas, *Federation Proc.* **13,** 200 (1954).

tion, but molecules which have no such group also serve as substrates. A number of structural analogs of substrates have been found to be competitive inhibitors of chymotrypsin. These include compounds with the D-configuration, split products, simple peptides, and other compounds with aromatic rings. The effects obtained with modified structures of substrates and inhibitors have been interpreted as indicating adsorption sites that react with the hydrolyzable bond, the primary amino acid side chain, and the secondary peptide.[45] Many of the substrates were tested in dilute methanol because of their insolubility in water. Methanol increases the K_m and has no effect on k_3.

The role of certain inhibitors with free carboxyl groups reflects on our dependence on analytical methods for the interpretation of experimental results. These compounds have no apparent hydrolyzable group, and therefore cannot participate in a reaction involving a net change. However, it has been found that chymotrypsin catalyzes an exchange of oxygen between water and the carboxyl group.[46] The "inhibitors" with

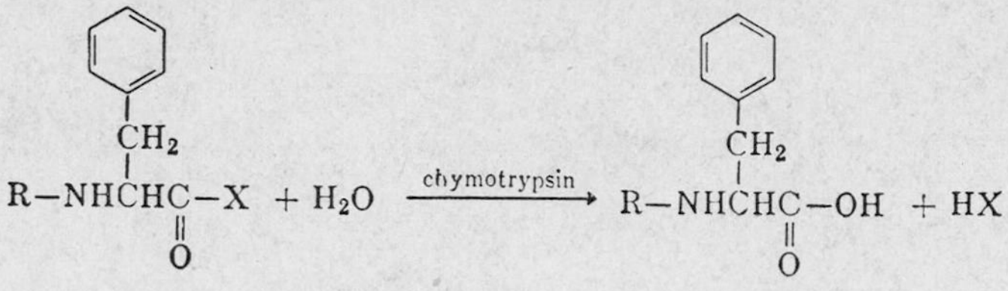

$$R-NHCHC-X + H_2O \xrightarrow{\text{chymotrypsin}} R-NHCHC-OH + HX$$

(III)

free carboxyl groups are not merely inert molecules that occupy the active sites of the enzyme and prevent substrates from reacting. They are themselves substrates, and compete with other substrates for the enzyme. The reaction of a typical substrate is shown in (III). If X is an —OH group, the reaction can only be measured isotopically as an exchange of oxygen between the carboxyl group and the medium. This has been demonstrated for carbobenzoxy-L-phenylalanine.

The nature of the "active center" of chymotrypsin and other hydrolytic enzymes (in general, esterases) has been explored with a specific type of inhibitor, dialkyl phosphate anhydrides, such as diisopropylfluorophosphate (DFP). These compounds react with chymotrypsin to add one, and only one, phosphorus per molecule and form a completely inactive compound.[47] The inactivated molecule yields phosphoserine on

[45] S. Kaufman and H. Neurath, *Arch. Biochem.* **21**, 437 (1949).
[46] D. B. Sprinson and D. Rittenberg, *Nature* **167**, 484 (1951); D. G. Doherty and F. J. Vaslow, *J. Am. Chem. Soc.* **74**, 931 (1952).
[47] E. F. Jansen, M. D. F. Nutting, R. Jang, and A. K. Balls, *J. Biol. Chem.* **179**, 189 (1949); E. F. Jansen, M. D. F. Nutting, and A. K. Balls, *ibid.* **179**, 201 (1949).

hydrolysis, but, since there are several serine residues per enzyme molecule, it is not clear which peculiar structure is responsible for the one reaction that does occur.

Trypsin. Like other pancreatic peptidases, trypsin is made by activation of a precursor, trypsinogen. Trypsinogen has been crystallized, and can be recrystallized only in the presence of an inhibitor for trypsin.[48] Activation may be brought about either by a rather specific, incompletely

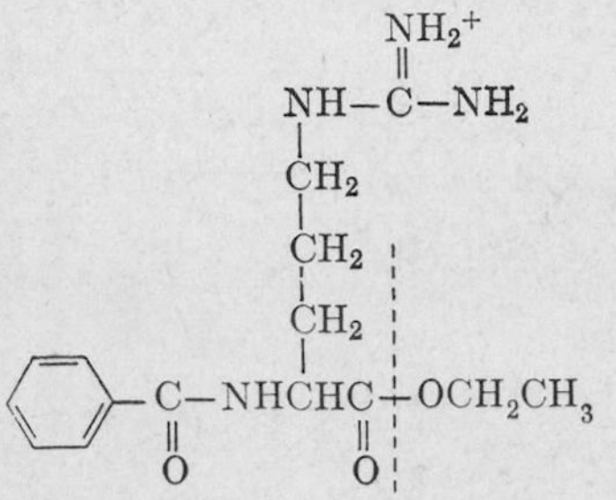

Benzoyl-L-arginine ethyl ester (250)

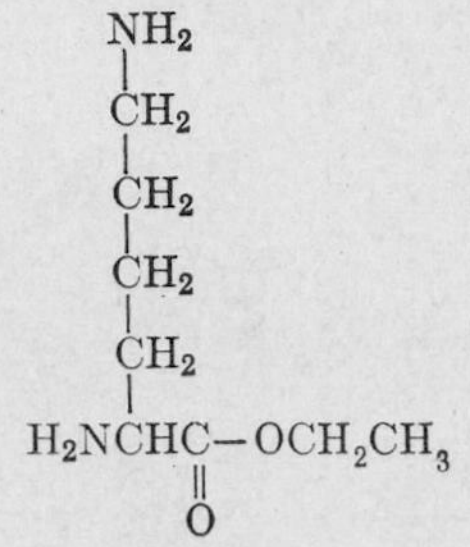

L-Lysine ethyl ester (110)

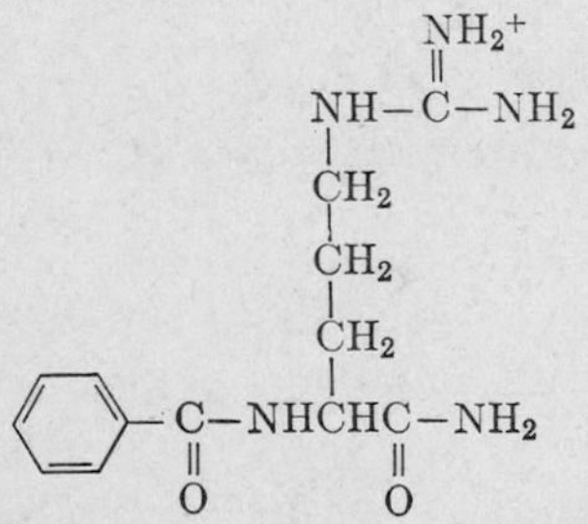

Benzoyl-L-arginine amide (3.8) 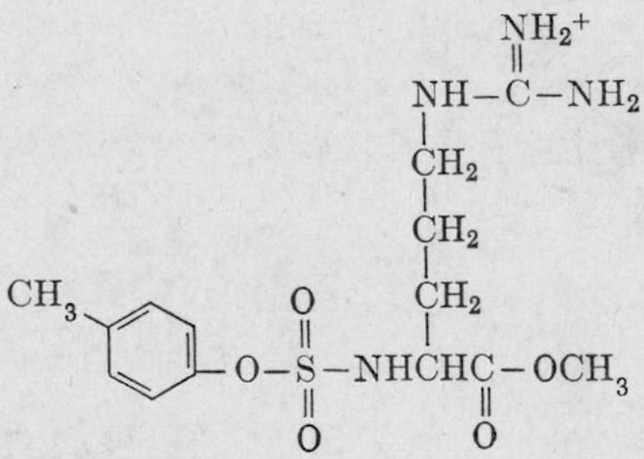   Toluenesulfonyl-L-arginine methyl ester (1800)

Representative Substrates for Trypsin

[The figures in parentheses are relative over-all reaction velocity constants (k_3).]

described, intestinal enzyme, enterokinase,[49] or by trypsin.[50] Since the product of activation is able to cause more activation, the rate of trypsin production increases with time until the concentration of trypsinogen becomes limiting. The appearance of trypsin with time thus follows an autocatalytic curve. Trypsin is also capable of converting trypsinogen to an inactive protein. The relative amounts of trypsin and inactive protein depend on pH, more inactive protein being formed as the pH increases above 6. The formation of inactive protein and the autodigestion of trypsin are inhibited by Ca^{++}. The activation of trypsinogen is

[48] F. Tietze, *J. Biol. Chem.* **204,** 1 (1953).
[49] N. P. Schepowalnikow, *Maly's Jahresber.* **29,** 378 (1899).
[50] M. Kunitz and J. H. Northrop, *J. Gen. Physiol.* **19,** 991 (1936).

brought about by removal of an N-terminal peptide, valine (aspartic acid)$_4$ lysine.[51]

Trypsin has more rigid substrate requirements than most proteases. Only bonds in which the carbonyl is contributed by an acid with the side chain of lysine or arginine are split.[52] This acid may have a free or substituted α-amino group or an α-hydroxy group in the L-configuration. Trypsin was the first enzyme shown to split esters as well as amides.[53] Trypsin, like chymotrypsin, is most active between pH 7 and 9.

Trypsin is inhibited by DFP in the same manner as chymotrypsin; one P per mole (M.W., 24,000).[54] It is also inhibited by naturally occurring proteins, several of which have been crystallized as trypsin inhibitors.[55] These inhibitors act by forming stoichiometric compounds with trypsin irreversibly.

It is often attempted to destroy enzymes in solution by heating. This procedure may result in serious misinterpretations, as experiments with trypsin and other of the pancreatic enzymes have shown. Trypsin is rapidly converted to a denatured protein at temperatures higher than 50°C., as shown by altered solubility in salt solutions and loss of enzymatic activity. On standing at low temperatures, however, both the solubility and activity become characteristic of the original trypsin.[56] This type of reversible denaturation depends very much on pH. Similar reversible inactivations have been reported for other enzymes, including myokinase and hyaluronidase.

Papain. Papain is representative of a group of plant proteases. It is found in the latex of papaya as 6–7 per cent of the total protein. Enzymes of this group in general are activated by reducing agents, and the inclusion of cysteine in the extract stabilizes the enzyme and permits its purification to proceed smoothly. Papain has been crystallized from salt solutions and from 70 per cent ethanol.[57,58] Highly purified crystals of the inactive mercury salt have been obtained that are fully active on removal of the mercury. The mercury–enzyme compound has two enzyme molecules (M.W., 20,700) per mercury atom.

For many years a controversy over the nature of papain activation confused the literature. Apparently cyanide, cysteine, glutathione, and

[51] E. W. Davie and H. Neurath, *J. Biol. Chem.* **212**, 515 (1955).

[52] K. Hofmann and M. Bergmann, *J. Biol. Chem.* **138**, 243 (1941).

[53] G. W. Schwert, H. Neurath, S. Kaufmann, and J. E. Snoke, *J. Biol. Chem.* **172**, 221 (1948).

[54] E. F. Jansen and A. K. Balls, *J. Biol. Chem.* **194**, 721 (1952).

[55] See reference 50, also M. Kunitz, *J. Gen. Physiol.* **30**, 31, 291 (1947); A. K. Balls and T. L. Swenson, *J. Biol. Chem.* **106**, 409 (1934).

[56] M. L. Anson and A. E. Mirsky, *J. Gen. Physiol.* **17**, 393 (1939).

[57] A. K. Balls and H. Lineweaver, *J. Biol. Chem.* **130**, 669 (1939).

[58] J. R. Kimmel and E. L. Smith, *J. Biol. Chem.* **207**, 515 (1954).

H_2S could bring about different degrees of activation, various buffers indicated different pH optima, etc. These facts have been reconciled by a single, simple explanation.[58] Sulfhydryl groups are essential for the activity of many enzymes, which may be inactivated either by oxidation or by heavy metal binding. The various reducing agents and buffers studied vary in their capacity to bind metals. If a reducing agent and a sequestering (chelating) agent (the one used most often is Versene, ethylenediamine tetraacetate) are both present, the same activities are found with various buffers, and other "activators" have no influence.

When assayed in the presence of Versene and sulfhydryl compounds, papain has a broad pH optimum between pH 5 and 7.5. It attacks a very wide range of substrates, including those characteristic of pepsin, trypsin, and chymotrypsin. It is an esterase also, but is not inhibited by DFP. Although most peptide bonds are attacked by papain, it apparently requires an α-substitution, as γ-glutamyl bonds are not attacked. The great variation in rates observed with different substrates led to questioning the homogeneity of the enzyme, on the suspicion that the low activity with certain substrates could be attributed to a contaminant. Competitive inhibition studies with carbobenzoxyglutamic acid indicate that a single enzyme is responsible for all of the reactions attributed to papain.

Cathepsins. Cathepsins are intracellular proteases of animal origin.[59] The occurrence of several such enzymes has been demonstrated in various tissues, including spleen, pituitary gland, kidney, thymus, etc. It is obvious that there is no reason to anticipate that all cathepsins will have similar properties to each other or to any other proteases. Cathepsins have been designated by both Roman numerals and by letters. Some of these enzymes have been identified with enzymes purified independently, as cathepsin III with leucine aminopeptidase. Several are activated by sulfhydryl compounds, some by metals. The isolation of the various cathepsins and studies of their substrate specificities are subjects currently under investigation, but because of the lower concentration of enzyme in the source materials and the number of related enzymes present, this area of investigation has not reached the development of the study of digestive enzymes.

There is very little information about many of the bacterial or fungal proteases, which are produced in large quantities by many organisms. Recently several have been crystallized. Crude preparations of mold and of bacterial media have been used commercially for their protease activity. A crystalline protease from *Bacillus subtilis*, subtilisin, has been

[59] J. S. Fruton and M. Bergmann, *J. Biol. Chem.* **130**, 19 (1939). Following the fractionation of cathepsins by these investigators, numerous other studies have described proteolytic enzymes from many sources.

used in the specific removal of a small peptide from crystalline ovalbumin to yield another well-defined, crystalline protein, plakalbumin.[60]

Hydrolysis of Large Polypeptides. Studies carried out with synthetic substrates give some insight into the catalytic abilities of various proteinases, but they have not yet reached a point from which the reaction between a protease and a protein can be predicted. Two types of observation remain to be explained. One is the resistance of many proteins to digestion by various enzymes. When these proteins are denatured by physical means, they become good substrates.[61] The nature of the process that converts a protein into a molecule susceptible to enzymatic hydrolysis is not understood. It is possible that the native protein has all of its sensitive bonds protected from access by the enzyme, and that the rupture of hydrogen bonds, specific salt bonds, disulfide bridges or weakening of van der Waals forces permits the protein to "unfold" and expose the hydrolyzable bonds. Also to be explained is the observation that bonds in large molecules are split whereas comparable bonds in small peptides resist hydrolysis. This was first observed by Katchalski,[62] who found synthetic polylysine to be split by chymotrypsin whereas small lysine peptides are not hydrolyzed. More recently Sanger found bonds in insulin to be split by pepsin contrary to expectation.[37] This is the only documented case with a native protein; insulin was the first protein whose amino acid sequence was completely determined, but additional information on susceptibility of bonds in proteins may come from studies with more recently analyzed proteins, including glucagon and ribonuclease.

Early studies with many types of enzymes failed to find evidence for intermediate products between large polymers and the ultimate products of an enzymatic degradation. From these studies a concept developed of "explosive" reaction, in which a large molecule was thought to be adsorbed to an enzyme, and held until completely digested. Currently, there is no basis for such a hypothesis. In some cases it has been found that the affinity of smaller molecules for the enzyme is greater than that of the parent compound, so that accumulation of intermediates is minimized, but where adequate methods (as chromatography) are available, it can usually be shown that each bond split is caused by an independent reaction.[63]

Transpeptidation. All of the emphasis on enzyme specificity so far has been restricted to properties of one of the reactants: the peptide.

[60] N. Egg-Larsen, K. Linderstrøm-Lang, and M. Ottesen, *Arch. Biochem.* **19**, 340 (1948).

[61] M. L. Anson and E. Mirsky, *J. Gen. Physiol.* **17**, 399 (1939).

[62] E. Katchalski, *Advances in Protein Chem.* **6**, 123 (1951).

[63] M. Rovery, P. Desnuelle, and G. Bonjour, *Biochim. et Biophys. Acta* **6**, 166 (1950); J. W. Williams, R. L. Baldwin, W. M. Saunders, and P. G. Squire, *J. Am. Chem. Soc.* **74**, 1542 (1952).

The other reactant has been considered to be water, which is present in large excess. Recently it has been found that other compounds may substitute for water, and, indeed, may react with sufficient velocity to compete successfully with water.[64] Hydroxylamine, amino acids, peptides, and other compounds act as acyl acceptors for appropriate substrates of all of the proteases tested, including pepsin, trypsin, chymotrypsin and papain. When amino acids or peptides act as acceptors for the acyl group of a substrate, the reaction is designated transpeptidation. Transpeptidation also occurs with specific peptidases, such as the one that hydrolyzes the γ-glutamyl bond of glutathione. It has often been suggested that such reactions are involved in the formation of proteins from peptides, but as yet no positive evidence has been obtained on this question. In general transpeptidation seems to involve transfer of an active acyl group, but examples are known that seem to indicate amino transfer.[65]

SYNTHETIC REACTIONS OF PROTEOLYTIC ENZYMES

Among the early studies of Bergmann and his collaborators were examples of synthetic reactions catalyzed by proteolytic enzymes. Two types of reaction were observed.[66] In one, an insoluble product was

$$
\begin{array}{lll}
APG & + GL & \rightarrow APGGL + H_2O \\
APGGL & + H_2O \rightarrow APGG & + L \\
APGG & + H_2O \rightarrow APG & + G \\
\hline
\text{Sum: } GL & \rightarrow G & + L
\end{array}
$$

(IV)

formed by condensation of a substituted amino acid and an amine, such as carbobenzoxyglycine $+$ aniline $\xrightarrow{\text{papain}}$ carbobenzoxyglycinanilide. In the other type of reaction the evidence for synthesis is indirect. An enzyme that was incapable of attacking either of two substrates was found to hydrolyze one in the presence of the other. The explanation suggested is that there is a condensation between the two substrates to form a molecule that is first split at one of the preexisting bonds. The synthesized bond is subsequently split, as shown in (IV). This is an example of the hydrolysis of glycyl-L-leucine (GL) by papain in the presence of acetyl-phenylalanylglycine (APG). In both cases the observations support the contention that enzymes catalyze reversible reactions and, when a means is available for "pulling" the reaction by removing a product (precipitation or hydrolysis in the cases above), thermodynamically unfavorable

[64] This subject has been summarized by C. S. Hanes, G. E. Connell, and G. H. Dixon, *in* "Phosphorus Metabolism" (W. D. McElroy and B. Glass, eds.), Vol. II, p. 95. Johns Hopkins Press, Baltimore, 1952; M. Bergmann and H. Fraenkel-Conrat, *J. Biol. Chem.* 119, 707 (1937).

[65] Y. P. Dowmont and J. S. Fruton, *J. Biol. Chem.* 197, 271 (1952).

[66] O. K. Behrens and M. Bergmann, *J. Biol. Chem.* 129, 587 (1937).

reactions proceed. Incubation of concentrated solutions of enzymatically hydrolyzed proteins with proteolytic enzymes results in the formation of new peptides, called plasteins.[67] These are random condensation products, much smaller than the original proteins; they do not represent protein synthesis by reversal of hydrolysis.

MECHANISM OF ENDOPEPTIDASE ACTIVITY

An outline of the reaction mechanism used by enzymes such as chymotrypsin may be constructed from available information, but it must be emphasized that all such schemes must remain conjectural until more is learned about the structure of the enzymes. The pronounced preference for compounds in which the acyl group is contributed by an aromatic amino acid indicates the presence of similar groups that interact with the side chain in such a way as to place the acyl group near a reactive group. Other groups more distant from the sensitive bond also influence the binding. Kinetic studies of DFP inhibition as a function of pH suggest that the initial reaction may be influenced by an imidazole group ($pK_a \sim 7$), a histidine residue in the enzyme.[68] The carbon of the acyl group may be attracted by a basic nitrogen with an unshared pair of electrons. The acylated group of the substrate is effectively displaced through the combination of the carbonyl carbon with the enzyme, but not necessarily with an imidazole group. Experiments with O^{18} support this concept.[69] In nonenzymatic reactions OH from the medium appears to attack the sensitive bond and may displace either the acyl oxygen or the acylated group. This permits oxygen from the medium to appear in the unreacted substrate as shown in the following scheme:

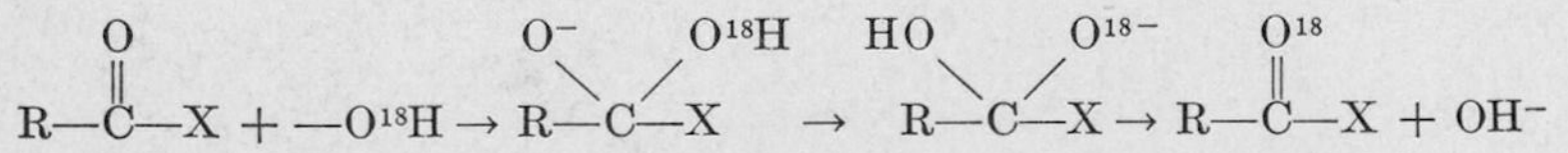

$$
\begin{array}{ccccc}
\text{O} & \text{O}^- \quad \text{O}^{18}\text{H} & \text{HO} \quad \text{O}^{18-} & \text{O}^{18} & \\
\| & \diagdown \diagup & \diagdown \diagup & \| & \\
\text{R—C—X} + \text{—O}^{18}\text{H} \rightarrow \text{R—C—X} & \rightarrow & \text{R—C—X} \rightarrow \text{R—C—X} + \text{OH}^-
\end{array}
$$

This phenomenon does not occur in the enzyme-catalyzed reaction. Therefore it might be concluded that when OH^- or any other acyl acceptor attacks the bound substrate, the acylated group has already been removed. The failure of the carbonyl oxygen to exchange with the medium is not conclusive evidence, however, since the OH^- that adds to the carbonyl group might be the only one to be removed in the reverse reaction. The stereospecificity of such reactions will be discussed later. The ability of OH^- (or H_2O) to attack what is essentially an acylated

---

[67] Early work of Danilewski and others is reviewed by H. Wasteneys and H. Borsook, *Physiol. Revs.* **10**, 110 (1930).

[68] I. B. Wilson and F. Bergmann, *J. Biol. Chem.* **186**, 683 (1950); T. Wagner-Jauregg and B. E. Hackley, Jr., *J. Am. Chem. Soc.* **75**, 2125 (1953).

[69] M. Bender, R. D. Ginger, and K. C. Kemp, *J. Am. Chem. Soc.* **76**, 3350 (1954).

enzyme is shown by the incorporation of isotope into free carboxyl groups in enzyme-catalyzed reactions.[46] The reaction of acyl carbon with the enzyme is a function of its carbonyl properties, as shown by the reaction

$$
\text{C}_6\text{H}_5\text{—CH}_2\text{CH}_2\overset{\text{O}}{\underset{\|}{\text{C}}}\text{—CH}_2\overset{\text{O}}{\underset{\|}{\text{C}}}\text{OC}_2\text{H}_5 + \text{H}_2\text{O} \longrightarrow \text{C}_6\text{H}_5\text{—CH}_2\text{CH}_2\overset{\text{O}}{\underset{\|}{\text{C}}}\text{OH} + \text{CH}_3\overset{\text{O}}{\underset{\|}{\text{C}}}\text{OC}_2\text{H}_5
$$

(V)

of ethyl-5-phenyl-3-ketovalerate with chymotrypsin (V).[45] The enzyme-catalyzed reaction may therefore be indicated thus:

$$
\text{R—}\overset{\text{O}}{\underset{\|}{\text{C}}}\text{—X} + \text{Enzyme} \rightarrow \text{R—}\overset{\text{O}}{\underset{\|}{\text{C}}}\text{—Enzyme} \overset{+\text{Y}}{\longrightarrow} \text{R—}\overset{\text{O}}{\underset{\|}{\text{C}}}\text{—Y} + \text{Enzyme}
$$
$$
\underset{\text{X}}{+}
$$

Direct experimental confirmation of this concept of enzyme action has recently been obtained with the isolation of acetyl-chymotrypsin as an intermediate in the hydrolysis of *p*-nitrophenyl acetate.[70] Kinetic studies have separated the initial reaction, in which acetyl enzyme is formed, from the subsequent hydrolysis of the intermediate.[71] The initial acylation is about 100 times as fast as the hydrolysis. In DFP studies the acylated group appears to be one specific serine hydroxyl group.[72] It is suggested that the pH dependency of the chymotrypsin reaction may be a function of an imidazole residue that has a role in catalysis as an attractive group because of the unshared electrons on the uncharged ring, but does not serve as an acyl acceptor. This scheme, derived from studies with chymotrypsin, may apply to other proteolytic enzymes, but modifications must be expected, as in the case of pepsin, whose activity depends greatly on the group contributing the amino portion of the sensitive bond.

GENERAL REFERENCES

Bergmann, M. (1942). *Advances in Enzymol.* **2**, 49.

Green, N. M., and Neurath, H. (1954). *In* "The Proteins" (H. Neurath and K. Bailey, eds.), Vol. II, Part B, p. 1057. Academic Press, New York.

Johnson, M. J., and Berger, I. (1942). *Advances in Enzymol.* **2**, 69.

Northrop, J. H., Kunitz, M., and Herriott, R. M. (1955). "Crystalline Enzymes." Columbia Univ. Press, New York.

Smith, E. L. (1951). *In* "The Enzymes" (J. B. Sumner and K. Myrbäck, eds.), Vol. I, Part 2, p. 793. Academic Press, New York.

[70] A. K. Balls and H. N. Wood, *J. Biol. Chem.* **219**, 245 (1956).

[71] G. H. Dixon, W. J. Dreyer, and H. Neurath, *J. Am. Chem. Soc.* **78**, 4810 (1956).

[72] N. K. Schaffer, S. C. May, Jr., and W. H. Summerson, *Federation Proc.* **11**, 282 (1952).

Chapter III

FERMENTATION AND OXIDATION OF MAJOR METABOLIC FUELS

Glycolysis

The special reactions that enable organisms to carry out their various functions involve the expenditure of energy. Although many of the conversions found in biological systems may proceed unassisted, many more require the participation of other reactions with which they couple, so that the thermodynamically unfavorable reaction of biological importance occurs as only one component of a thermodynamically favorable system. The mechanism of coupling may consist of adding a reagent or removing a product of a reaction. Another mechanism employed by biological systems is to alter a reaction sequence through the introduction of an extraneous component:

$$\text{Over-all reaction:} \quad A \rightarrow B$$
$$\text{Altered sequence:} \quad A + X \rightarrow Q$$
$$Q \rightarrow B + Y$$

When the extraneous component is eliminated later in the sequence of reactions, the products may be indistinguishable from those in a simpler system, but the change in state of the extraneous component has been coupled. When two over-all reactions are coupled, it is possible to determine which sequence is thermodynamically favorable and to ascribe to that system the function of supporting the other coupled reaction. While this is often a convenient mental aid, it can be misleading. The reaction that supplies the energy in one direction is the recipient of assistance when the system operates in the reverse direction. The assignment of "driving power" to part of a coupled system also carries an implication that the favorable reaction is free to couple or not, to use its "energy" or squander it. In general this is not the case; coupled reactions not only permit unfavorable reactions to occur, but mechanistically the favorable and unfavorable aspects of a system are so tied together that neither can proceed without the other.

In the metabolism of a cell or of complicated organisms, processes of great quantitative significance occur that result in the synthesis of molecules that are necessary for the formation of biological compounds and for the performance of physiological functions. These processes are the energy-yielding reaction sequences that couple with the reactions that make the improbable structures of biology. One of the over-all conversions widely used among biological systems to couple with synthetic or functional processes is the rearrangement of glucose to form smaller molecules of greater stability. The reaction sequences vary in different systems as to final products, but the intermediate reactions have been found to be remarkably similar in microorganisms, plants, and animals. The pathway taken in these various organisms for the degradation of glucose without the net utilization of other molecules in the formation of the carbon-containing products is called *glycolysis*. Glycolysis is an example of *fermentation*, which currently means a metabolic sequence for the degradation of any compound that does not involve the participation of extraneous oxidizing agents, such as molecular oxygen.

The over-all reaction of glycolysis in muscle is sometimes written glucose ($C_6H_{12}O_6$) $\rightarrow$ 2 lactic acid ($C_3H_6O_3$) $+$ 58 kcal. Such a designation is misleading on two counts. First, it ignores the participation of other compounds that are required for the actual reactions of glycolysis, and second, it implies that the standard free energy of a reaction is a tangible quantity, to be used as is convenient. The free energy of biological reactions will be considered later. At this point only the descriptive chemistry of glycolytic reactions will be considered.

During glycolysis in muscle each glucose molecule is indeed converted to two molecules of lactic acid. In yeast the products of glucose degradation are ethanol and CO_2. The products of fermentation of some microorganisms include butanol, acetone, butyric acid, acetic acid, acetoin, butylene glycol, propionic acid, hydrogen, and other compounds in addition to the products formed by muscle and yeast. In general, all of these products result from variations in the terminal reactions of very similar sequences.

Cofactors of Glycolysis

In these reactions, inorganic phosphate and certain cofactors are involved. Adenine nucleotides will appear repeatedly in considerations of various enzyme reactions, since they are the compounds most used in coupling glycolytic and other energy-yielding processes with energy-requiring systems. In addition to the simple nucleotides, an essential component of glycolyzing systems is a more complicated nucleotide, the cofactor originally isolated by Harden and Young.[1] Since the complex of enzymes responsible for glycolysis had been named *zymase*, the dia-

[1] A. Harden and W. J. Young, *J. Chem. Soc.* (proceedings) **21**, 189 (1905).

lyzable cofactor was named *cozymase*.[2] The nucleotide cofactors will be discussed before the individual reactions of glycolysis, which can be appreciated much better when the cofactors are well-known chemical entities than when they are alphabetical symbols of the mystery of biochemistry.

Adenine Nucleotides. Adenosine triphosphate (ATP) was isolated by Lohmann in 1929.[3] It has the structure shown below. The molecule may

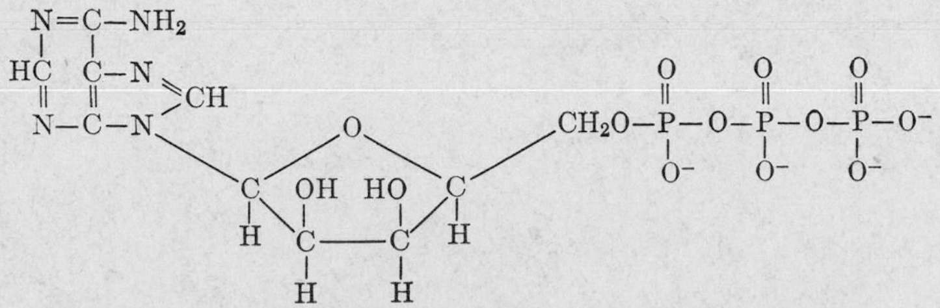

ATP

be considered as the condensation product of the purine base, adenine (Ad), the pentose, ribose (R), and three phosphate (P) groups. Compounds composed of a base, sugar, and one or more phosphates are called nucleo*ti*des. The dephosphorylated compounds, composed of only base and sugar, are termed nucleo*si*des. For many years ATP occupied a unique position in biochemistry. Recently its position has been changed to one of preeminence in a family of biologically occurring nucleoside triphosphates, in which uracil, guanine, hypoxanthine, cytosine, and perhaps other bases replace adenine. In such compounds the pyrophosphate bonds are easily hydrolyzed by acid, whereas the ester phosphate is relatively stable. Specific enzymes also occur that hydrolyze either the terminal or the two pyrophosphate bonds. The resulting diphosphate (ADP) or monophosphate (AMP) also occur naturally.

$$\text{AdRPPP (ATP)} \xrightarrow{\text{ATPase}} \text{AdRPP (ADP)} + \text{P}$$
$$\text{AdRPPP} \xrightarrow{\text{Apyrase}} \text{AdRP (AMP)} + 2\text{P}$$

The adenine nucleotides are equilibrated with each other by a widely distributed enzyme originally found in muscle, therefore named *myokinase*.[4] The general term *kinase* is now usually restricted to enzymes that transfer phosphate groups to specific acceptors. Myokinase has been renamed *adenylate kinase*, but the older name is still found in current

---

[2] H. von Euler and K. Myrbäck, *Z. physiol. Chem.* **131**, 179 (1923).
[3] K. Lohmann, *Naturwissenschaften* **17**, 624 (1929).
[4] S. P. Colowick and H. M. Kalckar, *J. Biol. Chem.* **148**, 117 (1943).

literature. The reaction catalyzed is:

$$ATP + AMP \rightleftharpoons 2ADP$$

This reaction is freely reversible (K_{eq} near 1). Similar reactions have been found for the phosphorylation of other nucleoside monophosphates. An enzyme, nucleoside diphosphate kinase, transfers a terminal phosphate from any of the nucleoside triphosphates to any of the diphosphates.[5] All of the pyrophosphate bonds are very similar, and the equilibrium constants of these transfers are near unity.

The biosynthesis of adenine and of ribose will be discussed later. In biochemistry there is no starting point for syntheses; instead there are

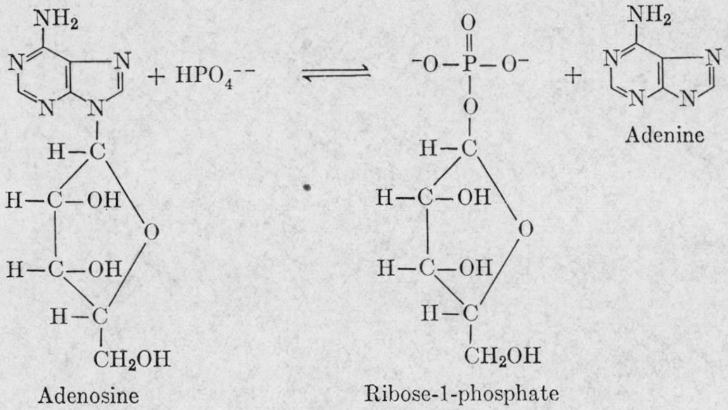

Adenosine   Ribose-1-phosphate
(I)

multiple cycles, in which the substrate and product roles are indistinguishable. Assuming, therefore, the existence of certain components, the synthesis of others may be considered. Two types of reaction are known to form adenine-ribose bonds. These are: (a) replacement of phosphate from ribose-1-phosphate; and (b) replacement of pyrophosphate from 5-phosphoribose-1-pyrophosphate.

Reaction (a) is catalyzed by nucleoside phosphorylase.[6] The reaction was originally detected by Kalckar as the phosphate-dependent splitting of nucleosides (I), and the detection of the resulting ribose-1-phosphate depended upon the availability of the Lowry-Lopez method for determination of inorganic phosphate in the presence of labile phosphate esters.[7] Although the enzyme initially described from rat liver was not active with adenine, the corresponding beef enzyme was found to use

[5] P. Berg and W. K. Joklik, Nature 172, 1008 (1953).
[6] H. M. Kalckar, J. Biol. Chem. 167, 477 (1947).
[7] O. H. Lowry and J. A. Lopez, J. Biol. Chem. 162, 421 (1946).

this compound in addition to other bases.[8] The reaction has an equilibrium constant of 16. If arsenate is present, so-called *arsenolysis* results in the formation of the products of hydrolysis: base and ribose. This is a reaction that occurs with many analogous systems, and presumably occurs by formation of unstable arsenate compounds in place of phosphate compounds (II).

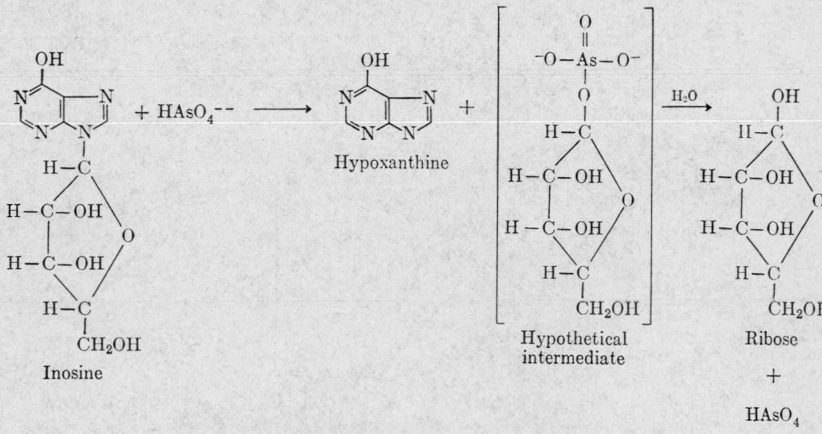

Presumed Mechanism of Arsenolysis
(II)

A reaction of deoxynucleosides was described by MacNutt[9] in which a transglycosidation results in the exchange of one base for another with-

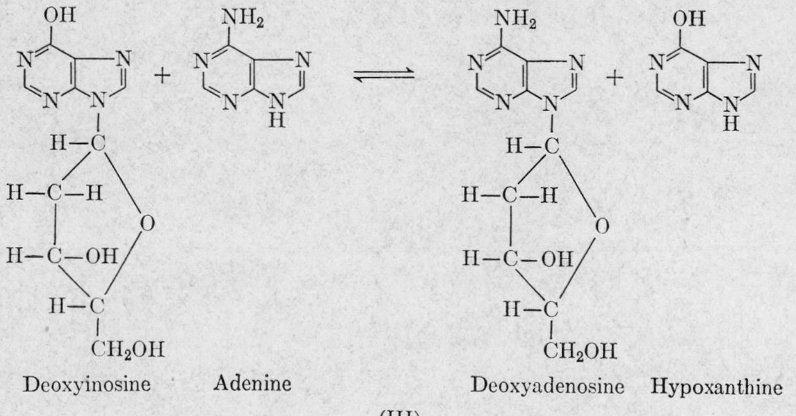

(III)

out the intermediate formation of a sugar phosphate (III). This type of reaction has not been described for ribosides (in contrast to deoxyribo-

[8] E. Korn and J. Buchanan, *J. Biol. Chem.* **217**, 183 (1955).
[9] W. S. MacNutt, *Biochem. J.* **50**, 384 (1952).

sides), but the existence of the analogous reaction encourages the speculation that a transfer of ribose from one base to another may take place.

If adenosine (adenine riboside) is formed by either of the two mechanisms discussed above, the formation of adenylic acid (AMP) and the pyrophosphates depends upon the introduction of a phosphate group. Kinases for the formation of riboside 5'-phosphates from the riboside and ATP are known.[10] Another reaction sequence (reaction b) has also been found for the formation of nucleotides (IV).[11] Individual enzymes have been found for handling orotic acid (a pyrimidine precursor) and adenine

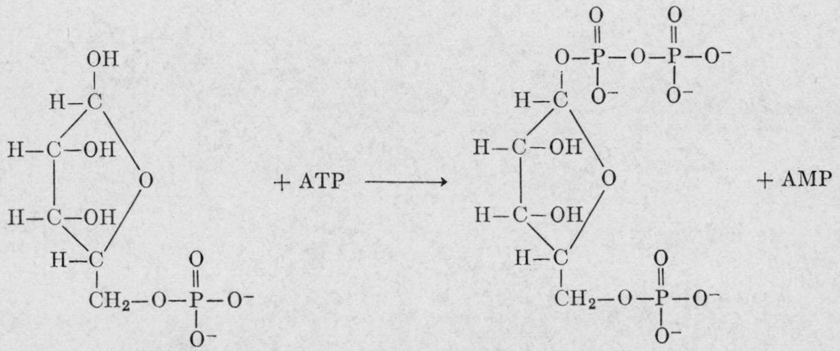

Ribose-5-phosphate   5-Phosphoribose-1-pyrophosphate (PRPP)

$$\text{PRPP} + \text{base} \longrightarrow \underset{\text{nucleotide}}{\text{base-R-P}} + \text{PP}$$

(IV)

(a purine). These enzymes appear to be specific with respect to the one base found to react with each. Other such enzymes have been found that react with other bases.

Pyridine Nucleotides. Nucleoside phosphorylase is capable of using the base nicotinamide.[12a] The product of the reaction of this base with ribose-1-phosphate is nicotinamide riboside. The formation of the corresponding nicotinamide mononucleotide is catalyzed by a typical kinase, using ATP. A specific enzyme purified from human erythrocytes has been shown to form nicotinamide mononucleotide by a second mechanism[12b] in which PRPP and nicotinamide react to form inorganic pyrophosphate and

---

[10] R. Caputto, *J. Biol. Chem.* **189,** 801 (1951). The carbon atoms of the ribose in nucleosides and nucleotides are numbered 1' through 5' to distinguish them from the numbers used to designate the atoms of the base linked to carbon 1'.

[11] A. Kornberg, I. Lieberman, and E. S. Simms, *J. Biol. Chem.* **215,** 389, 417 (1955).

[12a] J. W. Rowen and A. Kornberg, *J. Biol. Chem.* **193,** 497 (1951).

[12b] J. Preiss and P. Handler, *J. Biol. Chem.* **225,** 759 (1957).

the nucleotide. This nucleotide reacts with ATP to form a dinucleotide, in which the mononucleotides of nicotinamide and adenine are joined through a pyrophosphate bond.[13] In the condensation, inorganic pyrophosphate is formed from the terminal phosphates of ATP. The condensation is freely reversible. The dinucleotide is called diphosphopyridine nucleotide (DPN). It is the cofactor originally detected by Harden and

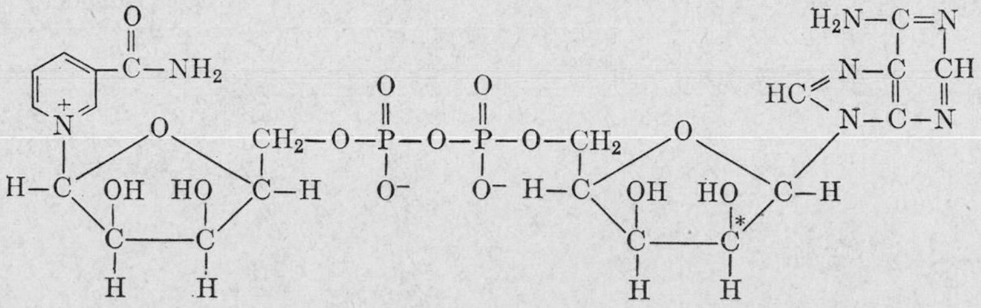

*Site of third phosphate of TPN

DPN

Young as an essential component of glycolyzing yeast extracts, and called cozymase. Another very similar pyridine nucleotide containing a third phosphate group was isolated and analyzed at the same time that the structure of DPN was being determined.[14] This is called triphosphopyridine nucleotide, TPN. DPN and TPN have also been called coenzyme I and coenzyme II, respectively. These nucleotides are ubiquitous in biology and will be mentioned repeatedly in connection with oxidative reactions. The reactive group of DPN in biological oxidations is the pyridine ring. This can accept two electrons and a proton to form a quinoid structure shown in (V). One electron neutralizes the positive

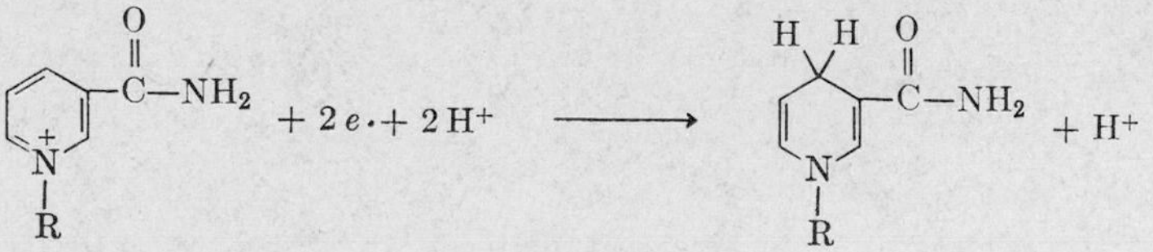

(V)

change on the nitrogen of the pyridine ring, and the other participates in the second carbon-hydrogen bond at position 4.[15] Ordinarily the substrate loses two protons as well as two electrons; the second proton remains in the medium as a hydrogen ion. Reduced DPN is designated DPNH; the corresponding reduced TPN is TPNH. DPN and TPN are

[13] A. Kornberg, *J. Biol. Chem.* **182,** 779 (1950).
[14] O. Warburg and W. Christian, *Biochem. Z.* **287,** 291 (1936).
[15] M. E. Pullman, A. San Pietro, and S. P. Colowick, *J. Biol. Chem.* **206,** 129 (1954).

cofactors for a large group of oxidative enzymes, named dehydrogenases. The reactions of these enzymes are conveniently followed spectrophotometrically.[16] The sensitivity, convenience and rapidity of the spectrophotometric method has made this analytical device one of the favorite tools of modern enzyme chemists. The ultraviolet-absorption spectra of oxidized and reduced pyridine nucleotides are shown in Fig. 8. The change in absorption on oxidation or reduction is most marked at 340 mμ; at this wavelength the molar extinction coefficient, ϵ, of reduced pyridine nucleotides is 6.22×10^3.[17] That is, a $1\ M$ solution in a vessel with a 1 cm. light path would have an optical density of 6220, and more dilute solutions have proportionally small optical densities ($10^{-4}\ M$, 0.1 μM per ml., DPNH or TPNH has an optical density of 0.622).

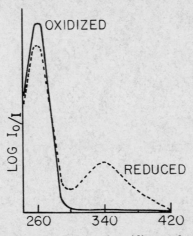

FIG. 8. Absorption spectra of pyridine nucleotides.[17a]

Spectrophotometry is well adapted for rate measurements, since readings do not interfere with the progress of a reaction. It has been found convenient to assay several enzymes that do not use pyridine nucleotides by coupling their reactions with those of dehydrogenases. In some studies several enzymes carry out intermediate reactions between the one being assayed and the one catalyzing pyridine nucleotide oxidation or reduction. Yet, if sufficient quantities of all of the other enzymes are present, the first enzyme in a sequence can be assayed accurately by measuring the rate of change in optical density at 340 mμ. This type of assay has been particularly useful in the study of glycolytic enzymes, and examples will be shown later. The measurement of the amount of

[16] O. Warburg, W. Christian, and A. Griese, *Biochem. Z.* **282,** 157 (1935).
[17] B. L. Horecker and A. Kornberg, *J. Biol. Chem.* **175,** 385 (1948).
[17a] F. Schlenk, *Symposium on Respirat. Enzymes*, p. 104 (1942).

change at 340 mμ can also be used as a quantitative determination of many substrates. The sensitivity of the spectrophotometric determination of oxidation or reduction has allowed this method to be used in a wide variety of experimental conditions. For special purposes an ultra-microanalytical technique was developed that distinguishes between oxidized and reduced pyridine nucleotides by their fluorescence. This method can measure smaller amounts of material than any other method in current use for enzyme determinations.[17b]

REACTION OF GLYCOLYSIS

In glycolysis carbohydrate metabolism is tightly coupled with phosphate metabolism. In the initial reaction sugar is phosphorylated, and the succeeding steps are concerned with the metabolism of various phosphate esters. Table 2 is a flowchart which lists the reactions of glycolysis.

TABLE 2
REACTIONS OF GLYCOLYSIS

Reaction	Enzyme
Glucose + ATP → glucose-6-P + ADP	Hexokinase
Glucose-6-P $\rightleftharpoons$ fructose-6-P	Phosphohexose isomerase
Fructose-6-P + ATP → fructose-1,6-diP	Phosphofructokinase
Fructose-1,6-diP $\rightleftharpoons$ dihydroxyacetone-P + 3-phosphoglyceraldehyde	Aldolase
Dihydroxyacetone-P $\rightleftharpoons$ 3-phosphoglyceraldehyde	Triose phosphate isomerase
3-Phosphoglyceraldehyde + DPN + P $\rightleftharpoons$ 1,3-diphosphoglyceric acid + DPNH	Triose phosphate dehydrogenase
1,3-Diphosphoglycerate + ADP $\rightleftharpoons$ 3-phosphoglycerate + ATP	Phosphoglycerate kinase
3-Phosphoglycerate $\rightleftharpoons$ 2-phosphoglycerate	Phosphoglyceromutase
2-Phosphoglycerate $\rightleftharpoons$ phospho(enol)pyruvate + H_2O	Enolase
Phosphopyruvate + ADP → pyruvate + ATP	Pyruvate kinase
Pyruvate + DPNH $\rightleftharpoons$ lactate + DPN	Lactic dehydrogenase (muscle)
Pyruvate → acetaldehyde + CO_2	Carboxylase ⎫ (yeast)
Acetaldehyde + DPNH $\rightleftharpoons$ ethanol + DPN	Alcohol dehydrogenase ⎬

These reactions have been designated the Embden-Meyerhof scheme, in recognition of the pioneer discoveries that led to the acceptance of this pathway, but workers in many other laboratories have contributed heavily to the study of the glycolytic enzymes. The reactions indicated as unidirectional are in fact reversible, but under ordinary circumstances the reverse reactions do not occur to significant extents. In order to reverse the glycolytic system, additional reactions by-pass the "irreversi-

[17b] O. H. Lowry, N. R. Roberts, and C. Lewis, *J. Biol. Chem.* **220,** 879 (1956).

ble" steps. Other reactions not included in the Embden-Meyerhof scheme also will be considered as they add or remove compounds involved in glycolysis.

Hexokinase. Glucose enters the glycolytic pathway by reacting with ATP in a reaction catalyzed by *hexokinase*. Hexokinase was used by Meyerhof to designate the yeast protein used to supplement muscle extracts to permit glucose fermentation.[18] The reaction was defined by

$$
\begin{array}{ccc}
\text{Glucose} + \text{ATP} & \longrightarrow & \text{Glucose-6-phosphate} + \text{ADP}
\end{array}
$$

(VI)

Colowick and Kalckar[19] as shown in (VI). Hexokinase activity has been determined in a variety of animal tissues as well as in microorganisms. The yeast enzyme has been crystallized.[20a] It is a protein with a molecular weight of 96,600, an isoelectric point at pH 4.5–4.8, and 3 atoms of phosphorus per mole. The enzyme phosphorylates glucose, fructose, and mannose, but not galactose, arabinose, xylose, or disaccharides. The affinity for glucose is much greater than for fructose ($K_m = 1.5 \times 10^{-4}$ and 1.5×10^{-3}, respectively) but V_{max} with fructose is twice that with glucose. All of the hexokinases require Mg^{++} or similar cations for activity, and it is probable that the true substrate is not ATP but its mono Mg^{++} complex. The hexokinase reaction proceeds quantitatively to completion, but experiments with labeled substrates have shown that measurable amounts of back-reaction do occur.[20b] The equilibrium constant of the reaction was determined with the use of isotope dilution techniques, and values of 3.86×10^2 and 1.55×10^2 were obtained in the absence and presence of added Mg^{++} at pH 6.0. The equilibrium constant increases by approximately a factor of 10 per pH unit in the region pH 6 to 8.[20c]

The determination of hexokinase activity may be made in several ways. In general the transfer of a phosphate from ATP to a hydroxyl group (including that of water) results in the formation of a new dis-

[18] O. Meyerhof, *Biochem. Z.* **183**, 176 (1927).

[19] S. P. Colowick and H. Kalckar, *J. Biol. Chem.* **148**, 117 (1943).

[20a] M. Kunitz and M. MacDonald, *J. Gen. Physiol.* **29**, 393 (1946); L. Berger, M. W. Slein, S. P. Colowick, and C. F. Cori, *ibid.* **29**, 379 (1946).

[20b] J. L. Gamble and V. A. Najjar, *Science* **120**, 1023 (1954); S. Kaufman, *J. Biol. Chem.* **216**, 153 (1955).

[20c] E. A. Robbins and P. D. Boyer, *J. Biol. Chem.* **224**, 121 (1957).

sociable acid group (at neutral pH values) (VII). The appearance of this group can be measured manometrically by evolution of CO_2 from a bicarbonate buffer, electrometrically with the glass electrode, or colorimetrically by measuring the change in light absorption of a pH indicator. Other methods for following hexokinase activity are determination of free glucose as reducing sugar after precipitation of salts of the phosphate esters and determination of the phosphate ester by a specific enzyme reaction (Zwischenferment) to be discussed later.

$$R—OH + \text{adenine-ribose}—O—\overset{\overset{O}{\|}}{\underset{\underset{O^-}{|}}{P}}—O—\overset{\overset{O}{\|}}{\underset{\underset{O^-}{|}}{P}}—O—\overset{\overset{O}{\|}}{\underset{\underset{O^-}{|}}{P}}—O^- \longrightarrow$$

$$R—O—\overset{\overset{O}{\|}}{\underset{\underset{O^-}{|}}{P}}—O^- + \text{adenine-ribose}—O—\overset{\overset{O}{\|}}{\underset{\underset{O^-}{|}}{P}}—O—\overset{\overset{O}{\|}}{\underset{\underset{O^-}{|}}{P}}—O^- + H^+$$

(VII)

Several of these methods were used by Sols and Crane in studies of hexokinase in brain and heart muscle preparations.[21] By using many structural analogs of D-glucose, they found that calf brain hexokinase does not have an absolute requirement for any of the groups as they exist in the glucose molecule. A terminal primary hydroxyl group is required, but it need not be in position 6; pentoses and heptoses also are phosphorylated. As shown in Table 3, compounds lacking the hydroxyl groups on positions 1, 3, and 4 or altered in steric configuration at these positions remain substrates, but the affinity of the enzyme is greatly influenced by these substituents. In contrast, the hydroxyl group on position 2 has very little influence on substrate binding. However a large substituent on position 2, an acetylamine group, converts a substrate to an inhibitor. The ring form of the sugars appears to be essential, also. Methyl groups on carbon 1, 2, or 3 convert glucose to inhibitors.

Essentially all of the brain enzyme is associated with particles, from which the enzyme has been separated only with difficulty and in poor yield. Both soluble and particulate fractions from heart muscle contain hexokinase activity. These preparations are grossly similar, but vary somewhat in response to pH changes. All of these enzymes are inhibited by glucose-6-phosphate and some analogous 6-phosphates. A recommended substrate for the assay of hexokinase is 2-deoxyglucose, because it does not form an inhibitory ester, and its phosphorylation product is not further metabolized by other enzymes.

[21] A. Sols and R. K. Crane, *J. Biol. Chem.* **210**, 581, 597 (1954); *Federation Proc.* **13**, 301 (1954).

TABLE 3
EFFECT OF STRUCTURE ON HEXOKINASE ACTIVITY[a]

Modified at carbon	Compound	K_m (moles/l.)	Relative maximal rate
	Glucose	8×10^{-6}	1.0
1	1,5-Sorbitan	3×10^{-2}	1.0
1, 2	1,5-Mannitan	2×10^{-2}	0.9
1	Glucoheptulose	2×10^{-4}	0.006
1, 2	Mannoheptulose	5×10^{-5}	0.015
1, 2	Fructose	1.6×10^{-3}	1.5
1, 2	2,5-Sorbitan	0.15	0.08
1, 2	Arabinose	2	0.1
1	Methyl-α-glucoside	—	—
1	Methyl-β-glucoside	—	—
1	1-Thioglucose	—	—
1,2	Glucal	—	—
1	α-Glucose-1-phosphate	—	—
1	Glucono-1,5-lactone	—	—
2	Mannose	5×10^{-6}	0.4
2	2-Deoxyglucose	2.7×10^{-5}	1.0
2	Glucosamine	8×10^{-5}	0.6
2	Glucosone	1×10^{-5}	0.08
2	N-Acetylglucosamine	8×10^{-5}	—
2	N-Methylglucosamine	2×10^{-4}	—
2	2-O-Methylglucose	—	—
3	Allose	7×10^{-3}	0.5
3, 2	Altrose	3×10^{-3}	0.11
3	3-Deoxyglucose	1.5×10^{-2}	0.2
3	3-O-Methylglucose	—	—
4	Galactose	0.1	0.02
5, 4	1,4-Sorbitan	—	—
5, 1	L-Sorbose	—	—
6	6-Deoxyglucose	2×10^{-3}	—
6	Xylose	2×10^{-3}	—
6, 2	Lyxose	1.3×10^{-3}	—
6, 3	Ribose	—	—
	Sorbitol	—	—
	Glucoguloheptose	—	—
	i-Inositol	—	—

[a] From Sols and Crane, J. Biol. Chem.[21]

Galactokinase. Galactose is phosphorylated by a specific galactokinase, formed in liver and yeast.[22] The product is galactose-1-phosphate (VIII). Galactose phosphate is converted to glucose phosphate by reactions to be discussed later (p. 242).

[22] H. W. Kosterlitz, *Biochem. J.* **37**, 322 (1943); R. E. Trucco, R. Caputto, L. Leloir, and N. Mittelman, *Arch. Biochem.* **18**, 137 (1948).

$$
\begin{array}{ccc}
& & \overset{\displaystyle O}{\underset{\displaystyle |}{\parallel}} \\
& & ^-O\!-\!\overset{|}{P}\!-\!O^- \\
& & | \\
& & O \\
\text{OH} & & | \\
| & & \\
\text{HC}\!\!-\!\!\rule{0pt}{2ex} & & \text{HC}\!\!-\!\!\rule{0pt}{2ex} \\
| & & | \\
\text{HCOH} & & \text{HCOH} \quad O \\
| & & | \\
\text{HOCH} \quad O & \xrightarrow[\text{galactokinase}]{+\ \text{ATP}} & \text{HOCH} \\
| & & | \\
\text{HOCH} & & \text{HOCH} \\
| & & | \\
\text{HC}\!\!-\!\!\rule{0pt}{2ex} & & \text{HC}\!\!-\!\!\rule{0pt}{2ex} \\
| & & | \\
\text{CH}_2\text{OH} & & \text{CH}_2\text{OH} \quad + \text{ADP} \\
\text{Galactose} & & \text{Galactose-1-phosphate}
\end{array}
$$

VIII

Phosphoglucomutase. Glucose-6-phosphate may react in several ways. An oxidative reaction will be discussed later (p. 116). The reactions leading to polysaccharide formation are usually considered as part of glycolysis, since the reverse process is used in the fermentation of the polysaccharide. In this sequence, glucose-6-phosphate is first converted to glucose-1-phosphate by phosphoglucomutase.[23] This enzyme was crystallized by Najjar.[24] It was shown early that the reaction mechanism involves a transfer of phosphate from carbon 6 to carbon 1 without exchange with inorganic phosphate. When Leloir and his collaborators[25] found that glucose-1,6-diphosphate is an essential cofactor for this enzyme, a reaction mechanism was proposed that seemed adequate and straightforward (IX). According to this scheme the phosphate on one

$$
\begin{array}{ccccccc}
1\ \text{C}-\text{P} & 1\ \text{C} & & 1\ \text{C}-\text{P} & & 1\ \text{C}-\text{P} \\
| & | & & | & & | \\
\text{C} & \text{C} & & \text{C} & & \text{C} \\
| & | & & | & & | \\
\text{C} & \text{C} & & \text{C} & & \text{C} \\
| \quad + & | & \rightleftharpoons & | \quad + & & | \\
\text{C} & \text{C} & & \text{C} & & \text{C} \\
| & | & & | & & | \\
\text{C} & \text{C} & & \text{C} & & \text{C} \\
| & | & & | & & | \\
6\ \text{C}-\text{P} & 6\ \text{C}-\text{P} & & 6\ \text{C} & & 6\ \text{C}-\text{P}
\end{array}
$$

(IX)

position of the cofactor is transferred to the opposite position of a glucose monophosphate. The cofactor thus becomes a product of the reaction, a monophosphate with phosphate on the opposite end from that of the

[23] E. W. Sutherland, S. P. Colowick, and C. F. Cori, *J. Biol. Chem.* **140,** 309 (1941).
[24] V. Najjar, *J. Biol. Chem.* **175,** 281 (1948).
[25] C. E. Cardini, A. C. Paladini, R. Caputto, L. F. Leloir, and R. E. Trucco, *Arch. Biochem.* **22,** 87 (1949).

original substrate. The substrate that accepts the phosphate is converted to a diphosphate, which donates its original phosphate to another substrate molecule. In this manner a phosphate group in position 6 is shifted to position 1, but of a different carbon chain.

Using stoichiometric amounts of enzyme, Najjar and Pullman[26a] have shown that the enzyme becomes phosphorylated, and proposed a reaction sequence:

$$\text{Enzyme-P} + \text{glucose-1-phosphate} \rightleftharpoons \text{Enzyme} + \text{glucose-1,6-diphosphate}$$
$$\text{Enzyme} + \text{glucose-1,6-diphosphate} \rightleftharpoons \text{Enzyme-P} + \text{glucose-6-phosphate}$$

At equilibrium the ratio of glucose-6-phosphate to glucose-1-phosphate is about 16–20 to 1, varying somewhat with temperature, but being independent of pH in the range 6.2 to 7.5.

The transfer of phosphate groups between aldehyde and primary alcohol groups is not peculiar to esters of glucose. Mannose, ribose, N-acetylglucosamine, glucosamine, and galactose phosphate esters all undergo similar reactions, and are probably substrates for the same enzyme, since glucose-1,6-diphosphate stimulates the reaction with each ester. It has been found, however, that additional enzymes of this type occur, and relatively specific phosphoribose and phosphoacetylglucosamine mutases have been described.[26b] Fructose phosphates do not react with phosphomutases.

The phosphoglucomutase reaction involves continual utilization and resynthesis of the cofactor, but it cannot cause the net synthesis of glucose-1,6-diphosphate. Two mechanisms have been found to increase the amount of this cofactor; in each case glucose-1-phosphate is the phosphate acceptor but the phosphate donors are different for the two enzymes involved. Glucose-1-phosphate kinase[26c] requires ATP to carry out the reaction:

$$\text{glucose-1-phosphate} + \text{ATP} \rightarrow \text{glucose-1,6-diphosphate} + \text{ADP}$$

Glucose-1-phosphate transphosphorylase[26d] uses glucose-1-phosphate as the phosphate donor as well as the acceptor.

$$2 \text{ glucose-1-phosphate} \rightarrow \text{glucose-1,6-diphosphate} + \text{glucose}$$

[26a] V. Najjar and M. E. Pullman, *Science* **119**, 631 (1954).
[26b] A. J. Guarino and H. Z. Sable, *Federation Proc.* **13**, 222 (1954), J. L. Reissig, *J. Biol. Chem.* **219**, 753 (1956).
[26c] A. C. Paladini, R. Caputto, L. F. Leloir, R. E. Trucco, and C. E. Cardini, *Arch. Biochem.* **23**, 55 (1949).
[26d] L. F. Leloir, R. E. Trucco, C. E. Cardini, and R. Caputto, *Arch. Biochem.* **24**, 65 (1949); J. B. Sidbury, Jr., L. L. Rosenberg, and V. A. Najjar, *J. Biol. Chem.* **222**, 89 (1956).

The transphosphorylation reaction is unusual in its ability to use a sugar phosphate as a phosphate donor. Except for the reaction in which phosphoglucomutase participates as a phosphate acceptor, no similar reactions of sugar phosphates have been described. The kinase has been demonstrated in extracts of yeast and rabbit muscle; the transphosphorylase has been found in *E. coli* and rabbit muscle.

Phosphohexose Isomerase. The reactions of glucose-1-phosphate will be discussed later. In glycolysis, glucose-6-phosphate is converted to fructose-6-phosphate. This reaction, catalyzed by phosphohexose isomerase, was found by Lohmann to come to an equilibrium when 70 per cent of the sugar is glucose.[27] Slein[28] has distinguished a phosphomannose isomerase from phosphoglucose isomerase. The two enzymes apparently react with fructose-6-phosphate to form different stereoisomers. This type of reaction occurs nonenzymatically in alkaline solutions, presumably through the intermediate formation of the ene-diol common to glucose, fructose, and mannose. It is not known whether this structure is an intermediate in the enzymatic reactions.

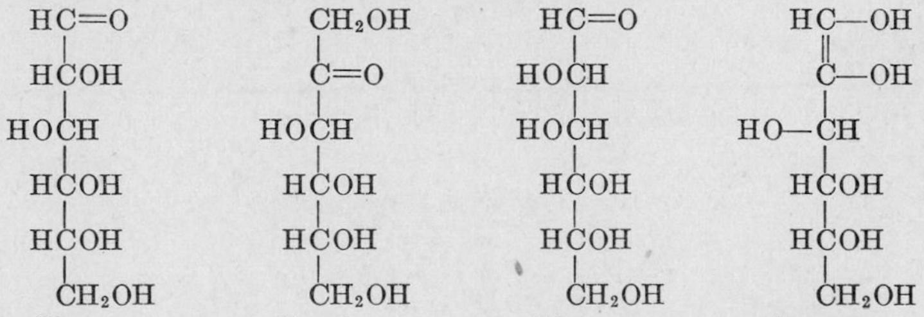

HC=O	CH$_2$OH	HC=O	HC—OH
HCOH	C=O	HOCH	C—OH
HOCH	HOCH	HOCH	HO—CH
HCOH	HCOH	HCOH	HCOH
HCOH	HCOH	HCOH	HCOH
CH$_2$OH	CH$_2$OH	CH$_2$OH	CH$_2$OH
Glucose	Fructose	Mannose	Common ene-diol

Phosphofructokinase. Fructose-6-phosphate is phosphorylated on position 1 by the action of phosphofructokinase (X),[29] which can use ATP, UTP, or ITP.[30] As with other kinases, Mg^{++} is an essential cofactor. The enzyme contains essential SH groups. This is the second essentially irreversible reaction of glycolysis, since the equilibrium lies far to the side of hexose diphosphate formation.

The further metabolism of hexose diphosphate (HDP) was elucidated by Meyerhof and his collaborators.[31] Initially they found that crude enzyme preparations converted HDP to dihydroxyacetone phosphate, and named the enzyme system involved *zymohexase*. This term is no

[27] K. Lohmann, *Biochem. Z.* **262**, 137 (1933).
[28] M. W. Slein, *Federation Proc.* **13**, 299 (1954).
[29] P. Ostern, J. Guthke, and J. Terszakowec, *Z. physiol. Chem.* **243**, 9 (1936).
[30] K.-H. Ling and W. L. Byrne, *Federation Proc.* **13**, 253 (1954).
[31] O. Meyerhof and K. Lohmann, *Biochem. Z.* **271**, 89 (1939).

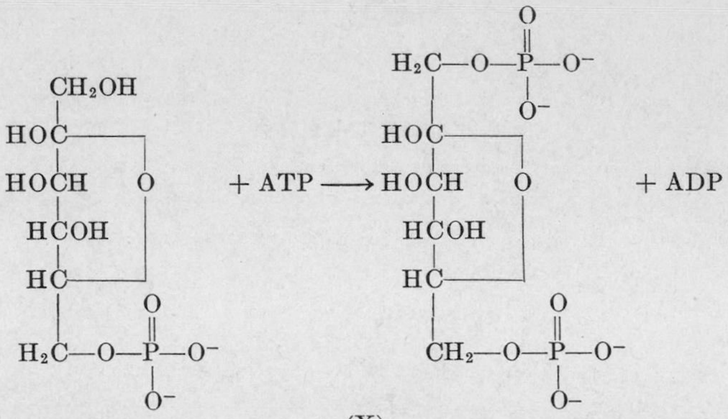

$$
\begin{array}{ccc}
\text{CH}_2\text{OH} & & \text{H}_2\text{C—O—P—O}^- \\
\text{HOC——} & & \text{HOC——} \quad \overset{\text{O}}{\underset{\text{O}^-}{\|}} \\
\text{HOCH} \quad \text{O} & + \text{ATP} \longrightarrow & \text{HOCH} \quad \text{O} \quad + \text{ADP} \\
\text{HCOH} & & \text{HCOH} \\
\text{HC——} & & \text{HC——} \\
\text{H}_2\text{C—O—P—O}^- & & \text{CH}_2\text{—O—P—O}^- \\
\text{O}^- & & \text{O}^-
\end{array}
$$

(X)

longer in common usage, as the individual enzymes, *aldolase* and *phosphotriose isomerase*, are known to carry out the over-all reaction.

Aldolase. Aldolase splits HDP to triose phosphates. These can be identified by their alkali lability, by their conversion to methyl glyoxal in acid, by derivatives of their carbonyl groups, and by enzymatic reactions to be discussed below. The availability of synthetic triose phosphates made individual identification possible, and it was shown that the initial reaction products are equivalent amounts of D-3-phosphoglyceraldehyde and dihydroxyacetone phosphate, and that these react back to form HDP. The condensation is an example of the well-known aldol condensation, hence the name aldolase (XI).

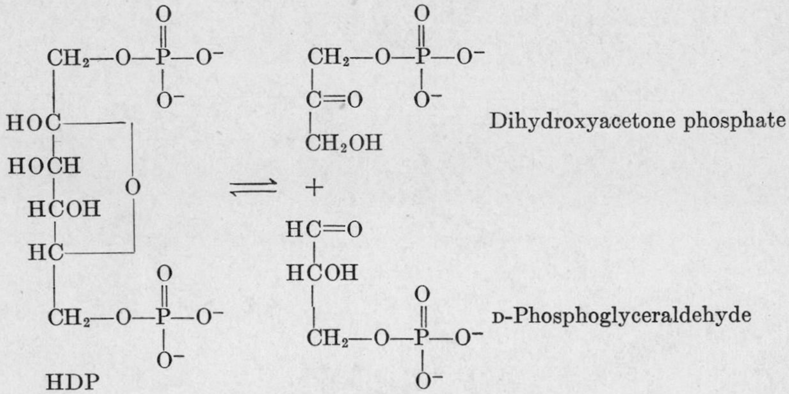

$$
\begin{array}{lll}
\text{CH}_2\text{—O—P—O}^- & \text{CH}_2\text{—O—P—O}^- & \\
\text{HOC——} & \text{C=O} & \text{Dihydroxyacetone phosphate} \\
\text{HOCH} & \text{CH}_2\text{OH} & \\
\text{HCOH} \quad \rightleftharpoons + & & \\
\text{HC——} & \text{HC=O} & \\
\text{CH}_2\text{—O—P—O}^- & \text{HCOH} & \\
\text{O}^- & \text{CH}_2\text{—O—P—O}^- & \text{D-Phosphoglyceraldehyde} \\
\text{HDP} & \text{O}^- &
\end{array}
$$

(XI)

Aldolase has been crystallized from muscle by various methods.[32] The crystalline enzyme is quite specific for the dihydroxyacetone portion of the molecule, but very nonspecific for the aldehyde. The only structure

[32] J. F. Taylor, A. A. Green, and G. T. Cori, *J. Biol. Chem.* **173**, 591 (1948).

found to substitute for dihydroxyacetone is the corresponding methyl compound, acetol phosphate. With dihydroxyacetone the hydroxyl groups assume a *trans* configuration about the new bond. This probably reflects an asymmetric activation of the substrate, but may also measure the tendency of the spontaneous reaction to form predominantly *trans* configurations. The apparent requirement for *trans* hydroxyl groups is not absolute, since aldolase has been found to split tagatose-1,6-diphosphate, in which the 3 and 4 hydroxyls have a *cis* configuration.[33] In place of phosphoglyceraldehyde, formaldehyde, acetaldehyde, glyceraldehyde, and many other aldehydes have been found to react. In general, the equilibriums of these reactions lie far to the side of condensation. This is true to a lesser extent of fructose diphosphate.

The equilibrium constant of the aldolase reaction depends greatly on temperature. At low temperatures the condensation is more favored, whereas the amount of triose at equilibrium increases with rising temperatures. The equilibrium constant is evaluated as (dihydroxyacetone phosphate)(phosphoglyceraldehyde)/(HDP) = K_{eq} = 6 × 10^{-5} at 28°C. At first glance it appears that this implies that very little triose exists at equilibrium at this temperature. This is an example of reactions in which one compound is converted to two, and closer examination shows that the percentage conversion in such cases is a function of the absolute concentration. Thus, with 1 M HDP, only a fraction of 1 per cent is split at equilibrium, whereas at 10^{-4} M, approximately half is converted to triose phosphates. The equilibrium constant is markedly affected by temperature. Lower temperatures favor the condensation to HDP, while higher temperatures cause the reaction to shift toward increased formation of triose phosphates.

Muscle aldolase crystallizes as a simple protein with a molecular weight near 150,000. The corresponding enzyme from yeast (molecular weight, 122,000) has recently been crystallized as an inactive mercury salt, which requires removal of the mercury and addition of divalent zinc, iron, or cobalt for activation.[34] Similar metal requirements have been found for aldolases from other microorganisms.

Phosphotriose Isomerase. Early work on aldolase was made difficult by the presence of phosphotriose isomerase. This enzyme is usually present in much larger effective concentrations than aldolase, and catalyzes the interconversion of phosphoglyceraldehyde and dihydroxyacetone phosphate.[35a] Since the equilibrium of this reaction lies to the

[33] T. C. Tung, K.-H. Ling, W. L. Byrne, and H. A. Lardy, *Biochim. et Biophys. Acta* **14,** 488 (1954).
[34] O. Warburg and K. Gawehn, *Z. Naturforsch.* **9b,** 206 (1954).
[35a] O. Meyerhof and W. Kiessling, *Biochem. Z.* **276,** 239 (1935).

side of dihydroxyacetone phosphate ($K_{eq} = 22$), this compound at first appeared to be the exclusive product of aldolase.

The two triose phosphates may be distinguished chemically. Both sugar phosphates are rapidly split by alkali at room temperatures. The oxidation of the aldehyde to phosphoglyceric acid by iodine has been used in the assay of mixtures. Only the remaining dihydroxyacetone phosphate is alkali labile. The phosphoglyceric acid may be determined by its high rotation in molybdate.

Each of the two triose phosphates participates in a different enzymatic oxidation-reduction system. Dihydroxyacetone phosphate may be reduced to glycerol phosphate and phosphoglyceraldehyde may be oxidized to phosphoglyceric acid. Both of these are enzyme-catalyzed reactions in which the pyridine nucleotide, DPN, participates. The two reactions may thus be linked by the coenzyme, and in some crude extracts glycerol phosphate and phosphoglyceric acid may accumulate as end-products of fermentation.

Stereospecificity of Dihydroxyacetone Phosphate Reactions. In both the aldolase and triose phosphate isomerase reactions a carbon-hydrogen bond of the hydroxymethyl group is broken. Isotope exchange experiments with tritium have shown that both enzymes catalyze an equilibration between one hydrogen of the substrate and the hydrogen of water.[35b,c] The two enzymes do not attack the same hydrogen atom; each is specific for only one position. In the projection shown (XII),

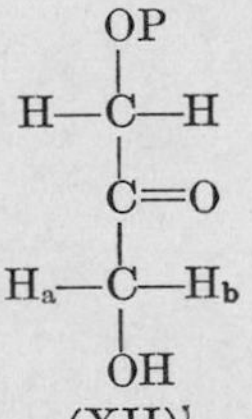

$$
\begin{array}{c}
\text{OP} \\
| \\
\text{H---C---H} \\
| \\
\text{C=O} \\
| \\
\text{H}_a\text{---C---H}_b \\
| \\
\text{OH} \\
\text{(XII)}
\end{array}
$$

If H_a is the hydrogen atom released through aldolase action, H_b would be labilized by the isomerase. The absolute positions of the exchangeable atoms have not been determined. It has been suggested that the enzyme action in each case involves the formation of an ene-diol structure complexed with the enzyme, and that in one case the *cis-* and in the other the *trans-* ene-diol is formed.[35c]

α-Glycerophosphate Dehydrogenase. Dihydroxyacetone phosphate and DPNH react in the presence of α-glycerophosphate dehydrogenase to form L-α-glycerophosphate. The enzyme has been found widely distrib-

[35b] I. A. Rose and S. V. Rieder, *J. Am. Chem. Soc.* **77**, 5764 (1955).
[35c] B. Bloom and Y. J. Topper, *Science* **124**, 982 (1956).

uted and was crystallized from rabbit muscle by Baranowski.[36] At neutral pH values the equilibrium constant is about 10^4, which means that for practical analytical purposes the reaction goes to completion in the direction of glycerophosphate production and DPNH oxidation. The reaction is more completely written as shown in (XIII). Since en-

$$
\begin{array}{ccc}
& \overset{\displaystyle O}{\underset{\displaystyle CH_2-O-\overset{\|}{\underset{\displaystyle |}{P}}-O^-}{}} & \\
CH_2-O-\overset{\overset{\displaystyle O}{\|}}{\underset{\underset{\displaystyle O^-}{\|}}{P}}-O^- & & \\
\end{array}
$$

CH₂—O—P—O⁻ (with =O and O⁻)

C=O + DPNH + H⁺ ⇌ HCOH + DPN

CH₂OH   CH₂OH

Dihydroxyacetone phosphate

L-α-Glycerophosphate

(XIII)

zyme reactions are usually carried out in buffered solutions, equilibrium constants are often calculated for given pH values and the hydrogen ion is ignored in these cases. The equilibrium constants thus evaluated are functions of pH; acid solutions, which supply H^+, tend to drive reactions such as (XIII) to the right, whereas DPN reduction is favored by more alkaline buffers. This is generally true for reactions involving pyridine nucleotides; reactions that appear to be irreversible oxidations of reduced pyridine nucleotides at pH 7 are found to react readily in the opposite direction at pH 10.

α-Glycerophosphate oxidation proceeds with difficulty and only to a very limited extent in the isolated reaction system at neutral pH, but when DPNH is removed by a thermodynamically favorable reaction, such as reduction of certain dyes, dihydroxyacetone phosphate can be made to accumulate. Another preparation, also from rabbit muscle, appears to oxidize α-glycerophosphate without the participation of pyridine nucleotides.[37] This preparation is particulate and uses either dyes or an enzyme electron transport system, the cytochromes, to support the oxidation. In each case the substrate for oxidation must be L-α-glycerophosphate, not the D-isomer. The product of the oxidation is dihydroxyacetone phosphate. Before the configuration of the glycerophosphates was established by Fischer and Baer in 1939,[38] it was postulated that oxidation occurred at the free primary alcohol group to form phosphoglyceraldehyde. It is now known that natural α-glycerophosphate has the opposite configuration of natural phosphoglyceraldehyde, and direct oxidation to the aldehyde would yield the inactive

[36] T. Baranowski, *J. Biol. Chem.* **180**, 535 (1949).
[37] D. E. Green, *Biochem. J.* **30**, 629 (1936).
[38] E. Baer and H. O. L. Fischer, *J. Biol. Chem.* **128**, 491 (1939).

L-phosphoglyceraldehyde. D-Phosphoglyceraldehyde is derived from L-α-glycerophosphate by oxidation to the optically inactive dihydroxyacetone phosphate and isomerization by triose phosphate isomerase.

Triose Phosphate Dehydrogenase. The oxidation of phosphoglyceraldehyde is catalyzed by a DPN-requiring enzyme that has been named triose phosphate dehydrogenase and nicknamed oxidizing enzyme, in acknowledgment of its role in glycolysis. The nature of the reaction catalyzed was established in Warburg's laboratory, when the enzyme was crystallized from yeast.[39] It was found that the reaction requires inorganic phosphate, and Negelein and Brömel isolated the product, 1,3-diphosphoglyceric acid.[40] The complete reaction is usually followed by measuring the absorption of DPNH at 340 mμ (XIV).

$$
\begin{array}{c}
\text{HC}=\text{O} \\
| \\
\text{HCOH} \\
| \\
\text{H}_2\text{C}-\text{O}-\overset{\overset{\text{O}}{\|}}{\text{P}}-\text{O}^- \\
| \\
\text{O}^-
\end{array}
\quad + \text{DPN} + \text{H}-\text{O}-\overset{\overset{\text{O}}{\|}}{\text{P}}-\text{O}^- \rightleftharpoons
\begin{array}{c}
\text{C}-\text{O}-\overset{\overset{\text{O}}{\|}}{\text{P}}-\text{O}^- \\
| \qquad\qquad | \\
\quad\qquad\quad \text{O}^- \\
\text{HCOH} \\
| \\
\text{H}_2\text{C}-\text{O}-\overset{\overset{\text{O}}{\|}}{\text{P}}-\text{O}^- \\
| \\
\text{O}^-
\end{array}
\quad + \text{DPNH} + \text{H}^+
$$

3-Phospho-D-glyceraldehyde

1,3-Diphospho-D-glyceric acid

(XIV)

Before the phosphate requirement for this reaction was known, it was generally accepted that the model of Wieland for the oxidation of aldehydes would apply.

$$
\text{R}-\overset{\overset{}{|}}{\underset{\underset{\text{H}}{|}}{\text{C}}}=\text{O} + \text{HOH} \rightleftharpoons \text{R}-\overset{\overset{\text{OH}}{|}}{\underset{\underset{\text{H}}{|}}{\text{C}}}-\text{O}\,\lfloor\text{H}\rfloor \longrightarrow \text{R}-\overset{\overset{\text{OH}}{|}}{\text{C}}=\text{O} + 2(\text{H})
$$

Wieland's Mechanism

An analogous mechanism was proposed by Warburg for the phosphate-requiring system.

$$
\text{R}-\overset{\overset{}{|}}{\underset{\underset{\text{H}}{|}}{\text{C}}}=\text{O} + \text{HOPO}_3^- \rightleftharpoons \text{R}-\overset{\overset{\text{OPO}_3^-}{|}}{\underset{\underset{\text{H}}{|}}{\text{C}}}-\text{OH} \longrightarrow \text{R}-\overset{\overset{\text{OPO}_3^-}{|}}{\text{C}}=\text{O} + 2(\text{H})
$$

Warburg's Mechanism

Attempts to demonstrate the primary reaction of phosphate with the aldehyde failed to support Warburg's proposed mechanism. More recent evidence strongly supports a mechanism in which phosphate participates

[39] O. Warburg and W. Christian, *Biochem. Z.* **301,** 221 (1939).

[40] E. Negelein and H. Brömel, *Biochem. Z.* **303,** 132 (1939).

in a secondary reaction, following the oxidation of the aldehyde. The development of this alternate description of the enzyme reaction has evoked one of the great polemics of biochemistry, in which Warburg has defended his original speculations with vigorous attacks.

Triose phosphate dehydrogenase has been studied both as a chemical molecule and as a catalyst. These studies were made possible by the availability of gram quantities of crystalline enzyme from rabbit muscle, by the procedure of Cori, Slein, and Cori,[41] and from yeast, by the procedure of Warburg and Christian.[39] Krebs[42] has isolated four fractions from yeast with equivalent specific activity, one of which is the enzyme crystallized by Warburg and Christian. The four components make up about 5 per cent of the total extractable protein of the yeast.

Composition of Triose Phosphate Dehydrogenase. The animal and yeast enzymes are similar but not identical in certain respects. The amino acid compositions are similar, and both have N-terminal valine residues.[43] The two are distinct immunologically.[44] The most striking difference is the unusual binding of DPN in the rabbit enzyme. After repeated recrystallizations, each mole of this protein contains two moles of firmly bound DPN.[45] This DPN can be removed by treatment of the enzyme with charcoal. The charcoal-treated enzyme is less stable and more soluble than the original enzyme.

The bound DPN is fully active enzymatically, and participates in coupled reactions with other DPN-dehydrogenases. After DPN has been removed, much larger quantities must be added back to saturate the enzyme, and 3 DPN molecules are bound to each enzyme molecule. Bound DPN is also displaced from the enzyme by PCMB. This reaction may be followed spectrophotometrically, as the DPN-enzyme compound has a broad absorption band with a peak at 360 mμ.[46] When DPN is removed from the enzyme, there is a decrease in optical density in a broad region about 360 mμ. The recombination of DPN with the enzyme restores this band. It has been known for many years that cysteine or another —SH compound is required for activation of the enzyme as usually obtained. Besides PCMB, iodoacetate also inhibits the enzyme, but this inhibition is prevented by phosphoglyceraldehyde. These observations support the concept that the essential —SH groups of the enzyme are involved in the binding of DPN.

[41] G. T. Cori, M. W. Slein, and C. F. Cori, *J. Biol. Chem.* **173,** 605 (1948).
[42] E. G. Krebs, G. W. Rafter, and J. M. Junge, *J. Biol. Chem.* **200,** 479 (1953).
[43] S. F. Velick and S. Udenfriend, *J. Biol. Chem.* **203,** 575 (1953).
[44] E. G. Krebs and V. Najjar, *J. Exptl. Med.* **88,** 569 (1948).
[45] J. F. Taylor, S. F. Velick, C. F. Cori, G. T. Cori, and M. W. Slein, *J. Biol. Chem.* **173,** 619 (1948).
[46] E. Racker and I. Krimsky, *J. Biol. Chem.* **198,** 731 (1952).

Digestion of triose phosphate dehydrogenase with proteolytic enzymes or other methods has resulted in the liberation of almost two equivalents per mole of the tripeptide, glutathione.[47] Since the γ-glutamyl structure of glutathione is not ordinarily found in proteins, the inference has been drawn that the tripeptide is a firmly bound prosthetic group. Iodoacetate-treated enzyme yields very little glutathione, implying that IAA reacts with glutathione in the enzyme. It has been suggested, however, that the glutathione associated with this enzyme may be an artifact of isolation, and may be merely tripeptide bound in disulfide linkages with the enzyme through metal-catalyzed oxidation.[48]

Reactions of Triose Phosphate Dehydrogenase. This enzyme is not specific in its reaction with phosphoglyceraldehyde. The nonphosphorylated compound is oxidized also, but only at 0.1 per cent of the rate found with the natural substrate.[49] Acetaldehyde, propionaldehyde and butyraldehyde are also oxidized, but at still slower rates.[50] The corresponding acyl phosphates are formed when these substrates are oxidized in the presence of inorganic phosphate. If arsenate is substituted for phosphate, the reactions proceed not to equilibrium, but to completion, with the formation of free acids. The formation of free acids is presumed to be the result of rapid spontaneous hydrolysis of unstable acyl arsenates, as proposed for other arsenolysis reactions.

The oxidation to form acyl phosphates is reversible. Acyl groups may also be transferred by this enzyme. Acetyl groups are transferred from acetyl phosphate to the sulfhydryl groups of compounds such as glutathione and CoA; acetyl phosphate is hydrolyzed in the presence of arsenate; and the phosphate of acetyl phosphate is exchanged with inorganic phosphate (measured by labeling with P^{32}).[51] These reactions also occur with other acyl phosphates. If DPNH were present, it might be concluded that these reactions involve reduction to the aldehyde and reoxidation with substitutes for inorganic phosphate forming the products found. However, the experiments were carried out with enzymes free of DPNH; peculiarly, DPN is required for all activities.

Kinetic studies reveal a K_m for phosphoglyceraldehyde of 5×10^{-5}. Ultracentrifuge studies have demonstrated specific binding of phosphoglyceraldehyde.[52] Under conditions in which no net reaction occurred,

[47] I. Krimsky and E. Racker, *Federation Proc.* **13**, 245 (1954).

[48] O. J. Koeppe, P. D. Boyer, and M. P. Stulberg, *J. Biol. Chem.* **219**, 569 (1956).

[49] O. Warburg and W. Christian, *Biochem. Z.* **303**, 40 (1939).

[50] J. Harting, *Federation Proc.* **10**, 195 (1951).

[51] J. Harting and S. Velick, *J. Biol. Chem.* **207**, 866 (1954); P. Oesper, *ibid.* **207**, 421 (1954).

[52] S. F Velick and J. E. Hayes, Jr., *J. Biol. Chem.* **203**, 545 (1953).

a dissociation constant of the phosphoglyceraldehyde-enzyme complex was measured and found to agree with the K_m. In contrast, the enzyme was not found to bind any phosphate specifically. The binding of phosphoglyceraldehyde was found to be independent of inorganic phosphate.

Mechanism of Action. The following scheme (XV) resembles one proposed by Racker to reconcile the various observations made with the oxidizing enzyme. It is suggested that the enzyme initially contains DPN bound through addition of a sulfhydryl group.[46] It is this addition compound that is thought to exhibit the broad 360 mμ absorption. The addition of an aldehyde splits the pyridine-sulfur bond (aldehydolysis),

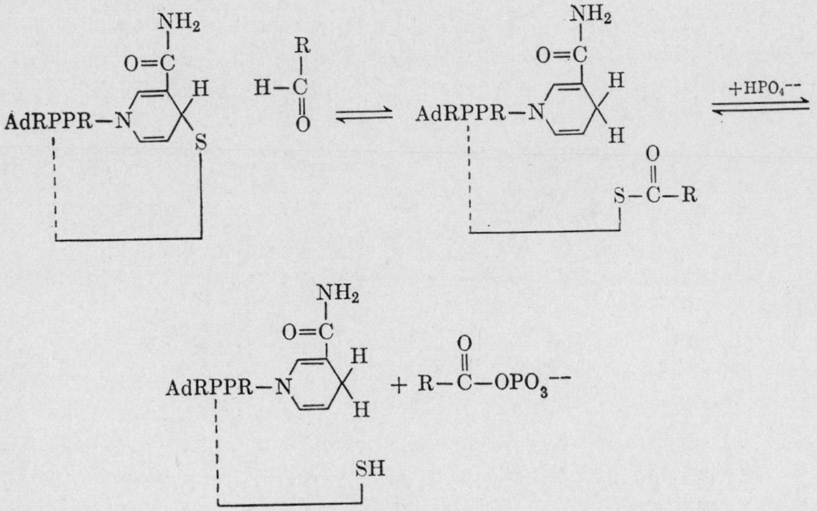

(XV)

forming enzyme-bound DPNH and a thioester. The DPNH may transfer electrons to an acceptor molecule or may exchange with free DPN. The acyl group may be transferred to phosphate, but in the absence of phosphate, slower reactions with sulfhydryl compounds or water may occur.

This mechanism does not explain all of the observations made with this enzyme. The role of DPN in the exchange reactions remains obscure. It is always possible to consider that DPN is not concerned directly with exchange reactions, but is necessary for maintenance of enzyme configuration. Careful kinetic measurements have eliminated the possibility that traces of DPNH could be involved in a reversible oxidation. The role of sulfhydryl compounds is also not obvious. Glutathione, which activates and protects the enzyme, permits acyl exchange reactions after the enzyme has been inhibited by iodoacetate, yet inhibits the hydrolysis of acetyl phosphate. These problems, however, concern details of a pattern whose most prominent feature is the formation of an acyl enzyme. An objection to this essential point was made[53] on the grounds that the initial reaction with phosphate present exceeded the initial rate, as well as the extent of the reaction in the absence of phosphate. Using glyceraldehyde as substrate, experiments have now shown that the initial rates of DPN reduction are indeed the same in the presence and absence of inorganic phosphate, and that the kinetics observed agree well with the theory for the establishment of an equilibrium:[48]

$$\text{R—CHO + DPN—Enzyme} \rightleftharpoons \text{R—}\overset{\overset{\displaystyle O}{\|}}{\text{C}}\text{—Enzyme—DPNH}$$

Phosphoglycerate Kinase. The acyl phosphate of 1,3-diphosphoglyceric acid is transferred to ADP by phosphoglycerate kinase (XVI). This

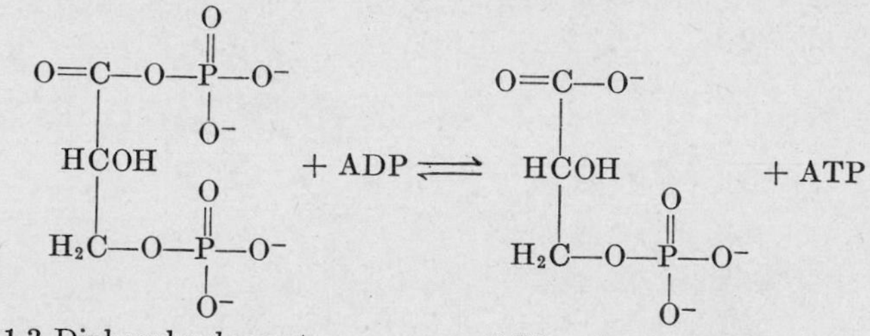

1,3-Diphosphoglycerate   3-Phosphoglycerate

(XVI)

enzyme was crystallized from yeast by Bücher,[54] and appears to be completely specific for all components of the system: ATP, ADP, 3-phosphoglycerate, and 1,3-diphosphoglycerate. Like other kinases, this enzyme also requires Mg^{++}, probably complexed with the adenine nucleotides.

The equilibrium constant $= (ATP)(PGA)/(diPGA)(ADP) = 3.3 \times 10^3$

The initial rates of reaction in both directions have been measured, and, for a given amount of enzyme the forward reaction goes 8.8 times as fast as the backward. K_m values for all the substrates have also been determined. They are:

ADP, 2×10^{-4} diPGA, 1.3×10^{-6}
ATP, 1.1×10^{-4} PGA, 2×10^{-4}

Substituting these values into the equation relating V_{max} and K_{eq}:

$$K_{eq} = \frac{V_{max}{}^1}{V_{max}{}^2} \times \frac{K_m{}^2}{K_m{}^1}$$

[53] O. Warburg, H. Klotsch, and K. Gawehn, Z. Naturforsch. **9b**, 391 (1954).
[54] T. Bücher, Biochim. et Biophys. Acta **1**, 292 (1947).

the equilibrium constant is calculated as:

$$\frac{8.8}{1} \times \frac{2 \times 10^{-4} \times 1.1 \times 10^{-4}}{2 \times 10^{-4} \times 1.8 \times 10^{-6}} = 5.4 \times 10^2$$

While the agreement between calculated and observed values is not perfect, the influence of K_m can be seen, and the discrepancy can probably be explained by the uncertainty of actual substrate concentrations introduced by lack of knowledge about the concentrations of Mg complexes.

Phosphoglyceromutase. The 3-phosphoglyceric acid formed by the transfer of phosphate by phosphoglycerate kinase is equilibrated with 2-phosphoglyceric acid by phosphoglyceromutase (XVII). The two

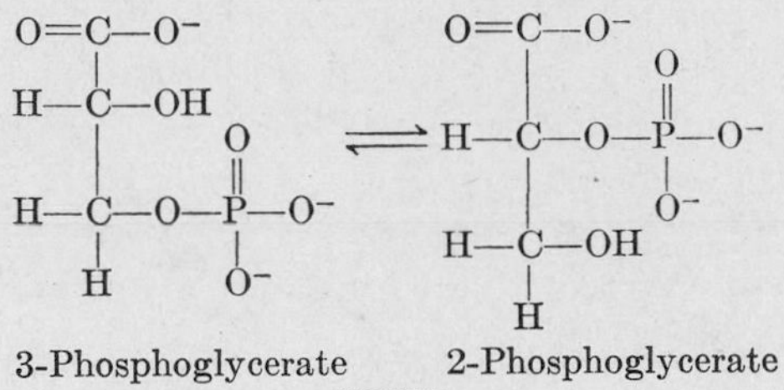

3-Phosphoglycerate   2-Phosphoglycerate
(XVII)

esters are distinguished by their different rotation of polarized light in molybdate solution.[55] The mechanism of this reaction was one of the first to be investigated with isotopes; Meyerhof and his collaborators[56] found that inorganic phosphate did not enter either ester during the reaction. After the mechanism of the phosphoglucomutase reaction was shown to involve a diphosphate, glucose-1,6-diphosphate, as a cofactor, a similar mechanism involving 2,3-diphosphoglycerate was shown to be used in the interconversion of the monophosphoglycerates.[57] Thus although the phosphorus atoms originally esterified remain in the original molecule, they are transferred from one carbon chain to another during the mutase reaction, and do not "flip" to an adjacent position on the same molecule. The cofactor had been isolated by Greenwald from erythrocytes many years before a metabolic function was found.[58] Phosphoglyceromutase has recently been crystallized.[59]

Enolase. Enolase is the name given to the enzyme that dehydrates 2-phosphoglycerate (XVIII).[60] The product is the phosphate ester of

[55] O. Meyerhof and W. Schulz, *Biochem. Z.* **297**, 60 (1938).
[56] O. Meyerhof, P. Ohlmeyer, W. Gentner, and H. Maier-Leibnitz, *Biochem. Z.* **298**, 396 (1938).
[57] E. W. Sutherland, T. Z. Posternack, and C. F. Cori, *J. Biol. Chem.* **179**, 501 (1949).
[58] I. Greenwald, *J. Biol. Chem.* **63**, 339 (1925).
[59] V. W. Rodwell, J. C. Towne, and S. Grisolia, *Biochim. et Biophys. Acta* **20**, 394 (1956).
[60] O. Meyerhof and W. Kiessling, *Biochem. Z.* **280**, 99 (1935).

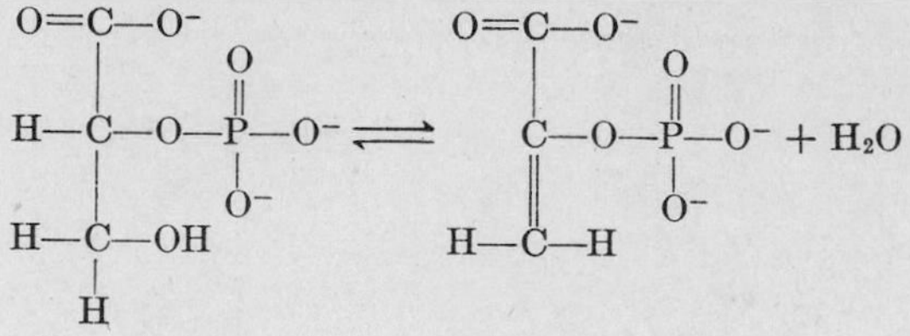

$$O=C—O^- \qquad\qquad O=C—O^-$$
$$\qquad\qquad\qquad O \qquad\qquad\qquad\qquad O$$
$$\qquad\qquad\qquad \|\qquad\qquad\qquad\qquad\qquad \|$$
$$H—C—O—P—O^- \rightleftharpoons \quad C—O—P—O^- + H_2O$$
$$\qquad\qquad\qquad O^- \qquad\qquad\qquad\qquad O^-$$
$$H—C—OH \qquad\qquad H—C—H$$
$$H$$

2-Phosphoglycerate   Phosphoenolpyruvate

(XVIII)

the enol group of pyruvic acid, called phosphoenolpyruvate (PEP), but often referred to as simply phosphopyruvate. The occurrence and the nature of the reaction were established in Meyerhof's laboratory, and the effect of fluoride as an inhibitor of glycolysis was shown to be caused by inhibition of enolase. Enolase was crystallized as an inactive mercury salt by Warburg and Christian,[61] who showed that all of the activity could be recovered when the mercury was removed by precipitation as a sulfide. The mercury-free enzyme requires activation by a divalent cation, Mg^{++}, Mn^{++}, or Zn^{++}. Warburg and Christian also explained the fluoride inhibition as the formation of a Mg-F-PO_4 complex that competes with the fluoride-free metal for the enzyme. Fluoride does not inhibit enolase when Mn^{++} is used as the activator or when phosphate is absent, as the inhibitory complex is formed only with Mg^{++} and phosphate. The metal does not combine with the substrate, but becomes a component of the enzyme.[62]

The equilibrium constant of the enolase reaction is about 1 (omitting water from the equation). The equilibrium constant of the reaction catalyzed by phosphoglyceromutase is about 10, varying somewhat with temperature. Since these constants are small, it is apparent that the three phosphorylated 3-carbon acids are freely interconverted, and that at equilibrium most of the ester will be 3-PGA, but several per cent of each of the others will also be present.

Pyruvate Kinase. The phosphate of PEP is transferred to ADP by pyruvate kinase (XIX).[63] The equilibrium of this reaction is far to the

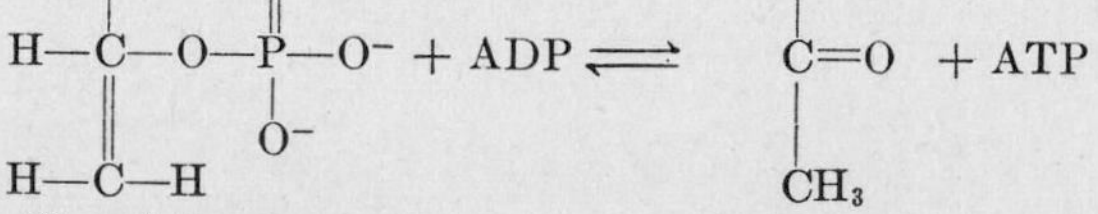

$$O=C—O^- \qquad\qquad\qquad O=C—O^-$$
$$\qquad\qquad\qquad O$$
$$\qquad\qquad\qquad \|$$
$$H—C—O—P—O^- + ADP \rightleftharpoons \quad C=O \ + ATP$$
$$\qquad\qquad\qquad O^-$$
$$H—C—H \qquad\qquad\qquad\qquad CH_3$$

Phosphopyruvate   Pyruvate

(XIX)

[61] O. Warburg and W. Christian, *Biochem. Z.* **310,** 389 (1942).
[62] B. G. Malmstrom, *Arch. Biochem. and Biophys.* **49,** 335 (1954).
[63] K. Lohmann and O. Meyerhof, *Biochem. Z.* **273,** 60 (1934).

side of ATP formation, and the reverse reaction is very difficult to demonstrate. The kinase from muscle requires K^+ in addition to the usual Mg^{++}.[64] The corresponding enzyme from yeast requires only Mg^{++}.[65] Pyruvate kinase has been crystallized from human muscle. Very recently it has been found that the enzyme from rabbit muscle is quite nonspecific in its nucleotide requirement, and that inosine diphosphate, uridine diphosphate, cytidine diphosphate, and guanosine diphosphate can all accept phosphate from phosphopyruvate.[66] This system has been used for the detection of nucleoside diphosphates in a spectrophotometric assay that measures the pyruvate formed.

Lactic Dehydrogenase. Muscle glycolysis is completed by the reduction of pyruvate to L-lactate. The pyridine nucleotide-requiring enzyme,

$$O{=}C{-}O^-$$
$$| $$
$$C{=}O \; + DPNH + H^+ \rightleftharpoons HO{-}C{-}H \; + DPN$$
$$| $$
$$CH_3 \qquad\qquad CH_3$$
$$\text{Pyruvate} \qquad\qquad \text{L-Lactate}$$

lactic dehydrogenase, was crystallized from heart muscle by Straub,[67] and similar enzymes have been crystallized from skeletal muscle and liver. The DPNH produced in the triose phosphate dehydrogenase reaction is thus used, permitting DPN to function catalytically in the oxidative reactions of glycolysis. At pH values near 7, the equilibrium is far to the side of lactate formation. The participation of H^+ in the DPN reaction causes this reaction to be reversed more readily in more alkaline solutions. Unlike triose phosphate dehydrogenase, lactic dehydrogenase is capable of reacting with TPN, but at rates much less than found with DPN.[68] The K_m for TPNH is similar to that for DPNH. Lactic dehydrogenase converts pyruvate quantitatively to lactate at neutral pH if equivalent amounts of DPNH are present. It has therefore been used extensively to determine pyruvate by measuring the decrease in optical density at 340 mμ. The reaction is not specific, however, as many other α-keto acids also react with lactic dehydrogenase.[69] The affinities for both pyruvate and DPNH are high, and this enzyme has been used in coupled reactions to assay small amounts of pyruvate, such as are formed in phosphopyruvate kinase reactions, and to trap very small amounts of DPNH,

[64] J. Parnas, P. Ostern, and T. Mann, *Biochem. Z.* **272,** 64 (1934); P. D. Boyer, H. A. Lardy, and P. Phillips, *J. Biol. Chem.* **149,** 529 (1943).
[65] J. Muntz, *J. Biol. Chem.* **171,** 653 (1947).
[66] J. L. Strominger, *Biochim. et Biophys. Acta* **16,** 616 (1955).
[67] F. B. Straub, *Biochem. J.* **34,** 483 (1940).
[68] A. H. Mehler, A. Kornberg, S. Grisolia, and S. Ochoa, *J. Biol. Chem.* **174,** 961 (1948).
[69] A. Meister, *J. Biol. Chem.* **184,** 117 (1950).

as produced sluggishly by illuminated chloroplast fragments.[70] The binding of pyruvate appears to depend on prior combination of the enzyme with DPN.[71] Lactic dehydrogenase reacts slowly with PCMB, losing activity and showing the ultraviolet absorption changes characteristic of the mercaptide,[72] but no other sulfhydryl reagents were found to affect the activity. DPN protects against PCMB inactivation. It has recently been reported that Zn^{++} is a firmly bound cofactor of muscle lactic dehydrogenase,[73] but this metal does not seem to be associated with the corresponding liver enzyme.[74]

Lactic dehydrogenase is the name of the enzyme that reacts with pyridine nucleotides. Several other types of enzyme have been described that oxidize lactic acid. These include flavoprotein enzymes, iron porphyrin-proteins (cytochrome b), an enzyme with no known cofactor, and an enzyme that decarboxylates during the oxidation to form acetate and CO_2. The various lactic oxidases are not capable, however, of reducing pyruvate. Thus many organisms that oxidize lactate cannot carry out the muscle type of glycolysis, which depends upon DPN-linked dehydrogenases to form lactate. Bacterial lactic dehydrogenases have been found that use pyridine nucleotides but differ from the animal enzyme in forming D-lactate or the racemic mixture.

Carboxylase. Muscle glycolysis is completed by reduction of pyruvate. In other systems pyruvate undergoes a variety of reactions. In yeast fermentation pyruvate is decarboxylated to acetaldehyde and CO_2.[75]

$$\underset{\text{Pyruvic acid}}{CH_3\overset{\overset{\displaystyle O}{\|}}{C}-COOH} \longrightarrow \underset{\text{Acetaldehyde}}{CH_3CHO} + CO_2$$

Similar reactions are involved in pyruvate metabolism in plants, animals, and other microorganisms. These various systems differ in their handling of the acetaldehyde.

The yeast enzyme was first studied in the laboratory of Neuberg, who named it *carboxylase,* and this name has been retained for the enzyme that decarboxylates pyruvate, even though many enzymes that decarboxylate other substrates have subsequently been discovered. The reaction is usually followed manometrically at pH 6 or lower by measuring

[70] W. Vishniac and S. Ochoa, *Nature* **167,** 768 (1951); L. J. Tolmach, *ibid.* **167,** 946 (1951).

[71] G. W. Schwert and Y. Takenaka, *Federation Proc.* **15,** 351 (1956).

[72] P. D. Boyer, *J. Am. Chem. Soc.* **76,** 4331 (1954).

[73] B. L. Vallee and W. E. C. Wacker, *J. Am. Chem. Soc.* **78,** 1771 (1956).

[74] H. Terayama and C. S. Vestling, *Biochim. et Biophys. Acta* **20,** 586 (1956).

[75] C. Neuberg and L. Karczag, *Biochem. Z.* **36,** 68 (1911).

the increase in pressure as CO_2 is produced. Carboxylase has an absolute requirement for an organic cofactor and a divalent cation[76] (Mg^{++} is usually found in the enzyme), but Mn^{++} can be used in place of Mg^{++}, and Co^{++}, Cd^{++}, Zn^{++}, Ca^{++}, and Fe^{++} also activate, although less efficiently. The cofactor was identified by Lohmann and Schuster[77] as thiamine pyrophosphate, and is often referred to as cocarboxylase.

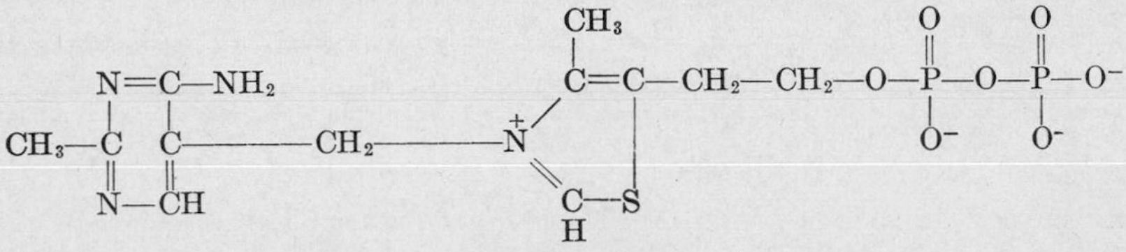

Cocarboxylase

Thiamine monophosphate and triphosphate have also been isolated, but they have little if any cofactor activity. The name cocarboxylase continues to be used, but there is a tendency to prefer the name thiamine pyrophosphate (ThPP or TPP), especially since this cofactor was found to participate in reactions that do not involve CO_2 or carboxyl groups (see p. 120).

Highly purified carboxylase contains 1 mole each of Mg^{++} and cocarboxylase per 75,000 gram of protein.[78] A molecular weight of 141,000 has been found for this preparation.[79] The cofactors are very firmly bound at pH 5–6, where the enzyme is most active. Resolution of the enzyme into protein (apocarboxylase) and cofactors has been accomplished by precipitation with ammoniacal ammonium sulfate and by brief treatment with pyrophosphate pH 8.1. Activity is restored to the apoenzyme by the addition of larger amounts of Mg^{++} and cocarboxylase.

Carboxylase attacks a large series of α-keto acids, pyruvate most rapidly, and the others at rates that decrease with chain length and other substitutions.[80] The first product of the decarboxylation is the corresponding aldehyde, but in general there is also formation of an acyloin. Yeast carboxylase does not form acetoin from acetaldehyde, but forms

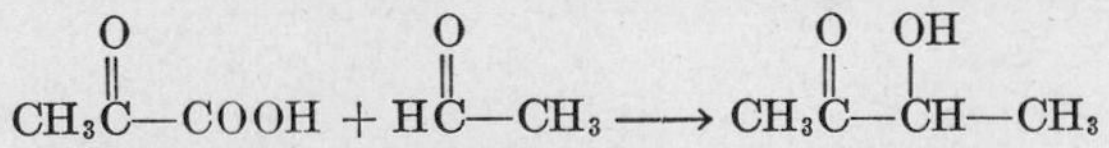

$$CH_3\overset{O}{\overset{\|}{C}}-COOH + H\overset{O}{\overset{\|}{C}}-CH_3 \longrightarrow CH_3\overset{O}{\overset{\|}{C}}-\overset{OH}{\overset{|}{CH}}-CH_3$$

Pyruvic acid   Acetaldehyde   Acetoin

[76] E. Auhagen, *Z. physiol. Chem.* **209**, 149 (1931).

[77] K. Lohmann and P. Schuster, *Biochem. Z.* **294**, 188 (1937).

[78] F. Kubowitz and W. Lüttgens, *Biochem. Z.* **307**, 170 (1941).

[79] J. L. Melnick and K. G. Stern, *Enzymologia* **8**, 129 (1940).

[80] S. Kobayasi, *J. Biochem.* **33**, 301 (1941); D. E. Green, D. Herbert, and V. Subrahmanyan, *J. Biol. Chem.* **138**, 327 (1941).

acetoin and CO_2 from pyruvate + acetaldehyde. This reaction was originally attributed to a "carboligase," but there is no evidence for the existence of such an enzyme distinct from carboxylase. Instead, acyloin formation appears to be a general property of carboxylases. The enzymes from various sources differ with respect to their affinities for acetaldehyde. Thus, the enzyme of pig heart[81] not only forms acetoin from acetaldehyde, but forms this product exclusively from pyruvate, as does an enzyme from *Aerobacter aerogenes*.[82] A wheat germ carboxylase has an intermediate position; it forms both acetaldehyde and acetoin from pyruvate and also forms acetoin from acetaldehyde. An interesting stereochemical phenomenon was noticed by Singer[83] in the reaction of plant carboxylases. In distinction from other carboxylases that yield optically active (−) acetoin, the plant enzymes formed a partially racemic mixture.

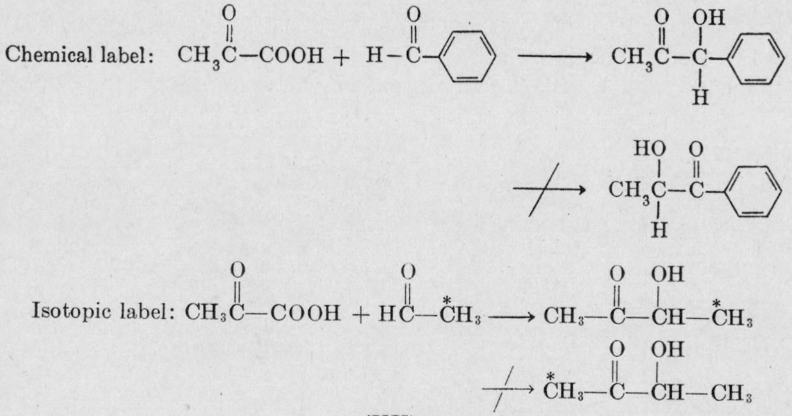

(XX)

During the formation of optically active acetoin, the two-carbon fragment derived from pyruvate must remain associated with the enzyme until it reacts with the carbonyl group of an acceptor molecule. The acceptor molecule may or may not be bound to the enzyme, but in either case it can approach the two-carbon fragment only in a single oriented manner, to yield an asymmetric product. The wheat germ carboxylase appears to hold its two-carbon fragment in an exposed position, so that aldehydes can approach from two directions; one of these leads to (+) and the other to (−) acetoin. The two approaches are not equivalent,

[81] D. E. Green, W. W. Westerfeld, B. Vennesland, and W. E. Knox, *J. Biol. Chem.* **145**, 69 (1942).
[82] M. Silverman and C. H. Werkman, *J. Biol. Chem.* **138**, 35 (1941).
[83] T. P. Singer, *Biochim. et Biophys. Acta* **8**, 108 (1952).

however, for one of the approaches is sufficiently hindered to prevent the formation of the isomers at equal rates.

The two-carbon fragment derived from pyruvate always contributes the acyl portion of the acyloin. This has been shown with both chemical[84] and isotopic[85] labels (XX).

Certain microorganisms synthesize acetoin by a modified pathway. The enzymes involved have been partially purified from extracts of *Aerobacter aerogenes*.[86] The first step resembles the carboxylase reaction of pig heart, in that acetaldehyde is not found, but all of the two-carbon fragment is transferred to another carbonyl group, forming an acyloin. The only known acceptor is pyruvate, and the product is d-α-acetolactate.

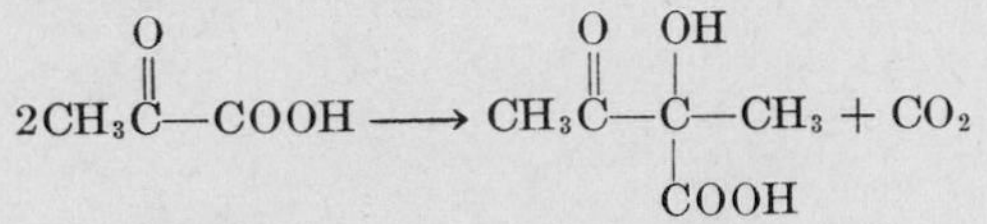

$$2CH_3\overset{O}{\overset{\|}{C}}{-}COOH \longrightarrow CH_3\overset{O}{\overset{\|}{C}}{-}\overset{OH}{\underset{\underset{COOH}{|}}{\overset{|}{C}}}{-}CH_3 + CO_2$$

α-Acetolactic acid

A second enzyme decarboxylates α-acetolactic acid. The enzyme is

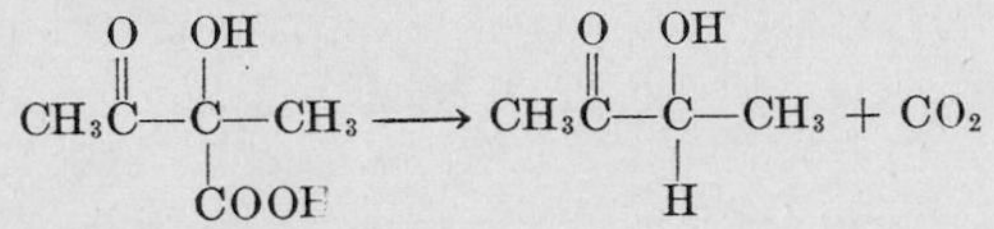

$$CH_3\overset{O}{\overset{\|}{C}}{-}\overset{OH}{\underset{\underset{COOH}{|}}{\overset{|}{C}}}{-}CH_3 \longrightarrow CH_3\overset{O}{\overset{\|}{C}}{-}\overset{OH}{\underset{\underset{H}{|}}{\overset{|}{C}}}{-}CH_3 + CO_2$$

α-Acetolactic acid   Acetoin

specific for the d-isomer of acetolactate, and it does not attack other β-keto acids, such as oxalacetic and acetoacetic acids. The decarboxylase requires a divalent cation, Mg^{++} or Mn^{++}, but not thiamine pyrophosphate.

Alcohol Dehydrogenase. Yeast glycolysis is concluded by reduction of

$$CH_3CHO + DPNH + H^+ \rightleftharpoons CH_3CH_2OH + DPN$$

acetaldehyde to ethanol by alcohol dehydrogenase. The equilibrium of alcohol oxidation by DPN lies far to the side of alcohol at neutral pH. The reaction can be measured directly in the direction of alcohol oxidation, however, if sufficient substrate is used, and it can be coupled with reactions that oxidize reduce DPN, and thus "pulled." Similar enzymes occur in mammalian liver, bacteria, and many other organisms. All use pyridine nucleotides, and in general DPN, but a TPN alcohol dehydrogenase has been described in *Leuconostoc mesenteroides*. These enzymes

---

[84] C. Neuberg and J. Hirsch, *Biochem. Z.* **115**, 282 (1921); C. Neuberg and H. Ohle, *ibid.* **127**, 327 (1922).
[85] N. H. Gross and C. H. Werkman, *Arch. Biochem.* **15**, 125 (1947).
[86] E. Juni, *J. Biol. Chem.* **195**, 715 (1952).

are active with many of the lower alcohols, and the liver enzyme has even been found to react with vitamin A alcohol.

Alcohol dehydrogenase has been crystallized from both yeast[87] and liver.[88] The two proteins differ in size and chemical composition. The yeast enzyme has a molecular weight of 153,000; the liver enzyme, 73,000. Yeast alcohol dehydrogenase contains 4 atoms of zinc per molecule.[89] The liver enzyme reacts with a larger variety of substrates than the yeast enzyme. The affinity of liver alcohol dehydrogenase for the coenzyme is much greater than that of the yeast enzyme, but in both cases the binding of DPNH is much stronger than the binding of DPN. In the case of the liver enzyme the enzyme-DPNH complex can be visualized spectrophotometrically when high molar concentrations of enzyme (10^{-5}–10^{-4} M, comparable to the DPNH concentration) are used. The absorption maximum shifts from 340 to 325 mμ and decreases in height.[90] When the enzyme-DPN compound is used as a reaction component in place of DPN with catalytic amounts of enzyme, there is a shift in the equilibrium of the reaction with ethanol and acetaldehyde that favors alcohol oxidation.[91] These observations have been interpreted as showing that the binding of DPN to alcohol dehydrogenase makes the coenzyme a more powerful oxidizing agent, and that the equilibrium of the reaction is therefore shifted. An alternate explanation is not concerned with the oxidizing ability of bound pyridine nucleotide, but deals only with the known properties of the original reaction and the combination of enzyme and coenzyme. The equilibrium constant at a given pH is expressed by the equation

$$K'_{eq} = \frac{(\text{acetaldehyde}) \, (\text{DPNH})}{(\text{ethanol}) \, (\text{DPN})}$$

This equation applies to those molecules in solution, regardless of the presence of other molecules, such as enzyme-bound compounds. If DPNH is preferentially removed from the solution by combination with the enzyme, the reaction must proceed in the direction of DPNH formation to satisfy the equation. It is perfectly possible to calculate an equilibrium involving enzyme-bound DPN, but this is a property of a different system. Such an additional system cannot alter the thermodynamic properties of the system of free compounds. The preferential binding of DPNH pulls the reaction in the same manner as any other coupled reac-

[87] E. Negelein and H. Wulff, *Biochem. Z.* **293**, 351 (1937).
[88] R. K. Bonnichsen and A. M. Wassen, *Arch. Biochem.* **18**, 361 (1948).
[89] B. L. Vallee and F. L. Hoch, *J. Am. Chem. Soc.* **77**, 821 (1955).
[90] H. Theorell and R. Bonnichsen, *Acta Chem. Scand.* **5**, 1105 (1951).
[91] H. Theorell and B. Chance, *Acta Chem. Scand.* **5**, 1127 (1951).

tion that removes one component from a reaction mixture, and is not concerned with the reaction mechanism. Alcohol dehydrogenase in substrate quantities[92] should have the same effect on the equilibrium of the triose phosphate dehydrogenase system (in which it has no catalytic function) that it has on the oxidation of ethanol. This interpretation of the phenomena observed when large amounts of enzyme are used is not a denial of the first interpretation; it is intended to illustrate the nature of the property, oxidation-reduction potential (see p. 163), which is a thermodynamic property, and related to other functions of reactions, such as equilibrium constants, that can be expressed in terms of mass action.

Popularly, enzymes are thought to occur at very low concentrations, so that the combination of enzyme and substrate is infinitesimally small compared with the amount of substrate present. This condition is true for most experimental conditions, but it is not true *in vivo*. The concentration of alcohol dehydrogenase in liver is of the order of 10^{-5} M (calculated for uniform distribution, therefore probably a minimum value). The concentration of DPN is higher, but, when allowance is made for the binding of DPN to other enzymes, it appears that a very large fraction of the DPN of liver is bound to enzymes.

GENERAL REFERENCES

Dickens, F. (1951). *In* "The Enzymes" (J. B. Sumner and K. Myrbäck, eds.), Vol. 2, p. 624. Academic Press, New York.
Harden, A. (1932). "Alcoholic Fermentation." Longmans, Green, London.
Lardy, H. A. (1949). *In* "Respiratory Enzymes," p. 179. Burgess Publishing, Minneapolis.
Nord, F. F., and Weiss, S. (1951). *In* "The Enzymes" (J. B. Sumner and K. Myrbäck, eds.), Vol. 2, p. 684. Academic Press, New York.
Racker, E. (1955). *Physiol. Revs.* **35**, 1.
Stumpf, P. K. (1954). *In* "Chemical Pathways of Metabolism" (D. M. Greenberg, ed.), Vol. I, p. 67. Academic Press, New York.

Pyruvate Oxidation and Acetyl Coenzyme A Formation

Glycolysis is an efficient mechanism for coupling the fermentation of carbohydrate to ATP formation, but it is inefficient in utilization of the total available substrate. More efficient utilization of a given amount of carbohydrate is accomplished by oxidation to carbon dioxide and

[92] When a catalytic component, enzyme, coenzyme or intermediate, is present in a system in amounts that permit direct measurement of that component, it is said to be present in substrate quantities.

water. There are several oxidative pathways known. Quantitatively the most important pathway follows the glycolytic sequence until the formation of pyruvate. In animal tissues and in some other organisms, at this point an oxidation converts pyruvate to CO_2 and a two-carbon fragment. The two-carbon fragment, "active acetate," has been identified in most cases as acetyl coenzyme A (acetyl CoA). In addition to formation in the oxidation of pyruvate, acetyl CoA is formed by several other routes, and it also participates in a number of biosyntheses. The metabolism of acetyl CoA is therefore of particular biological interest, since it is a metabolite common to many systems including those involved in carbohydrate, fat, steroid, carotenoid, and amino acid metabolism.

Several microbial systems have been found to oxidize pyruvate by somewhat different reactions. The reactions of animal preparations and *E. coli* are similar and involve, in addition to cocarboxylase and DPN, the cofactors CoA and lipoic acid.

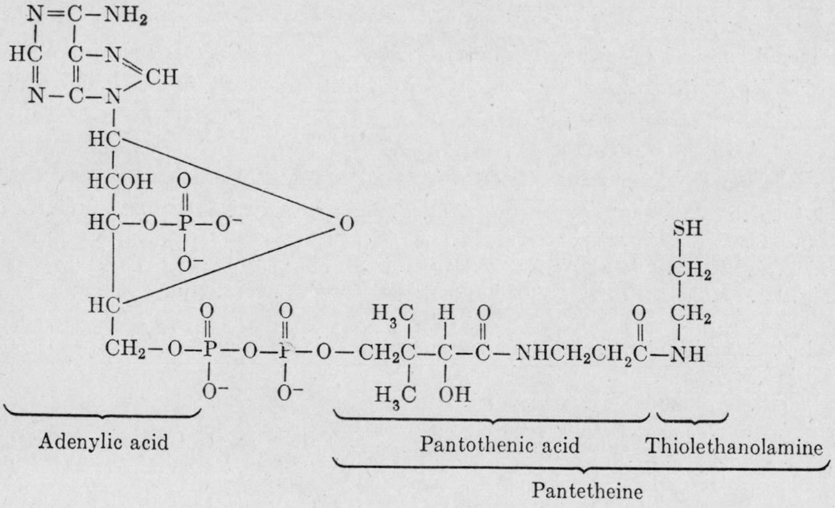

Coenzyme A

Coenzyme A. CoA was discovered as a factor needed for the acetylation of choline and of aromatic amines.[1] The isolation and identification of this material was one of the major advances of modern biochemistry. Not only is CoA of intrinsic interest as an active biochemical reagent, but it is essential for many diverse reactions that could not be studied until CoA and its many acylated forms became available. The study of

[1] F. Lipmann, *J. Biol. Chem.* **160**, 173 (1945); D. Nachmansohn and M. Berman, *ibid.* **165**, 551 (1946).

CoA involved many simultaneous approaches. Many years were spent in unraveling the tangle of information obtained from studies on microbial nutrition, enzyme reactions (in addition to amine acetylation) that require CoA, yeast oxidation of acetate, and degradation and synthesis of the cofactor. The results of these studies are summarized in the following sections.

Structure of Coenzyme A. The elucidation of the structure of CoA depended heavily on degradation by specific enzymes. The phosphate on carbon 3' of the adenosine was shown to be a monoester phosphate by hydrolysis with prostate phosphomonoesterase. The localization of the monoester at the 3' position was established by its sensitivity to a "b" nucleotidase that attacks only nucleoside 3'-phosphates, not 2'- or 5'-phosphates.[2] The original CoA molecule or the phosphatase product, dephospho CoA, can be split by nucleotide pyrophosphatases from potato[3] or snake venom.[2] These reactions permitted the identification of the adenosine phosphate portion of the molecule. The position of the phosphate on pantothenic acid cannot be determined enzymatically, but was established by studies on the synthesis of CoA from synthetic phosphorylated pantetheines.[4] Pantetheine is split to thiolethanolamine and pantothenic acid by an enzyme found in liver and kidney.[5] This enzyme also attacks larger molecules, including CoA.

Synthesis of Coenzyme A. Pantothenic acid is not synthesized by higher animals, but must be ingested by them as a vitamin. The synthesis

$$
\begin{array}{cccc}
\underset{\substack{\diagdown\ \diagup \\ HC \\ | \\ HCNH_2 \\ | \\ COOH}}{CH_3\ \ CH_3} & \underset{\substack{\diagdown\ \diagup \\ HC \\ | \\ C=O \\ | \\ COOH}}{CH_3\ \ CH_3} & \underset{\substack{CH_2OH \\ | \\ CH_3-C-CH_3 \\ | \\ C=O \\ | \\ COOH}}{} & \underset{\substack{CH_2OH \\ | \\ CH_3-C-CH_3 \\ | \\ HCOH \\ | \\ COOH}}{} \\
\text{Valine} & \alpha\text{-Ketovaline} & \alpha\text{-Ketopantoic} & \text{Pantoic acid} \\
 & & \text{acid} &
\end{array}
$$

(I)

of pantothenic acid by microorganisms may be considered starting with the amino acid valine. This is converted to its α-keto analog (see Chapter VII), which is the acceptor of an "active formaldehyde" (see Chapter VII).[6] The six-carbon compound is then reduced to pantoic acid (I). Pantoic acid condenses with β-alanine in the presence of ATP to form

[2] T. P. Wang, L. Schuster, and N. O. Kaplan, *J. Biol. Chem.* **206**, 299 (1954).

[3] G. D. Novelli, N. O. Kaplan, and F. Lipmann, *Federation Proc.* **9**, 209 (1950).

[4] J. Baddiley and E. M. Thain, *J. Chem. Soc.* p. 2253 (1951).

[5] G. M. Brown, J. A. Craig, and E. E. Snell, *Arch. Biochem.* **27**, 473 (1950); E. E. Snell, G. M. Brown, V. J. Peters, J. A. Craig, E. L. Wittle, J. A. Moore, V. M. McGlohon, and O. D. Bird, *J. Am. Chem. Soc.* **72**, 5349 (1950).

[6] M. Purko, W. D. Nelson, and W. A. Wood, *J. Biol. Chem.* **207**, 51 (1954).

$$
\begin{array}{ccccc}
\text{CH}_2\text{OH} & \text{NH}_2 & & \text{CH}_2\text{OH} & \\
| & | & & | & \\
\text{CH}_3-\text{C}-\text{CH}_3 + \text{CH}_2 & + \text{ATP} \rightleftharpoons \text{CH}_3-\text{C}-\text{CH}_3 & & +\text{AMP}+\text{PP} \\
| & | & & | & \\
\text{HCOH} & \text{CH}_2 & & \text{HCOH} & \\
| & | & & | & \\
\text{COOH} & \text{COOH} & & \text{C}-\text{NHCH}_2\text{CH}_2\text{COOH} & \\
& & & \| & \\
& & & \text{O} &
\end{array}
$$

Pantoic acid   β-Alanine   Pantothenic acid

(II)

pantothenic acid (II).[7] The remaining reactions have been demonstrated with preparations from livers of various animals. A second amide bond is formed when pantothenic acid is condensed with cysteine to form pantothenylcysteine. ATP is consumed in this process also. Pantothenyl-cysteine is decarboxylated to yield pantetheine (III).[8] A pantetheine

$$
\begin{array}{ccc}
\text{CH}_2\text{OH} & \text{CH}_2-\text{SH} & \text{CH}_2\text{OH} \\
| & | & | \\
\text{CH}_3-\text{C}-\text{CH}_3 \quad \text{HN}-\text{CH} & \longrightarrow & \text{CH}_3-\text{C}-\text{CH}_3 \quad \text{HN}-\text{CH}_2-\text{CH}_2-\text{SH} \\
| \qquad\qquad | & & | \qquad\qquad | \\
\text{HCOH} \quad \text{O}{=}\text{C} \quad \text{COOH} & & \text{HCOH} \quad \text{O}{=}\text{C} \\
| & & | \\
\text{C}-\text{N}-\text{CH}_2-\text{CH}_2 & & \text{C}-\text{N}-\text{CH}_2-\text{CH}_2 \\
\| \ \text{H} & & \| \ \text{H} \\
\text{O} & & \text{O}
\end{array}
$$

Pantothenylcysteine   Pantetheine

(III)

kinase places the terminal phosphate of ATP on the primary hydroxyl group of pantetheine.[9] Phosphopantetheine is linked to adenylic acid through a pyrophosphate bridge in a reaction with ATP in which pyrophosphate is split from ATP.[8] This reaction is analogous with the pyro-

$$
\text{phosphopantetheine} + \text{ATP} \underset{\text{enzyme}}{\overset{\text{condensing}}{\rightleftharpoons}} \text{dephospho CoA} + \text{PP}
$$

phosphorylase involved in the synthesis of DPN (p. 43). The condensation product, dephospho CoA, is converted to CoA by dephospho CoA kinase, which uses ATP and Mg^{++}.[10]

The reactions listed above were important in elucidating and establishing the structure of CoA, but information obtained by studies of microbial nutrition with natural and synthetic compounds was also extremely useful in the determination of the chemistry of CoA. Some of the microbial evidence indicates possible variations in the sequence of reactions, but those listed are the ones that have been demonstrated with isolated enzymes.

[7] W. Maas, *Federation Proc.* **13**, 256 (1954).
[8] M. B. Hoagland and G. D. Novelli, *J. Biol. Chem.* **207**, 767 (1954).
[9] L. Levintow and G. D. Novelli, *J. Biol. Chem.* **207**, 761 (1954).
[10] T. P. Wang and N. O. Kaplan, *J. Biol. Chem.* **206**, 311 (1954).

Lipoic Acid. The formation of acetyl CoA from pyruvate involves an additional cofactor, lipoic acid. This compound was studied independently as a growth factor for the protozoa *Tetrahymena* (protogen),[11] a growth factor for *Lactobacillus casei* (acetate replacing factor)[12] and a pyruvate oxidation factor for *Streptococcus faecalis*.[13] Pure synthetic compounds were made and called thioctic acid and lipoic acid.[14] The officially accepted name for the structure given below is lipoic acid. Lipoic acid is

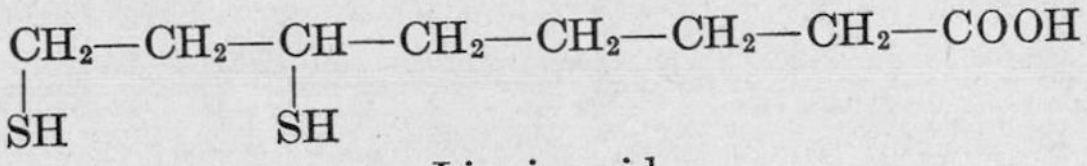

$$CH_2—CH_2—CH—CH_2—CH_2—CH_2—CH_2—COOH$$
$$\quad | \qquad\qquad\; |$$
$$\quad SH \qquad\quad SH$$

Lipoic acid

easily oxidized to a disulfide, in which the S-S bond closes a 5-membered ring.

Pyruvic Oxidase. The oxidation of pyruvate takes place in several integrated steps. It has been proposed that in the first reaction pyruvate

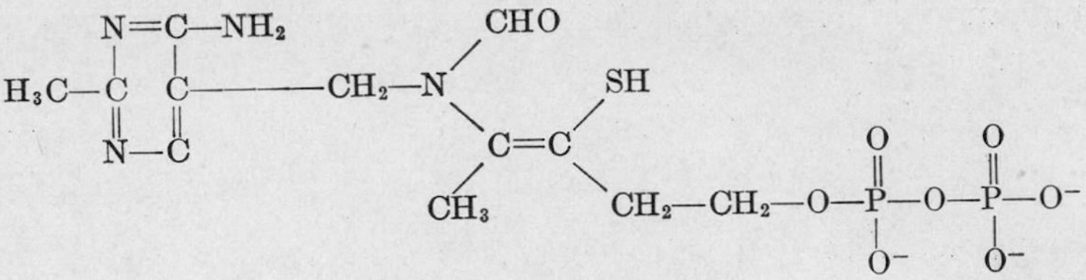

Thiamine pyrophosphate

$$+ \begin{array}{c} CH_3 \\ | \\ C=O \\ | \\ COOH \end{array} \longrightarrow R—S—\begin{array}{c} CH_3 \\ | \\ C—OH \\ | \\ COOH \end{array}$$

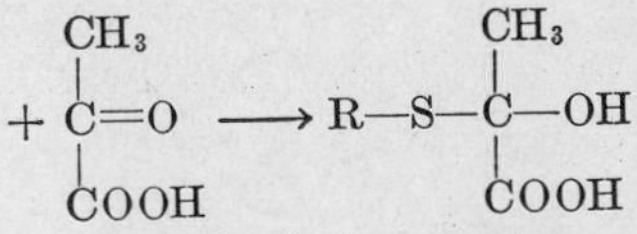

Hypothetical pyruvate
addition compound
(R is the remainder of
thiamine pyrophosphate)

(IV)

reacts with thiamine to form a bound aldehyde and CO_2 in a reaction common to both oxidative and glycolytic pathways. The precise mechanism of this reaction is far from clear. It has been speculated that the thiazole ring opens hydrolytically, forming a sulfhydryl group that can condense with carbonyl groups (IV). Decarboxylation of the condensa-

[11] G. W. Kidder and V. C. Dewey, *Arch. Biochem.* **8**, 293 (1945); E. L. R. Stokstad, C. E. Hoffmann, M. A. Regan, D. Fordham, and T. H. Jukes, *ibid.* **20**, 75 (1949).

[12] B. M. Guirard, E. E. Snell, and R. J. Williams, *Arch. Biochem.* **9**, 361 (1946).

[13] D. J. O'Kane and I. C. Gunsalus, *J. Bacteriol.* **56**, 499 (1948).

[14] L. J. Reed, B. G. DeBusk, I. C. Gunsalus, and C. S. Hornberger, *Science* **114**, 93 (1951); M. W. Bullock, J. A. Brockman, E. L. Patterson, J. V. Pierce, and E. L. R. Stokstad, *J. Am. Chem. Soc.* **74**, 3455 (1952).

tion product would leave the acetaldehyde addition compound. In the next step a transfer of the aldehyde to the disulfide form of lipoic acid would result in the formation of a thioester and a sulfhydryl group (V).

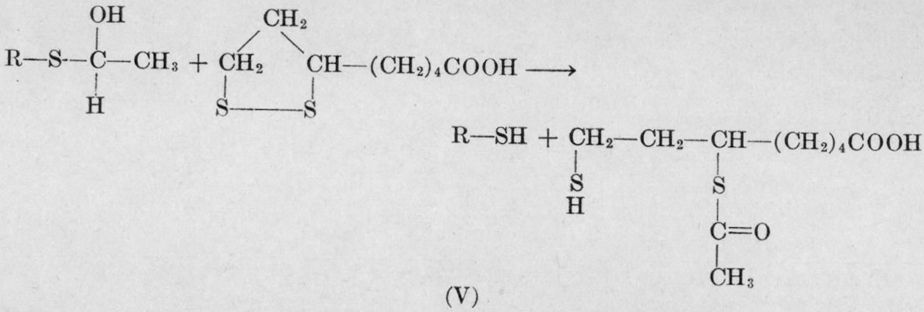

$$
\begin{array}{c}
\underset{\underset{H}{\mid}}{\overset{\overset{OH}{\mid}}{R-S-C-CH_3}} + \underset{\underset{S---S}{\mid\qquad\mid}}{\overset{\overset{CH_2}{\diagup\;\diagdown}}{CH_2\qquad CH-(CH_2)_4COOH}} \longrightarrow
\end{array}
$$

$$
R-SH + \underset{\underset{H}{\overset{\mid}{S}}}{CH_2}-CH_2-\underset{\underset{\underset{\underset{CH_3}{\mid}}{C=O}}{\mid}}{CH}-(CH_2)_4COOH
$$

(V)

It has been suggested that not only do these reactions all occur on one protein, but that the cofactors, cocarboxylase and lipoic acid, are combined through an amide linkage involving the carboxyl of lipoic acid and the amino group of the pyrimidine portion of thiamine. The condensed molecule has been called lipothiamide.[15] This proposal has not been established convincingly; it has been questioned, among other grounds, on the basis that cocarboxylase can be removed and restored to the enzyme without disturbing the more firmly bound lipoic acid.

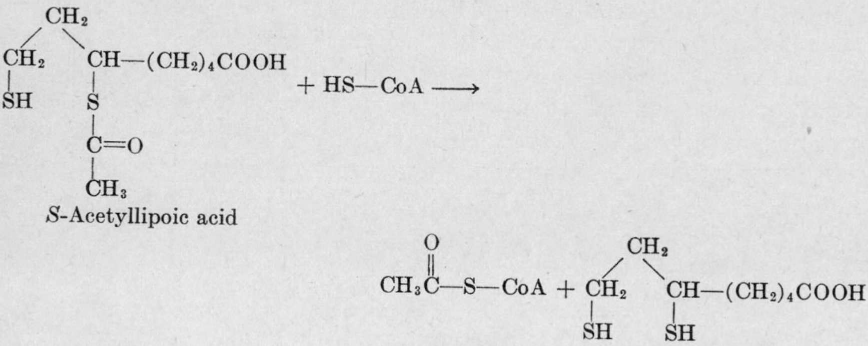

$$
\underset{\underset{SH}{\mid}}{CH_2} \quad \underset{\underset{\underset{\underset{CH_3}{\mid}}{C=O}}{\mid}}{CH}-(CH_2)_4COOH \qquad + HS-CoA \longrightarrow
$$

S-Acetyllipoic acid

$$
\underset{}{\overset{\overset{O}{\parallel}}{CH_3C}}-S-CoA + \underset{\underset{SH}{\mid}}{CH_2} \quad \underset{\underset{SH}{\mid}}{CH}-(CH_2)_4COOH
$$

Acetyl CoA   Reduced lipoic acid
(VI)

Although it is not clear exactly how acetyllipoic acid is formed, this structure is apparently an intermediate in pyruvate oxidation.[16] An enzyme has been separated from E. coli that transfers the acetyl group from lipoic acid to the sulfur of CoA (VI).[16] Another distinct enzyme

[15] L. J. Reed and B. G. DeBusk, J. Am. Chem. Soc. 74, 3964 (1952).
[16] Evidence for these reactions is presented in the review by Gunsalus (see General References).

$$\text{(CH}_2)_4\text{COOH} \qquad \qquad \text{(CH}_2)_4\text{COOH}$$

$$+ \text{ DPN} \longrightarrow \qquad\qquad + \text{ DPNH} + \text{H}^+$$

$$\text{SH} \quad \text{SH} \qquad\qquad\qquad \text{S} \quad \text{S}$$

$$\text{(VII)}$$

oxidizes lipoic acid with DPN as electron acceptor (VII).[17]
These reactions are summarized in (VIII).

pyruvate + thiamine PP → addition compound + CO_2
addition compound + oxidized lipoic acid → thiamine PP + acetyllipoic acid
acetyllipoic acid + CoA → lipoic acid + acetyl CoA
lipoic acid + DPN → oxidized lipoic acid + DPNH + H$^+$

Sum: pyruvate + CoA + DPN → acetyl CoA + DPNH + CO_2 + H$^+$

(VIII)

Three bacterial systems have been described that differ from the
E. coli and animal enzyme systems in that they oxidize pyruvate without
participation of lipoic acid, as shown by studies in the presence of specific
inhibitors. One of the classical specific inhibitors of α-keto acid oxidation
has been inorganic arsenite.[18] This inhibitor acts by binding lipoic acid
through combination with both sulfhydryl groups. The nonlipoic acid
systems are completely insensitive to arsenite.

The oxidation of pyruvate by *Lactobacillus delbruckii* was found to
require cocarboxylase, a flavin (FAD) (see Chapter IV), a divalent
cation (Mg^{++}, Mn^{++}, or Co^{++}), and inorganic phosphate. The electron
acceptor usually employed in studying this system is molecular oxygen,
but other acceptors, including ferricyanide, methylene blue, tetrazolium
chloride, and riboflavin, have also been found to support pyruvate oxidation. The reaction with oxygen satisfies the stoichiometry of the equation:

$$\text{CH}_3\overset{\text{O}}{\overset{\|}{\text{C}}}\text{—COOH} + \text{HO—}\overset{\text{O}}{\overset{\|}{\text{P}}}\text{—O}^- + \text{O}_2 \longrightarrow \text{CH}_3\overset{\text{O}}{\overset{\|}{\text{C}}}\text{—O—}\overset{\text{O}}{\overset{\|}{\text{P}}}\text{—O}^- + \text{CO}_2 + \text{H}_2\text{O}_2$$

$$\underset{\text{O}^-}{} \qquad\qquad\qquad \underset{\text{O}^-}{}$$

Pyruvate   Acetyl phosphate

This has been called the phosphoroclastic reaction. CoA and DPN are
not involved in this reaction, and do not influence the rate.[19]

A pyruvate oxidase from *Proteus vulgaris* catalyzes the simplest of the
α-keto acid oxidations. The only known requirements are cocarboxylase,

[17] L. P. Hager and I. C. Gunsalus, *J. Chem. Soc.* **75**, 5767 (1953); S. Korkes, A. del
Campillo, I. C. Gunsalus, and S. Ochoa, *J. Biol. Chem.* **193**, 721 (1951); H. A. Krebs
and W. A. Johnson, *Biochem. J.* **31**, 645 (1937).

[18] S. Ochoa, *J. Biol. Chem.* **138**, 751 (1941).

[19] L. P. Hager, D. M. Geller, and F. Lipmann, *Federation Proc.* **13**, 734 (1954).

divalent cations, and an electron acceptor. Pyruvate is converted to acetate and CO_2. The nature of the oxidizing group of the enzyme has not been determined.[20,21]

An anaerobic organism, *Clostridium butyricum*, contains a pyruvic oxidase that requires CoA and Fe^{++} in addition to cocarboxylase.[21] The function of the Fe^{++} is not clear; it may be in the electron transport. Crude extracts of this organism carry out a dismutation of pyruvate:

$$CH_3\overset{O}{\underset{}{C}}-COOH + HO-\overset{O}{\underset{O^-}{P}}-O^- \longrightarrow CH_3\overset{O}{\underset{}{C}}-O-\overset{O}{\underset{O^-}{P}}-O^- + CO_2 + H_2$$

In the presence of appropriate electron acceptors such as tetrazolium chloride, however, instead of H_2 gas appearing, the oxidant is reduced. The dismutation of pyruvate by the *Clostridium* is similar to a reaction found in *E. coli*, in which the products include formate in place of CO_2.[22]

$$CH_3\overset{O}{\underset{}{C}}-\overset{O}{\underset{}{C}}-O^- + HO-\overset{O}{\underset{O^-}{P}}-O^- \longrightarrow CH_3\overset{O}{\underset{}{C}}-O-\overset{O}{\underset{O^-}{P}}-O^- + H\overset{O}{\underset{}{C}}-O^-$$

The oxidation of pyruvate demonstrates the variety of oxidation reactions that can be found for the metabolism of a given substrate, and emphasizes the danger in attempting to carry over information from one system to another. All of the pyruvate oxidases, however, have one outstanding feature in common: they all require cocarboxylase and a divalent cation. These cofactors presumably play the same role both in the nonoxidative carboxylase reaction and in pyruvate oxidation, but the mechanism of thiamine action remains conjectural.

Work with bacterial enzymes has been partially paralleled by studies with animal and plant preparations. Particles with "molecular" weights over 10^6 have been shown to contain lipoic acid and to carry out oxidation of pyruvate, reduction of DPN, and formation of acetyl CoA.[23] The number of protein molecules involved and their relation to each other remain to be established.

Aldehyde Dehydrogenases. Acetaldehyde may also participate in oxidation reactions in addition to reduction and condensation. In liver the dismutation of acetaldehyde to ethanol and acetate was shown by

[20] P. K. Stumpf, *J. Biol. Chem.* **159**, 529 (1945).
[21] D. J. O'Kane, *Federation Proc.* **13**, 739 (1954).
[22] G. Kalnitsky and C. H. Werkman, *Arch. Biochem.* **2**, 113 (1943).
[23] V. Jagannathan and R. S. Schweet, *J. Biol. Chem.* **196**, 551 (1952).

Racker[24] to be caused by two pyridine nucleotide dehydrogenases: alcohol dehydrogenase and aldehyde dehydrogenase. The product of the oxidation is acetate, and the reaction has not been reversed. A similar

$$CH_3CHO + DPN + H_2O \rightarrow CH_3COOH + DPNH + H^+$$

activity was isolated by Black from yeast.[25] This enzyme requires cysteine and K^+ or a related cation, and is inhibited by Na^+, Li^+, and Cs^+. This yeast enzyme acts about ten times as rapidly with DPN as with TPN. Both the yeast and liver enzymes react with various aldehydes, but at different rates. Benzaldehyde is oxidized by the yeast enzyme but not by the liver enzyme, whereas salicylaldehyde is a substrate for the liver dehydrogenase but not for the yeast preparation. A TPN-specific aldehyde dehydrogenase was purified from yeast by Seegmiller.[26] The TPN-enzyme is activated by divalent cations and reacts with formaldehyde, glycolaldehyde, and propionaldehyde in addition to acetaldehyde. An aldehyde dehydrogenase was purified from *Acetobacter suboxydans* by King and Cheldelin.[27] This enzyme appears to react four times more rapidly with TPN than with DPN, it reacts only with acetaldehyde, propionaldehyde, and butyraldehyde, it has sulfhydryl groups that make the enzyme sensitive to traces of heavy metals, and it is not stimulated by cations.

A different type of aldehyde dehydrogenase was found by Burton and Stadtman in *Clostridium kluyveri*.[28] This DPN enzyme requires CoA and forms acetyl CoA, not free acetate. The reaction is freely reversible.

$$CH_3CHO + DPN + HS-CoA \rightleftharpoons CH_3\overset{\overset{\textstyle O}{\|}}{C}-S-CoA + DPNH + H^+$$

Reactions were observed with glycolaldehyde and higher homologs of acetaldehyde, but not with formaldehyde.

GENERAL REFERENCES

Gunsalus, I. C. (1954). *In* "Mechanism of Enzyme Action" (W. D. McElroy and B. Glass, eds.), p. 545. Johns Hopkins Press, Baltimore.
Ochoa, S. (1954). *Advances in Enzymol.* **15**, 183.
Symposium on Metabolic Role of Lipoic Acid (1954). *Federation Proc.* **13**, 695.

[24] E. Racker, *J. Biol. Chem.* **177**, 883 (1949).
[25] S. Black, *Arch. Biochem. and Biophys.* **34**, 86 (1951).
[26] J. E. Seegmiller, *J. Biol. Chem.* **201**, 629 (1953).
[27] T. E. King and V. Cheldelin, *J. Biol. Chem.* **220**, 177 (1956).
[28] R. M. Burton and E. R. Stadtman, *J. Biol. Chem.* **202**, 873 (1953).

Free Energy and the Concept of Bond Energy

Every compound contains energy in the forces that hold atoms together and in kinetic energy. When chemical compounds react, part of that energy may be released to do work in other systems. It is of interest to evaluate the amount of work that can be obtained from the reaction of given compounds. Three thermodynamic properties are concerned in this evaluation: heat content, H; free energy, F; and entropy, S. Thermodynamic properties are related to reactions, and are absolute only when referred to conventional standards. The heat content of the elements is defined as zero, and the H of any compound is the heat of formation from the elements. S is a measure of randomness of a system, and increases as a system becomes less organized.[1] At a given temperature, T, a portion of the total energy may be used for work. This is the free, or available, energy, F. These properties are related to each other.

$$F = H - TS$$

In every spontaneous reaction there is a loss in free energy because work is performed in the reaction process; there is less capacity for a system to do work after a spontaneous reaction than there was before the reaction. The total energy of the system maintained at constant temperature may increase or decrease during the reaction, and an entropy change also may occur. But since the fundamental relationship of these quantities applies at all times the change in free energy is related to the change in H and S by the equation: $\Delta F = \Delta H - T\Delta S$.

ΔH is the actual heat of reaction, and can be measured calorimetrically. ΔS is a less tangible quantity, and is evaluated indirectly. Therefore other relationships are useful in evaluating ΔF. In chemical reactions ΔF may be related to equilibrium constants by the equation

$$\Delta F_0 = -RT \ln K_{eq}$$

in which R is a constant (1.98 cal./degree) and T is the absolute temperature. The reaction for which ΔF is evaluated is the theoretical one in which 1 mole of reagents at unit activity (to a first approximation, at 1 molar concentration) react to produce products also at unit activity. Since reactions proceed only with the expenditure of energy, for every

---

[1] A complete discussion of entropy is not required for an understanding of the properties of free energy as related to biochemical reactions. For more information about the properties and significance of entropy, refer to physical chemistry texts or "Chemical Thermodynamics" by I. M. Klotz, Prentice-Hall, New York (1950).

reaction that proceeds spontaneously ΔF must be negative. The value for the theoretical reaction, however, may be either positive or negative. The ΔF so evaluated is defined as ΔF_0, the standard free energy change of a reaction. When conditions other than standard are used (such as pH 7 instead of pH 0), a modified standard free energy change, $\Delta F_0'$, is used.

In the reaction $A + B \rightleftharpoons C + D$, the reaction starting with A and B may proceed far toward completion. Nevertheless, a mixture of $C + D$ will also react to some extent, until equilibrium is reached. The equilibrium mixture contains less free energy than a mixture of reagents from

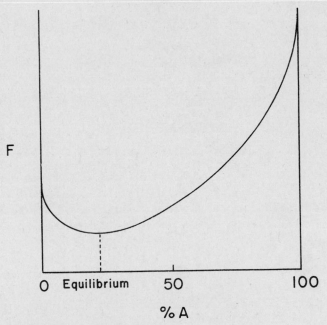

FIG. 9. Free energy of a system $A \rightleftharpoons B$ as a function of composition.

either side of the equation, but the difference in free energy between the equilibrium mixture and $A + B$ is much greater than the difference between the equilibrium mixture and $C + D$. This difference is expressed in ΔF_0 (Fig. 9). A reaction with a negative ΔF_0 will proceed toward an equilibrium containing a larger proportion of products than reagents, while a reaction with a positive ΔF_0 will proceed only to an equilibrium where most of the original reagents remain. The extent of the reaction in either case is given by the value of ΔF_0. A large negative number means that the reaction proceeds far toward completion; a large positive number means that the reaction does not proceed significantly. When

ΔF_0 is zero, the equilibrium constant is 1, and the reaction proceeds half way from either side.

Thermodynamic properties are statistical properties of large numbers of molecules. While the molecules of a given system can be assigned a certain free energy content, there is a great variation in the energy of the individual molecules. At equilibrium, then, reactions in both directions continue, always with a decrease in free energy of the reacting particles, but there is no change in the free energy of the entire system.

There can be many reasons why a reaction does not proceed to equilibrium. The standard free energy, then, does not define the amount of work that will be obtained from a reaction; it is the maximum energy available under defined conditions. When the standard free energy is known, it can be used to determine the equilibrium constant of a reaction. This, it must be remembered, measures the extent to which a reaction may proceed, but it does not indicate the speed of a reaction or even that a reaction will occur at all. ΔF is related indirectly to the relative rates of enzyme-catalyzed reactions by the Haldane relationship (p. 12), but the absolute rate of reaction is determined by the amount of enzyme and substrate in a given system.

The preceding chapters have described some of the reactions involved in the syntheses of ATP and acetyl CoA. These compounds are of particular interest because of their participation in many of the synthetic reactions of living systems. Studies on muscular contraction showed that the function of glycolysis in muscle could be interpreted as the synthesis of ATP, which serves to supply energy for contraction.[2] Another compound, phosphocreatine,[3] was found to act as a reservoir for phosphate that could be transferred to form ATP as it is consumed. The reaction is catalyzed by the enzyme, creatine phosphokinase,[4] which has recently been obtained as highly purified crystals.[5]

$$ADP^{-3} + {}^{-}O-\overset{\overset{O}{\|}}{\underset{\underset{O^{-}}{|}}{P}}-NH-\overset{\overset{NH_2^{+}}{\|}}{\underset{\underset{CH_3}{|}}{C}}-N-CH_2-COO^{-} \rightleftharpoons$$

Phosphocreatine

$$ATP^{-4} + H_2N-\overset{\overset{NH_2^{+}}{\|}}{\underset{\underset{CH_3}{|}}{C}}-N-CH_2-COO^{-}$$

Creatine

[2] S. Ochoa, Ann. N. Y. Acad. Sci. **47**, 835 (1947).
[3] C. H. Fiske and Y. SubbaRow, J. Biol. Chem. **81**, 629 (1929).
[4] K. Lohmann, Biochem. Z. **271**, 264 (1934); H. Lehmann, ibid. **281**, 271 (1935).
[5] S. A. Kuby, L. Noda, and H. A. Lardy, J. Biol. Chem. **209**, 191 (1954).

It became apparent that ATP is a uniquely important molecule, and that it can be synthesized by phosphate transfer from certain compounds, e.g., 1,3-diphosphoglycerate, phosphopyruvate, phosphocreatine, but not by many other compounds, including sugar phosphates, 2- or 3-phospho-glycerate, and glycerol phosphate. This led to the proposal by Lipmann that the terminal phosphates of ATP be considered to be linked in "high-energy" bonds, and that those compounds that can donate phosphate groups to ADP similarly be considered as having high-energy bonds.[6] The "high-energy bond" was designated ~P. Other phosphate compounds, such as simple esters, were considered to have low-energy bonds. The "bond energy" was evaluated as the standard free energy of hydrolysis with the sign reversed. At the time this concept was being developed, Kalckar[7] showed that thermodynamic stability can account for the large free-energy change characteristic of hydrolysis of certain phosphate anhydrides; some phosphate compounds are much less stable than their hydrolysis products because of such phenomena as opposing resonance. When a substituted pyrophosphate is hydrolyzed at neutral pH, for example, the structure

$$
\begin{array}{ccc}
& \overset{\displaystyle O}{\|} & \overset{\displaystyle O}{\|} \\
R\!-\!O\!-\!\!\underset{\underset{O^-}{|}}{P}\!-\!O\!-\!\!\underset{\underset{O^-}{|}}{P}\!-\!O^-
\end{array}
$$

in which three of the oxygens of the terminal phosphate moiety contribute equally to resonance forms, and two of the oxygen atoms of the penultimate phosphate also are equivalent, is converted to

$$
\begin{array}{ccc}
\overset{\displaystyle O}{\|} & & \overset{\displaystyle O}{\|} \\
R\!-\!O\!-\!\!\underset{\underset{O^-}{|}}{P}\!-\!O^- + HO\!-\!\!\underset{\underset{O^-}{|}}{P}\!-\!O^-
\end{array}
$$

in which an additional oxygen atom (the newly created dissociated hydroxyl group) also participates in resonating structures. This causes increased stability of the products, as stability is increased by increased resonance forms. Recently other factors, such as electrostatic repulsion, have also been invoked to explain the increased stability of hydrolysis products. Teleological reasons may be found to explain the prominent position of phosphate in intermediary metabolism, but it must be recognized that there is nothing unusual about phosphate in a chemical sense, and the free energies of reactions involving phosphate are determined, as in all cases, by the relative stabilities of reactants on the two sides of the equation. The thermodynamic stabilities that determine the equilib-

[6] F. Lipmann, *Advances in Enzymol.* **1**, 99 (1941).
[7] H. M. Kalckar, *Chem. Revs.* **28**, 71 (1941).

rium position must not be confused with kinetic stabilities; that is, the less stable molecules in the thermodynamic sense may be extremely stable in a biological system. In considering this aspect of chemical reactions, again a distinction must be maintained between the extent of reaction predicted from thermodynamic properties and the rate of reaction, which depends on a mechanism that may be influenced by many factors. Compounds such as ATP are useful in biological systems because they are not kinetically unstable (in the absence of enzymes that attack them), but are thermodynamically much less stable than their reaction products.

If a reaction involving transfer of a group is written as two partial reactions, one a hydrolysis and the other a dehydration, the free energy of the over-all reaction can be calculated as the sum of the free energies of the partial reactions.

(1) 3-phosphoglycerate + H_2O → glycerate + phosphate; $\Delta F_0' = -3000$ cal.
(2) glycerate + phosphate   → 2-phosphoglycerate + H_2O;

$$\Delta F_0' = \quad 4050 \text{ cal.}$$

(1) + (2) 3-phosphoglycerate   → 2-phosphoglycerate; $\Delta F_0' = \quad 1050$ cal.

If one considers reactions involving a "high-energy" bond, the ΔF for the over-all reaction may be a large negative number.

(3)   ATP + H_2O → ADP + P; $\Delta F_0' = -7800$ cal.
(4)   glucose + P → glucose-6-phosphate + H_2O; $\Delta F_0' = +3000$ cal.

(3) + (4) ATP + glucose → glucose-6-phosphate + ADP; $\Delta F_0' = -4800$ cal.

Since the equilibrium constant at 37°C. for a reaction with a ΔF of -4800 cal. is calculated

$$\log_{10} K_{eq} = \frac{-4800}{-2.303 \times 1.98 \times 310} = 3.4$$

$$K_{eq} = 2.5 \times 10^3$$

the phosphorylation of glucose proceeds essentially to completion. According to the high-energy bond concept it is the bond energy of the terminal phosphate group of ATP that "drives" the reaction. It is this concept that has been dangerously misleading in modern biochemistry. If free energies of hydrolysis are used correctly to determine the extent to which reactions *may* proceed, they are valuable tools in analyzing and predicting reactions. In order to have the concept of "bond energy" remain useful, however, it must be severely limited.[8]

The most serious error in the extension of the bond energy concept has been the treatment of ΔF_0 as a tangible object, a packet of energy to

[8] R. J. Gillespie, G. A. Maw, and C. A. Vernon, *Nature* 171, 1147 (1953).

be taken from one compound and applied to another. The illegitimate extension of the bond energy concept has led to the notion that reactions with positive values of ΔF cannot occur unless coupled with more favorable reactions. The example of equations (1) and (2) is of a familiar reaction that is known to proceed easily in either direction. The positive ΔF merely informs us that under the conditions of $\Delta F_0'$ the reaction reaches equilibrium with 6 times as much 3-phosphoglycerate as 2-phosphoglycerate. If one were to extend the concept of bond energy and driving force to the aldolase reaction, a bond energy would be assigned the function of tending to split the molecule into trioses. But, as was discussed previously, the equilibrium composition of the aldolase reaction shifts from one extreme to the other, depending on absolute concentrations. This illustrates the necessity of using ΔF only as a statistical property of molecules, not as an intrinsic property of single molecules.

Particularly in relation to oxidative reactions, the possible synthesis of ATP from ADP and inorganic P has been considered as a function of the free energy of the reaction. If the standard free energy of an oxidation is 10 kcal. or more, it is concluded that 1 ATP may be generated in a coupled reaction. If the free energy of the reaction is much larger, it is considered that the maximum number of ATP moles formed per mole of reaction is the total free energy divided by the free energy of hydrolysis of ATP. This sort of arithmetic is a completely invalid application of free energy. The free energy of a reaction does not tell whether a reaction (such as ATP synthesis) proceeds or not. It merely tells how far a reaction will proceed under given circumstances. Suppose ATP synthesis were coupled to a reaction in which the $-\Delta F$ were smaller than that of ATP hydrolysis. Such a system, starting with ADP and P and similar concentrations of the coupled system, would come to equilibrium with ADP exceeding ATP. If the product of the coupled reaction (for example, DPNH) were removed efficiently, the reaction would proceed to completion with all the ADP converted to ATP. In reactions to be considered in the next section it will be shown that ATP is used in the synthesis of acetyl phosphate. In this case there is a $\Delta F_0'$ of about $+3000$ cal. If the number of "active acetyl" groups to be formed by a coupled reaction with ATP hydrolysis were calculated on the basis used for calculating the number of ATP molecules that can be formed in a coupled oxidation, it would be concluded that acetate cannot be activated at all.

The proper use of free energy in calculations of biological reactions can give useful information. The ΔF values used in ordinary work are $\Delta F_0'$ values, whereas ordinarily reactions are carried out with very dilute solutions, and with the concentrations of the various reactants not equal to each other. The extent to which these actual reactions proceed may be calculated from ΔF_0 values only by determining the ΔF of dilution for each component.[9]

Free energy has been defined as the maximum amount of useful work that can be obtained from a reaction. In order for the energy of a reaction to be used, a transfer mechanism must be available. The useful work of systems such as glycolysis is only the production of materials such as ATP required by other systems. Such requirements are only for the specific reagents the other systems can handle. The free energy of hydrolysis of the hypothetical arsenate addition compound of phosphoglyceric acid may be much greater than that of the corresponding phosphate, but its value to an organism is negative. In the case of arsenate interference, a new reaction, hydrolysis, occurs, and the large free-energy change appears as an increase in entropy and a large ΔH, which is energy dissipated. In the absence of agents, such as arsenate, that open new pathways, energy cannot be so dissipated, as there is no mechanism for a pathway other than the reactions coupled with ATP formation. In living systems the availability of adenine nucleotides is less than that of primary substrates. Glycolysis thus depends upon continued utilization of ATP. In the presence of excess substrate, ATP may be used in processes that store food reserves (fat, polysaccharides). Kinetic limitations of the utilization system then determine the rate of glycolysis. This was demonstrated in the studies of Meyerhof who found that fermentation in yeast autolysates became limited by available ADP; in order to permit glycolysis to continue an ATPase (apyrase) was added. The hydrolysis of ATP eliminated a log-jam by forming ADP to accept phosphate from 1,3-diphosphoglycerate and phosphopyruvate, and permitted these compounds to be moved along the glycolytic sequence.

The energy-yielding reactions of living systems must proceed to completion in order to build up the concentration of materials needed for synthesis. A maximum yield of product of a coupled reaction (e.g., ATP) can thus be calculated for the over-all system, but such a calculation tells nothing at all about the mechanism of the coupling. The reaction involved in the actual coupling may be rather unfavorable, but, as part of a total system, the unfavorable reaction is "pulled" by steps with larger free-energy changes. That is, the equilibrium of the over-all system is in the opposite direction from that of the unfavorable step.

The term "bond energy" was created to provide a convenient means for expressing the standard free-energy changes of reactions that can be coupled biologically. Since the words "bond energy" have often had false connotations, it is suggested that another term be used that might

---

[9] The free energy of dilution is evaluated by the equation $\Delta F = -RT \ln \dfrac{[a_1]}{[a_2]}$.

focus attention on the relation of this property to a reaction, hydrolysis, rather than to a compound. The quantity evaluated is $\Delta F_0'$, the standard free-energy change of hydrolysis at pH 7.0 (selected for convenience). The term F_h is suggested as an abbreviation for $\Delta F_0'$ of hydrolysis.

TABLE 4

F_h VALUES OF VARIOUS REACTIONS

Reaction	F_h (cal.)
1,3-Diphosphoglycerate + H_2O → 3-phosphoglycerate + phosphate	−12,800
Acetoacetyl CoA + H_2O → acetoacetate + CoA	−12,000
Phosphopyruvate + H_2O → pyruvate + phosphate	−11,800
Creatine phosphate + H_2O → creatine + phosphate	−11,200
Acetyl phosphate + H_2O → acetate + phosphate	−11,000
Succinyl CoA + H_2O → succinate + CoA	− 9,100
Acetyl CoA + H_2O → acetate + CoA	− 8,000
ATP + H_2O → ADP + phosphate	− 7,800
ADP + H_2O → AMP + phosphate	− 7,300
Glucose-1-phosphate + H_2O → glucose + phosphate	− 4,800
Glutamine + H_2O → glutamate + NH_3	− 3,600
Glucose-6-phosphate + H_2O → glucose + phosphate	− 3,000
α-Glycerol phosphate + H_2O → glycerol + phosphate	− 2,500

The F_h values in Table 4 have been compiled from many sources, and have been adjusted to give appropriate differences from the value assigned to the ATP reaction. Recent measurements by independent methods agree within a few hundred calories on the −7800 cal. value. In all cases the concentration of water has been omitted from the calculations, thus the values can be added algebraically for coupled reactions. The negative signs are used to agree with the convention that the free-energy change of reactions which proceed spontaneously is negative.

The values included are intended to indicate the range found in biological reactions, and are not necessarily the most accurate.

Reactions of Coenzyme A

Formation of Acetyl Coenzyme A. Acetyl CoA is generated in the oxidation of pyruvate and acetaldehyde. As will be developed in a later section, oxidation of fatty acids also leads to the formation of acetyl CoA. This compound is the "active acetate" of many biological systems, and participates in many reactions. In bacteria, but not in animal tissues, there is an enzyme discovered by Stadtman, named phosphotransacetylase, often referred to as simply transacetylase.[1] This is the enzyme responsible for the utilization of acetyl phosphate. The equilibrium of the exchange reaction

$$CH_3COOPO_3^= + HS—CoA \rightleftharpoons CH_3CO—SCoA + HPO_4^=$$

lies somewhat to the side of acetyl CoA ($K_{eq} = 60$, $\Delta F = -3000$ cal.). The enzyme also catalyzes the arsenolysis of acetyl P. Since acetyl P is

$$CH_3\overset{\overset{O}{\|}}{C}—O—\underset{\underset{O^-}{|}}{\overset{\overset{O}{\|}}{P}}—O^- + H_2O \xrightarrow[\substack{HAsO_4^- \\ CoA}]{transacetylase} CH_3\overset{\overset{O}{\|}}{C}—O^- + HO—\underset{\underset{O^-}{|}}{\overset{\overset{O}{\|}}{P}}—O^- + H^+$$

readily synthesized, and is completely hydrolyzed in the presence of arsenate and transacetylase, the disappearance of acetyl P or the appearance of inorganic P is a measure of the enzyme. The reaction has an absolute requirement for CoA, and the rate is proportional to CoA concentration. Therefore, with constant levels of enzyme, the rate of the transacetylase reaction can be used to assay CoA. Since the role of CoA is catalytic, this kinetic assay is very sensitive. The sulfhydryl group of CoA is easily oxidized, and therefore must be protected by the presence of other SH compounds, such as glutathione, cysteine, or H_2S.

Acetate-ATP Reactions. Two mechanisms have been found for "activating" free acetate. A widely distributed enzyme forms acetyl CoA from acetate, CoA, and ATP.[2] This so-called acetate-ATP reaction appears to be catalyzed by a single enzyme, but occurs in two steps. The over-all reaction is:

$$ATP + acetate + CoA \rightleftharpoons AMP + PP + acetyl CoA$$

The enzyme catalyzes exchange reactions (determined with isotopic compounds) in which acetate exchanges with acetyl CoA and inorganic pyrophosphate exchanges with ATP. These reactions were once interpreted as showing compounds made of enzyme + AMP and enzyme + CoA. Recently Berg has shown acetyl AMP to be formed in the absence of CoA, and demonstrated a requirement for acetate in the ATP-PP exchange.[3] These experiments eliminate the necessity for postulating an enzyme–substrate compound as an intermediate in acetate activation and support the sequence shown in (I). Although acetyl AMP has not been detected in incubation mixtures, this sequence is supported by the finding that synthetic acetyl AMP rapidly reacts with either PP or CoA

[1] E. R. Stadtman, *J. Biol. Chem.* **196**, 527 (1952).
[2] M. E. Jones, S. Black, R. M. Flynn, and F. Lipmann, *Biochim. et Biophys. Acta* **12**, 141 (1953).
[3] P. Berg, *J. Biol. Chem.* **222**, 991 (1956); F. Lipmann, *ibid.* **155**, 55 (1944).

in the presence of the acetate-activating enzyme. Apparently the equilibrium of each partial reaction is reached with extremely low concentrations of acetyl AMP, which in effect may not dissociate from the enzyme.

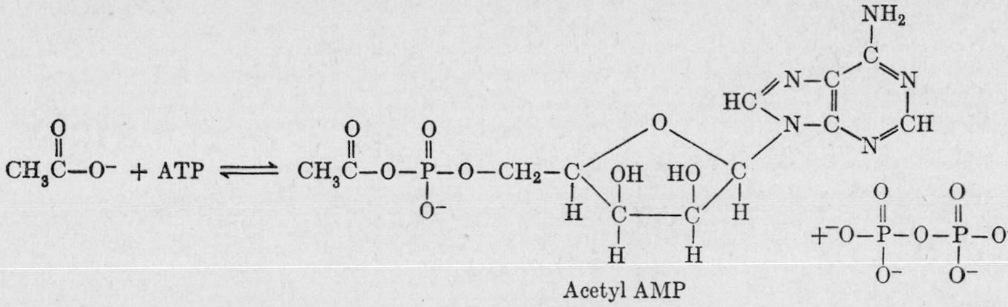

Acetyl AMP

acetyl AMP + CoA $\rightleftharpoons$ acetyl CoA + AMP

(I)

Acetokinase. A second method for activating acetate has been found in certain bacteria.[4] This involves formation of acetyl P in a kinase reac-

acetate + ATP $\rightleftharpoons$ acetyl P + ADP

tion. Besides ATP, ITP is also effective. The equilibrium constant (ADP) (acetyl P)/(ATP)(acetate) = approximately 0.01. Despite the unfavorable equilibrium, this reaction has been used to determine acetate by coupling the formation of acetyl P with the nonenzymatic reaction with hydroxylamine. In general, acid anhydrides such as acyl phosphates react rapidly at room temperature with NH_2OH to form hydroxamic acids.

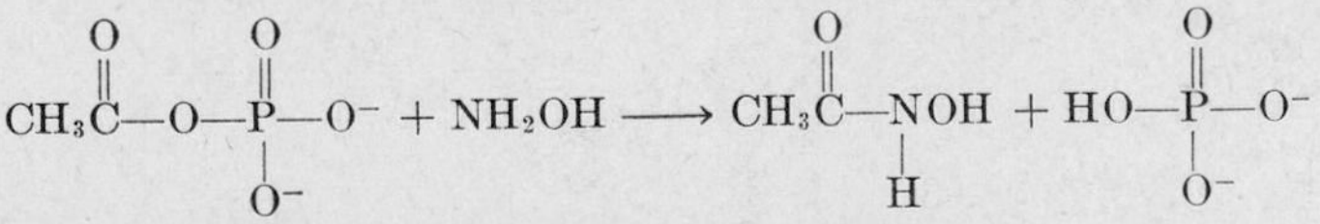

Acetyl phosphate   Hydroxyl-   Acethydroxamic
amine   acid

The formation of hydroxamic acids is essentially irreversible and pulls the reaction to completion. ATP and similar phosphate compounds do not react with NH_2OH. Reaction with hydroxylamine has long been used to determine acid derivatives such as esters, since hydroxamic acids form intensely colored complexes with ferric iron in acid solution. Acid anhydrides react much more readily than most other derivatives, and the hydroxamic reaction was introduced into biochemistry by Lipmann as a means for detecting acyl phosphate.[5] Thioesters also react readily with

[4] I. A. Rose, M. Grunberg-Manago, S. R. Korey, and S. Ochoa, *J. Biol. Chem.* **211**, 737 (1954).

[5] F. Lipmann and L. C. Tuttle, *J. Biol. Chem.* **159**, 21 (1945).

NH_2OH. Acetyl phosphate and acetyl CoA are differentiated by the heat lability of the former; heat-stable hydroxamic-forming material is thioester and the remainder of the total hydroxamic-forming material is acid anhydride. Many bacteria that contain the acetyl P-forming enzyme, acetokinase, also contain transacetylase, so that acetyl phosphate can serve as an acetyl donor.

Coenzyme A Transferases. In addition to enzymes that incorporate CoA into acyl thioesters, a group of enzymes have been found to form the CoA derivatives of certain acids by transfer of CoA from a preexisting thioester. The first example of this type of reaction to be found was the CoA transphorase of *C. kluyveri*.[6] This enzyme transfers CoA among various small fatty acids. Bacterial enzymes have been implicated in CoA transfers between propionate and succinate,[7] acetate and formate,[8]

$$CH_3\overset{\text{O}}{\underset{\|}{C}}-S-CoA + CH_3CH_2CH_2\overset{\text{O}}{\underset{\|}{C}}-O^- \rightleftharpoons CH_3\overset{\text{O}}{\underset{\|}{C}}-O^- + CH_3CH_2CH_2\overset{\text{O}}{\underset{\|}{C}}-S-CoA$$

and acetate and malonate.[9] A transferase from pig heart was found to transfer CoA specifically from succinyl CoA to acetoacetate and similar β-keto acids.[10] A similar enzyme was found in *Pseudomonas* to be involved in the oxidation of β-ketoadipate.[11] There is presumptive evidence for a CoA transferase in the activation of oxalate.[12]

Amine Acetylation. The first acetylation reaction to be studied intensively was the acetylation of aromatic amines, of which sulfanilamide was used as a representative (II).[13] This system was used for the first assays of CoA. The amine-acetylating enzyme has been partially purified.

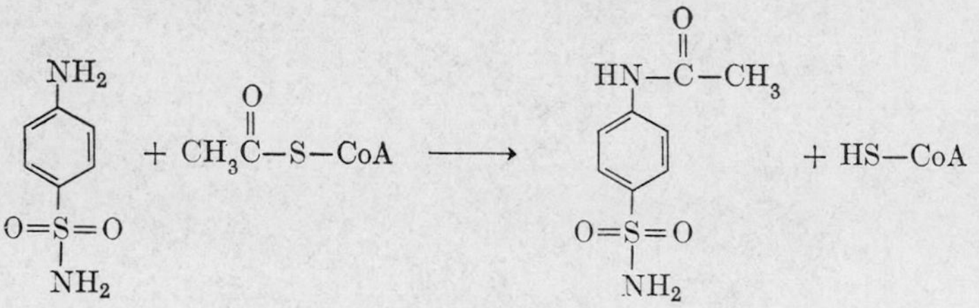

(II)

[6] E. R. Stadtman, *J. Biol. Chem.* **203**, 501 (1953).

[7] H. R. Whiteley, *Proc. Natl. Acad. Sci. U. S.* **39**, 772 (1953); E. A. Delwiche, E. F. Phares, and S. F. Carson, *Federation Proc.* **12**, 194 (1953).

[8] I. Lieberman, *Arch. Biochem. and Biophys.* **51**, 350 (1954).

[9] O. Hayaishi, *J. Biol. Chem.* **215**, 125 (1955).

[10] J. R. Stern, M. J. Coon, and A. del Campillo, *Nature* **171**, 28 (1953); D. E. Green, D. S. Goldman, S. Mii, and H. Beinert, *J. Biol. Chem.* **202**, 137 (1953).

[11] M. Katagiri and O. Hayaishi, *Federation Proc.* **15**, 285 (1956).

[12] W. B. Jakoby, E. Ohmura, and O. Hayaishi, *J. Biol. Chem.* **222**, 435 (1956).

[13] F. Lipmann, *J. Biol. Chem.* **160**, 173 (1945).

Early studies used acetyl P as substrate with CoA and transacetylase, or used the acetate-ATP system, and showed complete disappearance of free amine as measured colorimetrically.[14] It was believed that acetyl CoA was formed and subsequently donated its acetyl group to the amine. Yet when acetyl CoA became available in substrate concentrations and was used as the acetyl donor, the reaction went only part way.

The unexpected difference between the observed extent of reaction and the extent anticipated from independent measurements of acetyl CoA has been explained by the occurrence of a competing reaction that

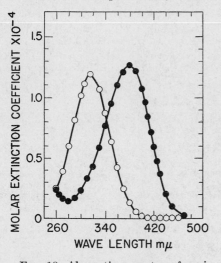
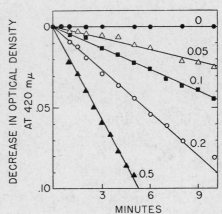

FIG. 10. Absorption spectra of *p*-nitroaniline (●) and *p*-nitroacetanilide (○).[15]

FIG. 11. Rate of acetylation of *p*-nitroaniline with the amounts of enzyme indicated.

removes acetyl CoA. The rate of amine acetylation was followed spectrophotometrically, as there is a large shift in the absorption peak of aromatic amines on acetylation (Figs. 10, 11). The spectrophotometric assay showed that the system has a high affinity for acetyl CoA. The reaction is very rapid, and when it stops the addition of more acetyl CoA allows the acetylation to proceed to completion.

The loss of acetyl CoA was found to be caused by a nonenzymatic transfer of acetyl groups to other sulfhydryl compounds added to the system. Ordinarily sulfhydryl compounds should not be required in reactions of acetyl CoA, as opposed to systems using free CoA, whose —SH group needs protection. But the amine-acetylating enzyme also has essential —SH groups that require protection. The dilemma was

[14] A. C. Bratton and E. K. Marshall, Jr., *J. Biol. Chem.* **128**, 537 (1939).
[15] H. Tabor, A. H. Mehler, and E. R. Stadtman, *J. Biol. Chem.* **204**, 127 (1953).

resolved when the reaction was run at lower pH values to minimize the nonenzymatic exchange and Versene (ethylenediamine tetraacetate) was added to protect the enzyme from both heavy metal inhibition and oxidation catalyzed by heavy metals. Under these conditions the enzyme uses acetyl CoA and the various acetate-activating systems equally well.[15] In addition to catalyzing a net acetylation of amines, this enzyme also catalyzes a slow exchange of acetate from one amine to another.[16] The

$$
\underset{\underset{Ar^1}{|}}{HN}-\overset{\overset{O}{\|}}{C}-CH_3 + \underset{\underset{Ar^2}{|}}{NH_2} \longrightarrow \underset{\underset{Ar^1}{|}}{NH_2} + \underset{\underset{Ar^2}{|}}{HN}-\overset{\overset{O}{\|}}{C}-CH_3
$$

enzyme is not absolutely specific for acetyl CoA, but also reacts slowly with other low molecular weight acyl CoA compounds. The long-chain palmityl CoA is an inhibitor of the amine-acylating enzyme.[15]

Citrate Formation. Among the condensation reactions of acetyl CoA is the condensation of two acetyl groups to form acetoacetyl CoA.[17] This is an example of a type of reaction called thiolase. These reactions will be discussed in connection with both fatty acid oxidation (p. 144) and tryptophan degradation (Chapter VII). Quantitatively, the most important reaction that uses acetyl CoA is the one that results in citrate formation.[18] The reaction is shown in (III). The enzyme that catalyzes this reaction

$$
\underset{\text{Acetyl CoA}}{CH_3\overset{\overset{O}{\|}}{C}-S-CoA} + \underset{\text{Oxalacetate}}{\underset{\underset{COO^-}{|}}{\underset{\underset{CH_2}{|}}{\underset{\underset{C=O}{|}}{COO^-}}}} + H_2O \rightleftharpoons \underset{\text{Citrate}}{\underset{\underset{H_2C-\overset{\overset{O}{\|}}{C}-O^-}{|}}{\underset{\underset{HOC-\overset{\overset{O}{\|}}{C}-O^-}{|}}{H_2C-\overset{\overset{O}{\|}}{C}-O^-}}} + HS-CoA + H^+
$$

(III)

is the citrate condensing enzyme, often called simply condensing enzyme. This enzyme was crystallized from pig heart.[19] The importance of this reaction lies in the fact that it is the means for introducing products of carbohydrate, fat, and protein metabolism into the tricarboxylic acid cycle, the site of complete oxidation. The reaction catalyzed by this enzyme differs from most reactions of acyl CoA derivatives, in that condensation involves the methyl group rather than the carboxyl of acetate. Two steps occur in citrate formation: the condensation and the hydrolysis

[16] S. P. Bessman and F. Lipmann, *Arch. Biochem. and Biophys.* **46**, 252 (1953).
[17] F. Lynen, L. Wessely, D. Wieland, and L. Rueff, *Z. angew. Chem.* **64**, 687 (1952).
[18] J. R. Stern and S. Ochoa, *J. Biol. Chem.* **191**, 161 (1951).
[19] S. Ochoa, J. R. Stern, and M. C. Schneider, *J. Biol. Chem.* **193**, 691 (1951).

of the thioester. There is no evidence for any separation of these steps. The specificity of the condensing enzyme has not been explored extensively. It appears that certain analogs of acetyl CoA can react with this enzyme, including fluoroacetyl CoA, which results in the formation of fluorocitrate[20] (see p. 96).

The condensation reaction proceeds essentially to completion. The equilibrium constant (citrate)(CoA)/(acetyl CoA)(OAA) was evaluated by measuring the equilibrium of a system including other reactions whose equilibria were known. A value of 5×10^5 was calculated, equivalent to a $\Delta F_0'$ of -8000 cal.[21] It is of interest to note that this reaction with a very large $\Delta F_0'$ can be reversed, and, in the presence of transacetylase and inorganic phosphate, gives rise to acetyl P. Since acetyl P transfers P to ADP in the acetokinase reaction, a mechanism for forming ATP by coupling with a thermodynamically unfavorable reaction is demonstrated with known enzymes. Of course, this reaction will not proceed measurably unless the citrate cleavage is coupled with reactions that remove the products efficiently.

GENERAL REFERENCES

Baddiley, J. (1955). *Advances in Enzymol.* **16,** 1.
Lipmann, F. (1954). *Science* **120,** 855.
Novelli, G. D. (1953). *Physiol. Revs.* **33,** 525.
Snell, E. E., and Brown, G. M. (1953). *Advances in Enzymol.* **14,** 49.
Symposium on Chemistry and Functions of Coenzyme A (1953). *Federation Proc.* **12,** 673.

The Krebs Citric Acid Cycle

The citric acid cycle, or tricarboxylic acid cycle, is a series of reactions by which certain organic acids are oxidized to carbon dioxide and water. The oxidations of the cycle are linked to reactions that transfer electrons ultimately to molecular oxygen, and, in the presence of oxygen, proceed to completion. These reactions involve esterification of inorganic phosphate at both the level of substrate oxidation and electron transport, and together the cycle and subsequent electron transport are an efficient mechanism for producing energy for other biological systems. For many organisms, plant, animal, and bacterial, this is the major or only oxidative pathway.

[20] R. A. Peters, R. W. Wakelin, D. E. A. Rivelt, and L. C. Thomas, *Nature* **171,** 1111 (1953).
[21] J. R. Stern, B. Shapiro, E. R. Stadtman, and S. Ochoa, *J. Biol. Chem.* **193,** 703 (1951).

The citric acid cycle is one of several biochemical cycles, in which one reagent is regenerated after a series of reactions, and thus acts catalytically during the metabolism of other reagents. The regenerated molecule may or may not be composed of the same atoms in successive cycles. In the citric acid cycle, the reaction that introduces new carbon is the condensation of oxalacetate and acetyl CoA, to form citrate. The cycle involves dehydration, hydration, oxidation and decarboxylation, and ultimately yields oxalacetate in addition to carbon dioxide and water. As will be seen in the following discussion, the atoms of the oxalacetate that is produced are not identical with those that initiated the cycle.

The present concept of a cyclic pathway for terminal metabolism developed from the observations of Thunberg, who found that only a few organic acids, including succinate, fumarate, malate, and citrate, were oxidized actively by tissue homogenates.[1] Szent-György developed the hypothesis that 4-carbon dicarboxylic acids were involved in an electron-transport system.[2] This function for the dicarboxylic acids is no longer advocated, as evidence has accumulated to show that all of the carbon atoms of the dicarboxylic acids are converted to CO_2. Using homogenates of pigeon breast muscle, Krebs and his collaborators demonstrated that the organic acids found by Thunberg to cause reduction of methylene blue stimulated oxygen consumption when added in catalytic amounts. The nature of the catalytic effect of these acids on respiration was explained by Krebs as a result of their participation in a cyclic series of reactions needed for the oxidation of pyruvate. Under anaerobic conditions citrate was shown to accumulate when pyruvate and oxalacetate were incubated with pigeon muscle homogenates. Malonate inhibition of respiration caused both citrate and α-ketoglutarate to accumulate. These and other observations were used, together with some known reactions and some subsequently discovered, to support the hypothesis described below. Subsequent work in many laboratories has only added detailed information without modifying the basic concept of the cycle as proposed by Krebs[3] (Fig. 12).

The reactions of the citric acid cycle include:

(1) condensation of oxalacetate and acetate from acetyl CoA to form citrate;
(2) rearrangement of citrate to form isocitrate;

[1] T. Thunberg, *Skand. Arch. Physiol.* **40**, 1 (1920).
[2] An informative and entertaining personal account of the development of his theories was published by A. Szent-Györgyi as a small monograph: "Studies on Biological Oxidation." Karl Reaugi, Leipzig, 1937.
[3] H. A. Krebs, *Harvey Lectures Ser.* **44**, 165 (1948–49).

(3) oxidative decarboxylation of isocitrate to form α-ketoglutarate;

(4) oxidative decarboxylation of α-ketoglutarate to form succinate;

(5) oxidation of succinate to fumarate;

(6) hydration of fumarate to malate;

(7) oxidation of malate to oxalacetate, which closes the cycle by entering reaction (1).

When all of the necessary enzymes are present, this cycle can be initiated by the addition of any of the compounds listed. For continuous recycling, acetyl CoA must be supplied, but the addition of a large

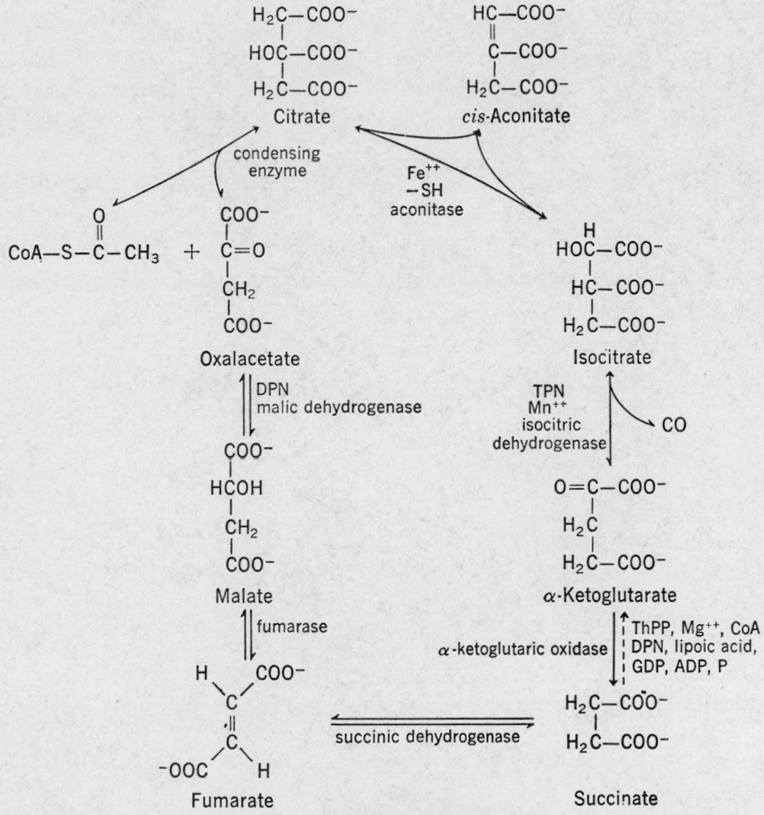

FIG. 12. The citric acid cycle.

amount of a citric acid cycle intermediate does not force the system to retain large amounts of carrier compounds, since reactions exist for converting the 4-carbon compounds, oxalacetate and malate, to pyruvate, thus permitting complete oxidation of the components of the cycle. In addition, the cycle may be started by reactions that synthesize dicarboxylic acids by carbon dioxide fixation. Thus homeostatic mechanisms are available to increase or decrease the concentrations of Krebs cycle intermediates.

The citric acid cycle is usually considered in terms of oxidation of substrate and generation of ATP, but it is also of value in the production and utilization of the carbon skeletons for many other essential compounds. Reactions will be described later that demonstrate the utilization of citric acid cycle intermediates for the fixation of inorganic nitrogen into amino acids. The amino acids thus formed are used in the syntheses of other amino acids, including those required for the production of urea. The carbon of citric acid cycle acids also appears in porphyrins along with the amino acid glycine. Indeed, it has been proposed that during rapid growth the principal function is to supply cell material.

Hydration and Dehydration

Two of the enzymes of the tricarboxylic acid cycle, aconitase and fumarase, catalyze reactions in which water is added reversibly to an unsaturated polycarboxylic acid. Both enzymes exhibit rigid stereospecificity; fumarase forms only L-malate from fumarate and forms only fumarate (*trans*) and not maleate (*cis*-ethylenedicarboxylic acid),[4] and aconitase reacts with only *cis*-, not *trans*-aconitate, and with D-, not L-isocitrate.[5] Citrate is a symmetrical molecule, with no optical isomers, but it will be shown that steric factors also enter into the reaction of this substrate with aconitase. The enzymes of the tricarboxylic acid cycle, in contrast to the glycolytic enzymes, are associated with intracellular granules known as mitochondria. Studies of the individual enzymes have depended to a large extent on the separation of soluble activities from these particles. Aconitase and fumarase are released from the particles very rapidly under mild conditions; often in the preparation of cell-free homogenates these activities are largely solubilized, and special care must be taken to demonstrate their origin in mitochondria.

The history of the purification of fumarase demonstrates the danger of equating crystallization with homogeneity of proteins. Laki and Laki purified fumarase extensively, and eventually obtained very active crystalline precipitates.[6] Crystals of lower specific activity have been obtained during some preparations, and, on recrystallization, these lose specific activity.[7] The crystalline preparations obtained finally by

---

[4] H. D. Dakin, *J. Biol. Chem.* **52**, 183 (1922).
[5] C. Martius, *Z. physiol. Chem.* **247**, 104 (1937); **257**, 29 (1938).
[6] E. Laki and K. Laki, *Enzymologia* **9**, 139 (1941).
[7] E. M. Scott, *Arch. Biochem.* **18**, 131 (1948).

Massey[8] and by Frieden *et al.*,[9] using different procedures, are identical with each other, and satisfy physical criteria for a homogeneous protein.

Assay of Aconitase and Fumarase. Several types of assay have been used for these enzymes. The simplest is the measurement of the appearance of an unsaturated compound detected spectrophotometrically at 240 mμ.[10] More specific assays employ enzymes that oxidize the appropriate hydroxy acid, malic or isocitric, with pyridine nucleotides. In these, the unsaturated compound is hydrated in the presence of excess isocitric dehydrogenase (p. 103) or "malic enzyme" (p. 105), which oxidize the hydroxy acids quantitatively as they are formed and allow the hydration to be followed as an increase in optical density at 340 mμ. Other assays depend upon the change in rotation of polarized light on formation of the optically active hydroxy acids (increased by molybdate),[5] titration of the double bond with permanganate,[11] fluorometric determination of malic acid[12] and colorimetric determination of citric acid.[13] These methods are less suitable for kinetic studies than the spectrophotometric methods, and the system that does not require secondary reactions is to be preferred for studying influences on the rate of reaction. Unfortunately the large amount of light absorption at 240 mμ contributed by many reagents, as well as by proteins and other naturally occurring compounds, limits this assay; only solutions with optical densities less than 2 should be used unless very precise optical equipment is available.

Equilibria of Aconitase and Fumarase. The reactions catalyzed by fumarase and aconitase have small free-energy changes. Omitting water from the calculation, the equilibrium constant for fumarase, K_{eq} = (malate)/(fumarate) = about 3–4 at room temperature and pH 7.[14] This reaction is temperature dependent, forming relatively more fumarate at higher temperatures.[15] The aconitase equilibrium must include two separate reactions: the equilibration of *cis*-aconitate with

[8] V. Massey, *Biochem. J.* **51**, 490 (1952).

[9] C. Frieden, R. M. Bock, and R. A. Alberty, *J. Am. Chem. Soc.* **76**, 2982 (1954).

[10] E. Racker, *Biochim. et Biophys. Acta* **4**, 211 (1950). 240 mμ has been used as the lowest wavelength convenient for use with ordinary spectrophotometers for the measurement of α,β-unsaturated acids, but the absorption peaks of these compounds are near 210 mμ, where the extinction coefficients are much larger than those at 240 mμ.

[11] F. B. Straub, *Z. physiol. Chem.* **236**, 43 (1935).

[12] The qualitative reaction of Eegriwe [*Z. anal. Chem.* **89**, 121 (1939)] was developed for quantitative analysis by Dr. John Speck of the University of Chicago.

[13] G. W. Pucher, C. C. Sherman, and H. B. Vickery, *J. Biol. Chem.* **113**, 235 (1936).

[14] K. P. Jacobsohn, *Biochem. Z.* **274**, 167 (1934).

[15] H. A. Krebs, D. H. Smyth, and E. A. Evans, Jr., *Biochem. J.* **34**, 1091 (1940).

citrate and isocitrate. This system does not vary with temperature, and contains about 89.5 per cent citrate, 6–7 per cent isocitrate, and 3–4 per cent cis-aconitate at equilibrium.[16] When aconitate is used as substrate, the equilibration with isocitrate occurs more rapidly than the citrate reaction. The equilibration of aconitate with isocitrate leads to the formation of more isocitrate than is present in the over-all equilibrium mixture.[17] This overshoot of the final equilibrium concentration is not a violation of thermodynamics; the partial reaction involving only aconitate and isocitrate does not go beyond its own equilibrium. It is the removal of tricarboxylic acid from the aconitate–isocitrate system as citrate is formed that causes the concentration of isocitrate to fall after an initial increase. In spite of this apparent dissociation into two reactions, there is no evidence that more than one enzyme is involved in the aconitase reaction in animal preparations. Evidence has been presented, however, that in the mold Aspergillus niger two separate enzymes exist.[18]

Nature of Aconitase. Many attempts to purify aconitase following its discovery in 1937 failed to accomplish more than modest increases in specific activity. The difficulty in retaining activity on purification was explained by the finding of Dickman and Cloutier that both Fe^{++} and a sulfhydryl compound are required for aconitase activity.[19] Morrison subsequently achieved a high degree of purification of the enzyme and concluded that one equivalent each of iron and reducing agent is bound to the enzyme.[20]

Some insight into the mechanism of aconitase action was gained as a result of investigation into the basis for the pharmacological effects of fluoroacetic acid. Fluoroacetic acid is a highly toxic compound that is capable of interfering with oxidation of acetate. Recent work indicates the possibility of inhibition at several sites, but one site has been shown to be affected in a highly specific manner. In fluoroacetate poisoning citrate accumulates. In the laboratory of Peters it was found that one of the isomeric fluorocitrates accumulated when fluoroacetate was incubated with homogenates. This compound and synthetic fluorocitrate have been shown to be effective inhibitors of aconitase, and this inhibition is presumably the cause of citrate accumulation.[21] Inhibition of both citrate and isocitrate reactions strengthens the conclusion that only a single aconitase exists.

[16] H. A. Krebs and L. V. Eggleston, *Biochem. J.* **37**, 334 (1943).
[17] C. Martius and H. Leonhardt, *Z. physiol. Chem.* **278**, 208 (1943).
[18] N. E. Neilson, *Biochim. et Biophys. Acta* **17**, 139 (1953).
[19] S. R. Dickman and A. A. Cloutier, *J. Biol. Chem.* **188**, 379 (1951).
[20] J. F. Morrison, *Biochem. J.* **56**, xxxvi (1954).
[21] J. F. Morrison and R. A. Peters, *Biochem. J.* **56**, xxxvi (1954).

The equations usually written to describe the action of aconitase show *cis*-aconitate as an intermediate between citrate and isocitrate. Experiments with deuterated water have recently indicated that a more correct formulation is the interconversion of all of the three tricarboxylic acids through a common intermediate (I).[22] The most convincing experiments

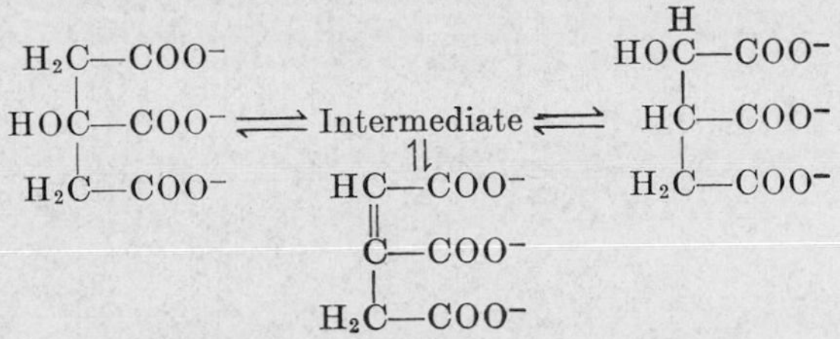

$$
\begin{array}{ccccc}
H_2C-COO^- & & & & \overset{H}{HOC}-COO^- \\
| & & & & | \\
HOC-COO^- & \rightleftharpoons & \text{Intermediate} & \rightleftharpoons & HC-COO^- \\
| & & \updownarrow & & | \\
H_2C-COO^- & & HC-COO^- & & H_2C-COO^- \\
& & \parallel & & \\
& & C-COO^- & & \\
& & | & & \\
& & H_2C-COO^- & & \\
& & (I) & &
\end{array}
$$

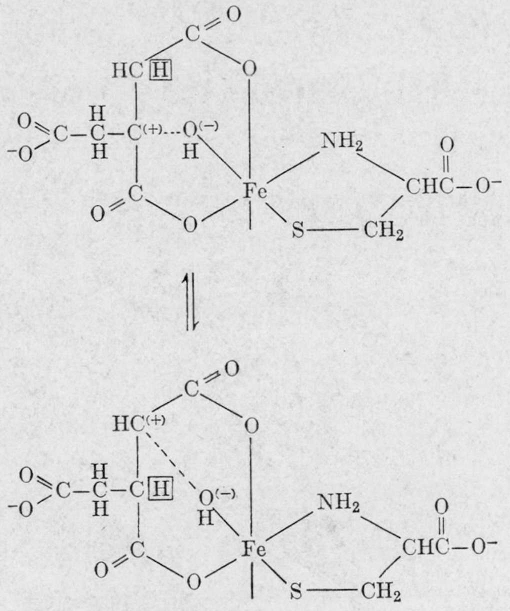

(II)

show that one atom of deuterium is incorporated into citrate when *cis*-aconitate is the substrate (the other position that becomes deuterated is the hydroxyl group, which exchanges its hydrogen with water during isolation of the product, and therefore appears as unlabeled). When isocitrate is the substrate, however, only traces of deuterium are found in the citrate formed. If *cis*-aconitate were an intermediate in the latter case,

22 J. F. Speyer and S. R. Dickman, *J. Biol. Chem.* **220**, 193 (1956).

the isotope incorporation should be equivalent to that found when *cis*-aconitate is the primary substrate. The small incorporation of label in citrate produced from isocitrate is consistent with the formation of a small amount of *cis*-aconitate during the isocitrate–citrate reaction. Kinetic analyses are also consistent with the formulation of a common intermediate. Speyer and Dickman have suggested the structures shown in (II) as representing the mode of action of the enzyme. These structures show a tricarboxylic acid held by the enzyme in a complex with Fe^{++} and cysteine. In the complex a hydride ion (H^-) shifts from a carbon atom to an oxygen associated with the iron. The transfer of the hydride ion to a carbonium (C^+) ion allows interconversion of citrate and isocitrate without equilibration with hydrogen (H^+) ions. Aconitate is formed according to this proposal by elimination of a proton from the position marked $\boxed{H}$ with the simultaneous formation of a double bond in the tricarboxylic acid and a water molecule associated with the Fe^{++}. The uncharged water molecule can exchange with water in the medium. The common intermediate exists only as part of an enzyme complex. When the complex dissociates it may split with the oxygen-carbon bonds shown in dotted lines, forming citrate or isocitrate, or the oxygen and both potential hydride ions may remain with the iron, leaving *cis*-aconitate.

Kinetics of Fumarase Activity. The kinetics of the fumarase reaction have been studied intensively by Alberty and his collaborators.[23] They have found that interaction of enzyme with phosphate can cause activation at low phosphate concentrations, but that at high concentrations, phosphate acts as a competitive inhibitor. An unusual effect was noted when the effect of fumarate concentration on the rate of hydration was measured. At low substrate concentrations the Lineweaver-Burk plots are linear, but at higher concentrations the rate is faster than anticipated. This phenomenon was interpreted as indicating an interaction of fumarate with the enzyme at sites other than the catalytic site, to form a more active enzyme. At very high substrate concentrations (0.1 M) there is inhibition of the reaction, and the theoretical V_{max} is never attained.

Fumarase has been studied intensively with respect to the effects of pH changes and certain other environmental changes on enzyme activity. These effects will be considered in detail, as the principles involved apply to all enzymes. Changes in the composition of incubation mixtures influence the rate of enzyme-catalyzed reactions by altering the substrate or by modifying the enzyme. Even though effects caused by changes in pH may sometimes appear complicated, they can all be

[23] R. A. Alberty, V. Massey, C. Frieden, and A. R. Fuhlbrigge, *J. Am. Chem. Soc.* **76**, 2485 (1954); R. A. Alberty, *ibid.* **76**, 2494 (1954); C. Frieden and R. A. Alberty, *J. Biol. Chem.* **212**, 859 (1955).

explained in terms of simple acid–base dissociations. The dissociation curve of every monovalent compound, whether it be of an acidic or basic group, exhibits the same shape; the relation between concentration of one form of a compound and pH is given by the Henderson-Hasselbach equation:

$$\mathrm{pH} = \mathrm{p}K_a + \log\frac{A}{B}$$

in which A represents the concentration of the component that has one fewer proton than the component whose concentration is given as B

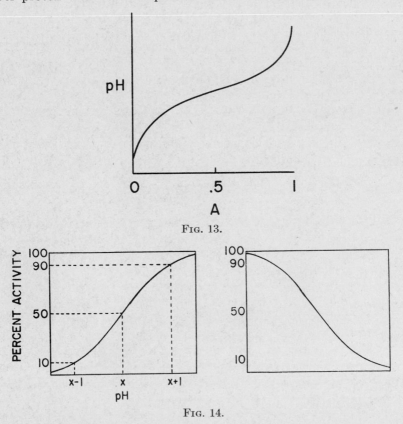

Fig. 13.

Fig. 14.

(Fig. 13). A may be an anion in case of an acidic group or an undissociated molecule in the case of basic compounds. In the simplest case, either A or B represents an active form, and the other is inactive. The curves of activity versus pH then assume S shapes (Fig. 14). If two groups in the protein are involved in the determination of activity of an enzyme, one limiting the reaction in more acidic solutions and the other in more basic solutions, the composite activity curve may assume the shape shown in Fig. 15. It is not necessary, however, that the pK's of the relevant groups be close together; if they are several pH units apart, the curve may include a broad plateau of maximum activity (Fig. 16). It is not necessary that only a single dissociation influence the activity in a given region of pH values. When several groups influence the activity, composite curves may be obtained that do not show the symmetry of the single dissociations.

When an enzyme interacts with an ionizable substrate, in general it attacks only one molecular species. If the enzyme is not sensitive to pH in the region of the pK of the substrate, it is possible to demonstrate changes in the rate of reaction as the fraction of the substrate available to the enzyme changes with pH. In effect this type of change is equivalent to a change in substrate concentration. If the total amount of substrate

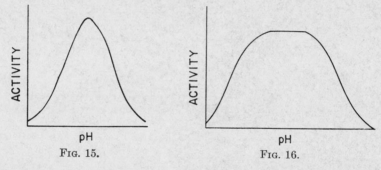

FIG. 15.   FIG. 16.

is increased as the pH becomes less favorable for the reaction, the effect of pH can be compensated for. To determine the effect of pH on an enzyme, comparisons of rates at different pH values should therefore be made on the basis of V_{max}, not on the basis of rates with arbitrary substrate concentrations. On the other hand, if the rate with limiting amounts of substrate changes with pH at the same values as the substrate dissociates, the pH–activity curve can be used to determine the active form of the substrate.

The kinetic studies of Alberty and his collaborators on homogeneous, crystalline fumarase showed a number of complicating factors. These include activation by high concentrations of fumarate, stimulation by low concentrations of phosphate, and competitive inhibition by high phosphate concentrations. It has been concluded that binding of certain ions including fumarate to the enzyme at sites other than the catalytic site change the enzyme to a more active catalyst. The effect of fumarate in changing the catalytic properties of the enzyme is seen in Lineweaver-Burk plots, which show striking curvatures at high fumarate concentrations (Fig. 17). The curve is straightened markedly by stimulating concentrations of phosphate. At very high fumarate concentrations there is inhibition of fumarase, so that V_{max} cannot be measured experimentally. For determination of pH–activity curves, therefore, V_{max} must be calculated by extrapolation of Lineweaver-Burk plots.

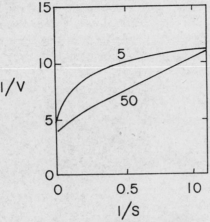

FIG. 17. Lineweaver-Burk plots of fumarase in phosphate buffer pH 6.5 at the millimolar phosphate concentrations indicated. From Alberty et al.[23]

In the pH range that fumarase exhibits activity, phosphate exists as the monobasic and dibasic salts. Since the two forms may influence the enzyme differently, pH curves were established with tris(hydroxymethyl)aminomethane (Tris) buffer. The most striking result of these studies is that the curves obtained with malate and fumarate as substrates are quite different. The maximum rate of hydration of fumarate occurs at pH 6.5, whereas the dehydration of malate is most rapid near pH 8 (Fig. 18).

$$
\begin{array}{ccccccc}
& E^{(n+1)} & & E^{(n+1)}F & & E^{(n+1)}M & & E^{(n+1)} \\
& \updownarrow & & \updownarrow & & \updownarrow & & \updownarrow \\
F + & E^{(n)} & \rightleftharpoons & E^{(n)}F & \rightleftharpoons & E^{(n)}M & \rightleftharpoons & E^{(n)} & + M \\
& \updownarrow & & \updownarrow & & \updownarrow & & \updownarrow \\
HF & E^{(n-1)} & & E^{(n-1)}F & & E^{(n-1)}M & & E^{(n-1)} & HM
\end{array}
$$

(III)

The Michaelis constants of the two substrates also show striking variations with pH (Fig. 19). These values of K_m and V_{max} have been combined with the dissociation constants of the substrates to permit calculation of pH-independent values for V_{max} and K_m. These values apply to a hypothetical enzyme in its active ionic form reacting with the divalent ions of fumarate and malate. The general description of the hypothesis is shown in (III).

As the diagram indicates, the active form of the enzyme $(E^{(n)})$ can be converted to a more acid $(E^{(n-1)})$ or more alkaline $(E^{(n+1)})$ form, but only one ionic species has catalytic activity. Similar dissociations are proposed for the enzyme–substrate complexes. This picture of enzyme action is supported by the calculation of the dissociation constants of the free enzyme. When these constants are calculated from data on the forward and reverse reactions, essentially the same values are obtained.

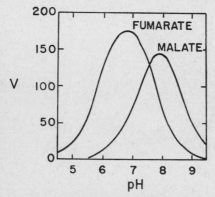

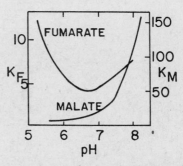

FIG. 18. From Alberty *et al.*[23]   FIG. 19. From Frieden and Alberty.[23]

From the symmetrical bell-shaped curves obtained, pK values of 6.2 and 6.8 were obtained for the ionization of the groups essential for enzyme activity.

The elaborate studies of fumarase have provided the most critical test of the Haldane relationship. For fumarase this is expressed

$$K_{eq} = \frac{V_{max} \text{ fumarate} \times K_m \text{ malate}}{V_{max} \text{ malate} \times K_m \text{ fumarate}}$$

Experimental verification of this relationship was found when K_{eq} was calculated from V_{max} and K_m values at different pH values. In experiments carried out in various concentrations of phosphate and at varying pH values, the ratio V_{max} fumarate/V_{max} malate varied by a factor of 65, but the K_{eq} calculated by using the Haldane equation was found to be remarkably constant, averaging 4.4 at 25°C.

Isotopic Studies with Fumarase. The kinetic data reported above have led to preliminary attempts to interpret the nature of the binding of substrate to enzyme in terms of specific amino acid side chains. Another approach to the nature of the reaction was made in studies with D_2O.[24] It was found that the hydrogens of fumarate do not equilibrate with

[24] H. F. Fisher, C. Frieden, J. S. M. McKee, and R. A. Alberty, *J. Am. Chem. Soc.* **77**, 4436 (1956).

water, whereas one stable (methylene) hydrogen is incorporated into malate. These experiments measure incorporation of deuterium only into C—H bonds, as the hydrogens in O—H bonds in both alcoholic and carboxylic acid groups exchange freely with the hydrogen of water. Therefore deuterium in these positions is lost during the isolation of the reaction products. The failure of fumarate to incorporate deuterium shows that only a definite one of the hydrogen atoms of the methylene group of malate may be removed by the enzyme. The atoms shown in heavy type in (IV) are not labilized by fumarase. This reaction furnishes

$$
\begin{array}{ccc}
\text{COO}^- & & \text{COO}^- \\
| & & | \\
\text{H—C—H} & & \text{C—H} \\
| & \rightleftharpoons & \| \\
\text{H—C—OH} & & \text{H—C} \\
| & & | \\
\text{COO}^- & & \text{COO}^-
\end{array}
$$

(IV)

an example of stereochemical specificity discussed in more detail in Chapter IV (p. 159).

Recent experiments by Englard and Colowick[24a] have shown that the exchange of hydrogen of citrate with deuterium of the medium catalyzed by aconitase occurs at a rate consistent with the formation of aconitate and hydration back to citrate. The same type of mechanism seems to account for the exchange of the reactive hydrogen of malate.

Isocitric Dehydrogenase. The oxidation of isocitric acid is catalyzed by pyridine nucleotide enzymes. What was originally thought to be a citric dehydrogenase was shown to be isocitric dehydrogenase, acting after aconitase produced the proper hydroxy acid.[25a] The enzymes first described from various plant and animal sources were shown to be completely specific for TPN.[25b,26] Recently another type of isocitric dehydrogenase has been found to react exclusively with DPN.[27] Both the TPN- and DPN-requiring enzymes have been found in yeast and heart muscle.

The reaction catalyzed by isocitric dehydrogenase involves both oxidation and decarboxylation (V). This was the first isolated enzymatic reaction to be shown to fix CO_2 into an organic compound, and, in so doing, restored CO_2 to the family of normally reactive molecules.[26] A concept of CO_2 as an inert end-product of metabolism had pervaded biochemical thinking; it was considered that only unusual reactions such as might participate in photosynthesis could fix CO_2. It is now quite

[24a] S. Englard and S. P. Colowick, *J. Biol. Chem.* **226**, 1047 (1957).
[25a] C. Martius and F. Knoop, *Z. physiol. Chem.* **296**, 5 (1937).
[25b] E. Adler, H. von Euler, G. Gunther, and M. Plass, *Biochem. J.* **33**, 1028 (1939).
[26] S. Ochoa, *J. Biol. Chem.* **174**, 133 (1948).
[27] A. Kornberg and W. E. Pricer, Jr., *J. Biol. Chem.* **189**, 123 (1951).

$$
\begin{array}{ll}
\begin{array}{l}
\mathrm{H} \\
| \\
\mathrm{HO-C-COO^-} \\
| \\
\mathrm{H-C-COO^-} \\
| \\
\mathrm{H-C-COO^-} \\
| \\
\mathrm{H}
\end{array}
+ \; \mathrm{TPN} \; \rightleftharpoons \;
\begin{array}{l}
\mathrm{O=C-COO^-} \\
| \\
\mathrm{H-C-H} \\
| \\
\mathrm{H-C-COO^-} \\
| \\
\mathrm{H}
\end{array}
+ \; \mathrm{TPNH} + \mathrm{CO_2}
\end{array}
$$

D-Isocitrate   α-Ketoglutarate
(V)

clear that decarboxylations are reversible as all reactions are, and in favorable thermodynamic circumstances net fixation of CO_2 occurs.

It now appears that both the oxidation and decarboxylation occur simultaneously, that is, without the appearance of an intermediate that dissociates from the enzyme. The mechanism by which these processes occur is not apparent. A possible intermediate in the over-all reaction is oxalosuccinate (VI). This compound has been synthesized and found to

$$
\begin{array}{l}
\mathrm{H} \\
| \\
\mathrm{HO-C-COO^-} \\
| \\
\mathrm{H-C-COO^-} \; \longleftarrow \; \mathrm{TPNH} + \\
| \\
\mathrm{H-C-COO^-} \\
| \\
\mathrm{H}
\end{array}
\quad
\begin{array}{l}
\mathrm{O=C-COO^-} \\
| \\
\mathrm{H-C-COO^-} \; \longrightarrow \\
| \\
\mathrm{H-C-COO^-} \\
| \\
\mathrm{H}
\end{array}
\quad
\begin{array}{l}
\mathrm{O=C-COO^-} \\
| \\
\mathrm{H-C-H} \\
| \\
\mathrm{H-C-COO^-} \\
| \\
\mathrm{H}
\end{array}
+ \; \mathrm{CO_2}
$$

Reaction of Oxalosuccinate
(VI)

react with isocitric dehydrogenase in both an oxidation and decarboxylation.[27] The decarboxylation has been measured both manometrically and spectrophotometrically. Purification of the enzyme has shown no separation of the carboxylase activity from the over-all oxidative decarboxylation of isocitrate. Both reactions require the presence of a divalent cation; Mn^{++} is the most effective. The decarboxylation of oxalosuccinate is inhibited by isocitrate, but not by other tricarboxylic acids. These observations all indicate that oxalosuccinate reacts with isocitric dehydrogenase. When the enzyme is incubated with TPNH and oxalosuccinate, there is rapid disappearance of the absorption at 340 mμ. It is possible that this represents a direct reduction of oxalosuccinate, but it has not been excluded that the reduction follows formation of α-ketoglutarate and CO_2.

The DPN-isocitric dehydrogenase presents a mysterious situation. Catalyzing the same over-all reaction, yielding α-ketoglutarate and CO_2, the enzyme does not reverse the oxidation under conditions that promote reversal readily with the TPN enzyme. DPN-isocitric dehydrogenase has no oxalosuccinic carboxylase activity. The DPN enzyme requires adenylic acid. The possibility exists that the oxidation involves an obscure coupling. Thermodynamically there is essentially no difference between the DPN and TPN systems. Another possibility, however, is that substrate inhibition of the reverse reaction prevents its demonstration with the DPN-isocitric dehydrogenase.

Malic Dehydrogenase. The oxidation of the hydroxy dicarboxylic acid, malic acid, is catalyzed by two different enzymes: a simple dehydrogenase, employing pyridine nucleotides to oxidize malate to oxalacetate (VII), and an enzyme that catalyzes an oxidative decarboxylation (VIII).

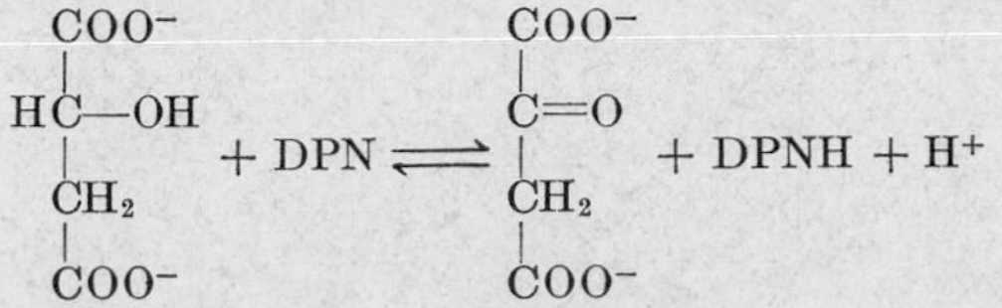

$$
\begin{array}{ccc}
\text{COO}^- & & \text{COO}^- \\
| & & | \\
\text{HC--OH} & + \text{DPN} \rightleftharpoons & \text{C=O} \quad + \text{DPNH} + \text{H}^+ \\
| & & | \\
\text{CH}_2 & & \text{CH}_2 \\
| & & | \\
\text{COO}^- & & \text{COO}^- \\
\text{L-Malate} & & \text{Oxalacetate} \\
& & \text{(VII)}
\end{array}
$$

The former is called malic dehydrogenase.[28] Although it was reported early that the enzyme is specific for DPN, in fact malic dehydrogenase from several sources has been found to react with TPN at 5–7 per cent of the rate of the DPN reaction.[29] The affinities for DPN and TPN are quite similar, K_m equals about 10^{-5} at neutral pH. The equilibrium of the reaction at neutral pH values lies far to the side of malate and DPN, but, as discussed previously, at higher pH values the equilibrium of reactions in which H^+ participates is shifted, and near pH 10 the oxidation of malate proceeds to a considerable extent and at a good rate. Even at low pH values, however, the oxidation of malate is readily carried out when coupled with an effective system for oxidizing DPNH or removing oxalacetate (as citrate formation).

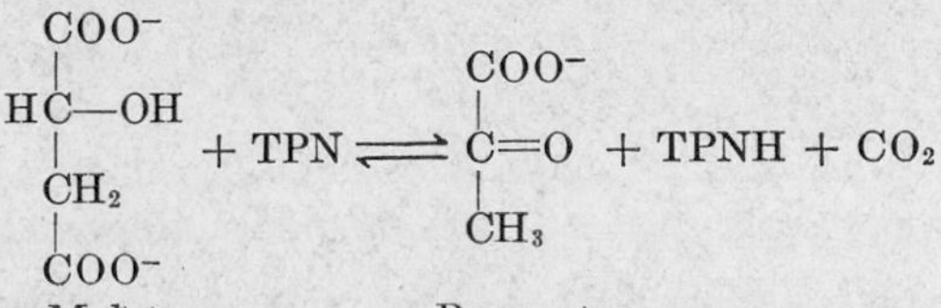

$$
\begin{array}{ccc}
\text{COO}^- & & \\
| & & \text{COO}^- \\
\text{HC--OH} & + \text{TPN} \rightleftharpoons & | \\
| & & \text{C=O} \quad + \text{TPNH} + \text{CO}_2 \\
\text{CH}_2 & & | \\
| & & \text{CH}_3 \\
\text{COO}^- & & \\
\text{L-Malate} & & \text{Pyruvate} \\
& & \text{(VIII)}
\end{array}
$$

Malic Enzyme. Another enzyme that oxidizes malate resembles the TPN-isocitric dehydrogenase in many respects. This is the malic enzyme, which catalyzes an oxidative decarboxylation (VIII).[30] Malic enzymes

[28] F. B. Straub, *Z. physiol. Chem.* **275**, 63 (1942).
[29] A. H. Mehler, A. Kornberg, S. Grisolia, and S. Ochoa, *J. Biol. Chem.* **179**, 961 (1948).
[30] S. Ochoa, A. H. Mehler, and A. Kornberg, *J. Biol. Chem.* **174**, 979 (1948).

from several animal and plant sources require TPN,[30,31] but a DPN enzyme has been found to be formed adaptively in *Lactobacillus arabinosus*.[32] Malic enzymes are capable of decarboxylating oxalacetate, and in this respect resemble TPN-isocitric dehydrogenase. In this case also there is no evidence that oxalacetate is formed as an intermediate. Unlike isocitric dehydrogenase, which can reduce oxalosuccinate, the malic enzyme does not reduce oxalacetate under conditions in which pyruvate $+ CO_2$ are reduced. The decarboxylation of oxalacetate is an intrinsic property of the enzyme, and some day may provide information leading to a more complete understanding of the nature of the enzyme. The intimate association of oxalacetic carboxylase activity with the activity concerned with the oxidative decarboxylation of malate is emphasized by the finding that the decarboxylation is specifically stimulated 3-fold by the presence of TPN when malic enzyme of pigeon liver is used.[30,33] Some malic enzymes do not show this stimulation. Oxalacetic carboxylase activity, however, can only be demonstrated at acid pH values, whereas no malate oxidation occurs under these conditions. At their respective pH optima, 4.5 and 7.5, oxalacetate and malate decarboxylations proceed at the same rate with the enzyme purified from pigeon liver.[34] Preparations from other sources have other ratios of the two activities.

Malic Oxidase. Recently another enzymatic mechanism for oxidizing malate was discovered in *Micrococcus lysodeikticus*.[35] The chemical nature of this enzyme has not been determined as yet, but it does not use pyridine nucleotides. Ferricyanide was used as the electron acceptor, and in the reaction oxalacetate accumulated. Evidence for a similar activity has also been found in extracts of pigeon liver.

Phosphopyruvate Carboxylases. In addition to the malic enzyme, other oxalacetic decarboxylases are known.[36] Enzymes purified from various

[31] B. Vennesland, M. Gollub, and J. F. Speck, *J. Biol. Chem.* **178,** 301 (1949); E. Conn, B. Vennesland, and L. M. Kraemer, *Arch. Biochem.* **23,** 179 (1949).

[32] M. L. Blanchard, S. Korkes, A. del Campillo, and S. Ochoa, *J. Biol. Chem.* **187,** 875 (1950).

[33] B. Vennesland, E. A. Evans, Jr., and K. I. Altman, *J. Biol. Chem.* **171,** 675 (1947).

[34] I. Harary, S. R. Korey, and S. Ochoa, *J. Biol. Chem.* **203,** 595 (1953).

[35] D. V. Cohn, *J. Biol. Chem.* **221,** 413 (1956).

[36] The decarboxylation of oxalacetic acid was postulated by Wood and Werkman to participate in the fermentation process of Propionobacteria. They postulated that a reversal of this decarboxylation resulted in the fixation of CO_2 by these bacteria, and the name, Wood-Werkman reaction, was used to identify this postulated mechanism for the first fixation of CO_2 in carbon chains to be demonstrated in heterotrophic organisms. [C. H. Werkman and H. G. Wood, *Advances in Enzymol.* **2,** 135 (1942).] The fixation of CO_2 in succinate observed by these workers may be explained by the reactions discussed on p. 148.

bacteria appear to be simple decarboxylases.[37] They have pH optima from 4.5 to 7.5, use divalent cations, and have never been reversed. The decarboxylation of oxalacetate proceeds to completion in any system, but experiments with isotopic CO_2 showed that during the decarboxylation by avian liver extracts there was significant back reaction, which was increased by ATP.[38] Recently the mechanism of this reaction has been established with a purified enzyme.[39] The reaction is shown in (IX).

$$
\begin{array}{cc}
\text{COO}^- & \text{COO}^- \\
| & | \quad\quad \text{O} \\
\text{C}=\text{O} & | \quad\quad\; \| \\
| \quad\quad + \text{GTP} \rightleftharpoons & \text{C}-\text{O}-\text{P}-\text{O}^- + \text{CO}_2 + \text{GDP} \\
\text{CH}_2 & \| \quad\quad | \\
| & \quad\quad\; \text{O}^- \\
\text{COO}^- & \text{CH}_2 \\
\text{Oxalacetate} & \text{Phosphopyruvate} \\
& \text{(IX)}
\end{array}
$$

ITP is the only nucleotide that is known to substitute for GTP. The reaction was thought for many years to use ATP, but the apparent utilization of ATP was the result of the presence of other nucleotides in ATP preparations and the presence of nucleoside diphosphate kinase in the enzyme preparations. Thus the active nucleotide reacted in catalytic quantities and the nucleoside diphosphate produced accepted the terminal phosphate of ATP to give the appearance of participation of ATP in the decarboxylation of oxalacetate. Another complication in the analysis of this reaction is the secondary reaction of phosphopyruvate with nucleoside diphosphates. Phosphopyruvate kinase, which is present in crude extracts, transfers phosphate from the product of decarboxylation to the nucleotide also formed in the reaction, so that no phosphate transfer is seen, and the reaction appears to be a simple decarboxylation. It

$$
\begin{array}{c}
\text{oxalacetate} + \text{GTP} \rightleftharpoons \text{phosphopyruvate} + \text{CO}_2 + \text{GDP} \\
\underline{\text{phosphopyruvate} + \text{GDP} \rightleftharpoons \text{pyruvate} + \text{GTP}} \\
\text{oxalacetate} \rightleftharpoons \text{pyruvate} + \text{CO}_2
\end{array}
$$

was not possible to demonstrate the reaction mechanism until the enzyme had been purified and interfering activities were removed.

A similar enzyme was found in wheat germ and other plant tissues to catalyze the carboxylation of phosphopyruvate (X).[40] This enzyme

[37] L. O. Krampitz and C. H. Werkman, *Biochem. J.* **35**, 595 (1941); D. Herbert, *ibid.* **47**, i (1950); G. W. E. Plaut and H. A. Lardy, *J. Biol. Chem.* **180**, 13 (1949).

[38] M. F. Utter and H. G. Wood, *J. Biol. Chem.* **164**, 455 (1946).

[39] M. F. Utter, K. Kurahashi, and I. A. Rose, *J. Biol. Chem.* **207**, 803 (1954); K. Kurahashi and M. F. Utter, *Federation Proc.* **14**, 240 (1955).

[40] R. S. Bandurski and C. M. Greiner, *J. Biol. Chem.* **204**, 781 (1953); T. T. Chen and B. Vennesland, *ibid.* **213**, 533 (1955).

appears to form inorganic phosphate and does not seem to require nucleotide triphosphate.

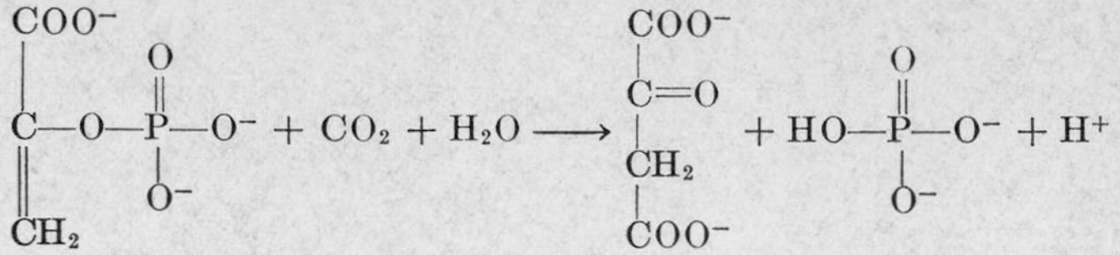

$$
\underset{\text{(X)}}{
\begin{array}{c}
\text{COO}^- \\
| \\
\text{C}-\text{O}-\overset{\overset{\text{O}}{\|}}{\text{P}}-\text{O}^- \\
\| \quad | \\
\text{CH}_2 \quad \text{O}^-
\end{array}
+ \text{CO}_2 + \text{H}_2\text{O} \longrightarrow
\begin{array}{c}
\text{COO}^- \\
| \\
\text{C}=\text{O} \\
| \\
\text{CH}_2 \\
| \\
\text{COO}^-
\end{array}
+ \text{HO}-\overset{\overset{\text{O}}{\|}}{\text{P}}-\text{O}^- + \text{H}^+
}
$$

The decarboxylation of the β-keto acids oxalosuccinate and oxalacetate catalyzed either by appropriate enzymes or by polyvalent cations can be measured spectrophotometrically by the appearance of absorption bands at higher wavelengths than those characteristic of the substrates alone.[41] These transient bands were once thought to represent complexes of metal with the enol form of the substrate. Studies with model compounds by Steinberger and Westheimer[42] showed, however, that a compound that cannot form an enol, β,β-dimethyloxalacetate, is also decarboxylated by metals to form a transient product that has an increased ultraviolet absorption. They identified the absorbing material with the enol of the product. The mechanism of β-keto acid decarboxylation is therefore as shown in (XI).

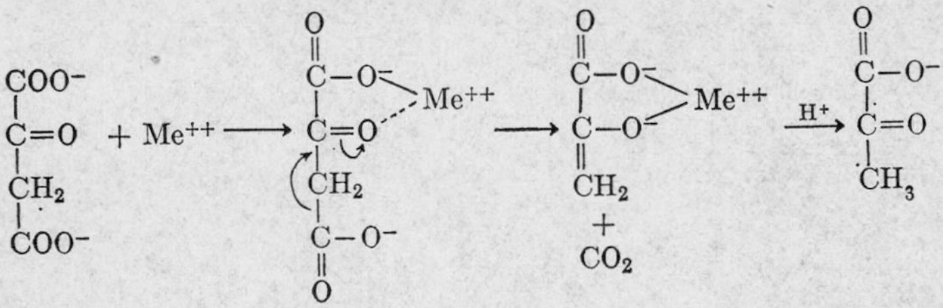

$$
\underset{\text{(XI)}}{
\begin{array}{c}
\text{COO}^- \\
| \\
\text{C}=\text{O} \\
| \\
\text{CH}_2 \\
| \\
\text{COO}^-
\end{array}
+ \text{Me}^{++} \longrightarrow
\cdots \longrightarrow
\cdots \xrightarrow{\text{H}^+}
\begin{array}{c}
\text{C}-\text{O}^- \\
| \\
\text{C}=\text{O} \\
| \\
\text{CH}_3
\end{array}
}
$$

Metabolic Functions of CO_2-Fixing Enzymes. The various enzymes that metabolize malate and oxalacetate often occur in the same tissue. The functions of these enzymes permit a high degree of control over the tricarboxylic acid cycle. In the simplest case, oxalacetate is formed by oxidation of malate by malic dehydrogenase and continues the cycle by condensing with acetyl CoA to form citrate. In the event that large quantities of a cycle intermediate are introduced into a system, two devices exist for permitting complete oxidation. These are decarboxylation of oxalacetate, which may be very slow, and oxidative decarboxylation of malate, which goes rapidly with amounts of enzyme found in many

[41] A. Kornberg, S. Ochoa, and A. H. Mehler, *J. Biol. Chem.* **174**, 159 (1948).
[42] R. Steinberger and F. H. Westheimer, *J. Am. Chem. Soc.* **71**, 4158 (1949).

cells. These two reactions form pyruvate, which can be oxidized to acetyl CoA and enter the cycle to be oxidized completely. The third of the decarboxylases, the one using ITP, provides a mechanism for converting cycle intermediates to carbohydrate. This mechanism for producing phosphopyruvate by-passes a glycolytic step that is difficult to reverse, the phosphopyruvate kinase reaction. Finally, the phosphopyruvate reactions and the malic enzyme provide mechanisms for generating 4-carbon dicarboxylic acids from phosphopyruvate and pyruvate, respectively, thus permitting the tricarboxylic acid cycle to continue to operate under circumstances in which it is used to provide the carbon skeleton for synthetic reactions.

α-Ketoglutaric Oxidase. In addition to the two oxidation steps involving hydroxy acids, the Krebs cycle includes an oxidation of an α-keto acid. The oxidation of α-ketoglutarate in most respects is similar to the oxidation of pyruvate. For many years only particulate preparations were found to be active. These showed requirements for cocarboxylase, Mg^{++}, and inorganic phosphate.[43] The products of the oxidation were found to be succinate and CO_2. A coupled reaction caused the formation of ATP from ADP and inorganic phosphate. Recently the other cofactors that participate in pyruvate oxidation were also found to be required in the α-ketoglutarate system. These include coenzyme A, lipoic acid, and DPN.

In animal and plant preparations the most purified enzymes obtained carry out the reaction

$$\begin{array}{l} O{=}C{-}COO^- \\ \ \ \ | \\ \ \ CH_2 \\ \ \ \ | \\ \ \ CH_2{-}COO^- \end{array} + DPN + CoA{-}SH \longrightarrow \begin{array}{l} O{=}C{-}S{-}CoA \\ \ \ \ | \\ \ \ CH_2 \\ \ \ \ | \\ \ \ CH_2{-}COO^- \end{array} + CO_2 + DPNH$$

α-Ketoglutarate

This reaction has been resolved into two steps in preparations from *E. coli*, and it is probable that the reaction mechanism follows the outline described for pyruvate.[44] One enzyme, using cocarboxylase and Mg^{++}, presumably decarboxylates the keto acid to an aldehyde derivative that splits the disulfide group of oxidized lipoic acid (XII). A second enzyme transfers the succinyl group to CoA, leaving reduced lipoic acid. A third enzyme oxidizes reduced lipoic acid with DPN, forming the 5-membered ring of oxidized lipoic acid and DPNH. A soluble preparation from pig heart was fractionated into an α-ketoglutaric dehydrogenase that forms succinyl CoA[45] and a transferring enzyme that couples the hydrolysis

[43] S. Ochoa, *J. Biol. Chem.* **155,** 87 (1944).
[44] L. P. Hager, J. D. Fortney, and I. C. Gunsalus, *Federation Proc.* **12,** 213 (1953).
[45] S. Kaufman, C. Gilvarg, O. Cori, and S. Ochoa, *J. Biol. Chem.* **203,** 869 (1953).

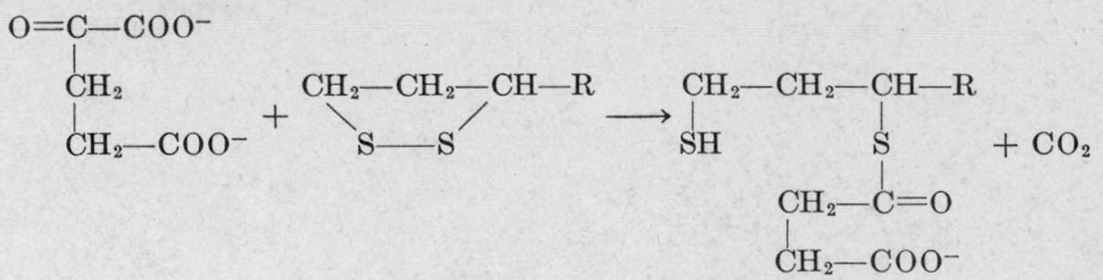

$$O=C-COO^-$$
$$CH_2$$
$$CH_2-COO^- \quad + \quad \begin{matrix} CH_2-CH_2-CH-R \\ \diagdown \quad \diagup \\ S-S \end{matrix} \quad \longrightarrow \quad \begin{matrix} CH_2-CH_2-CH-R \\ | \qquad\qquad | \\ SH \qquad\quad S \end{matrix} \quad + CO_2$$
$$CH_2-C=O$$
$$CH_2-COO^-$$

α-Ketoglutarate   Lipoic acid   Succinyllipoic acid

(XII)

of succinyl CoA with the formation of ATP from ADP and inorganic phosphate.[46]

Reaction of Succinyl CoA. The first enzyme-dissociated product of α-ketoglutarate oxidation is succinyl CoA. Succinyl CoA may be metabolized by three known routes; the thioester may be hydrolyzed; the CoA may be transferred to other acids; and a coupled reaction may result in organic phosphate incorporation into ATP. The last reaction is the only substrate-level phosphorylation of the tricarboxylic acid cycle.

Hydrolysis of succinyl CoA is catalyzed by a specific enzyme named succinyl CoA deacylase.[47] This is one of a family of deacylases of which three members are known; these hydrolyze succinyl CoA (XIII), acetyl CoA, and acetoacetyl CoA, respectively.

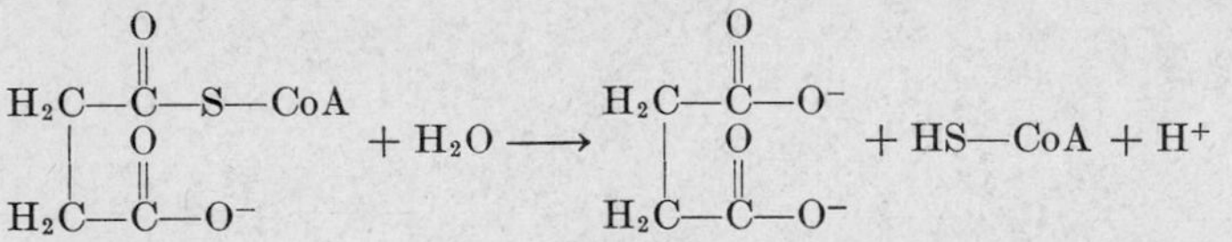

$$\begin{matrix} & O \\ & \| \\ H_2C-&C-S-CoA \\ | & O \\ & \| \\ H_2C-&C-O^- \end{matrix} \quad + H_2O \longrightarrow \quad \begin{matrix} & O \\ & \| \\ H_2C-&C-O^- \\ | & O \\ & \| \\ H_2C-&C-O^- \end{matrix} \quad + HS-CoA + H^+$$

(XIII)

The hydrolysis of succinyl CoA permits α-ketoglutarate oxidation to proceed to completion in the presence of catalytic quantities of CoA, but it results in a waste of the potentially useful thioester bond. One use for the thioester group is to permit metabolism of certain β-keto acids. A CoA transferase isolated from pig heart[48] is absolutely specific for succinate and succinyl CoA, but only relatively specific for the β-keto acid (XIV). The rate of reaction is greatest with acetoacetate, and decreases as the chain length is increased from β-ketovalerate to β-ketocaproate. Longer chains are not used. Branched chain β-keto acids can also serve as substrates, but α,β-unsaturated acids, β-hydroxy acids, and dicarboxylic acids cannot accept CoA in this transfer reaction. An enzyme

[46] S. Kaufman and S. G. A. Alivisatos, *J. Biol. Chem.* **216**, 141 (1955).
[47] J. Gergely, P. Hele, and C. V. Ramakrishnan, *J. Biol. Chem.* **198**, 323 (1952).
[48] J. R. Stern, M. J. Coon, and A. del Campillo, *Nature* **171**, 28 (1953); D. E. Green, D. S. Goldman, S. Mii, and H. Beinert, *J. Biol. Chem.* **202**, 137 (1953).

$$
\underset{\text{Succinyl CoA}}{
\begin{array}{l}
\text{H}_2\text{C}-\overset{\displaystyle O}{\overset{\|}{\text{C}}}-\text{S}-\text{CoA} \\
\ \ \ | \\
\text{H}_2\text{C}-\overset{\displaystyle O}{\overset{\|}{\text{C}}}-\text{O}^-
\end{array}}
\ + \ \underset{\text{Acetoacetate}}{\text{CH}_3\overset{\displaystyle O}{\overset{\|}{\text{C}}}-\text{CH}_2\overset{\displaystyle O}{\overset{\|}{\text{C}}}-\text{O}^-} \ \rightleftharpoons
$$

$$
\underset{\text{Succinate}}{
\begin{array}{l}
\text{H}_2\text{C}-\overset{\displaystyle O}{\overset{\|}{\text{C}}}-\text{O}^- \\
\ \ \ | \\
\text{H}_2\text{C}-\overset{\displaystyle O}{\overset{\|}{\text{C}}}-\text{O}^-
\end{array}}
\ + \ \underset{\text{Acetoacetyl CoA}}{\text{CH}_3\overset{\displaystyle O}{\overset{\|}{\text{C}}}-\text{CH}_2\overset{\displaystyle O}{\overset{\|}{\text{C}}}-\text{S}-\text{CoA}}
$$

(XIV)

found in *Pseudomonas* has a similar action, but, instead of reacting with simple β-keto acids, reacts with the dicarboxylic β-ketoadipic acid.

The free energy of the transfer reaction is a function of pH, because in addition to the carboxyl groups the carbonyl group of acetoacetate thioester can ionize (as an enol). Since this ionization is favored by alkaline media, the equilibrium is somewhat shifted towards acetoacetyl CoA formation at higher pH values. The reaction favors succinyl CoA formation at all pH values studied, however (F_0' at pH 7.0 = +3470 cal.). Another factor that influences the apparent equilibrium of the transferase is the presence of Mg^{++}. This ion forms a complex with the β-ketothioester. The enzyme has been assayed by measuring the appearance of the absorption band of the β-ketothioester at 303 mμ, where simple thioesters do not absorb. This band is shifted slightly and the extinction coefficient greatly increased by Mg^{++}, which has therefore been added to assay systems to increase the sensitivity. CoA transferase does not require Mg^{++} for activity, however.[49]

Phosphorylating (P) Enzyme. The formation of ATP associated with cleavage of succinyl CoA is catalyzed by the P enzyme. This enzyme has been purified from both heart muscle[50] and spinach.[46] In both cases a single protein appears to carry out the reaction:

succinyl CoA + XDP + P $\rightleftharpoons$ succinate + XTP + CoA (K_{eq} = 3.7)

The base in the nucleotide may be either guanine or hypoxanthine in the reaction with the heart muscle enzyme, while the spinach enzyme is specific for adenine. In crude heart muscle preparations nucleoside diphosphate kinase permits catalytic amounts of GTP or ITP to support α-ketoglutarate oxidation in the presence of ADP, which is converted to ATP.

[49] J. R. Stern, M. J. Coon, and A. del Campillo, *J. Am. Chem. Soc.* **75**, 1517 (1953).
[50] D. R. Sanadi, D. M. Gibson, and P. Ayengar, *Biochim. et Biophys. Acta* **14**, 434 (1954).

Studies on the action of the P enzyme have revealed several exchange reactions catalyzed by this enzyme.[51a] C^{14}-labeled succinate exchanges with succinyl CoA; this exchange depends upon inorganic phosphate, but is inhibited by ADP. The incorporation of inorganic phosphate into ATP requires all of the components of the system, and merely measures the sum of forward and reverse reactions, but the exchange of ADP^{32} with ATP is catalyzed by the P enzyme supplemented only with Mg^{++}. The over-all reaction and succinate exchange are inhibited by PCMB, but the ATP-ADP exchange is not. It is possible to write partial reactions involving succinyl phosphate to account for the properties of the P enzyme, but synthetic succinyl phosphate is inert with this enzyme. Another hypothetical intermediate, S-phosphoryl CoA (CoA-P), was rejected as an intermediate on kinetic grounds in studies with purified P enzyme, but this compound has been found as an intermediate in bacterial systems.[51b] It participates in the following reactions:

ATP + CoA $\rightleftarrows$ ADP + CoA-P   CoA thiolkinase

CoA-P + succinate $\rightleftarrows$ succinyl CoA + P   phosphoryl CoA transferase

It has not been determined whether similar reactions are catalyzed by animal or plant enzymes involved in α-ketoglutarate oxidation, or whether similar intermediates remain bound to the enzyme. The formation of CoA-P has been demonstrated in a thiolkinase reaction with extracts of acetone-dried brain, which also contain a transferase that forms acetyl CoA.[51c]

Succinic Dehydrogenase. The oxidation of succinate in general has been studied with insoluble particles that contain a series of enzymes capable of oxidizing succinate and transporting electrons efficiently to molecular oxygen. The particulate system has been named succinoxidase. This will be discussed in more detail in connection with the cytochromes, which are involved in the electron transport.

The failure of a generation of investigators to obtain soluble succinic dehydrogenase preparations was caused by the unusually fastidious requirements of this enzyme. Of the ordinary electron acceptors, including methylene blue and dichlorophenolindophenol, none supports the oxidation of succinate. Phenazine methosulfate was found to be reduced by solubilized preparations;[52] the enzyme is extracted from acetone powders and was undoubtedly present in many extracts, where it was not detected for lack of a suitable oxidant. A preparation from yeast was reported

[51a] S. Kaufman, *J. Biol. Chem.* **216**, 153 (1955).
[51b] R. A. Smith, I. F. Frank, and I. C. Gunsalus, *Federation Proc.* **16**, 251 (1957).
[51c] M. Wolleman and G. Feuer, *Acta Physiol. Acad. Sci. Hung.* **7**, 329, 343 (1955).
[52] T. P. Singer and E. B. Kearney, *Biochim. et Biophys. Acta* **15**, 151 (1954).

by Fisher and Eysenbach to be a fumaric reductase which could accept electrons from a reduced dye, leucosafranine (XV).[53] It is not known why this reaction was not studied in the reverse direction. Purified preparations of succinic dehydrogenase from beef heart muscle and yeast carry out the reaction in both directions.[54]

$$
\begin{array}{ccc}
\underset{\displaystyle \overset{\displaystyle \text{C}}{\|}}{\overset{\displaystyle \text{H} \quad \text{COO}^-}{}} \\
\underset{\displaystyle \text{-OOC} \quad \text{H}}{\text{C}}
\end{array}
+ \text{ leucosafranine} \longrightarrow
\begin{array}{c}
\text{COO}^- \\
| \\
\text{H—C—H} \\
| \\
\text{H—C—H} \\
| \\
\text{COO}^-
\end{array}
+ \text{ safranine}
$$

(XV)

Analyses of the composition of solublized succinic dehydrogenase from both animal and yeast preparations show the presence of both Fe^{++} and a flavin, FAD.[55] Both enzymes are very sensitive to —SH reagents. The animal dehydrogenase, but not that of yeast, is markedly stimulated by inorganic phosphate.

Inhibitors of the Citric Acid Cycle. Many compounds have been found to inhibit various steps of the citric acid cycle. Two inhibitors are of particular importance. Arsenite specifically inactivates α-ketoglutarate oxidation. This compound has a similar effect on pyruvate oxidation, and it is now believed that arsenite sensitivity is characteristic of systems containing disulfide groups, as found in lipoic acid. Malonate is a competitive inhibitor of succinate oxidation.[56] Besides its usefulness in studies in which it is desirable to prevent further oxidation, this inhibition was instrumental in developing our current concept of competitive inhibition, since the similarity of structures is striking, and the competitive nature of the reaction is easily demonstrated. The specific inhibition of aconitase by fluorocitrate has already been mentioned. Succinate oxidation is inhibited by naturally occurring C-4 dicarboxylic acids; in particular oxalacetate has been found to inhibit succinate oxidation in systems that permit oxalacetate to accumulate.

The Ogston Concept of Three-Point Attachment. When carbon-labeled acetate is oxidized by an inhibited tricarboxylic acid cycle, so that α-ketoglutarate accumulates, the isotope is found in the ketoglutarate. It was expected that the isotope should be equally distributed in the

[53] F. G. Fisher and H. Eysenbach, *Ann.* **530**, 99 (1937).
[54] T. P. Singer, V. Massey, and E. B. Kearney, *Biochim. et Biophys. Acta* **19**, 200 (1956); T. P. Singer, N. Z. Thimot, V. Massey, and E. B. Kearney, *Arch. Biochem. and Biophys.* **62**, 497 (1956).
[55] E. B. Kearney and T. P. Singer, *Biochim. et Biophys. Acta* **17**, 596 (1955); T. P. Singer, E. B. Kearney, and V. Massey, *Arch. Biochem. and Biophys.* **60**, 255 (1956).
[56] J. H. Quastel and A. H. M. Wheatley, *Biochem. J.* **25**, 117 (1931).

two carboxyls of ketoglutarate, since it was assumed that the symmetrical molecule, citric acid, was an intermediate (XVI). According to this

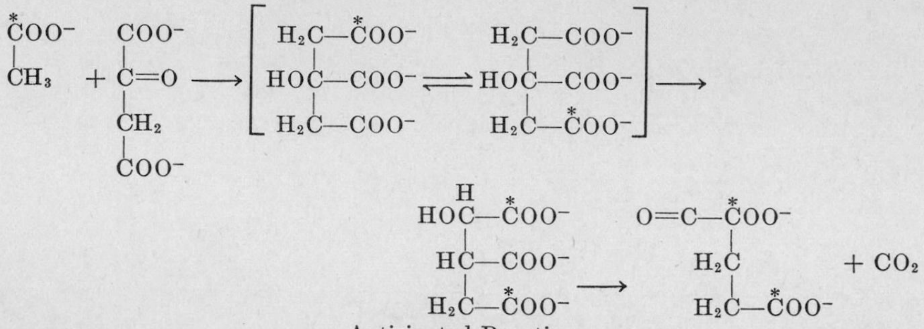

$$
\overset{*}{C}OO^- \quad COO^-
$$
$$
CH_3 \;+\; C{=}O \longrightarrow
\left[
\begin{array}{cc}
H_2C{-}\overset{*}{C}OO^- & H_2C{-}COO^- \\
HOC{-}COO^- \rightleftharpoons HOC{-}COO^- \\
H_2C{-}COO^- & H_2C{-}\overset{*}{C}OO^-
\end{array}
\right] \longrightarrow
$$
$$
CH_2
$$
$$
COO^-
$$

$$
\overset{H}{HOC}{-}\overset{*}{C}OO^- \qquad O{=}C{-}\overset{*}{C}OO^-
$$
$$
HC{-}COO^- \qquad\qquad H_2C \qquad\qquad + CO_2
$$
$$
H_2C{-}\overset{*}{C}OO^- \longrightarrow H_2C{-}\overset{*}{C}OO^-
$$

Anticipated Reaction
(XVI)

view, the two primary carboxyl groups should be equivalent, and should have equal chances of becoming the α- or γ-carboxyl of α-ketoglutarate.

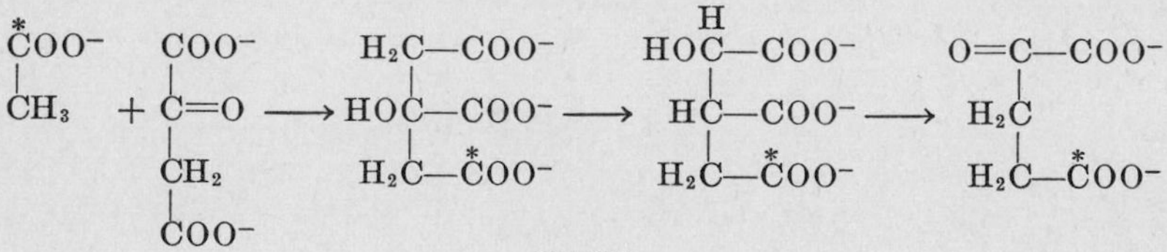

$$
\overset{*}{C}OO^- \quad COO^- \qquad H_2C{-}COO^- \qquad \overset{H}{HOC}{-}COO^- \qquad O{=}C{-}COO^-
$$
$$
CH_3 \;+\; C{=}O \longrightarrow HOC{-}COO^- \longrightarrow HC{-}COO^- \longrightarrow H_2C
$$
$$
CH_2 \qquad H_2C{-}\overset{*}{C}OO^- \qquad H_2C{-}\overset{*}{C}OO^- \qquad H_2C{-}\overset{*}{C}OO^-
$$
$$
COO^-
$$

(XVII)

A major disturbance in biochemistry occurred when it was found and confirmed that the carboxyls of ketoglutarate were very unsymmetrically labeled (XVII), and that, when label was introduced via CO_2, the pattern was reversed (XVIII).[57]

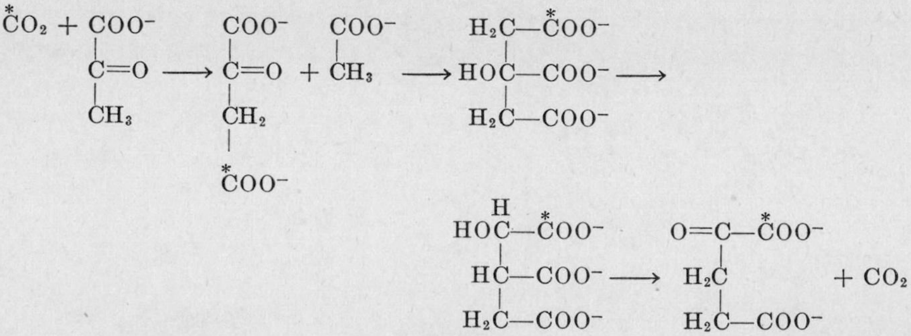

$$
\overset{*}{C}O_2 \;+\; COO^- \qquad COO^- \quad COO^- \qquad H_2C{-}\overset{*}{C}OO^-
$$
$$
C{=}O \longrightarrow C{=}O \;+\; CH_3 \longrightarrow HOC{-}COO^- \longrightarrow
$$
$$
CH_3 \qquad CH_2 \qquad H_2C{-}COO^-
$$
$$
\overset{*}{C}OO^-
$$

$$
\overset{H}{HOC}{-}\overset{*}{C}OO^- \qquad O{=}C{-}\overset{*}{C}OO^-
$$
$$
HC{-}COO^- \longrightarrow H_2C \qquad\qquad + CO_2
$$
$$
H_2C{-}COO^- \qquad H_2C{-}COO^-
$$

(XVIII)

[57] J. M. Buchanan, W. Sakami, S. Gurin, and D. W. Wilson, *J. Biol. Chem.* **159,** 695 (1945); S. Weinhouse, G. Medes, and N. F. Floyd, *J. Biol. Chem.* **166,** 691 (1946); H. G. Wood, *Physiol. Revs.* **26,** 198 (1946).

The two carboxyl groups of α-ketoglutarate can be distinguished from each other by either enzymatic or chemical oxidation to succinate and CO_2; the α-carboxyl appears as CO_2 and the γ- remains as a carboxyl group of succinate. Various explanations were offered, such as the idea that citrate is not the product of condensation, but is a side product, perhaps on a blind alley. Another suggestion considered the possibility that a derivative of citrate, such as citryl phosphate, not the symmetrical citrate, actually is the intermediate.

The dilemma was caused by a curious restriction in the thinking of biochemists throughout the world. It is perfectly obvious that the two methylene carbons of citrate are strictly equivalent chemically, but it is also clear that the carbinol carbon orients these substituents sterically, bearing, as it were, a right and left methylene carbon.

$$
\begin{array}{c}
OH \\
| \\
C \\
\diagup \ | \ \diagdown \\
\text{COO}^- \\
^-OOC\text{—}CH_2 \quad\ \ CH_2\text{—}COO^-
\end{array}
$$

Citric acid

If a reaction involves only the two carbons that participate in the double bond formed by aconitase, the right and left groups are equally susceptible. On the other hand, it was finally pointed out by Ogston,[58] it is highly probable that the enzyme binds the substrate at additional points. If only one more group is bound, the three binding sites of the protein orient the substrate so that only one of the potentially active sites is in a position to react. Considering the carbon bearing the hydroxyl group as the center of a tetrahedron, alignment of the molecule to place the hydroxyl group and methylene carbon ① in position to react with aconitase automatically places the tertiary carboxyl on one side and methylene ② carbon on the other. If either of these groups (or the carboxyl group ②) reacts with the enzyme in this position, aconitase may act. Alignment of the hydroxyl group and methylene carbon ②, on the other hand, reverses the positions of the tertiary carboxyl and methylene groups, so that they no longer can interact with the enzyme. Thus, structures that are chemically identical can be distinguished by their positions in a molecule; an asymmetric enzyme can impose an asymmetric reaction on a symmetrical molecule, possessing only one plane of symmetry. The validity of the Ogston "3-point attachment" hypothesis was thoroughly demonstrated by Potter and Heidelberger[59] who isolated

[58] A. G. Ogston, *Nature* **162,** 693 (1948).

[59] V. R. Potter and C. Heidelberger, *Nature* **169,** 180 (1947); P. E. Wilcox, C. Heidelberger, and V. R. Potter, *J. Am. Chem. Soc.* **72,** 5019 (1950).

biologically synthesized, labeled citric acid, and demonstrated that the chemically indistinguishable carboxyl groups were distinguished by the enzymes that make α-ketoglutarate. Citrate made from carboxyl-labeled acetate labeled the γ-carboxyl of ketoglutarate, whereas CO_2 entered the primary carboxyl of citrate that became the α-carboxyl. This work was neatly polished when the two types of citrate were synthesized chemically from optically resolved γ-chloro-β-carboxyl-β-hydroxybutyric acid. Similar studies were carried out by Martius and Schorre.[60]

The brilliant analysis of Ogston and experiments of Potter and Heidelberger not only preserved citric acid in the citric acid cycle, but greatly improved our concept of enzyme–substrate interaction. The same type of asymmetric reaction is observed with many different chemical structures, and the explanation of these stereospecific reactions is the same 3-point orientation. The advance in concept does not involve mere consideration of binding at additional sites, since this had been demonstrated for many years with various substrates, especially peptides. The unappreciated point was the fact that the binding of the substrate places particular atoms in position to react, and leaves other similar atoms in inert positions. Asymmetric syntheses are the rule, not the exception, in biochemistry, and stereochemical considerations as proposed by Ogston account for the large number of optically active compounds synthesized biologically.

GENERAL REFERENCES

Krebs, H. A. (1948–49). *Harvey Lectures Ser.* **44**, 165.
Ochoa, S. (1950-51). *Harvey Lectures Ser.* **46**, 153.
Ochoa, S. (1954). *Advances Enzymol.* **15**, 183.

Alternative Pathways of Carbohydrate Metabolism

THE PENTOSE PHOSPHATE PATHWAY

Glucose-6-phosphate Dehydrogenase. The establishment of the mechanism of glycolysis led to the acceptance of this sequence as the major pathway of carbohydrate metabolism. However, one of the first enzymes to be identified and studied catalyzes a reaction that takes hexose phosphate on a different route. Warburg and Christian found an oxidative enzyme that used a cofactor, later identified as TPN, to form 6-phosphogluconic acid from glucose-6-phosphate.[1] At the time these observations

[60] C. Martius and G. Schorre, *Z. Naturforsch.* **5b**, 170 (1950).

[1] O. Warburg and W. Christian, *Biochem. Z.* **242**, 206 (1931); O. Warburg, W. Christian, and A. Griese, *ibid.* **282**, 157 (1935).

were made the concept of individual specific dehydrogenases as components of complex metabolic pathways had not been established, and the enzyme was distinguished from the so-called "respiratory enzyme" (that reacts with oxygen) by the designation *Zwischenferment*. Zwischenferment is now called glucose-6-phosphate dehydrogenase.

For many years the reaction catalyzed by Zwischenferment was written glucose-6-phosphate + TPN → 6-phosphogluconate + TPNH, and was believed to proceed to completion. Recently this reaction has been resolved into two components. The form of glucose-6-phosphate that reacts with the dehydrogenase is the 6-membered pyranose ring. This ring persists during the oxidation, and the product of the reaction is the δ-lactone of 6-phosphogluconic acid.[2] In spite of the rapid, spontaneous hydrolysis of the lactone, it has been possible to demonstrate the reversibility of the oxidation (I).[3]

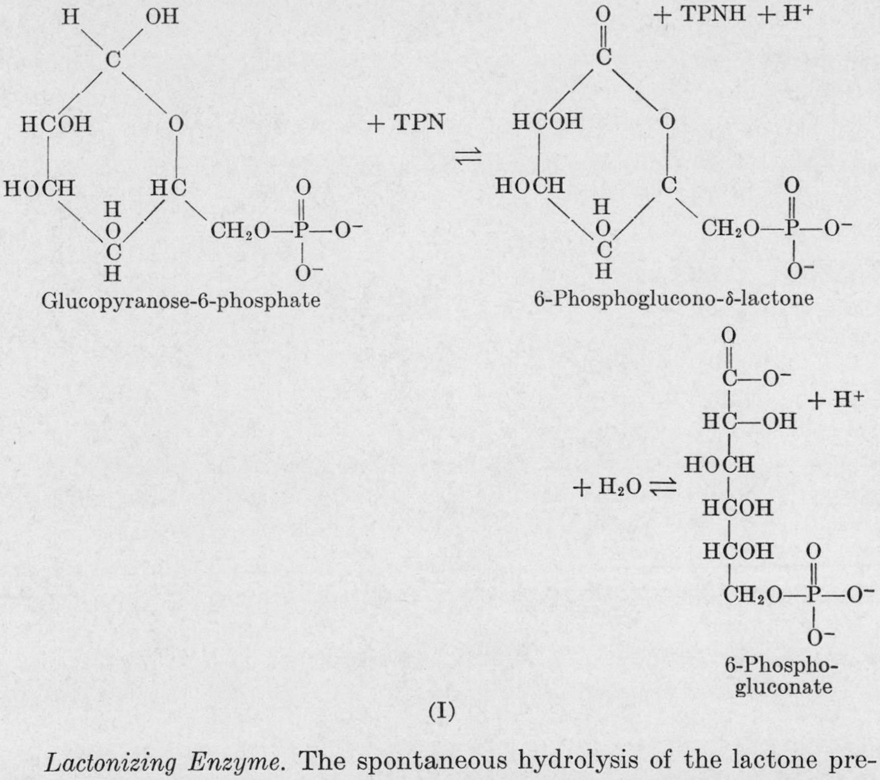

Glucopyranose-6-phosphate

6-Phosphoglucono-δ-lactone

6-Phosphogluconate

(I)

Lactonizing Enzyme. The spontaneous hydrolysis of the lactone prevented its detection for many years, but is not fast enough to keep pace with the rate of formation of the lactone by naturally occurring levels

[2] O. Cori and F. Lipmann, *J. Biol. Chem.* **194,** 417 (1952).
[3] B. L. Horecker and P. Z. Smyrniotis, *Biochim. et Biophys. Acta* **12,** 98 (1953).

of the dehydrogenase. Therefore a hydrolytic enzyme was postulated, and subsequently discovered in both bacteria and animals by Brodie and Lipmann.[4] This so-called lactonizing enzyme reacts with δ-glucono-lactone as well as with the 6-phosphate derivative. Theoretically this reaction is reversible, but, at neutral pH values, the equilibrium lies far to the side of hydrolysis, and therefore the over-all reaction: glucose-6-phosphate $+ \text{TPN} + H_2O \rightarrow$ 6-phosphogluconate $+ \text{TPNH} + 2H^+$ goes to completion. This is a very useful system for the analysis of TPN, the determination of glucose-6-phosphate, and the preparation of TPNH.

6-Phosphogluconic Dehydrogenase. The further oxidation of 6-phosphogluconate in yeast extracts was observed many years ago, and TPN was implicated in the oxidation. The enzyme involved was purified recently by Horecker and Smyrniotis,[5a] who identified the product of the oxidation as the ketose, ribulose-5-phosphate (II).[5b] The oxidation of

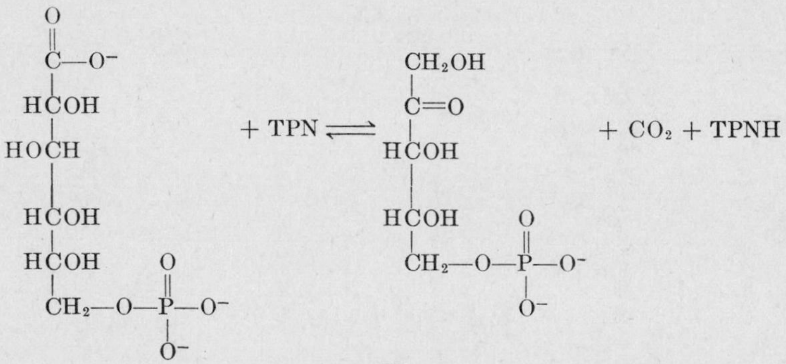

(II)

the β-, rather than the α-, carbon establishes a relation between phosphogluconic dehydrogenase and the malic enzyme and isocitric dehydrogenase. This enzyme also requires a divalent cation, Mg^{++} or Mn^{++}. It is probable that with this enzyme also the β-keto acid does not exist as a dissociable intermediate, but that the reaction is an oxidative decarboxylation without free intermediates. This hypothesis is supported by the relatively easy reversal of the decarboxylation. If the reversal were to proceed in two steps, with the intermediate formation of a β-keto acid by CO_2 fixation, the over-all equilibrium would be the same as in the case of simultaneous carboxylation and reduction. The rate would be much slower, however. The rate of a reaction is determined by the amount of enzyme, so that in the presence of sufficient enzyme even very

[4] A. F. Brodie and F. Lipmann, *J. Biol. Chem.* **212**, 677 (1955).
[5a] B. L. Horecker and P. Z. Smyrniotis, *J. Biol. Chem.* **193**, 371 (1951).
[5b] B. L. Horecker, P. Z. Smyrniotis, and J. E. Seegmiller, *J. Biol. Chem.* **193**, 383 (1951).

unfavorable reactions can be measured. If the K_m values of substrates and products are similar, the ratio of the rates of forward and reverse reactions is the equilibrium constant. The equilibrium constants of β-keto acid decarboxylations are large numbers, so the rates of the reverse reactions are probably several orders of magnitude less than the rates of decarboxylation. Even with the reaction maintained at maximum rates by reduction of the product, the two-step process would require extremely high concentrations of enzyme for rapid CO_2 fixation. The efficient reversal of oxidative decarboxylations is evidence for the two steps, oxidation and decarboxylation, occurring with the substrate bound to a single enzyme.

Pentose Phosphate Isomerase. Ribulose-5-phosphate may undergo several reactions. One is isomerization to the corresponding aldose, ribose 5-phosphate (III).[6] The widely distributed phosphopentose

```
        CH2OH                      CHO
         |                          |
         C=O                       HCOH
         |                          |
        HCOH          ⇌            HCOH
         |                          |
        HCOH     O                 HCOH     O
         |       ||                 |       ||
        CH2—O—P—O-                 CH2—O—P—O-
                 |                          |
                 O-                         O-
   D-Ribulose-5-phosphate        D-Ribose-5-phosphate
                          (III)
```

isomerase was purified highly from alfalfa by Axelrod and Jang,[7] who measured many physical properties of the system. They reported that at 37°C. and pH 7, $\Delta F_0' = -0.7$ kcal. in the direction of ribose-5-phosphate formation, whereas $\Delta H = 3.1$ kcal. Therefore $\Delta S = 12.1$ e.u.

Epimerase. Another rearrangement of the pentose phosphate molecule was recently discovered. This involves an isomerization at carbon 3, to form xylulose-5-phosphate (IV).[8a] The enzyme responsible has been named phosphoketopentose epimerase. The equilibrium of the epimerase favors slightly the formation of xylulose.[8b] The isomerization of the aldose and ketose forms may be presumed to proceed via an ene-diol, as was considered for the isomerization of hexoses. Such a mechanism may also be involved in the synthesis and utilization of arabinose. A comparable mechanism may be invoked for the inversion at carbon 3. Evidence for

[6] J. E. Seegmiller and B. L. Horecker, *J. Biol. Chem.* **194**, 261 (1952).

[7] B. Axelrod and R. Jang, *J. Biol. Chem.* **209**, 847 (1954).

[8a] G. Ashwell and J. Hickman, *J. Am. Chem. Soc.* **76**, 5889 (1954); P. K. Stumpf and B. L. Horecker, *J. Biol. Chem.* **218**, 753 (1956).

[8b] B. L. Horecker, J. Hurwitz, and P. Z. Smyrniotis, *J. Am. Chem. Soc.* **78**, 692 (1956).

the existence of a 3-keto pentose has been found;[9] if this is an intermediate it may revert to a 2-keto sugar with the hydroxyl group on carbon 3 on either side, to yield ribulose or xylulose.

$$
\begin{array}{ccc}
\text{CH}_2\text{OH} & & \text{CH}_2\text{OH} \\
\text{C}{=}\text{O} & & \text{C}{=}\text{O} \\
\text{HCOH} & \rightleftharpoons & \text{HOCH} \\
\text{HCOH} \quad \text{O} & & \text{HCOH} \quad \text{O} \\
\text{CH}_2{-}\text{O}{-}\text{P}{-}\text{O}^- & & \text{CH}_2{-}\text{O}{-}\text{P}{-}\text{O}^- \\
\text{O}^- & & \text{O}^-
\end{array}
$$

D-Ribulose-5-phosphate   D-Xylulose-5-phosphate

(IV)

Transketolase. The two pentose phosphates derived from ribulose phosphate, ribose-5-phosphate and xylulose-5-phosphate, participate in a transfer reaction in which two carbons are shifted from one to the other (V). Both the molecule split and the one formed are ketols, con-

$$
\begin{array}{ccccccc}
\text{CH}_2\text{OH} & & \text{CHO} & & \text{CH}_2\text{OH} \\
\text{C}{=}\text{O} & & \text{HCOH} & & \text{C}{=}\text{O} \\
\text{HOCH} & + & \text{HCOH} & \rightleftharpoons & \text{HC}{=}\text{O} & + & \text{HOCH} \\
\text{HCOH} \quad \text{O} & & \text{HCOH} \quad \text{O} & & \text{HCOH} \quad \text{O} & & \text{HCOH} \\
\text{CH}_2{-}\text{O}{-}\text{P}{-}\text{O}^- & & \text{CH}_2{-}\text{O}{-}\text{P}{-}\text{O}^- & & \text{CH}_2{-}\text{O}{-}\text{P}{-}\text{O}^- & & \text{HCOH} \\
\text{O}^- & & \text{O}^- & & \text{O}^- & & \text{HCOH} \quad \text{O} \\
& & & & & & \text{CH}_2{-}\text{O}{-}\text{P}{-}\text{O}^- \\
& & & & & & \text{O}^-
\end{array}
$$

Xylulose-5-   Ribose-5-   D-Phospho-   Sedoheptulose-7-
phosphate   phosphate   glyceraldehyde   phosphate

(V)

taining the structure —CO—CHOH—. Therefore the enzyme has been named transketolase. Transketolase has been purified from liver, spinach,[10] and yeast[11] and has been found to contain firmly bound cocarboxylase. The cofactor has been removed by the methods used earlier for carboxylase, by precipitation of the enzyme at low pH, and by alkaline dialysis. These treatments also reveal a requirement for Mg^{++}. The participation of cocarboxylase in a reaction that does not include a carboxyl group indicates that the cofactor was named for the wrong portion of the keto acid substrate. Apparently it acts by binding carbonyl

[9] G. Ashwell and J. Hickman, *J. Am. Chem. Soc.* **77**, 1063 (1955).
[10] B. L. Horecker and P. Z. Smyrniotis, *J. Am. Chem. Soc.* **75**, 1009 (1953).
[11] E. Racker, G. de la Haba, and I. Leder, *J. Am. Chem. Soc.* **75**, 1010 (1953).

groups, and participates in transfers to oxidizing systems or to other carbonyl groups (as in acyloin formation and in the transketolase reaction).

Transketolase requires an acceptor; it cannot split a ketol to form aldehydes. Presumably an intermediate "active glycolaldehyde" is formed by combination of the two-carbon unit with the enzyme. The enzyme will react with a number of aldehydes, and the reactions in many cases have been shown to be reversible. The donor requirements are not completely understood. Among the sugars only those with the hydroxyl on carbon 3 on the "L" side are substrates. Before the discovery of epimerase, ribulose-5-phosphate was thought to be a substrate, but it has been shown that a mixture of ribose and ribulose phosphates does not undergo the transketolase reaction until epimerase is added. Besides xylulose phosphate, fructose-6-phosphate, sedoheptulose-7-phosphate, and hydroxypyruvate are "glycolaldehyde" donors.

Transaldolase. When ribose-5-phosphate is the acceptor in the transketolase reaction, the product is the 7-carbon sugar, sedoheptulose-7-phosphate. The further metabolism of this compound was found to involve another type of transfer reaction, in which a dihydroxyacetone group is shifted to a phosphoglyceraldehyde molecule. The products of this transfer are fructose-6-phosphate and a 4-carbon sugar, erythrose-4-phosphate (VI). The transfer of a dihydroxyacetone group resembles the

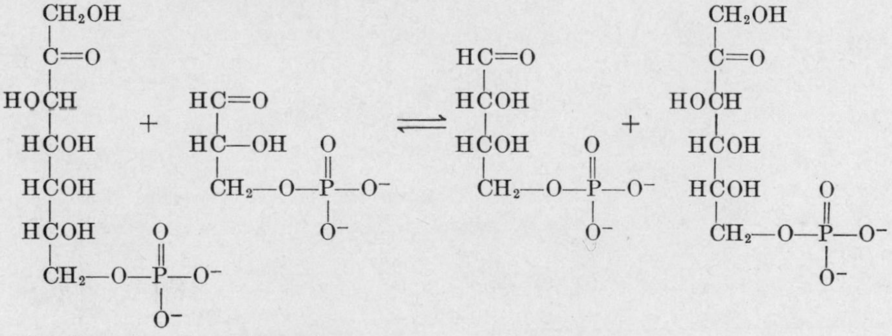

Sedoheptulose-7-phosphate   Phosphoglyceraldehyde   Erythrose-4-phosphate   Fructose-6-phosphate

(VI)

aldolase reaction, but in this case there is no equilibration with free dihydroxyacetone. Therefore the enzyme involved is called transaldolase.[12] The only compounds that have been found to react are fructose-6-phosphate, sedoheptulose-7-phosphate, and the corresponding aldehydes produced from them during the reaction. The equilibrium constant for the reaction is near 1.

[12] B. L. Horecker and P. Z. Smyrniotis, *Federation Proc.* **13**, 232 (1954).

Sedoheptulose Diphosphate. The transaldolase reaction requires phosphoglyceraldehyde, which has become available as a synthetic compound only recently. A convenient method for supplying triose phosphate is to add fructose diphosphate and aldolase. When this device was used in a study of the transaldolase reaction, the reaction products of the system containing both aldolase and transaldolase, and hexose diphosphate and sedoheptulose phosphate were expected to include tetrose phosphate, as shown in equation (VI) above. Tetrose phosphate failed to accumulate, however. Instead, sedoheptulose-1,7-diphosphate was found to accumulate as a result of condensation of the tetrose ester with dihydroxyacetone phosphate in the presence of aldolase.[13] This diphosphate reacts rapidly with aldolase, and it is not known whether it can react in any other systems, or only shuttles back and forth in response to changes in tetrose phosphate level.

Integration of the Pentose Phosphate Cycle. The two oxidative reactions catalyzed by glucose-6-phosphate and phosphogluconic dehydrogenase convert glucose-6-phosphate to a pentose phosphate and CO_2. A mechanism has been discovered for converting pentose phosphate quantitatively to hexose phosphate, thus establishing a cycle by which carbohydrate may be oxidized completely to CO_2 and water with only these two oxidative steps. The multi-cyclic process shown in Fig. 20 employs transketolase (TK), transaldolase (TA), and aldolase to transfer carbon atoms and several isomerases, pentose phosphate isomerase, epimerase, phosphotriose isomerase, and phosphohexose isomerase, to establish the requisite structures. The transfer reactions and the isomerizations are freely reversible processes and do not require the expenditure of energy. The cycle is driven by several irreversible steps. The oxidative reactions produce TPNH, which is oxidized by systems that ultimately transfer electrons to molecular oxygen; these reactions proceed irreversibly to completion. The hydrolysis of phosphogluconolactone is also essentially irreversible, and one additional reaction employed to permit utilization of all of the carbon by this cycle, the hydrolysis of hexose diphosphate, also proceeds in only one direction. This hydrolysis is catalyzed by a specific, Mg^{++}-requiring enzyme.[14a] The hydrolysis results in the liberation of inorganic phosphate, originally inserted into carbohydrate at the expense of ATP. No other phosphate is esterified or liberated during the operation of this cycle. The expenditure of the small amount of esterified phosphate is a small price for the maintenance of the cycle, since many more equivalents of phosphate are esterified during the oxidation of TPNH. The reactions shown in the diagram are summarized in (VII).

[13] P. Z. Smyrniotis and B. L. Horecker, *J. Biol. Chem.* **218**, 745 (1956).
[14a] G. Gomori, *J. Biol. Chem.* **148**, 139 (1943); B. M. Pogell and R. W. McGilvery, *ibid.* **208**, 149 (1954).

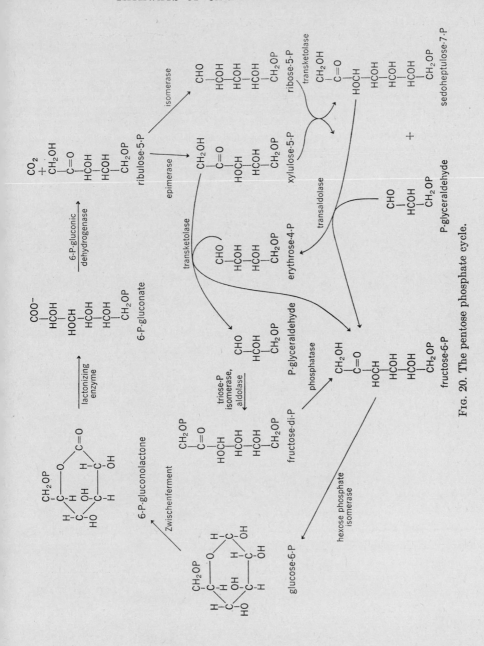

Fig. 20. The pentose phosphate cycle.

$$6 \text{ glucose-6-phosphate} + 12\text{TPN} \rightarrow 12\text{TPNH} + 6\text{CO}_2 + \underbrace{6 \text{ pentose phosphate}}$$

$$\begin{aligned} &4 \text{ xylulose phosphate} \\ &2 \text{ ribose phosphate} \end{aligned}$$

2 xylulose phosphate + 2 ribose phosphate →
$$2 \text{ triose phosphate} + 2 \text{ heptulose phosphate}$$
2 heptulose phosphate + 2 triose phosphate →
$$2 \text{ tetrose phosphate} + 2 \text{ fructose-6-phosphate}$$
2 xylulose phosphate + 2 tetrose phosphate →
$$2 \text{ triose phosphate} + 2 \text{ fructose-6-phosphate}$$
2 triose phosphate → 1 fructose diphosphate
1 fructose diphosphate → 1 fructose monophosphate + 1 inorganic phosphate
5 fructose-6-phosphate → 5 glucose-6-phosphate

$$\overline{\text{glucose-6-phosphate} + 12\text{TPN} \rightarrow 12\text{TPNH} + 6\text{CO}_2 + \text{inorganic phosphate}}$$

(VII)

The enzymes of the pentose phosphate pathway are widely distributed in plants, animals, and microorganisms. Recently attempts have been made to evaluate the activity of this system *in vivo*. Isotope experiments, in which the rate of conversion of C-1 of glucose to CO_2 is compared with the rate of CO_2 formation from other atoms of glucose, indicate that a major part of the oxidation in liver may proceed by the pentose pathway. In contrast, glycolysis accounts for essentially all of muscle carbohydrate metabolism. The presence of Zwischenferment, however, does not imply that this cyclic mechanism is operative, since the first steps may be used for the production of pentose phosphate for nucleotide synthesis, polysaccharides, or other purposes.

A series of nonoxidative reactions using enzymes of the pentose phosphate pathway permits pentose phosphates to be converted to hexose phosphates (VIII).

$$\text{xylulose-5-phosphate} + \text{ribose-5-phosphate} \xrightarrow{\text{TK}} \text{sedoheptulose-7-phosphate}$$
$$+$$
$$\text{phosphoglyceraldehyde}$$
$$\text{sedoheptulose-7-phosphate} + \text{triose phosphate} \xrightarrow{\text{TA}}$$
$$\text{tetrose-4-phosphate} + \text{fructose-6-phosphate}$$
$$\text{xylulose-5-phosphate} + \text{tetrose-4-phosphate} \xrightarrow{\text{TK}}$$
$$\text{fructose-6-phosphate} + \text{triose phosphate}$$

$$\overline{3 \text{ pentose-5-phosphate} \rightarrow 2 \text{ fructose-6-phosphate} + \text{triose phosphate}}$$

(VIII)

The utilization of pentose phosphate without conversion to hexose is effected by *Lactobacillus pentosus*. An enzyme isolated from this organism converts xylulose-5-phosphate in one step to acetyl phosphate and triose phosphate (IX).[14b] This enzyme has been named phosphopentoketolase.

[14b] E. C. Heath, J. Hurwitz, and B. L. Horecker, *J. Am. Chem. Soc.* **78**, 5449 (1956).

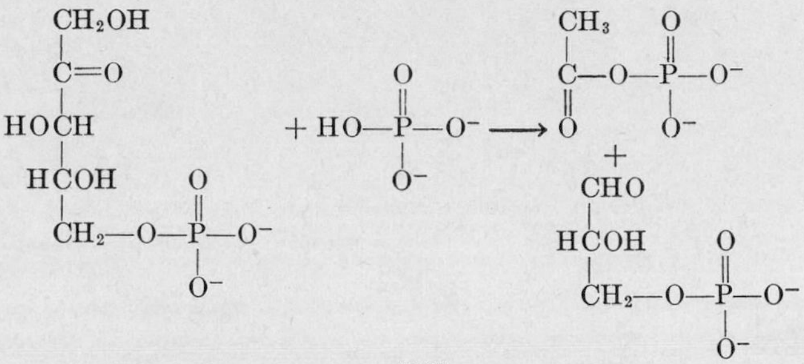

$$
\begin{array}{c}
CH_2OH \\
| \\
C=O \\
| \\
HOCH \\
| \\
HCOH \quad O \\
| \quad\quad \| \\
CH_2-O-P-O^- \\
| \\
O^-
\end{array}
\quad + \quad
\begin{array}{c}
O \\
\| \\
HO-P-O^- \\
| \\
O^-
\end{array}
\longrightarrow
\begin{array}{c}
CH_3 \quad O \\
| \quad\quad \| \\
C-O-P-O^- \\
\| \quad\quad | \\
O \quad\quad O^- \\
+ \\
CHO \\
| \\
HCOH \quad O \\
| \quad\quad \| \\
CH_2-O-P-O^- \\
| \\
O^-
\end{array}
$$

(IX)

The mechanism of this unusual reaction is under investigation. The enzyme requires thiamine pyrophosphate, Mg^{++}, and a sulfhydryl compound.

The Path of Carbon in Photosynthesis

Photosynthesis is the process in which most of the organic material of our world is produced. There are some organisms, the chemo-autotrophs, that synthesize organic matter by coupling CO_2 fixation with inorganic reactions, but the overwhelming majority of net organic synthesis depends upon solar radiation. The general equation for the formation of carbohydrate (glucose) is:

$$6CO_2 + 6H_2O \xrightarrow{h\nu} C_6H_{12}O_6 + 6O_2$$

This process has been studied intensively from many points of view, but the fundamental mechanisms by which light affects a chemical reaction are still unknown. Many properties of photosynthesizing systems are known, but these properties cannot be ascribed to defined chemical reactions. The enzyme chemistry of photosynthesis is an almost completely unknown area.

The Hill Reaction. The photochemical reaction has been studied with isolated chloroplasts and chloroplast fragments (grana). These evolve oxygen in the presence of suitable oxidants, such as various iron complexes, quinones, and dyes. These reactions are known as Hill reactions, in recognition of the pioneer studies of Hill in demonstrating and defining the properties of isolated chloroplasts (X).[15]

The balanced Hill reaction is a simplification of the actual process. Each individual light reaction, requiring a single photon, probably transfers one electron from a molecule of water through an enzymatic electron transport mechanism to an undefined electron acceptor. The remainder of the water molecule is used to produce oxygen (XI).

[15] R. Hill, *Proc. Roy. Soc.* **B127,** 192 (1939).

The identities of the reducing and oxidizing moieties of water are not known. It is possible that the chlorophyll molecule participates chemically in the Hill reaction in addition to absorbing the radiant energy on which all photosynthesis depends.

$$2 \text{ cytochrome } c^{+++} + H_2O \xrightarrow{\;h\nu\;} 2 \text{ cytochrome } c^{++} + \tfrac{1}{2}O_2 + 2H^+$$

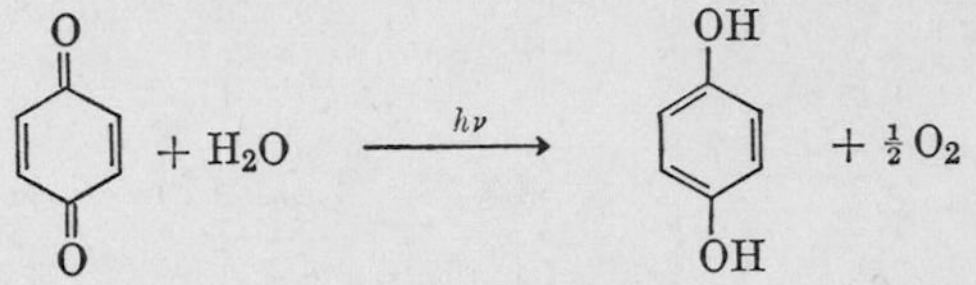

Representative Hill Reactions
(X)

$$4H_2O \xrightarrow{\;h\nu\;} 4[H] + 4[OH] \quad (\text{photolysis})$$
$$4[H] + 2X \to 2XH_2$$
$$\underline{4[OH] \to O_2 + 2H_2O}$$
$$2H_2O + 2X \to 2XH_2 + O_2$$

(XI)

Coupling of Light Reactions. In some photosynthetic organisms photolysis results not in oxygen liberation but in oxidation of other materials, such as organic acids or sulfur compounds. The energy requirements of these variants are much less than that of the more familiar process in which oxygen is liberated; nevertheless, the requirements for light quanta are similar, indicating that the basic mechanism of photolysis is essentially the same in all types of photosynthesis. The reduction process is capable of reducing pyridine nucleotides, which are thereby made available for reducing CO_2.[16] In the metabolism of CO_2, ATP is required, and this may be generated photosynthetically by enzymatic recombination of the products of photolysis. This may employ the mechanisms of oxidative phosphorylation characteristic of nonphotosynthetic oxidation (see p. 385).

The Phosphoglyceric Cycle. One area of knowledge has been revealed in detail in recent years: the path of carbon. Isotope studies have established phosphoglyceric acid as the first compound to become labeled with carbon dioxide during photosynthesis.[17] The following sequence of known reactions is a scheme for converting phosphoglycerate to the CO_2 acceptor that forms more phosphoglycerate.

[16] W. Vishniac and S. Ochoa, *Nature* **167**, 768 (1951); L. J. Tolmach, *ibid.* **167**, 946 (1951).

[17] A. A. Benson, J. A. Bassham, M. Calvin, T. C. Goodale, V. A. Haas, and W. Stepka, *J. Am. Chem. Soc.* **72**, 1710 (1950); E. W. Fager, J. L. Rosenberg, and H. Gaffron, *Federation Proc.* **9**, 535 (1950).

Reduction of phosphoglycerate by the combined action of phosphoglycerate kinase and triose phosphate dehydrogenase results in the formation of triose phosphate (XII). This reduction is presumably effected

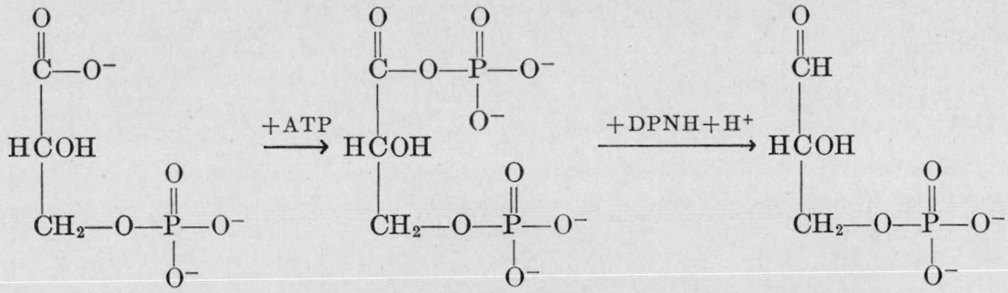

$$
\begin{array}{ccccc}
\overset{\displaystyle O}{\overset{\displaystyle \|}{C}}-O^- & & \overset{\displaystyle O}{\overset{\displaystyle \|}{C}}-O-\overset{O}{\overset{\|}{P}}-O^- & & \overset{\displaystyle O}{\overset{\displaystyle \|}{C}}H \\
| & \xrightarrow{+ATP} & | \quad\quad\; | & \xrightarrow{+DPNH+H^+} & | \\
HCOH & & HCOH \quad O^- & & HCOH \\
| & & | & & | \\
CH_2-O-\overset{O}{\overset{\|}{P}}-O^- & & CH_2-O-\overset{O}{\overset{\|}{P}}-O^- & & CH_2-O-\overset{O}{\overset{\|}{P}}-O^- \\
\quad\quad | & & \quad\quad | & & \quad\quad | \\
\quad\quad O^- & & \quad\quad O^- & & \quad\quad O^-
\end{array}
$$

(XII)

by the light reaction which supplies the electrons for reducing DPN and
also generates ATP by oxidative phosphorylation during reoxidation of
part of the photochemical reduction product. Triose phosphate isomerase
and aldolase together form hexose diphosphate. This can be hydrolyzed

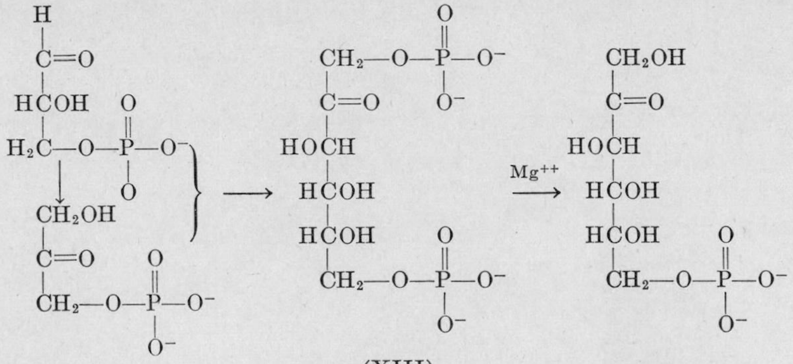

(XIII)

to the monophosphate by the specific, Mg^{++} requiring hexosediphos-
phatase (XIII). Hexose monophosphate, through the transketolase reac-

fructose-6-phosphate + triose phosphate $\xrightarrow{\text{TK}}$
   erythrose-4-phosphate + xylulose-5-phosphate

fructose-6-phosphate + erythrose-4-phosphate $\xrightarrow{\text{TA}}$ sedoheptulose-7-phosphate
   +
   triose phosphate

sedoheptulose-7-phosphate + triose phosphate $\xrightarrow{\text{TK}}$ xylulose-5-phosphate
   +
   ribose-5-phosphate

___

Sum: 2 fructose-6-phosphate + triose phosphate →
   2 xylulose-5-phosphate + ribose-5-phosphate

(XIV)

tion, enters the pentose phosphate system, and, with the participation of transaldolase, is converted to pentose (XIV). All of the pentose phosphate may be converted to ribulose-5-phosphate by epimerase and phosphopentose isomerase.

Ribulokinase. A specific ribulokinase converts ribulose monophosphate to a diphosphate (XV).[18]

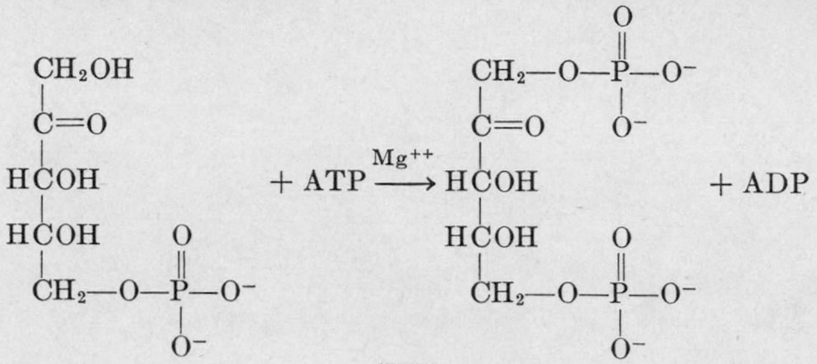

$$
\begin{array}{ccc}
\text{CH}_2\text{OH} & & \text{CH}_2\text{—O—P—O}^- \\
| & & | \qquad | \\
\text{C=O} & & \text{C=O} \quad \text{O}^- \\
| & \overset{\text{Mg}^{++}}{\xrightarrow{\quad}} & | \\
\text{HCOH} & + \text{ATP} & \text{HCOH} \qquad + \text{ADP} \\
| & & | \\
\text{HCOH} \quad \text{O} & & \text{HCOH} \quad \text{O} \\
| \qquad || & & | \qquad || \\
\text{CH}_2\text{—O—P—O}^- & & \text{CH}_2\text{—O—P—O}^- \\
\quad\;\; | & & \quad\;\; | \\
\quad\;\; \text{O}^- & & \quad\;\; \text{O}^-
\end{array}
$$

(XV)

The Carboxylation Reaction. The carboxylation reaction of green plants appears to be a single-step reaction in which ribulose diphosphate adds CO_2 to carbon 2, and in the process is split to form two molecules

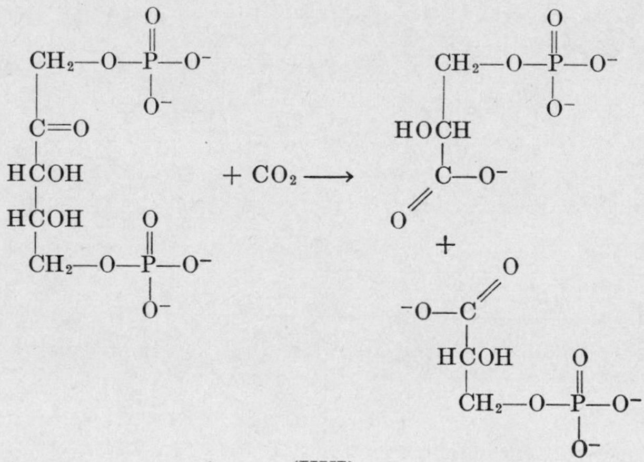

(XVI)

of phosphoglycerate (XVI).[19a] This unique reaction has not been reversed. The carboxylation enzyme has been purified only 15–20-fold from leaf extracts, and many diverse attempts to increase the specific activity have failed. It is therefore possible to consider that as much as

[18] J. Hurwitz, A. Weissbach, B. L. Horecker, and P. Z. Smyrniotis, *J. Biol. Chem.* **218,** 769 (1956).

[19a] A. Weissbach, B. L. Horecker, and J. Hurwitz, *J. Biol. Chem.* **218,** 795 (1956); W. B. Jakoby, D. O. Brummond, and S. Ochoa, *ibid.* **218,** 811 (1956).

5 per cent of the soluble protein of green leaves is concerned with the fixation of carbon dioxide in phosphoglycerate. Such an emphasis on the formation of phosphoglycerate is consistent with the isotope experiments which show that all of the $C^{14}O_2$ fixed in the first instant of photosynthesis appears in the carboxyl group of phosphoglycerate. The subsequent appearance of isotope in other compounds and in other positions of phosphoglycerate can be accounted for by the cyclic process initiated by the reduction of phosphoglycerate and completed by the carboxylation reaction.

Since its demonstration in photosynthetic organisms, the carboxylation enzyme has been demonstrated in *E. coli*[19b] and *Thiobacillus*.[19c] The physiological significance of this enzyme in nonphotosynthetic organisms is not clear, but its occurrence emphasizes that the carbon metabolism of photosynthesis is an enzymatic process distinct from the photochemical reaction.

Each turn of the cycle may be summarized as the formation of 3 pentose molecules from 5 triose molecules. The carboxylation reaction forms 6 phosphoglycerates from the 3 pentoses. The net increase in compounds of this cycle may be balanced by the withdrawal of substances for the synthesis of cell constituents or, for the most part, by storage of hexose as polysaccharides (Chapter V).

NONPHOSPHORYLATED CARBOHYDRATE OXIDATION

In certain microorganisms, glucose is oxidized by a pathway that begins with metabolism of nonphosphorylated compounds. Eventually kinases introduce phosphate into the intermediates in these systems also, but many steps remain obscure. An oxidation of nonphosphorylated glucose also occurs in animal tissues, but it is probable that this reaction is followed by a phosphorylation.

Glucose Oxidation. The first reaction of these sequences is oxidation of glucose. A glucose dehydrogenase has been purified from animal livers.[20] This enzyme oxidizes the β-pyranose form of D-glucose with DPN to form δ-gluconolactone (XVII). The enzyme also attacks D-galactose, D- and L-xylose, and D-arabinose. A similar reaction is catalyzed by a glucose oxidase of molds.[21] This enzyme uses molecular oxygen and also produces the δ-lactone. It is extremely specific for glucose. Additional properties of this enzyme will be discussed with the flavoproteins.

[19b] R. C. Fuller, *Bacteriol. Proc.* p. 112 (1956).

[19c] P. A. Trudinger, *Biochem. J.* **69,** 279 (1956); M. Santer and W. Vishniac, *Biochim. et Biophys. Acta* **18,** 157 (1955).

[20] D. C. Harrison, *Biochem. J.* **25,** 1016 (1931); H. J. Strecker and S. Korkes, *J. Biol. Chem.* **196,** 769 (1952).

[21] D. Keilin and E. F. Hartree, *Biochem. J.* **42,** 221 (1948).

Mutarotase. The substrate for glucose oxidation has been shown to be in the β-configuration. Glucose exists in aqueous solution as an equilibrium mixture of α and β forms, with less than 1 per cent open chain or other possible forms. The interconversion of α- and β-glucose, mutarotation, is a rapid reaction which is accelerated by certain inorganic ions. Even though the reaction is rapid, it was possible to demonstrate a lag in the oxidation of α-glucose by the oxidase from molds; this could be eliminated by another enzyme, mutarotase.[22] Mutarotase has been found in high concentrations in certain animal tissues,[23] but its function is still uncertain. It is active with D-glucose, D-galactose, D-xylose, lactose, and mannose, but not with a large number of other sugars.

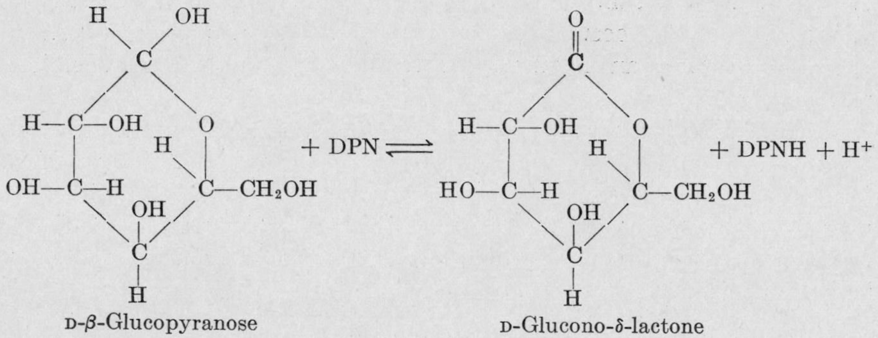

$$
\begin{array}{cc}
\text{D-}\beta\text{-Glucopyranose} & \text{D-Glucono-}\delta\text{-lactone}
\end{array}
$$

(XVII)

Gluconate Metabolism. Gluconate is oxidized to 2-ketogluconate by incompletely described bacterial enzymes. Kinases have been found for phosphorylating both gluconate and its 2-keto derivative.[24] 2,5-Diketogluconate has been described as a product in some systems. Eventually this pathway leads to the Krebs cycle, but it is not known how.

Pseudomonas Pathway. Certain strains of *Pseudomonas* have been found to possess different enzymes from those previously described. First, a glucose-6-phosphate dehydrogenase and a phosphogluconic dehydrogenase have been found that react with DPN in contrast to the enzymes from other sources that are specific for TPN.[25] A novel pathway was found in studies in Doudoroff's laboratory. A dehydrase converts phosphogluconate to 2-keto-3-deoxygluconate.[26] This enzyme has been separated from a second enzyme, which requires Fe^{++} and glutathione,

[22] D. Keilin and E. F. Hartree, *Biochem. J.* **50,** 341 (1952).
[23] A. S. Keston, *Science* **120,** 355 (1954).
[24] J. DeLey, *Biochim. et Biophys. Acta* **13,** 302 (1954); S. A. Narrod and W. A. Wood, *Bacteriol. Proc.* p. 108 (1954).
[25] W. A. Wood and R. F. Schwerdt, *J. Biol. Chem.* **206,** 625 (1954).
[26] J. MacGee and M. Doudoroff, *J. Biol. Chem.* **210,** 617 (1954).

and splits the dehydrated compound in an aldolase-like reaction to form pyruvate from carbons 1, 2, and 3, and phosphoglyceraldehyde from carbons 4, 5, and 6 (XVIII).[27]

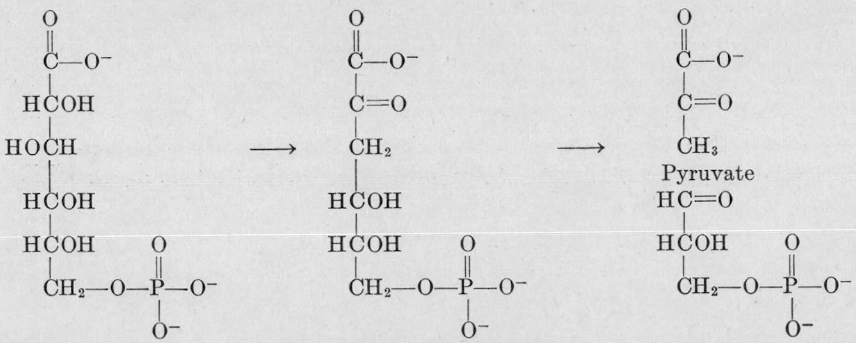

$$
\begin{array}{ccccc}
\overset{\displaystyle O}{\underset{\displaystyle \|}{C}}\!-\!O^- & & \overset{\displaystyle O}{\underset{\displaystyle \|}{C}}\!-\!O^- & & \overset{\displaystyle O}{\underset{\displaystyle \|}{C}}\!-\!O^- \\
HCOH & & C\!=\!O & & C\!=\!O \\
HOCH & \longrightarrow & CH_2 & \longrightarrow & CH_3 \\
& & & & \text{Pyruvate} \\
HCOH & & HCOH & & HC\!=\!O \\
HCOH \quad O & & HCOH \quad O & & HCOH \quad O \\
CH_2\!-\!O\!-\!\overset{\|}{P}\!-\!O^- & & CH_2\!-\!O\!-\!\overset{\|}{P}\!-\!O^- & & CH_2\!-\!O\!-\!\overset{\|}{P}\!-\!O^- \\
O^- & & O^- & & O^-
\end{array}
$$

6-Phosphogluconate   2-Keto-3-deoxy-6-phosphogluconate   Phosphoglyceraldehyde

(XVIII)

ADDITIONAL PATHWAYS

Certain microorganisms ferment glucose by pathways that have been very useful for isotope analysis, but which have not been separated into individual reactions. *Leuconostoc mesenteroides* degrades glucose as shown in (XIX).[28] *Pseudomonas lindneri* produces 2 ethanol and $2CO_2$ from

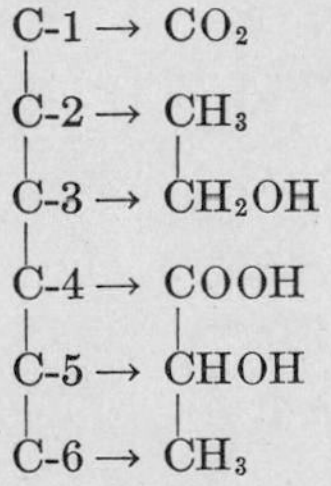

$$
\begin{array}{l}
C\text{-}1 \rightarrow CO_2 \\
C\text{-}2 \rightarrow CH_3 \\
C\text{-}3 \rightarrow CH_2OH \\
C\text{-}4 \rightarrow COOH \\
C\text{-}5 \rightarrow CHOH \\
C\text{-}6 \rightarrow CH_3
\end{array}
$$

(XIX)

glucose, but unlike yeast, forms CO_2 from C-1 and C-4.[29] The pathway is probably related to the 2-keto-3-deoxygluconic reaction described above.

Fructose-1-phosphate Metabolism. A kinase in mammalian liver has been found to phosphorylate fructose on carbon 1.[30] This is in contrast to

[27] N. Entner and M. Doudoroff, *J. Biol. Chem.* **196,** 853 (1952); R. Kovachevich and W. A. Wood, *J. Biol. Chem.* **213,** 745, 757 (1955).

[28] I. C. Gunsalus and M. Gibbs, *J. Biol. Chem.* **194,** 871 (1952).

[29] M. Gibbs and R. D. DeMoss, *J. Biol. Chem.* **207,** 689 (1954).

[30] M. W. Slein, G. T. Cori, and C. F. Cori, *J. Biol. Chem.* **186,** 763 (1950); R. E. Parks, *Federation Proc.* **13,** 271 (1954).

the phosphorylation on carbon 6 catalyzed by the less specific hexo-kinases. Fructose-1-phosphate is converted to glucose-6-phosphate by liver enzymes, but the conversion is indirect. There is no known mutase for the conversion of fructose-1-phosphate to fructose-6-phosphate; similarly there is no isomerization of fructose-1-phosphate to glucose-1-phosphate. The metabolism of fructose-1-phosphate appears to require cleavage to trioses. A liver aldolase differs from muscle aldolase in having a high affinity for and a rapid rate of reaction with fructose-1-phosphate.[31] The products are dihydroxyacetone phosphate and glyceraldehyde (XX).

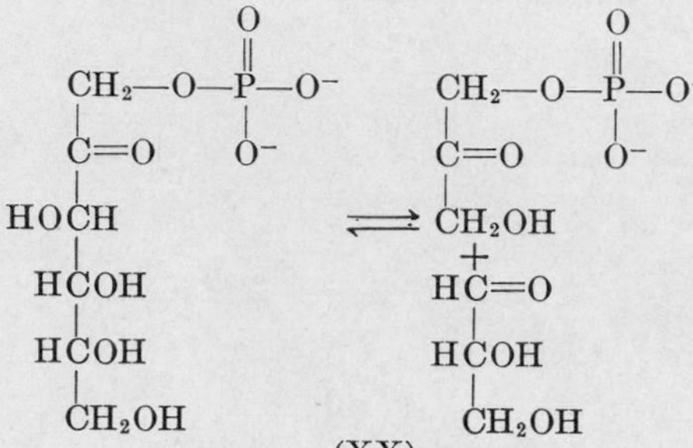

$$
\begin{array}{ccc}
\overset{\displaystyle O}{\underset{\displaystyle \parallel}{CH_2-O-P-O^-}} & & \overset{\displaystyle O}{\underset{\displaystyle \parallel}{CH_2-O-P-O^-}} \\
C=O \quad O^- & & C=O \quad O^- \\
HOCH & \rightleftharpoons & CH_2OH \\
HCOH & & + \\
HCOH & & HC=O \\
CH_2OH & & HCOH \\
& (XX) & CH_2OH
\end{array}
$$

Efficient utilization of the fructose requires phosphorylation of the glyceraldehyde. Tracer experiments show that the carbon 1 of fructose appears as both carbons 1 and 6 of glucose. This is the result of triose phosphate isomerization followed by (conventional) aldolase condensation to hexose diphosphate. The conversion of fructose diphosphate to glucose-6-phosphate requires a phosphatase and an isomerase, as discussed in the pentose phosphate pathway.

Triose Phosphate Dehydrogenases in Plants. Triose phosphate dehydrogenase, as described earlier, exists in many organisms. In green plants two additional enzymes have been reported.[32] One is a very similar enzyme that has TPN in place of DPN. The other also uses TPN, but has no requirement for inorganic phosphate or any other acyl acceptor. This enzyme apparently catalyzes an irreversible formation of phospho-glyceric acid.

Glyoxalase. Before the Embden-Meyerhof scheme was recognized as the major pathway for glucose metabolism, an enzyme was found that could produce lactic acid rapidly. It was reasonable, therefore, that this enzyme, glyoxalase, and its substrate, methyl glyoxal, were used in many speculations about the nature of glycolysis. Today they are not

[31] H. G. Hers and T. Kusaka, *Biochim. et Biophys. Acta* **11**, 427 (1953).
[32] M. Gibbs, *Nature* **170**, 164 (1952); D. I. Arnon, L. L. Rosenberg, and F. R. Whatley, *ibid.* **173**, 1132 (1954).

included in schemes of fermentation because they are not involved in the complete mechanism involving phosphorylated compounds, and because no enzymatic mechanism has been found for producing the substrate. Methyl glyoxal is formed nonenzymatically from triose and triose phosphates. The enzyme remains of interest because of the studies on its mode of action and because of its cofactor, glutathione.

Glyoxalase has been resolved into two proteins, glyoxalase I and II.[33] The two act independently in the formation and decomposition of an intermediate formed from methyl glyoxal and glutathione (XXI). The

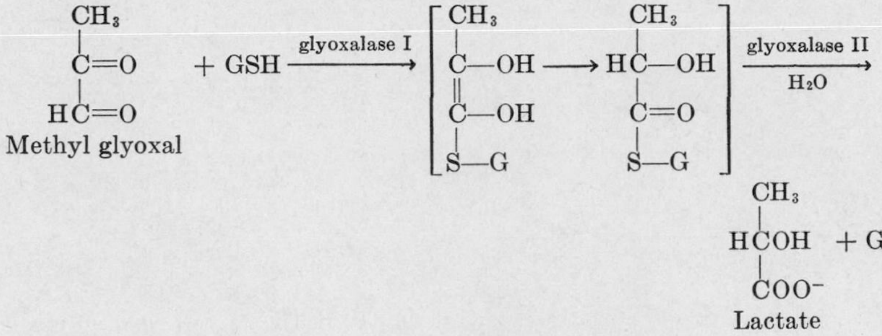

$$
\begin{array}{c}
\underset{|}{\overset{|}{C}}H_3 \\
C=O \\
| \\
HC=O \\
\text{Methyl glyoxal}
\end{array}
+ GSH \xrightarrow{\text{glyoxalase I}}
\left[
\begin{array}{cc}
CH_3 & CH_3 \\
| & | \\
C-OH \longrightarrow HC-OH \\
\| & | \\
C-OH & C=O \\
| & | \\
S-G & S-G
\end{array}
\right]
\xrightarrow[H_2O]{\text{glyoxalase II}}
$$

$$
\begin{array}{c}
CH_3 \\
| \\
HCOH \\
| \\
COO^-
\end{array}
+ GSH + H^+
$$

Lactate

(XXI)

intermediate can be detected spectrophotometrically by its absorption at 240 mμ, but its structure is not certain. The reaction is completely dependent on reduced glutathione, and has been used as a simple quantitative assay for this compound.

Glycerol Oxidation. Aerobacter aerogenes is capable of oxidizing glycerol by two pathways.[34] The free compound is oxidized by one strain to dihydroxyacetone; a similar oxidation with a DPN-requiring enzyme occurs in *B. subtilis*.[35] Another pathway, also found in *Acetobacter*,[36] involves phosphorylation to L-α-glycerophosphate, which is then oxidized to phosphoglyceraldehyde. A kinase, triokinase, phosphorylates dihydroxyacetone prior to its further metabolism.

Cleavage of Tricarboxylic Acids. Both citric and isocitric acids are reversibly split by enzymes found in various bacteria. Citritase splits citrate to oxalacetate and acetate (XXII).[37] CoA does not participate in this reaction. Unlike the reaction catalyzed by the citrate condensing enzyme, which greatly favors citrate synthesis, the standard free energy

[33] E. Racker, *Federation Proc.* **12,** 711 (1953).
[34] B. Magasanik, M. S. Brooke, and D. Karibian, *J. Bacteriol.* **66,** 611 (1953).
[35] J. M. Wiame, S. Bourgeois, and R. Lambion, *Nature* **174,** 37 (1954).
[36] J. G. Hauge, T. E. King, and V. H. Cheldelin, *Federation Proc.* **13,** 224 (1954).
[37] R. A. Smith, J. R. Stamer, and I. C. Gunsalus, *Biochim. et Biophys. Acta* **19,** 567 (1956).

of the citritase reaction is a very small number, and the relative amounts of reactants at equilibrium depends upon the absolute concentration, as is generally true when one compound yields two.

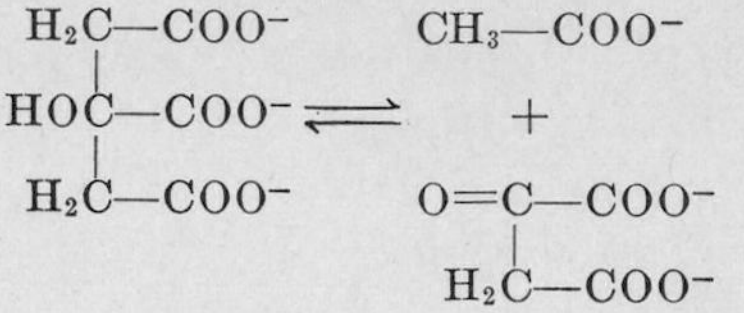

$$
\begin{array}{l}
H_2C—COO^- \\
HOC—COO^- \\
H_2C—COO^-
\end{array}
\rightleftharpoons
\begin{array}{l}
CH_3—COO^- \\
+ \\
O=C—COO^- \\
H_2C—COO^-
\end{array}
\qquad
K_{eq} = \frac{(citrate)}{(oxalacetate)(acetate)} = 1.5
$$

(XXII)

The cleavage of isocitrate is catalyzed by an enzyme designated iso-citritase.[38] The products are glyoxylate and succinate (XXIII). This reaction favors synthesis even more than the citritase reaction but at the

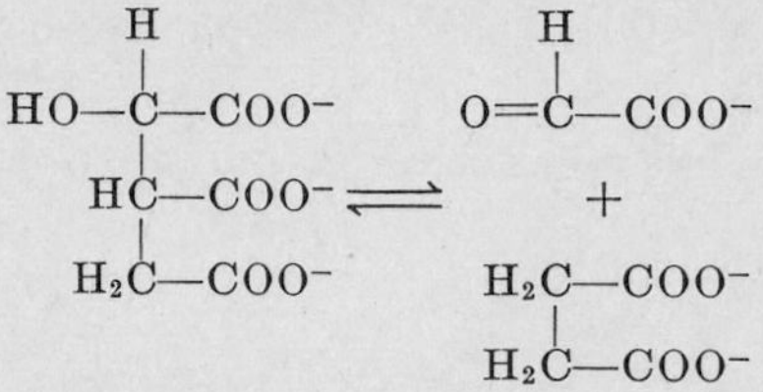

$$
\begin{array}{l}
H \\
HO—C—COO^- \\
HC—COO^- \\
H_2C—COO^-
\end{array}
\rightleftharpoons
\begin{array}{l}
H \\
O=C—COO^- \\
+ \\
H_2C—COO^- \\
H_2C—COO^-
\end{array}
\qquad
K_{eq} = \frac{(isocitrate)}{(glyoxalate)(succinate)} = 34
$$

(XXIII)

concentrations ordinarily used most of the isocitrate is split.

Hydrolysis of Oxalacetate. In certain fungi oxalate is a major metabolic product. The mechanism of oxalate formation appears to involve the hydrolytic cleavage of oxalacetate:[39]

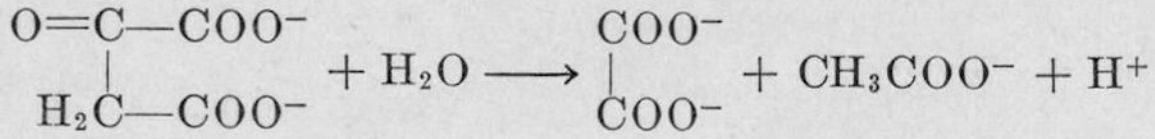

$$
\begin{array}{l}
O=C—COO^- \\
H_2C—COO^-
\end{array}
+ H_2O \longrightarrow
\begin{array}{l}
COO^- \\
COO^-
\end{array}
+ CH_3COO^- + H^+
$$

Polyols. Polyols, the reduction products of sugars, serve as oxidizable substrates for many organisms. Edson has identified three types of polyol dehydrogenases with varying substrate specificities. A DPN-specific enzyme has been named L-iditol dehydrogenase because both ends of L-iditol contain the required structure.[40] The configuration of carbon 3 is not significant in this reaction, but carbon 4 must have the configuration shown (XXIV). This type of enzyme has been purified from animal livers, seminal vesicles and coagulating glands.[41]

[38] H. J. Saz, *Biochem. J.* **58**, xx (1954); J. A. Olson, *Nature* **174**, 695 (1954); R. A. Smith and I. C. Gunsalus, *Nature* **175**, 774 (1955).

[39] O. Hayaishi, H. Shimazono, M. Katagiri, and Y. Saito, *J. Am. Chem. Soc.* **78**, 5126 (1956).

[40] J. McCorkindale and N. L. Edson, *Proc. Univ. Otago Med. School* **31**, 2 (1953).

[41] C. M. Todd, *Proc. Univ. Otago Med. School* **32**, 9 (1954); H. G. Williams-Ashman and J. Banks, *Arch. Biochem. and Biophys.* **50**, 513 (1954).

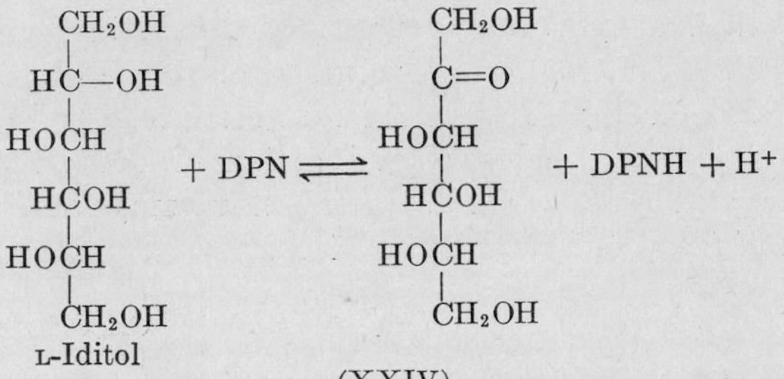

$$
\begin{array}{c}
\text{CH}_2\text{OH} \\
\text{HC—OH} \\
\text{HOCH} \\
\text{HCOH} \\
\text{HOCH} \\
\text{CH}_2\text{OH} \\
\text{L-Iditol}
\end{array}
\quad + \text{DPN} \rightleftharpoons \quad
\begin{array}{c}
\text{CH}_2\text{OH} \\
\text{C=O} \\
\text{HOCH} \\
\text{HCOH} \\
\text{HOCH} \\
\text{CH}_2\text{OH}
\end{array}
\quad + \text{DPNH} + \text{H}^+
$$

(XXIV)

Another dehydrogenase found in *Acetobacter suboxydans* requires that the substrate have the structure of one end of D-mannitol, and has been named D-mannitol dehydrogenase (XXV).[42] The reaction occurs in the position indicated if one of the hydrogens of carbon 1 is replaced by a methyl group. A similar oxidation is catalyzed by an enzyme with no cofactor requirement.

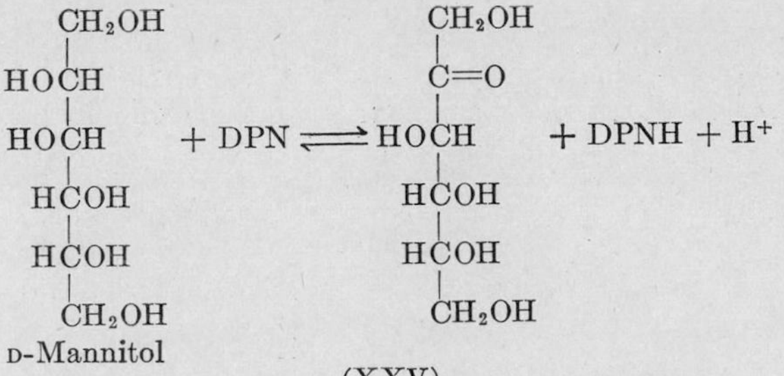

$$
\begin{array}{c}
\text{CH}_2\text{OH} \\
\text{HOCH} \\
\text{HOCH} \\
\text{HCOH} \\
\text{HCOH} \\
\text{CH}_2\text{OH} \\
\text{D-Mannitol}
\end{array}
\quad + \text{DPN} \rightleftharpoons
\begin{array}{c}
\text{CH}_2\text{OH} \\
\text{C=O} \\
\text{HOCH} \\
\text{HCOH} \\
\text{HCOH} \\
\text{CH}_2\text{OH}
\end{array}
\quad + \text{DPNH} + \text{H}^+
$$

(XXV)

A third type of polyol dehydrogenase is galactitol dehydrogenase.[43] This activity has been found in *Pseudomonas* grown on appropriate substrates. This enzyme requires that the hydroxyl groups on carbons 2 and 3 be in the *trans* configuration (XXVI).

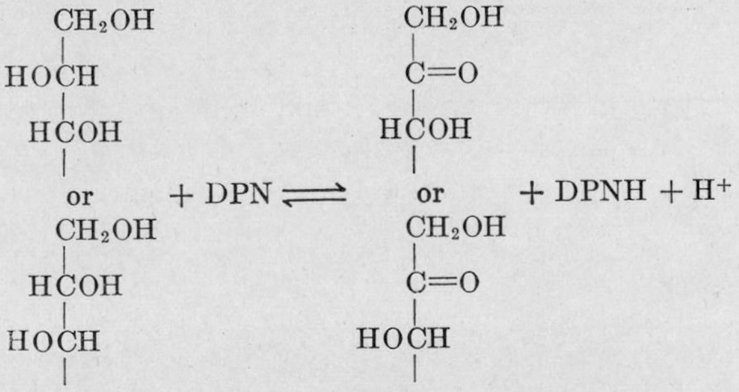

$$
\begin{array}{c}
\text{CH}_2\text{OH} \\
\text{HOCH} \\
\text{HCOH} \\
\text{or} \\
\text{CH}_2\text{OH} \\
\text{HCOH} \\
\text{HOCH}
\end{array}
\quad + \text{DPN} \rightleftharpoons
\begin{array}{c}
\text{CH}_2\text{OH} \\
\text{C=O} \\
\text{HCOH} \\
\text{or} \\
\text{CH}_2\text{OH} \\
\text{C=O} \\
\text{HOCH}
\end{array}
\quad + \text{DPNH} + \text{H}^+
$$

(XXVI)

[42] J. McCorkindale and N. L. Edson, *Biochem. J.* **57**, 518 (1954).
[43] D. R. P. Shaw, *Proc. Univ. Otago Med. School* **32**, 5 (1954).

Another polyol dehydrogenase has been found in seminal gland.[44] This requires TPN and converts sorbitol to glucose (XXVII). The formation of an aldose contrasts with the formation of ketose by the other polyol dehydrogenases. The combined action of the TPN and DPN enzymes concerned with sorbitol metabolism may account for the accumulation of fructose by seminal vesicles.

$$
\begin{array}{ccccc}
\text{CHO} & & \text{CH}_2\text{OH} & & \text{CH}_2\text{OH} \\
| & & | & & | \\
\text{HCOH} & & \text{HCOH} & & \text{C}{=}\text{O} \\
| & & | & & | \\
\text{HOCH} & \xrightleftharpoons[\text{TPN}]{\text{TPNH}} & \text{HOCH} & \xrightleftharpoons[\text{DPNH}]{\text{DPN}} & \text{HOCH} \\
| & & | & & | \\
\text{HCOH} & & \text{HCOH} & & \text{HCOH} \\
| & & | & & | \\
\text{HCOH} & & \text{HCOH} & & \text{HCOH} \\
| & & | & & | \\
\text{CH}_2\text{OH} & & \text{CH}_2\text{OH} & & \text{CH}_2\text{OH} \\
\text{Glucose} & & \text{Sorbitol} & & \text{Fructose} \\
& & \text{(XXVII)} & &
\end{array}
$$

An enzyme found in *E. coli* converts hexose phosphate to polyol phosphate (XXVIII).[45] Evidence has been found for additional enzymes of this type that oxidize phosphate esters of various polyols.

$$
\begin{array}{ccc}
\text{CH}_2\text{OH} & & \text{CH}_2\text{OH} \\
| & & | \\
\text{C}{=}\text{O} & & \text{HOCH} \\
| & & | \\
\text{HOCH} & & \text{HOCH} \\
| & +\ \text{DPNH} + \text{H}^+ \rightleftharpoons & | \\
\text{HCOH} & & \text{HCOH} \\
| & & | \\
\text{HCOH} \quad \text{O} & & \text{HCOH} \quad \text{O} \\
| \qquad \parallel & & | \qquad \parallel \\
\text{CH}_2{-}\text{O}{-}\text{P}{-}\text{O}^- & & \text{CH}_2{-}\text{O}{-}\text{P}{-}\text{O}^- \ +\ \text{DPN} \\
\qquad | & & \qquad | \\
\qquad \text{O}^- & & \qquad \text{O}^- \\
\text{Fructose-6-phosphate} & & \text{Mannitol-1-phosphate} \\
& \text{(XXVIII)} &
\end{array}
$$

In this chapter a number of variations employed for the metabolism of carbohydrates has been described. These illustrate the many mechanisms devised to support the metabolism of organisms at the expense of fermentation or oxidation of carbohydrate. Many more individual reactions are known. For example, several additional reactions for metabolizing various pentoses have been reported. The specificity of various enzyme systems is shown by oxidation of free arabinose by *Pseudomonas*, which follows the general mechanism described for glucose phosphate in this organism, but the dehydrogenase and lactonizing

[44] H. G. Hers, *Biochim. et Biophys. Acta* **22**, 202 (1956).
[45] J. B. Wolff and N. O. Kaplan, *J. Biol. Chem.* **218**, 849 (1956).

enzyme are specific for the substrate on which the cells are grown, glucose or arabinose. Additional specific enzymes will undoubtedly be found responsible for synthesis and degradation of the rarer sugars. Some additional reactions of sugars will be described in the chapters on polysaccharides and uridine nucleotides. Many compounds related to carbohydrates and the Krebs cycle acids are known to occur biologically, and enzymes for forming and destroying compounds, such as glycolic, glyoxylic, tartaric, citramalic, dihydroxyfumaric, hydroxypyruvic, oxalic, and other acids have already been found. The imminent description of additional enzymes responsible for the metabolism of many more compounds is to be anticipated.

GENERAL REFERENCES

Gunsalus, I. C., Horecker, B. L., and Wood, W. A. (1955). *Bacteriol. Revs.* **19**, 79.
Horecker, B. L. (1953). *Brewers Dig.* **28**, 214.
Horecker, B. L., and Mehler, A. H. (1955). *Ann. Rev. Biochem.* **24**, 207.
Racker, E. (1954). *Advances in Enzymol.* **15**, 141.

Oxidation of Fatty Acids

The fatty acids of biological importance include the homologous series of straight-chain acids, beginning with formic acid and continuing through acetic acid to compounds with more than twenty carbon atoms. The naturally occurring fatty acids are predominantly even-numbered. In addition to the straight-chain saturated compounds, long-chain acids with unsaturated bonds, side chains, and other substitutions occur. The metabolism of such specific modified fatty acids of various organisms has, in general, not been studied at the enzyme level. Within the last few years, on the other hand, the enzymatic mechanism of oxidation of the saturated series has been studied in great detail. It appears that all of the compounds of this series, with the exception of formic and propionic acids, are handled by one mechanism. These two exceptions will be considered separately.

Formation of Ketone Bodies. Fatty acids have been known for many years to be a principal fuel for biological oxidations, being converted to carbon dioxide and water. The so-called ketone bodies, acetone, aceto-

$$CH_3(CH_2)_{16}COOH + 26O_2 \rightarrow 18CO_2 + 18H_2O$$

acetic acid, and β-hydroxybutyric acid, were also found to be produced

$$\underset{\text{Acetone}}{CH_3-\overset{\overset{\displaystyle O}{\|}}{C}-CH_3} \qquad \underset{\text{Acetoacetic acid}}{CH_3-\overset{\overset{\displaystyle O}{\|}}{C}-CH_2COOH} \qquad \underset{\beta\text{-Hydroxybutyric acid}}{CH_3-\overset{\overset{\displaystyle OH}{|}}{C}H-CH_2COOH}$$

by fatty acid oxidation. An early form of labeling was used by Knoop to study this process in intact animals.[1] He found that ω-phenyl fatty acids were oxidized; even-numbered chains became phenylacetic acid, and odd-numbered chains yielded benzoic acid (I). These results were interpreted as evidence for oxidation at the β-carbon, with removal of successive 2-carbon units. This so-called β-oxidation product was considered to remove acetate groups until a 4-carbon unit remained. The situation became more complicated when it was learned that more than one molecule of acetoacetate could be formed from a fatty acid molecule.[2]

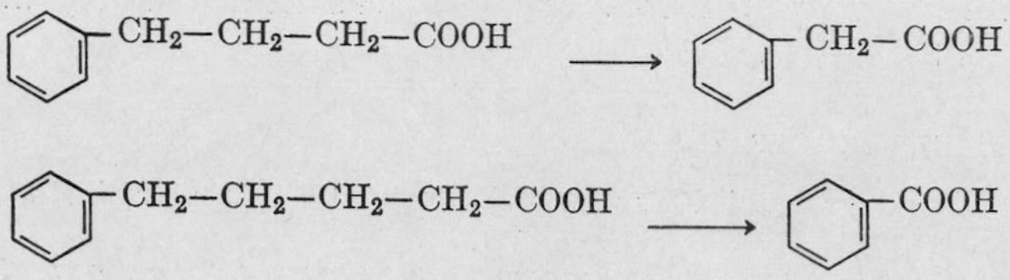

(I)

The mechanism of acetoacetate formation was then studied by isotopic methods. At first, experiments by Weinhouse and collaborators[3] using liver slices seemed to indicate a random recombination of 2-carbon fragments to result in acetoacetate formation. Octanoate labeled with C^{13} in the carboxyl group was found to produce acetoacetate with almost equal label in the carboxyl and carbonyl groups. Later work with liver

$$CH_3(CH_2)_6 C^{13}OO^- \rightarrow CH_3 \overset{\overset{\displaystyle O}{\|}}{C^{13}} \rightarrow CH_3 C^{13}OCH_2 C^{13}OO^-$$

homogenates indicated a differential handling of the carboxyl and methyl ends of a fatty acid molecule, with only partial randomization.[4] When C-7-labeled octanoate was studied, ratios of carbonyl: carboxyl label in acetoacetate as high as 3.8 were found.

The finding that carboxyl groups of fatty acids tend to remain carboxyl groups in acetoacetate and that methyl groups tend to remain methyl groups supported speculation that two types of "active acetate" were formed, with different propensities for reaction, but both capable of entering either end of acetoacetate. More recent studies on the chemistry of fatty acid oxidation have found evidence for only one 2-carbon

[1] F. Knoop, Beitr. chem. Physiol. Pathol. 6, 150 (1904).
[2] M. Jowett and J. H. Quastel, Biochem. J. 29, 2159 (1935).
[3] S. Weinhouse, G. Medes, and N. F. Floyd, J. Biol. Chem. 155, 143 (1944); J. M. Buchanan, W. Sakami, and S. Gurin, ibid. 169, 411 (1947).
[4] D. I. Crandall and S. Gurin, J. Biol. Chem. 181, 829 (1949); D. I. Crandall, R. O. Brady, and S. Gurin, ibid. 181, 845 (1949).

product, acetyl CoA, and have explained deviations from random recombination in kinetic terms.

Fatty Acid Oxidation by Cell-Free Preparations. An important development in the study of fatty acid oxidation was made by Munoz and Leloir,[5] who found that homogenates of liver oxidize fatty acids when supplemented with adenine nucleotides, inorganic phosphate, Mg^{++}, cytochrome c (a compound involved in electron transport), and a member of the tricarboxylic acid cycle. Others found that washed mitochondria contain all the necessary enzymes for oxidation of fatty acids to either CO_2 or acetoacetate, provided the supplements of Munoz and Leloir were added. Such systems are extremely delicate, and for many years it was considered that maintenance of the intact structure of mitochondria was essential for preservation of fatty acid oxidation.

A requirement for adenine nucleotides suggested the possibility that acyl phosphate formation was a step in activation of fatty acids, but synthetic fatty acid phosphates were found to be inert in fatty acid oxidizing systems.[6] Studies of acetate metabolism culminated in the demonstration of acetyl CoA as the reactive molecule in condensations involving acetate, including the formation of citrate that precedes further oxidation.

When it was found that two molecules of acetyl CoA participate in the formation of acetoacetate, that acyl phosphates support the oxidation of short-chain fatty acids catalyzed by extracts of *Clostridium kluyveri*, and that phosphotransacetylase and CoA transphorase were present in these extracts, Barker proposed a scheme of fatty acid oxidation in which acyl CoA compounds served as substrates.[7a] The reactions postulated by Barker for the conversion of butyrate to acetyl groups are essentially those that were subsequently found to participate in fatty acid oxidation in animals and other organisms. With the demonstration that CoA is involved in butyrate metabolism in *Clostridium kluyveri*, several laboratories simultaneously, independently and successfully studied the role of CoA in fatty acid metabolism.

The chemical nature of acetyl CoA as a thioester was discovered by Lynen and Reichert,[7b] who isolated this compound from yeast. This discovery provided great impetus to the study of the various reactions in which CoA is involved. Extracts of acetone-dried liver mitochondria were found to oxidize fatty acids when supplemented with ATP and CoA.[8]

[5] J. M. Munoz and L. F. Leloir, *J. Biol. Chem.* **147**, 355 (1943).

[6] A. L. Lehninger and E. P. Kennedy, *J. Biol. Chem.* **173**, 753 (1948).

[7a] H. A. Barker *in* "Phosphorus Metabolism" (W. D. McElroy and B. Glass, eds.), Vol. I, p. 204. Johns Hopkins Press, Baltimore, 1951.

[7b] F. Lynen and E. Reichert, *Angew. Chem.* **63**, 47 (1951).

[8] G. R. Drysdale and H. A. Lardy, *J. Biol. Chem.* **202**, 119 (1953).

Within 2 years the essential reactions of fatty acid oxidation were discovered. Known enzymes account not only for fatty acid oxidation, but also for the special metabolism of the ketone bodies.

Fatty Acid Oxidation Spiral. The individual reactions of fatty acid oxidation require the activation of the fatty acid by formation of a thioester with CoA. Subsequent oxidation, dehydration, oxidation, and cleavage all occur only with the acyl CoA compounds (II). The product

(1) $RCH_2CH_2CH_2COO^- + CoA + ATP \rightleftharpoons RCH_2CH_2CH_2COCoA + AMP + PP$

(2) $RCH_2CH_2CH_2COCoA + FAD \rightleftharpoons RCH_2CH=CHCOCoA + FADH_2$

(3) $RCH_2CH=CHCOCoA + H_2O \rightleftharpoons RCH_2CHOHCH_2COCoA$

(4) $RCH_2CHOHCH_2COCoA + DPN \rightleftharpoons RCH_2COCH_2COCoA + DPNH$

(5) $RCH_2COCH_2COCoA + CoA \rightleftharpoons RCH_2COCoA + CH_3COCoA$

(II)

of reaction (5) differs from the substrate of reaction (2) only in containing two less methylene groups. A second series of reactions (2) through (5) may therefore occur, and this series may be repeated until the fatty acid is completely converted to acetyl CoA. The generation of substrate for repetitive operation of a series of reactions resembles the Krebs cycle and the pentose phosphate pathway. These cyclic processes, however, regenerate identical substrates, whereas fatty acid oxidation offers a shorter chain at each repetition. It is more appropriate to describe fatty acid oxidation as a spiral, rather than cyclic, process (Fig. 21). The number of turns to the spiral is half the number of carbon atoms in the fatty acid.

Following this sequence, the acetyl CoA produced may enter any of the systems that use acetyl CoA. If the spiral oxidizes the 4-carbon butyric acid CoA thioester, or if reaction (5) is reversed with acetyl CoA as sole substrate, the product is acetoacetyl CoA. Two additional enzymes have been found to influence acetoacetate metabolism. One is a hydrolytic enzyme, the deacylase, that converts acetoacetyl CoA to acetoacetate and CoA.[9] The other is the specific CoA-transferring enzyme that forms acetoacetyl CoA from succinyl CoA and acetoacetate, acetoacetyl-succinic thiophorase.[9,10a]

Fatty Acid Activation. The activation of fatty acids [reaction (1)] is catalyzed by enzymes that are specific for various chain lengths. The first of these, the acetate-activating enzyme, acetic thiokinase,[10b] has

[9] J. R. Stern, M. J. Coon, and A. del Campillo, *Nature* **171**, 28 (1953).

[10a] D. E. Green, D. S. Goldman, S. Mii, and H. Beinert, *J. Biol. Chem.* **202**, 137 (1953).

[10b] The enzymes involved in fatty acid oxidation have each been given several names, as no two groups of investigators happened to invent the same name for a given activity. An international committee has devised a system of terminology for both systematic and trivial nomenclature. H. Beinert, D. E. Green, P. Hele, O. Hoffman-Ostenhof, F. Lynen, S. Ochoa, G. Popjak, and R. Ruyssen, *Science* **124**, 614 (1956).

already been discussed (p. 86). This enzyme has been partially purified from yeast, liver, and heart muscle, and activates propionate in addition to acetate. A second activating enzyme was isolated from beef liver.[11] This enzyme differs from the acetate activating enzyme primarily in its substrate requirements. It does not activate acetate or propionate; butyrate and higher acids through C_{12} are attacked with a maximum

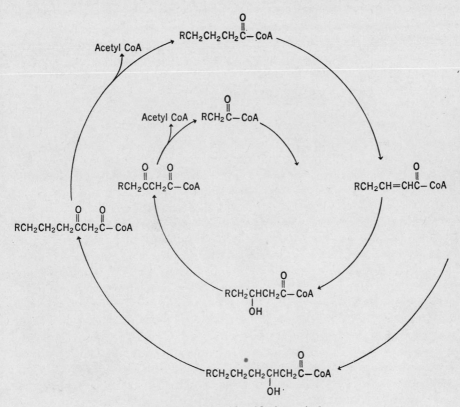

FIG. 21. Fatty acid oxidation spiral.

rate at C_7. Hydroxylated, unsaturated, methyl substituted, and aromatic acids are all activated. Higher fatty acids (C_5 to C_{22}) are converted to their CoA thioesters by a third enzyme, also found in liver.[12] An enzyme named CoA transphorase has been found in *Clostridium kluyveri;* this transfers CoA from one small-chain organic acid to another.[13]

[11] H. R. Mahler, S. J. Wakil, and R. J. Bock, *J. Biol. Chem.* **204,** 453 (1953).

[12] A. Kornberg and W. E. Pricer, Jr., *J. Biol. Chem.* **204,** 329 (1953).

[13] E. R. Stadtman, *J. Biol. Chem.* **203,** 501 (1953); D. E. Green, S. Mii, and H. R. Mahler, *ibid.* **206,** 1 (1953).

Acyl CoA Dehydrogenases. Reaction (2) is catalyzed by a family of related enzymes. The lower fatty acid derivatives are oxidized by an enzyme purified from beef liver mitochondria. The isolated enzyme, butyryl dehydrogenase, has been reported to contain both riboflavin and copper.[14] The nature of this enzyme will be discussed later in conjunction with other flavin enzymes. Chains of 3 to 8 carbons are oxidized in the presence of suitable electron acceptors, including 2,6-dichlorophenolindophenol, cytochrome c, and ferricyanide. The maximum rate is obtained with butyryl CoA. It has been reported that the initial oxidation of fatty acid-CoA and the reduction of the employed electron acceptors are carried out by two different flavoproteins, which react with each

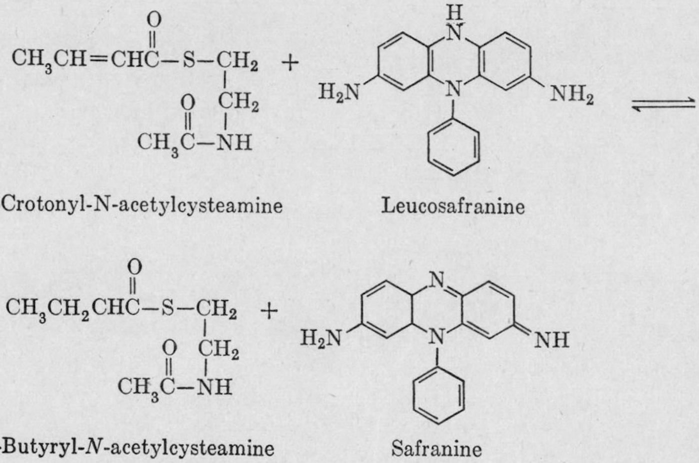

S-Crotonyl-N-acetylcysteamine   Leucosafranine

S-Butyryl-N-acetylcysteamine   Safranine
Ethylene Reductase Reaction
(III)

other in an intermediary step. The first step can be visualized as a bleaching of the dehydrogenase in the presence of its substrate. This reaction is not catalytic, since only an amount of acyl CoA equivalent to the enzyme can be oxidized until the enzyme finds an acceptor for its electrons. The addition of the second flavoprotein, ETF (electron-transferring flavoprotein), permits electrons to be passed to various acceptors,[15] and the dehydrogenase then acts catalytically. Oxidation of longer chain fatty acids is carried out by an enzyme similar to butyryl-dehydrogenase. This enzyme, yellow in color, was originally studied in the reverse reaction with leucosafranine as electron donor and was called ethylene reductase (III).[16] Because of the difficulty in obtaining substrate quanti-

[14] H. R. Mahler, *J. Biol. Chem.* **206**, 13 (1953).
[15] F. L. Crane and H. Beinert, *J. Biol. Chem.* **218**, 717 (1956).
[16] W. Seubert and F. Lynen, *J. Am. Chem. Soc.* **75**, 2787 (1953). A similar enzyme was also prepared by H. Beinert and F. L. Crane, *Federation Proc.* **13**, 181 (1954).

ties of CoA, this reaction was attempted with CoA analogs, and, re-markably, was found to proceed well with thioesters of N-acetylthio-ethanolamine. In all of these reactions oxidation or reduction of the dyes was measured spectrophotometrically.

Crotonase. The hydration of unsaturated acyl CoA thioesters is a readily reversible reaction, with almost equal concentrations of β-hydroxy and unsaturated compounds present at equilibrium.[17a] The enzyme that catalyzes this reaction was named unsaturated fatty acyl coenzyme A hydrase by one group, and crotonase by another. The systematic name enoyl hydrase has been recommended for enzymes of this type, with crotonase being retained as a trivial name for the enzyme most active with crotonyl CoA. Originally this enzyme was studied by measuring the oxidative reaction involving the β-hydroxy derivative formed. Later a direct assay was developed based on the decrease in absorption at 263 mμ on hydration. Originally the hydrase was reported to be specific for the *trans* configuration. A recent report suggests that enzymes exist for hydrating both *cis* and *trans* bonds, forming L- and D-β-hydroxyacyl CoA, respectively.[17b] Crotonase has been crystallized by Stern *et al.*,[18] who concluded that this enzyme is capable of catalyzing a *cis-trans* isomerization, and that the only product of hydration is the L-(+)-β-hydroxyacyl CoA.

Hydroxyacyl CoA Dehydrogenase. The oxidation of β-hydroxyacyl thioesters was studied in opposite directions in different laboratories,[19] hence was named both β-hydroxyacyl coenzyme A dehydrogenase and β-keto reductase. The systematic name β-hydroxyacyl dehydrogenase has been recommended. The enzyme is specific for DPN, and in general the change in absorption at 340 mμ is used to follow this reaction. The equilibrium constant for the reaction with CoA derivatives was reported to be 2.5–6.3×10^{-11} for 6 and 4 carbon acids, while a value of about 2×10^{-10} was obtained with the N-acetylthioethanolamine analog. It is not known whether these differences are real or represent experimental error. In any event, it is apparent that in the case of this dehydrogenase, as for others discussed previously, at neutral pH values reduction of the keto acid is favored, while at higher pH values oxidation of the hydroxy compound becomes more favorable. The enzyme is active above pH 10, so that the reaction can be studied in the direction of DPN reduction.

[17a] J. Stern and A. del Campillo, *J. Am. Chem. Soc.* **75**, 2277 (1953); S. J. Wakil and H. R. Mahler, *J. Biol. Chem.* **207**, 125 (1954).

[17b] S. J. Wakil, *Biochim. et Biophys. Acta* **19**, 497 (1956).

[18] J. R. Stern, A. del Campillo, and I. Raw, *J. Biol. Chem.* **218**, 971 (1956); J. R. Stern and A. del Campillo, *ibid.* **218**, 985 (1956).

[19] P. Lynen, L. Wessely, O. Wieland, and L. Rueff, *Angew. Chem.* **64**, 687 (1952); S. J. Wakil, D. E. Green, S. Mii, and H. R. Mahler, *J. Biol. Chem.* **207**, 631 (1954).

A β-hydroxybutyric acid dehydrogenase that uses DPN has been known for many years.[20a] It is specific for L-β-hydroxybutyrate, and does not act with the CoA derivative. The original description of the CoA-requiring system included a specific requirement for D-hydroxy acids. Later a "racemase" that converts L-β-hydroxy CoA compounds to D- was reported, but it appears that the "racemase" is the resultant of an L-dehydrogenase together with the previously described D-enzyme. Racemization requires DPN.[20b]

Thiolase. The reversible cleavage of β-ketothioesters is catalyzed by an enzyme named β-ketothiolase.[9,10,19] This enzyme was detected by measuring coupled reactions. Acetyl CoA was found to condense with itself to form a substrate for the keto reductase, and the combined reaction was followed as a decrease in DPNH (IV). Similarly, crotonyl CoA

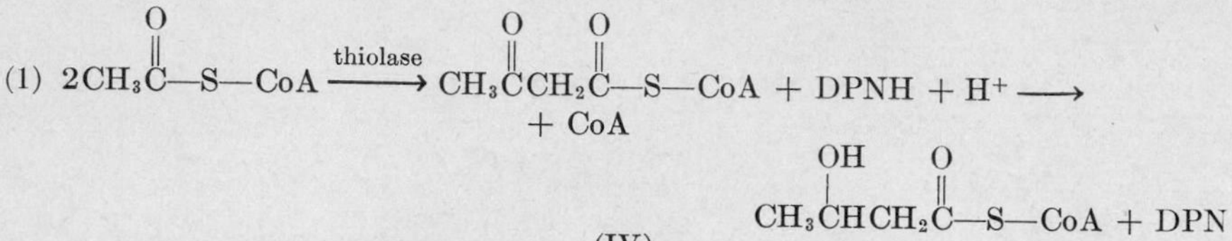

$$(1) \quad 2CH_3\overset{\displaystyle O}{\overset{\|}{C}}-S-CoA \xrightarrow{\text{thiolase}} CH_3\overset{\displaystyle O}{\overset{\|}{C}}CH_2\overset{\displaystyle O}{\overset{\|}{C}}-S-CoA + DPNH + H^+ \longrightarrow + CoA$$

$$CH_3\overset{\displaystyle OH}{\overset{|}{C}}HCH_2\overset{\displaystyle O}{\overset{\|}{C}}-S-CoA + DPN$$

$$(IV)$$

was hydrated and oxidized to a product that could be used ultimately to form citrate (V). More direct assays include measuring the decrease

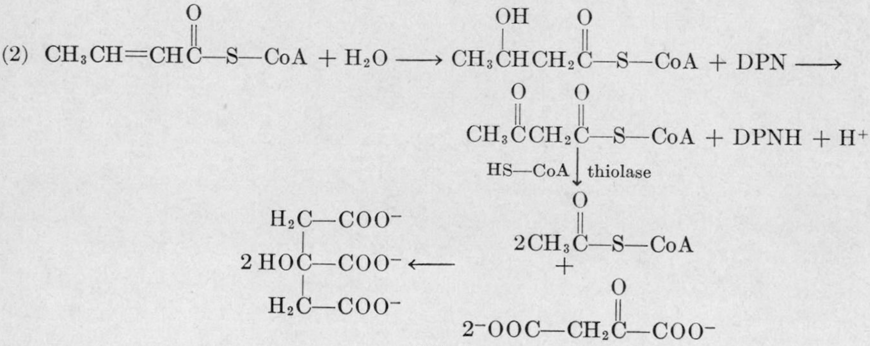

$$(2) \quad CH_3CH{=}CH\overset{\displaystyle O}{\overset{\|}{C}}-S-CoA + H_2O \longrightarrow CH_3\overset{\displaystyle OH}{\overset{|}{C}}HCH_2\overset{\displaystyle O}{\overset{\|}{C}}-S-CoA + DPN \longrightarrow$$

$$CH_3\overset{\displaystyle O}{\overset{\|}{C}}CH_2\overset{\displaystyle O}{\overset{\|}{C}}-S-CoA + DPNH + H^+$$

$$HS{-}CoA \downarrow \text{thiolase}$$

$$\begin{array}{c} H_2C{-}COO^- \\ | \\ 2\,HOC{-}COO^- \longleftarrow \\ | \\ H_2C{-}COO^- \end{array} \qquad \begin{array}{c} \overset{\displaystyle O}{\overset{\|}{ }} \\ 2CH_3C{-}S{-}CoA \\ + \\ \overset{\displaystyle O}{\overset{\|}{ }} \\ 2{}^-OOC{-}CH_2C{-}COO^- \end{array}$$

$$(V)$$

in absorption at 305 mμ as acetoacetyl CoA is split or the corresponding increase at 240 mμ caused by the increase in thioester. The failure of β-keto CoA compounds to yield a color in the hydroxamic acid reaction can also be used to distinguish acetoacetyl CoA from acetyl CoA, which yields a colored ferric complex of acethydroxamic acid. The equilibrium

[20a] D. E. Green, J. D. Dewan, and L. F. Leloir, *Biochem. J.* **31**, 934 (1937).
[20b] S. J. Wakil, *Biochim. et Biophys. Acta* **18**, 314 (1955).

of the cleavage reaction is far to the side of cleavage. Purified β-keto-thiolase is active only with short-chain compounds, and has been called acetoacetylthiolase, but crude preparations have been found to be active with higher homologs, implying the existence of one or more similar enzymes active with long-chain acyl CoA compounds.

Acetoacetate Metabolism. An active deacylase in liver is responsible for the formation of free acetoacetate from its CoA derivative. The β-hydroxybutyric dehydrogenase mentioned above and a decarboxylase are capable of converting acetoacetate into the other "ketone bodies," β-hydroxybutyrate, and acetone. Liver does not contain a mechanism for activating acetoacetate. Heart muscle has been found to contain a specific thiophorase that forms acetoacetyl CoA at the expense of succinyl CoA. Acetoacetate is thus used by peripheral tissues by activation through transfer, then reaction with either the enzymes of fatty acid synthesis or β-ketothiolase and the enzymes that use acetyl CoA.

Krebs Cycle and Fatty Acid Oxidation. A possible role of Krebs cycle intermediates in supporting fatty acid oxidation is now apparent. Complete oxidation to CO_2 requires oxalacetate to introduce acetyl CoA into the citric acid cycle. But even the formation of acetoacetate requires the continued generation of ATP to support the activation of fatty acids. The transfer of electrons from fatty acid to oxygen is coupled with phosphate esterification, so that fatty acid oxidation has the theoretical capacity to be self-supporting. In the crude systems that contain all of the essential factors for fatty acid oxidation, fatty acid activation must compete with other reactions for the available ATP, and maximum rates of oxidation occur only when additional ATP is generated through operation of the Krebs cycle.

Synthesis of Fatty Acids. All of the reactions of the fatty acid oxidation spiral are reversible, and theoretically this series of reactions could account for both synthesis and degradation of fatty acids. Experimentally it has not been possible to force the reversal through more than one loop of the spiral, and long-chain fatty acids have not been made from acetyl CoA. The incorporation of labeled acetate into fatty acids was observed with soluble preparations of pigeon liver;[20c] this finding supported earlier suggestions that the oxidizing and synthesizing systems are not identical. Another system has been found to form long chain fatty acids, but the individual steps in the overall synthesis have not been elucidated yet.[21] Whereas the enzymes of fatty acid oxidation occur in cell particles, the enzymes that incorporate acetate into long chains appear to be soluble. Soluble systems from mammary gland require CoA, ATP,

[20c] R. O. Brady and S. Gurin, *J. Biol. Chem.* **199**, 421 (1952).
[21] G. Popják and A. Tietz, *Biochem. J.* **56**, 46 (1954).

and DPNH for fatty acid synthesis.[22] The reaction has been followed spectrophotometically by measuring the disappearance of DPNH and C^{14}-labeled acetate has been used to show incorporation of acetate into both short- and long-chain fatty acids. A similar system from liver appears to use TPNH instead of DPNH.[23] A soluble fatty-acid–synthesizing system from pigeon liver has been separated into three protein fractions, and requires as cofactors ATP, Mg^{++}, CoA, DPN, TPN, lipoic acid, glutathione, a sugar phosphate, and isocitrate. This complicated system catalyzes the net synthesis of higher fatty acids from acetate.[24] It is possible that the enzymes of the soluble systems differ from those of mitochondria only in their affinities for various chain lengths, but it seems more probable that the reaction sequences are not identical.

The Thunberg Condensation. An alternate to the tricarboxylic acid cycle has been proposed many times as a mechanism for acetate (or acetyl CoA) oxidation. The suggestion was originally made by Thunberg that two acetates could condense to form succinate, which could be oxidized to acetate and CO_2. This Thunberg condensation has been invoked repeatedly to explain various observations, and it is indeed

$$\begin{array}{ccc} COO^- & & COO^- \\ | & & | \\ C^{13}H_3 & -2H & C^{13}H_2 \\ + & \longrightarrow & | \\ C^{13}H_3 & & C^{13}H_2 \\ | & & | \\ COO^- & & COO^- \end{array}$$

Formation of Doubly Labeled Succinate by Thunberg Mechanism

(VI)

possible that such a reaction will be found. However, the most promising examples have been shown to involve more devious routes. A critical test was applied by Krampitz and co-workers.[25] They made use of an elegant mass spectrophotometric technique developed by Wood[26] for distinguishing between succinate labeled in one methylene carbon atom with C^{13} and succinate doubly labeled. In this technique the succinate is converted to ethylene, and the individual components of a mixture, $C^{12}C^{12}$, $C^{12}C^{13}$, and $C^{13}C^{13}$ are determined in the mass spectrometer. If highly labeled $C^{13}H_3COOH$ is converted to succinate, a certain amount of the $C^{13}C^{13}$ species would be expected from the Thunberg condensation (VI). Instead, none of this molecule was found; only the distribution

[22] A. Tietz and G. Popják, *Biochem. J.* **60,** 155 (1955).

[23] R. E. Langdon, *J. Am. Chem. Soc.* **77,** 5190 (1955).

[24] S. J. Wakil, J. W. Porter, and A. Tietz, *Federation Proc.* **15,** 377 (1956).

[25] H. J. Saz and L. O. Krampitz, *J. Bacteriol.* **67,** 409 (1954); H. E. Swim and L. O. Krampitz, *ibid.* **67,** 426 (1954).

[26] H. G. Wood, *J. Biol. Chem.* **194,** 905 (1952).

$$
\begin{array}{c}
\text{COO}^- \\
| \\
\text{C}^{13}\text{H}_3 \\
+ \\
\text{COO}^- \\
| \\
\text{C}{=}\text{O} \\
| \\
\text{CH}_2 \\
| \\
\text{COO}^-
\end{array}
\longrightarrow
\begin{array}{c}
\text{H}_2\text{C}^{13}{-}\text{COO}^- \\
| \\
\text{HOC}{-}\text{COO}^- \\
| \\
\text{H}_2\text{C}{-}\text{COO}^-
\end{array}
\longrightarrow
\begin{array}{c}
\text{H}_2\text{C}^{13}{-}\text{COO}^- \\
| \\
\text{HC}{-}\text{COO}^- \\
| \\
\text{HOC}{-}\text{COO}^-
\end{array}
\longrightarrow
\begin{array}{c}
\text{H}_2\text{C}^{13}{-}\text{COO}^- \\
| \\
\text{H}_2\text{C} \\
| \\
\text{O}{=}\text{C}{-}\text{COO}^-
\end{array}
\longrightarrow
\begin{array}{c}
\text{COO}^- \\
| \\
\text{C}^{13}\text{H}_2 \\
| \\
\text{CH}_2 \\
| \\
\text{COO}^-
\end{array}
$$

Formation of Singly Labeled Succinate by Krebs Cycle
(VII)

expected from the operation of the Krebs cycle appeared (VII). A reaction was recently found in *E. coli* in which acetyl CoA condenses with glyoxylate to form malate (VIII).[27a] The importance of this reaction has

$$
\begin{array}{c}
\text{O} \\
\|\\
\text{C}{-}\text{S}{-}\text{CoA} \\
| \\
\text{CH}_3
\end{array}
+
\begin{array}{c}
\text{O} \\
\|\\
\text{C}{-}\text{O}^- \\
| \\
\text{HC}{=}\text{O}
\end{array}
\longrightarrow
\begin{array}{c}
\text{O} \\
\|\\
\text{C}{-}\text{O}^- \\
| \\
\text{HCOH} \\
| \\
\text{CH}_2 \\
| \\
\text{COO}^-
\end{array}
+ \text{CoA}
$$

(VIII)

yet to be evaluated. At the present time no mechanism is known for forming glyoxylate directly from acetate. If the glyoxylate were formed from isocitrate in the isocitritase reaction, it would contain carbon from preformed dicarboxylic acid, not from acetate. A modified tricarboxylic acid cycle has been proposed to include this reaction, by-passing the formation of α-ketoglutarate. In this scheme, acetyl CoA + glyoxylate form malate. Oxidation of malate forms oxalacetate, which condenses with another acetyl CoA to form citrate. Aconitase and isocitritase convert the citrate to succinate and glyoxylate. The net reaction is the formation of succinate and DPNH from two acetyl CoA.[27b]

Formate Oxidation. Formate is oxidized to CO_2. A DPN-specific dehydrogenase has been found in plants.[28] The nature of microbial formic dehydrogenase is not known. The only animal enzyme known to oxidize

$$
\begin{array}{c}
\text{O} \\
\|\\
\text{H}{-}\text{C}{-}\text{O}^-
\end{array}
+ \text{DPN} \longrightarrow CO_2 + \text{DPNH}
$$

formate is catalase (which will be discussed later, Chapter IV). A reaction ascribed to an enzyme called formic hydrogenlyase has been de-

[27a] D. T. O. Wong and S. J. Ajl, *J. Am. Chem. Soc.* **78**, 3230 (1956).
[27b] H. L. Kornberg and H. A. Krebs, *Nature* **179**, 988 (1957).
[28] D. C. Davison, *Biochem. J.* **49**, 520 (1951).

scribed in which formate is split to CO_2 and H_2.[29] Despite years of intensive investigation, it has not been established whether this is catalyzed by a discreet enzyme or by the sum of formic dehydrogenase and a hydrogenase. Formate may be metabolized by several other routes, involving condensation with various molecules. Some of these reactions will be considered later.

Propionate Metabolism. Propionate metabolism is not completely understood. It has been mentioned that some of the activating systems can form propionyl CoA. It has been suggested that the oxidation proceeds via lactate and pyruvate.[30] Another mechanism has been found in which an incompletely understood CO_2 fixation results in the formation of methyl malonic, then succinic acid (IX).[31] This reaction has been

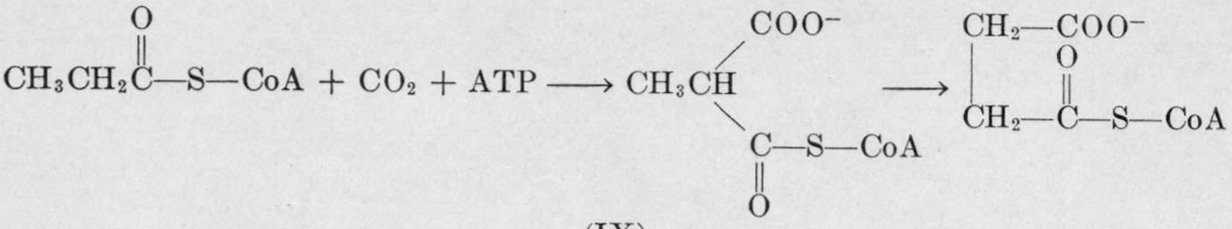

$$CH_3CH_2\overset{O}{\overset{\|}{C}}-S-CoA + CO_2 + ATP \longrightarrow CH_3CH\begin{array}{l} COO^- \\ \\ C-S-CoA \\ \| \\ O \end{array} \longrightarrow \begin{array}{l} CH_2-COO^- \\ | \quad\quad O \\ CH_2-\overset{\|}{C}-S-CoA \end{array}$$

(IX)

studied in the reverse direction as the process by which certain microorganisms convert succinate to propionate and CO_2. It is also a mechanism for introducing the terminal group of odd-numbered chains into the citric acid cycle.

A direct oxidation of the 3-carbon chain has been observed with extracts of *Clostridium propionicum*.[32] The thioester propionyl pantetheine was used as a more available substrate than propionyl CoA, and

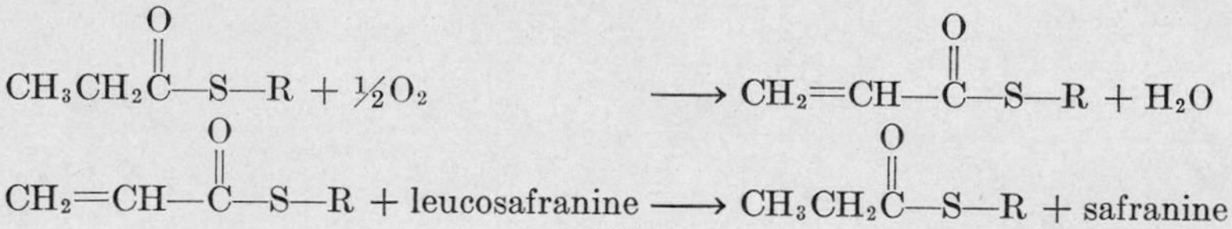

$$CH_3CH_2\overset{O}{\overset{\|}{C}}-S-R + \tfrac{1}{2}O_2 \longrightarrow CH_2{=}CH-\overset{O}{\overset{\|}{C}}-S-R + H_2O$$

$$CH_2{=}CH-\overset{O}{\overset{\|}{C}}-S-R + \text{leucosafranine} \longrightarrow CH_3CH_2\overset{O}{\overset{\|}{C}}-S-R + \text{safranine}$$

(X)

was oxidized in the presence of molecular oxygen. The oxidation product behaves like the very reactive acrylic acid derivative, adding sulfhydryl compounds nonenzymatically across the double bond. In the presence of ammonia the oxidation product is converted by extracts of *C. propioni-*

[29] M. Stephenson and L. H. Stickland, *Biochem. J.* **26**, 712 (1932).
[30] F. M. Huennekens, H. R. Mahler, and J. Nordmann, *Arch. Biochem. and Biophys.* **30**, 66, 77 (1951).
[31] H. A. Lardy, *Proc. Natl. Acad. Sci. U.S.* **38**, 1003 (1952); H. R. Whitely, *ibid.* **39**, 772, 779 (1953); E. A. Delwiche, E. F. Phares, and S. F. Carson, *Federation Proc.* **13**, 198 (1954).
[32] E. R. Stadtman, *Federation Proc.* **15**, 360 (1956).

cum to a β-alanyl thioester. Further evidence that propionyl thioester is dehydrogenated to the acrylyl compound comes from the demonstration of the reduction of acrylyl pantetheine by the bacterial extracts with leucosafranine as electron donor (X).

Branched Chain Acids. Three of the amino acids found as usual components of proteins, valine, leucine, and isoleucine, contain branched chains. The amino groups are lost in the first step of degradation (see Chapter VII), and the resulting α-keto acids are metabolized by specific pathways related to the pathways of straight-chain fatty acid metabolism.

Removal of the amino group of valine results in the formation of α-ketoisovalerate.[33] In a subsequent reaction that resembles the oxidation of pyruvate, isobutyryl CoA is formed.[34] Isobutyryl CoA is oxidized to

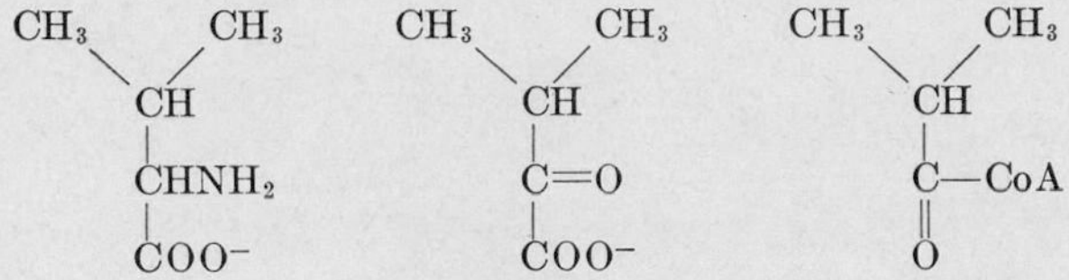

$$
\begin{array}{ccc}
CH_3 \diagdown \diagup CH_3 & CH_3 \diagdown \diagup CH_3 & CH_3 \diagdown \diagup CH_3 \\
CH & CH & CH \\
| & | & | \\
CHNH_2 & C{=}O & C{-}CoA \\
| & | & \| \\
COO^- & COO^- & O \\
\text{Valine} & \text{α-Ketoisovalerate} & \text{Isobutyryl CoA} \\
& \text{(XI)} &
\end{array}
$$

the unsaturated ester, methacrylyl CoA, presumably by the same flavoprotein that oxidizes intermediate length straight-chain acyl CoA derivatives.[35] The unsaturated ester is hydrated by crotonase, to become β-hydroxyisobutyryl CoA.[34] At this point, the CoA is removed by a specific hydrolytic enzyme, HIB CoA deacylase.[35]

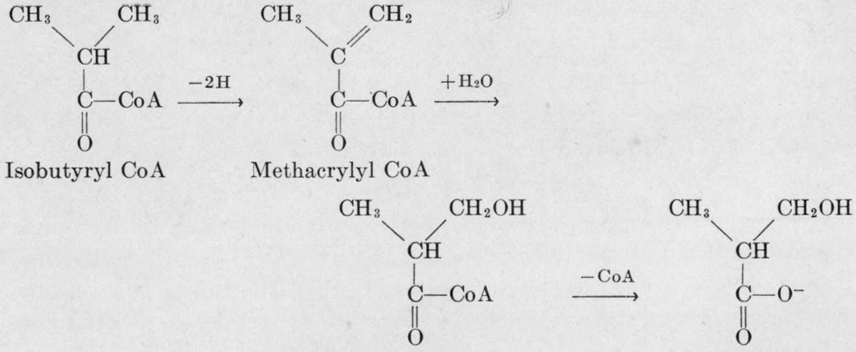

$$
\begin{array}{ccc}
CH_3 \diagdown \diagup CH_3 & CH_3 \diagdown \diagup CH_2 & \\
CH & C & \\
| & \| & \\
C{-}CoA \xrightarrow{-2H} & C{-}CoA \xrightarrow{+H_2O} & \\
\| & \| & \\
O & O & \\
\text{Isobutyryl CoA} & \text{Methacrylyl CoA} &
\end{array}
$$

$$
\begin{array}{cc}
CH_3 \diagdown \diagup CH_2OH & CH_3 \diagdown \diagup CH_2OH \\
CH & CH \\
| & | \\
C{-}CoA \xrightarrow{-CoA} & C{-}O^- \\
\| & \| \\
O & O \\
\text{β-Hydroxyisobutyryl CoA} & \text{β-Hydroxyisobutyrate} \\
\text{(XII)} &
\end{array}
$$

[33] D. S. Kinnory, Y. Takeda, and D. M. Greenberg, *J. Biol. Chem.* **212**, 385 (1955).
[34] W. G. Robinson, R. Nagle, B. K. Bachhawat, F. P. Kupiecki, and M. J. Coon, *J. Biol. Chem.* **224**, 1 (1957).
[35] G. Rendina and M. J. Coon, *J. Biol. Chem.* **225**, 523 (1957).

β-Hydroxisobutyrate is oxidized by a DPN-specific dehydrogenase; TPN is not attacked and the enzyme does not attack any known analogs of either β-hydroxyisobutyrate or the oxidation product, methylmalonate semialdehyde.[36] The further degradation of the branched chain has not

$$
\begin{array}{c}
CH_3 \quad\ CH_2OH \\
\diagdown\diagup \\
CH \\
\mid \\
COO^-
\end{array}
+ DPN \rightleftarrows
\begin{array}{c}
CH_3 \quad\ CHO \\
\diagdown\diagup \\
CH \\
\mid \\
COO^-
\end{array}
+ DPNH + H^+ \quad K_{eq} = 3.0 \times 10^{-4}
$$

(XIII)

been demonstrated enzymatically, but includes loss of the carboxyl group as CO_2 and conversion of the remainder of the molecule to a 3-carbon fragment at the oxidation level of propionate.[33] This process may involve the formation of methylmalonyl CoA, an intermediate in the interconversion of succinate and propionate.[37]

The pathways involved in the oxidation of leucine and isoleucine are similar in many respects, but result in the formation of quite different products.

$$
\begin{array}{c}
CH_3 \quad\quad NH_2 \\
\diagdown\quad\ \mid \\
CHCH_2CHCOO^- \\
\diagup \\
CH_3 \\
\text{Leucine}
\end{array}
\rightarrow
\begin{array}{c}
CH_3 \quad\quad O \\
\diagdown\quad\ \parallel \\
CHCH_2CCOO^- \\
\diagup \\
CH_3 \\
\alpha\text{-Keto-}\gamma\text{-methylvalerate}
\end{array}
\rightarrow
\begin{array}{c}
CH_3 \quad\quad O \\
\diagdown\quad\ \parallel \\
CHCH_2CCoA \\
\diagup \\
CH_3 \\
\text{Isovaleryl CoA}
\end{array}
$$

$$
\begin{array}{c}
\quad\quad NH_2 \\
\quad\quad \mid \\
CH_3CH_2CHCHCOO^- \\
\mid \\
CH_3 \\
\text{Isoleucine}
\end{array}
\rightarrow
\begin{array}{c}
\quad\quad O \\
\quad\quad \parallel \\
CH_3CH_2CHCCOO^- \\
\mid \\
CH_3 \\
\alpha\text{-Keto-}\beta\text{-methylvalerate}
\end{array}
\quad
\begin{array}{c}
\quad\quad O \\
\quad\quad \parallel \\
CH_3CH_2CHCCoA \\
\mid \\
CH_3 \\
\alpha\text{-Methylbutyryl CoA}
\end{array}
$$

(XIV)

Isovaleryl CoA is oxidized in a reaction similar to that found for other acyl CoA compounds to senecioyl CoA. This unsaturated thioester is a substrate for crotonase and is hydrated to β-hydrosyisovaleryl CoA (HIV CoA).[38] An unusual type of reaction converts this product to β-hydroxy-β-methyl glutaryl CoA (HMG CoA). Free CO_2 is not fixed directly, but is first activated in a reaction with ATP. The activating enzyme "H enzyme" was purified as a hydroxylamine and bicarbonate dependent ATP-splitting enzyme; the initial products are believed to be AMP-CO_2 and pyrophosphate.[39] This enzyme is different from a fluoride, bicarbonate dependent ATPase that has been shown to form

[36] W. G. Robinson and M. J. Coon, *J. Biol. Chem.* **225**, 511 (1957).
[37] M. Flavin, P. J. Ortiz, and S. Ochoa, *Nature* **176**, 823 (1955).
[38] B. K. Bachhawat, W. J. Robinson, and M. J. Coon, *J. Biol. Chem.* **219**, 539 (1956).
[39] B. K. Bachhawat, F. P. Kupiecki, and M. J. Coon, *Federation Proc.* **16**, 148 (1957).

fluorophosphate and ADP[40] CO_2 activated by the "H enzyme" reacts with HIV CoA.[38] The carboxylation product, HMG CoA, is split by an enzyme that requires Mg^{++} or Mn^{++} and a thiol compound. The cleavage

$$
\begin{array}{ccccccc}
\underset{\text{CH}_3}{\diagdown} \quad \underset{\text{CH}_3}{\diagup} & & \underset{\text{CH}_3}{\diagdown} \quad \underset{\text{CH}_3}{\diagup} & & \underset{\text{CH}_3}{\diagdown} \quad \underset{\text{CH}_3}{\diagup} & & \underset{\text{CH}_3}{\diagdown} \quad \underset{\text{CH}_2\text{COO}^-}{\diagup} \\
\text{CH} & \rightarrow & \text{C} & \rightarrow & \text{C—OH} & \xrightarrow{\text{``CO}_2\text{''}} & \text{COH} \\
| & & \| & & | & & | \\
\text{CH}_2 & & \text{CH} & & \text{CH}_2 & & \text{CH}_2 \\
| & & | & & | & & | \\
\text{COCoA} & & \text{COCoA} & & \text{COCoA} & & \text{COCoA} \\
\text{Isovaleryl CoA} & & \text{Senecioyl CoA} & & \text{HIV CoA} & & \text{HMG CoA} \\
& & & & \text{(XV)} & &
\end{array}
$$

enzyme is specific for both the HMG component and the CoA of the substrate; it does not attack HIV CoA or other thioesters of HMG. The products are acetoacetate and acetyl CoA.

$$
\begin{array}{ccc}
\underset{\text{OH}}{} \quad \underset{\text{O}}{} & \underset{\text{O}}{} & \underset{\text{O}}{} \\
\text{CH}_3\text{C—CH}_2\text{—C—O}^- \rightarrow & \text{CH}_3\text{C—CH}_2\text{—C—O}^- \; + & \text{CH}_3\text{CCoA} \\
| & & \\
\text{CH}_2\text{CCoA} & & \\
\| & & \\
\text{O} & & \\
\text{HMG CoA} & \text{Acetoacetate} & \text{Acetyl CoA} \\
& \text{(XVI)} &
\end{array}
$$

α-Methylbutyryl CoA is oxidized to tiglyl CoA and hydrated to α-methyl-β-hydroxybutyryl CoA (HMB CoA) in reactions that probably involve the same enzymes that carry out the corresponding reactions in leucine degradation. This hydroxy compound is oxidized by a DPN-

$$
\begin{array}{ccc}
\text{CH}_3\text{CH}_2\text{CHCOCoA} & \text{CH}_3\text{CH}{=}\text{CCOCoA} \rightarrow & \text{CH}_3\text{CH—CHCOCoA} \\
| & | & \quad | \quad \; | \\
\text{CH}_3 & \text{CH}_3 & \text{OH} \;\; \text{CH}_3 \\
\text{α-Methylbutyryl CoA} & \text{Tiglyl CoA} & \text{α-Methyl-β-hydroxybutyryl CoA} \\
& \text{(XVII)} &
\end{array}
$$

requiring dehydrogenase, which may be a nonspecific hydroxy acyl CoA dehydrogenase, to α-methylacetoacetyl CoA. The β-keto thioester is cleaved in a thiolase reaction to acetyl CoA and propionyl CoA.[41]

$$
\begin{array}{cccc}
\text{CH}_3\text{CH—CHCOCoA} + \text{DPN} \rightleftarrows & \text{CH}_3\text{C—CHCOCoA} & \underset{\rightleftharpoons}{\pm \text{CoA}} & \\
| \quad \;\; | & \| \quad \; | & & \\
\text{OH} \;\; \text{CH}_3 & \text{O} \;\; \text{CH}_3 & & \\
\text{HMB CoA} & \text{α-Methylacetoacetyl CoA} & &
\end{array}
$$

$$
\begin{array}{c}
\text{CH}_3\text{COCoA} + \text{CH}_3\text{CH}_2\text{COCoA} \\
\text{Acetyl CoA} \quad \text{Propionyl CoA} \\
\text{(XVIII)}
\end{array}
$$

[40] M. Flavin, H. Castro-Mendoza, and W. S. Beck, *Federation Proc.* **15**, 252 (1956).
[41] W. G. Robinson, B. K. Bachhawat, and M. J. Coon, *J. Biol. Chem.* **218**, 391 (1956).

The three series of reactions described above demonstrate how acetyl CoA, propionyl CoA, and acetoacetate are formed from the three-branched chain amino acids. Through the efforts of Coon and his collaborators, most of the individual reactions by which these products are obtained have been described, and it can be seen how the early steps follow the same pathway, and only at later steps, where the reactive groups differ, do variations appear in the pathways. Thus, at the level of β-hydroxy acyl CoA esters the three pathways diverge; the valine product, β-hydroxyisobutyryl CoA is hydrolyzed; the isoleucine derivative, α-methyl-β-hydroxybutyryl CoA, is oxidized to the corresponding β-keto compound; and the β-hydroxyisovaleryl CoA derived from leucine is carboxylated. Subsequent cleavage reactions result in the formation of two, three, and four carbon fragments. Since the three-carbon propionate can be converted to carbohydrate (via succinate or pyruvate) but acetyl CoA and acetoacetate appear as ketone bodies, these reactions explain the grossly different metabolic fates of the three similar branched chain amino acids.

GENERAL REFERENCES

Lynen, F. (1953). *Federation Proc.* **12,** 683.
Mahler, H. R. (1953). *Federation Proc.* **12,** 694.
Stadtman, E. R. (1954). *Record. Chem. Progr. (Kresge-Hooker Sci. Lib.)* **15,** 1.
Stern, J. R., del Campillo, A., and Raw, I. (1956). *J. Biol. Chem.* **218,** 971.

Chapter IV

BIOLOGICAL OXIDATION: TRANSFER OF OXYGEN, HYDROGEN, AND ELECTRONS

Pyridine Nucleotides

Pyridine nucleotides participate in most biological oxidations and reductions at the substrate level. The two naturally occurring pyridine nucleotides are DPN and TPN. The structure and enzymatic synthesis of DPN were discussed in the section on glycolysis. It was recognized during the original isolation of cofactors of oxidation that DPN and TPN closely resemble each other. Simultaneous studies on the two cofactors led to the establishment of the nicotinamide and adenine nucleotide structures as components of both DPN and TPN.[1] The complete structure of DPN was soon established,[2] but the position of the third phosphate of TPN remained uncertain. Early studies with crude yeast extracts demonstrated the conversion of DPN to TPN with the aid of ATP.[3] A similar reaction was observed in extracts of acetone-dried pigeon liver.[4] Hydrolysis of TPN to DPN was also observed in these systems. The precise localization of the third phosphate was finally made by identifying the fragment produced by specific enzymatic hydrolysis. Nucleotide pyrophosphatase formed nicotinamide mononucleotide and an adenine riboside, not the well-known ADP, containing two equivalents of phosphate. One of these phosphates was removed by a 5' nucleotidase. The remaining adenine nucleotide was identified with adenylic acid *a*, a product of ribonucleic acid degradation finally established as adenosine-2'-phosphate.[5a] Specific kinases for the formation of TPN from DPN have been isolated from both yeast and pigeon liver (I).[5b]

Desamino DPN. Muscle adenylic acid (adenosine-5'-phosphate) is converted to inosinic acid by specific deaminase.[6] Similar enzymes have been found to attack adenosine and adenine. An enzyme from *Aspergillus*

[1] O. Warburg, W. Christian, and A. Griese, *Biochem. Z.* **282**, 157 (1935); H. von Euler, H. Albers, and F. Schlenk, *Z. physiol. Chem.* **240**, 113 (1936).

[2] F. Schlenk and H. von Euler, *Naturwissenschaften* **24**, 794 (1936).

[3] E. Adler, S. Elliott, and L. Elliott, *Enzymologia* **8**, 80 (1940).

[4] A. H. Mehler, A. Kornberg, S. Grisolia, and S. Ochoa, *J. Biol. Chem.* **174**, 961 (1948).

[5a] A. Kornberg and W. E. Pricer, Jr., *J. Biol. Chem.* **186**, 557 (1950).

[5b] A. Kornberg, *J. Biol. Chem.* **182**, 805 (1950); T. P. Wang and N. O. Kaplan, *ibid.* **206**, 311 (1954).

[6] G. Schmidt, *Z. physiol. Chem.* **179**, 243 (1928).

(Taka-diastase), originally described as an adenosine deaminase, was found to deaminate a number of adenosine derivatives, including ADP, ATP, and DPN, but not TPN.[7] When TPN is dephosphorylated enzymatically, the resulting DPN is susceptible to deamination by the

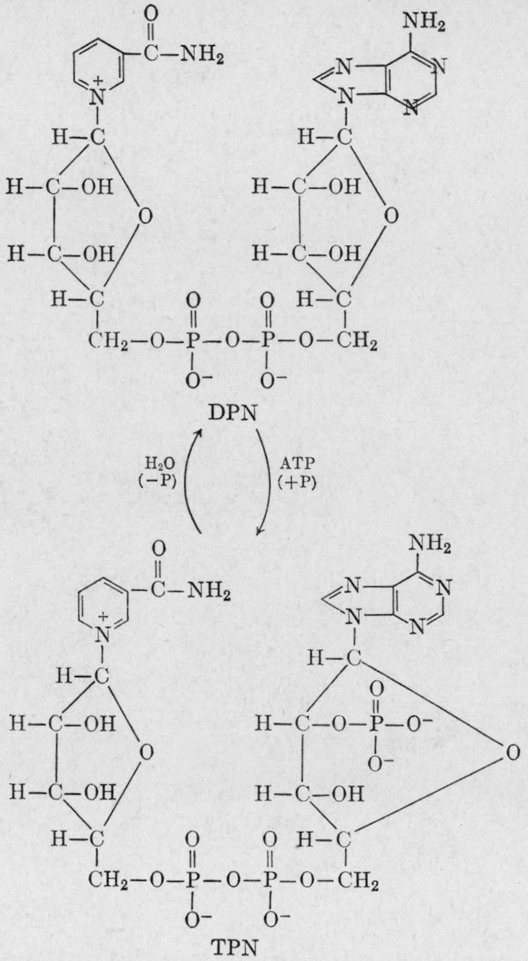

(I)

Taka-diastase deaminase. The product of deamination of DPN is called desamino DPN. This compound has also been prepared by treatment of DPN with nitrous acid.[8]

[7] N. O. Kaplan, S. P. Colowick, and M. M. Ciotti, *J. Biol. Chem.* **194**, 579 (1952).
[8] F. Schlenk, H. Hellstrom, and H. von Euler, *Ber.* **71**, 1471 (1938).

Coenzyme Specificity of Dehydrogenases. A number of dehydrogenases have been reported to require DPN, others are specific for TPN, and several react with both, although not necessarily at the same rate. Examples have already been cited of enzymes that catalyze identical reactions but possess different coenzyme requirements. An additional means for studying pyridine nucleotide reactions became available with desamino DPN.[9] Certain dehydrogenases (e.g., liver alcohol dehydrogenase) react at equal rates with DPN and desamino DPN. Others (as

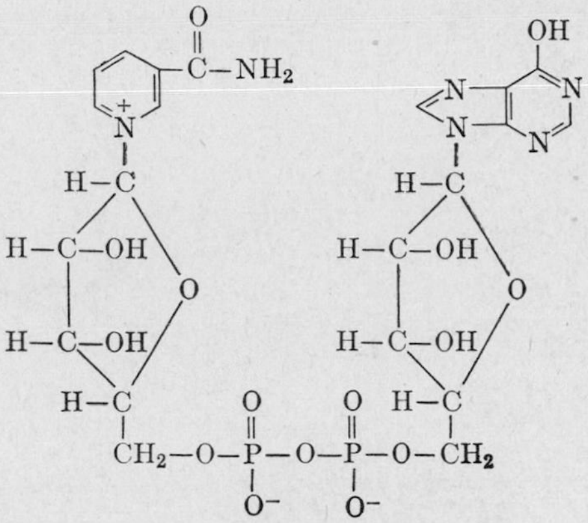

Desamino DPN

yeast alcohol dehydrogenase) react more slowly with the desamino compound, while no reaction is found with the desamino compound and β-hydroxybutyric dehydrogenase. Muscle lactic dehydrogenase provides an interesting case of an enzyme that reacts more rapidly in the direction of coenzyme reduction with DPN and more rapidly in the reverse direction with desamino DPN. The discrepancy in rates is, of course, compensated by a difference in Michaelis constants, since the same equilibrium is characteristic of both systems.

DPNases. Besides the amino group of adenine, two other groups in the pyridine nucleotides are known to be sensitive to enzymatic attack. One is the nicotinamide-ribose bond and the other is the pyrophosphate bond. Nicotinamide is split from DPN by a group of widespread DPNases.[10] The reaction catalyzed is a hydrolysis that results in the formation of free nicotinamide and a fragment abbreviated RPPRA. Some DPNases catalyze an exchange reaction in which nicotinamide from the medium enters the DPN molecule.[11] In these systems nico-

[9] M. E. Pullman, S. P. Colowick, and N. O. Kaplan, *J. Biol. Chem.* **194,** 593 (1952); P. J. G. Mann and J. H. Quastel, *Biochem. J.* **35,** 502 (1941).
[10] P. Handler and J. R. Klein, *J. Biol. Chem.* **150,** 447 (1943).
[11] L. J. Zatman, N. O. Kaplan, and S. P. Colowick, *J. Biol. Chem.* **200,** 197 (1953).

tinamide inhibits the hydrolysis of DPN, but since it apparently competes with water for RPPRA bound to the enzyme, the inhibition of hydrolysis is noncompetitive with DPN. Other DPNases, especially

$$NRPPRA + E \rightleftharpoons ERPPRA + N$$
$$\downarrow + H_2O$$
$$E + RPPRA$$

those from microorganisms, do not catalyze the exchange of nicotinamide. These are inhibited competitively by nicotinamide. Competitive inhibition in this case implies that the nicotinamide portion of DPN is bound to the enzyme, and that free nicotinamide also reacts with the same site on the enzyme and interferes with DPN binding. The ability to catalyze exchange reactions and sensitivity to nicotinamide have been associated with an enzyme that splits nicotinamide riboside; these properties are lost during purification and are restored by ergothioneine (2-thiolhistidine betaine). Ergothioneine also endows the *Neurospora* DPNase with the ability to catalyze the exchange reaction and makes it sensitive to nicotinamide. The mechanism by which the properties of these enzymes are changed drastically by the presence of small amounts (10^{-5} M) of ergothioneine is completely unknown.[12] Animal DPNases also differ in their sensitivities to a nicotinamide analog, isonicotinic acid hydrazide.[13] It is probable that the differences are related to sensitivity of the enzymes to the DPN analog formed by replacement of nicotinamide by isonicotinic acid hydrazide. The exchange reaction has also been used to prepare the 3-acetylpyridine analog of DPN.[14] This analog is of interest because of its ability to replace DPN in several dehydrogenase reactions.

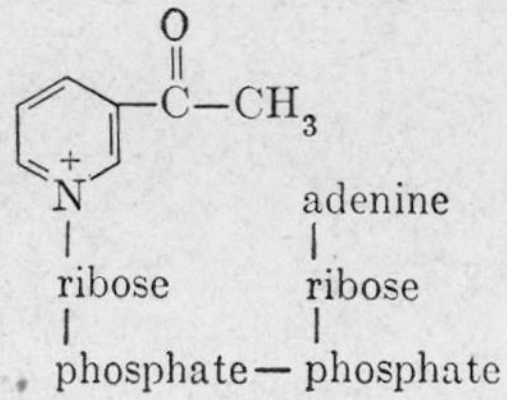

Acetylpyridine Analog of DPN

The DPNase of *Neurospora* has been purified and found to be a useful tool. This enzyme splits nicotinamide from only the oxidized forms of DPN and TPN, and does not attack the various pyridine analogs of

[12] L. Grossman and N. O. Kaplan, *J. Am. Chem. Soc.* **78,** 4175 (1956).
[13] L. J. Zatman, N. O. Kaplan, S. P. Colowick, and M. M. Ciotti, *J. Biol. Chem.* **209,** 453 (1954).
[14] N. O. Kaplan and M. M. Ciotti, *J. Am. Chem. Soc.* **76,** 7613 (1954).

DPN. The rate of hydrolysis of desamino DPN by this enzyme is much slower than the rate with DPN or TPN.[15]

Nucleotide Pyrophosphatase. Nucleotide pyrophosphatase was first detected in particles from animal cells, and was subsequently purified from potato tubers.[16] This enzyme, which has been found widely distributed, splits DPN, TPN, and other dinucleotides with a pyrophosphate bridge. As would be expected from the broad specificity, reduction

nicotinamide   adenine   nicotinamide   adenine
| | | |
ribose   ribose $+ H_2O \longrightarrow$ ribose $+$ ribose
| | | |
phosphate—phosphate   phosphate   phosphate
DPN   Nicotinamide   Adenylic acid
mononucleotide (NMN)

of the pyridine ring does not influence the pyrophosphatase. The pyrophosphate bond of DPN is also attacked by the specific pyrophosphorylase.[17] This enzyme catalyzes the reversible reaction

$$DPN + PP \rightleftharpoons NMN + ATP$$

The biologically significant reaction of the pyridine nucleotides is the reversible reduction of the pyridine ring. This causes a decrease in the absorption at 260 mμ and the appearance of a peak at 340 mμ.[1] The accepted value of the molar extinction coefficient of 6.22×10^3 was obtained by averaging the results obtained with several enzymes that quantitatively oxidize or reduce pyridine nucleotides in the presence of limiting amounts of substrates.[18] Standard solutions of pyruvate, acetaldehyde, and isocitrate were prepared and allowed to react with excess coenzyme in the presence of lactic dehydrogenase, alcohol dehydrogenase, or isocitric dehydrogenase. The lactic dehydrogenase system was used with both DPN and TPN. The advantages of this type of analysis include dependence on the purity of easily handled substrates instead of the coenzyme. Before these enzymatic reactions were used, large discrepancies were reported for extinction coefficients for the coenzymes, because the determinations depended on an independent assay of the amount of coenzyme present, and this could not be precise without coenzyme preparations of known purity.

Reduced Pyridine Nucleotides. The structure of the reduced pyridine ring was considered for many years to be an β-dihydro pyridine, probably

[15] N. O. Kaplan, S. P. Colowick, and A. Nason, *J. Biol. Chem.* **191,** 473 (1951)
[16] A. Kornberg and O. Lindberg, *J. Biol. Chem.* **176,** 665 (1948); A. Kornberg and W. E. Pricer, Jr., *ibid.* **182,** 763 (1950).
[17] A. Kornberg, *J. Biol. Chem.* **182,** 779 (1950).
[18] B. L. Horecker and A. Kornberg, *J. Biol. Chem.* **175,** 385 (1948).

substituted in position 6, with a second electron serving to eliminate the positive charge from the nitrogen of the ring. This picture was derived from analogy with the spectra of model compounds. It is now believed that reduced pyridine nucleotides bear two hydrogens on the γ-carbon.[19] Chemical reduction of pyridine rings in D_2O results in the formation of compounds with stably bound deuterium. DPND (with one atom of stable deuterium) has also been prepared by oxidation of deuterated substrates (see below). When these reduced rings are oxidized with neutral ferricyanide, the corresponding oxidized ring is obtained, partially labeled with deuterium. These reactions have been carried out with N-methylnicotinamide and DPN. The DPN was then converted to N-methylnicotinamide. Alkaline ferricyanide converts this compound to a mixture of 2- and 6-N-methylpyridone. These were isolated and assayed for deuterium. If either the 2 or 6 position were the site of reduction, the deuterium should be lost when that position is oxidized. The results showed no loss of isotope from either the 2 or 6 position; both pyridones retained all of the label. Therefore it was concluded that all of the label was bound at position 4. Positions 3 and 5 were eliminated, since addition of electrons to the nitrogen and a meta position would form a free radical structure. Reduction of N-substituted pyridine rings is therefore pictured as shown in (II).

$$2H^+ + 2e + \underset{\underset{R}{\overset{|}{N^+}}}{\bigcirc}\!-CONH_2 \rightleftharpoons \underset{\underset{R}{\overset{|}{N}}}{\overset{H\;H}{\bigcirc}}\!-CONH_2 + H^+$$

(II)

The addition of hydrogen to position 4 of the pyridine ring is only one of several addition reactions known. The addition of CN^- to DPN was reported in 1938,[20] and was later found to be a general reaction of N-substituted nicotinamide compounds. The spectrum of the cyanide addition compound resembles that of DPNH, and the appearance of the cyanide-compound spectrum has been used to assay N-substituted nicotinamide compounds.[21] Acetone and other carbonyl compounds also add to DPN to form similar compounds with similar spectra.[22] It was suggested that the binding of DPN to triose phosphate dehydro-

[19] M. E. Pullman, A. San Pietro, and S. P. Colowick, *J. Biol. Chem.* **206**, 129 (1954).
[20] O. Meyerhof, P. Ohlmeyer, and W. Mohle, *Biochem. Z.* **297**, 113 (1938).
[21] S. P. Colowick, N. O. Kaplan, and M. M. Ciotti, *J. Biol. Chem.* **191**, 447 (1951); V. A. Najjar, V. White, and D. B. Scott, *Bull. Johns Hopkins Hosp.* **74**, 378 (1944).
[22] R. M. Burton and N. O. Kaplan, *J. Biol. Chem.* **206**, 283 (1954).

genase involves addition of a sulfhydryl group of the enzyme to the pyridine ring, and that the formation of DPNH occurs by splitting the sulfur bond with an aldehyde.[23] In this process the hydrogen of the aldehyde is transferred to the pyridine ring. This hypothetical mechanism has strong support from an analysis of chemical reduction of DPN. It has long been known that during the reduction of DPN by hydrosulfite a yellow intermediate appeared. It has recently been shown that the yellow compound cannot be a radical, but is very probably an addition compound.[24] The hydrosulfite ion $S_2O_4^=$ is thought to hydrolyze to give sulfinate, $HOSO^-$, which adds to DPN as shown in (III). This compound

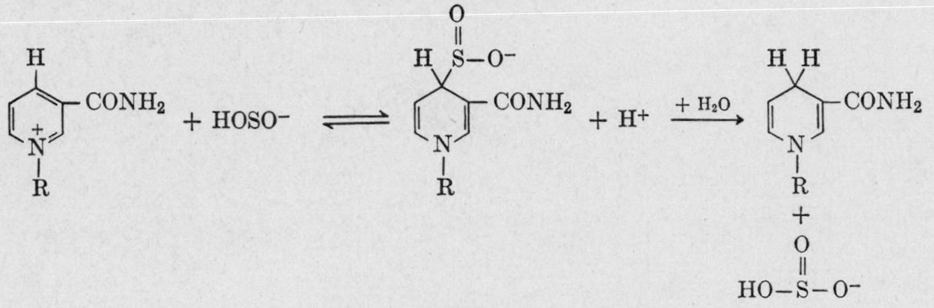

(III)

is formed in alkaline solution and is hydrolyzed at neutral pH values. Experiments in D_2O have shown that deuterium is incorporated into DPNH only in the second step, at neutral pH.

Stereochemistry of DPN Reduction. The carbon that bears two hydrogen atoms in DPNH is not optically active, but has the two hydrogens located on either side of the pyridine ring. The stereochemistry of DPNH has been studied with brilliant success by investigators in the laboratories of Westheimer and Vennesland. The first experiments of this group[25] involved the use of ethanol prepared by reduction of phenyl acetate by

$LiAlD_4$. Reduced DPN was prepared with yeast alcohol dehydrogenase either by reaction with CH_3CH_2OH in D_2O or by reaction with CH_3CD_2OH in H_2O (the hydroxyl H exchanges with water, and does not

[23] E. Racker and I. Krimsky, *J. Biol. Chem.* **198,** 731 (1952).

[24] M. B. Yarmolinsky and S. P. Colowick, *Biochim. et Biophys. Acta* **20,** 177 (1956).

[25] F. H. Westheimer, H. F. Fisher, E. E. Conn, and B. Vennesland, *J. Am. Chem. Soc.* **73,** 2403 (1951); H. F. Fisher, E. E. Conn, B. Vennesland, and F. H. Westheimer, *J. Biol. Chem.* **202,** 687 (1953).

retain a label). The reduced DPN was isolated and analyzed for stable D. None was found in the sample incubated with D_2O, but one atom was incorporated in the sample prepared with deuterated ethanol. Enzymatically prepared DPND could be distinguished from the chemically reduced compound (hydrosulfite in D_2O) by the reverse reaction with alcohol dehydrogenase and acetaldehyde. Whereas the chemically deuterated DPN is randomly labeled (on both sides of the pyridine

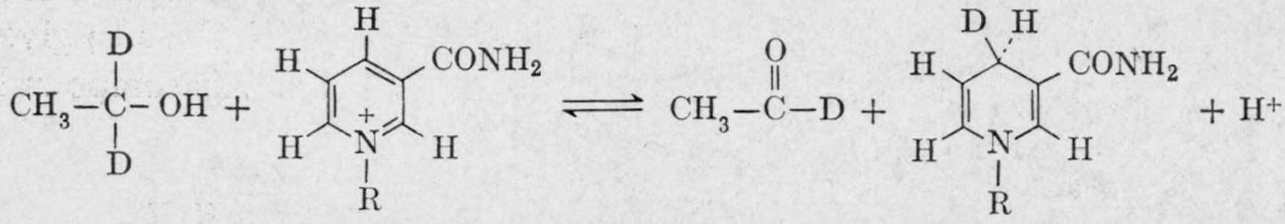

$$CH_3-\overset{\overset{\displaystyle D}{|}}{\underset{\underset{\displaystyle D}{|}}{C}}-OH \; + \; \text{[pyridinium ring]} \;\; \rightleftharpoons \;\; CH_3-\overset{\overset{\displaystyle O}{||}}{C}-D \; + \; \text{[dihydropyridine ring]} \; + \; H^+$$

(IV)

ring), the enzymatically deuterated DPN has all of the D in one position. The DPN formed by enzymatic oxidation of the chemically reduced material retained half of the deuterium, whereas the enzymatically reduced nucleotide lost all of its label. These experiments demonstrate the direct transfer of hydrogen from substrate to DPN, and also show that only a specific one of the two hydrogens on the γ-carbon of the pyridine ring is removed by alcohol dehydrogenase. In other words, the

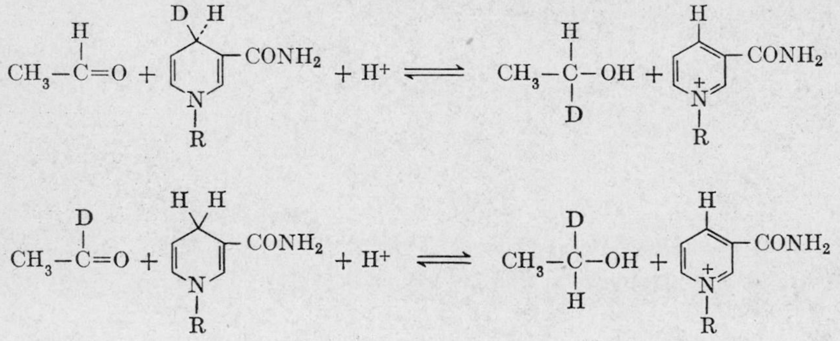

$$CH_3-\overset{\overset{\displaystyle H}{|}}{C}=O \; + \; \text{[ring]} \; + \; H^+ \;\; \rightleftharpoons \;\; CH_3-\overset{\overset{\displaystyle H}{|}}{\underset{\underset{\displaystyle D}{|}}{C}}-OH \; + \; \text{[ring]}$$

$$CH_3-\overset{\overset{\displaystyle D}{|}}{C}=O \; + \; \text{[ring]} \; + \; H^+ \;\; \rightleftharpoons \;\; CH_3-\overset{\overset{\displaystyle D}{|}}{\underset{\underset{\displaystyle H}{|}}{C}}-OH \; + \; \text{[ring]}$$

Formation of Isomers of Deutero-Labeled Ethanol
(V)

reaction that introduces or removes a hydrogen atom into or from DPN is stereospecific, and involves only one side of the pyridine ring (IV).

    Deuterated and normal DPNH have been used with acetaldehyde, either CH_3CHO or CH_3CDO, to form two types of mono-deuterated ethanol, CH_3CHDOH.[26] Only one of these transfers D to DPN: the

[26] F. A. Loewus, F. H. Westheimer, and B. Vennesland, *J. Am. Chem. Soc.* **75**, 5018 (1953).

ethanol obtained by reducing CH_3CHO with DPND. This result shows that the ethanol is attacked in the same stereospecific manner as DPN (V). The ability of an enzyme to distinguish between two hydrogen atoms attached to the same carbon atom is another illustration of the consequences of the three-point attachment hypothesis of Ogston (discussed in connection with citric acid p. 113).

The stereospecificity of malic and lactic dehydrogenases have been found to be the same as alcohol dehydrogenase, as far as the type of deuterated DPN formed and the activation of only a single hydrogen of the substrate (VI).[27] A dehydrogenase that attacks steroids with a hydroxyl group in the β configuration was found to add hydrogen to the opposite side of DPN (VII).[28]

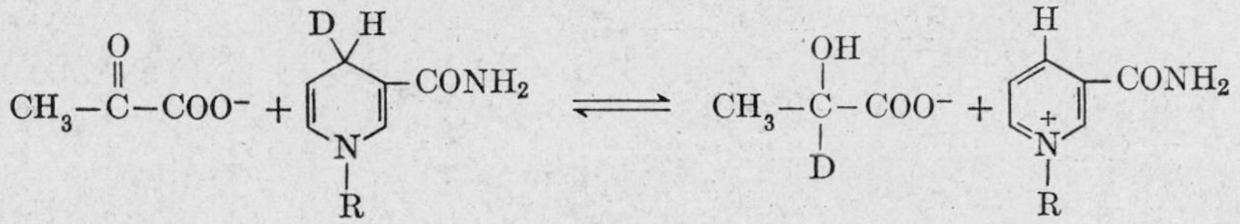

(VI)

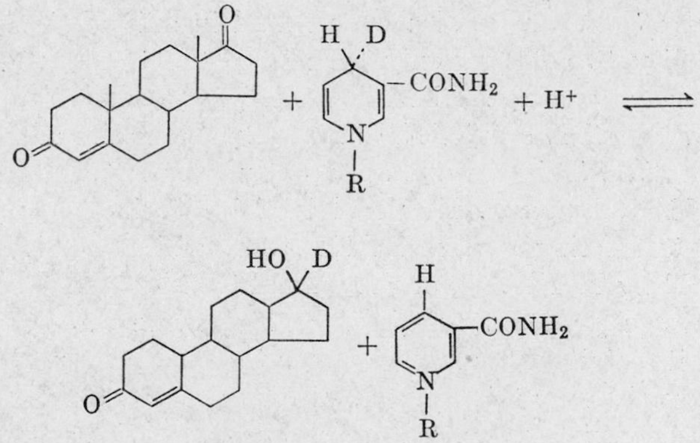

(VII)

Transhydrogenase. DPN and TPN are equivalent in terms of their ability to oxidize (or reduce) other molecules, since precisely the same structure is involved in each nucleotide. The possibility of a reaction equilibrating the two coenzymes by transferring hydrogen from one to the other had been considered for many years, and finally was found

[27] F. A. Loewus, P. Ofner, H. F. Fisher, F. H. Westheimer, and B. Vennesland, *J. Biol. Chem.* **202**, 699 (1953); F. A. Loewus, T. T. Tchen, and B. Vennesland, *ibid.* **212**, 787 (1955).

[28] P. Talalay, F. A. Loewus, and B. Vennesland, *J. Biol. Chem.* **212**, 801 (1955).

to be catalyzed by an enzyme purified from *Pseudomonas*.[29] This enzyme, transhydrogenase, acts with all of the known nucleotides of nicotinamide.

Not all of the theoretically possible combinations of oxidized and reduced nucleotides have been demonstrated yet because of some unusual properties of the enzyme. Although it catalyzes an efficient reaction between TPNH and DPN, the reverse reaction, with DPNH and TPN,

$$TPNH + DPN \rightarrow TPN + DPNH$$

proceeds at a negligible rate. The discrepancy is caused by an inhibition of the enzyme by TPN, not by TPNH. The addition of adenylic acid *a* antagonizes the inhibition and permits the reaction to proceed in both

$$TPNH + DPN \xrightleftharpoons{\text{adenylic acid}} TPN + DPNH$$

directions.[30] Reactions involving desamino DPN and NMN are also greatly stimulated by adenylic acid *a*, but other reactions, such as transfer of H from TPNH to desamino TPN, are not influenced by adenylic acid *a*. These phenomena are not completely explained by any model, but obviously involve many types of combination of enzyme and nucleotide.

Transhydrogenases have also been found in animal tissues.[31] These have the ability to catalyze only certain of the transfers effected by bacterial transhydrogenase. A beef heart enzyme catalyzes the reversible reaction between DPN and TPNH and does not require adenylic acid *a* in either direction. This enzyme apparently does not reduce desamino TPN by TPNH, however. Hog and beef brain, in contrast, contain enzymes for transfering H from DPNH to desamino DPN, but not for reducing DPN by TPNH. All of the animal transhydrogenases are associated with large particles. The beef heart enzyme was solubilized by treatment with digitonin, and subsequently could be purified. Other methods for solubilizing this enzyme failed.

The stereochemistry of transhydrogenase has been studied with deuterium.[32] Bacterial transhydrogenase specifically uses the same side of the ring as steroid dehydrogenase, both as donor and recipient. In this way the same hydrogen remains transferable by dehydrogenases, regardless of the number of transfers by transhydrogenase. The presence of transhydrogenase therefore has no influence on the distribution of

[29] S. P. Colowick, N. O. Kaplan, E. F. Neufeld, and M. M. Ciotti, *J. Biol. Chem.* **195**, 95 (1952).
[30] N. O. Kaplan, S. P. Colowick, E. F. Neufeld, and M. M. Ciotti, *J. Biol. Chem.* **205**, 17 (1953).
[31] N. O. Kaplan, S. P. Colowick, and E. P. Neufeld, *J. Biol. Chem.* **205**, 1 (1953).
[32] A. San Pietro, N. O. Kaplan, and S. P. Colowick, *J. Biol. Chem.* **212**, 941 (1955).

deuterium between nucleotide and substrate; if mono-deuterated reduced DPN is made by transfer of D from deuterated ethanol, the DPND will be able to transfer D to acetaldehyde, pyruvate, and oxalacetate. When this nucleotide is used to reduce steroids or other pyridine nucleotides, only H, no D, will be transferred.

DPNH-X. DPNH is rapidly destroyed by acid at room temperature. A similar destruction of DPNH was found to be catalyzed by triose phosphate dehydrogenase specifically, and by no other dehydrogenases tested.[33] The pH optimum for the destructive reaction is near 5, in contrast to the optimum for the oxidative reaction at 8.5. The reaction product is identified by an absorption peak at 290 mμ and by an altered optical rotation. This product is called DPNH-X, and is distinguished from the acid degradation product by its rotation and by its enzymatic activity. A kinase has been found that catalyzes the reaction:[34]

$$DPNH-X + ATP \rightarrow DPNH + ADP + P$$

Isotopic and spectrophotometric studies give results consistent with DPNH-X being a hydrated DPNH, with one molecule of water inserted into the pyridine ring.[35] The metabolic significance of this compound and the kinase that attacks it cannot be assessed at this time.

GENERAL REFERENCES

Kaplan, N. O. (1955). *Record Chem. Progr.* (*Kresge-Hooker Sci. Lib.*) **16,** 177.
San Pietro, A. (1956). *Am. Brewer* p. 31.
Schlenk, F. (1951). *In* "The Enzymes" (J. B. Sumner and K. Myrbäck, eds.), Vol. II, Part 1, p. 250. Academic Press, New York.
Singer, T. P., and Kearney, E. B. (1954). *Advances in Enzymol.* **15,** 79.
Vennesland, B., and Westheimer, F. H. (1954). *In* "The Mechanism of Enzyme Action" (W. D. McElroy and B. Glass, eds.), p. 357. Johns Hopkins Press, Baltimore, Maryland.

Oxidation-Reduction Potentials

The oxidation reactions already described have been discussed in terms of equilibria. From the equilibrium values it can be seen that some compounds, such as acetaldehyde, are oxidized by DPN or TPN quantitatively, whereas, at the same pH values, malate, lactate and ethanol react to a very slight extent. The equilibrium constants of oxidation-reduction reactions have been used to evaluate a property of members

[33] G. W. Rafter, S. Chaykin, and E. G. Krebs, *J. Biol. Chem.* **208,** 799 (1954).
[34] J. O. Meinhart, S. Chaykin, and E. G. Krebs, *J. Biol. Chem.* **220,** 821 (1956).
[35] J. O. Meinhart and M. C. Hines, *Federation Proc.* **16,** 425 (1957).

of a reacting system called the oxidation-reduction potential. This property, being derived from equilibrium constants, is also related to the thermodynamic quantity, free energy. Equilibrium values are functions of an entire reaction system. Oxidation-reduction potentials are characteristics of half-reactions; that is, the components on the two sides of a reaction that are interconverted by the addition or substraction of electrons. A complete oxidation is thus the sum of two half-reactions. Some examples of half-reactions are:

$$DPN^+ + H^+ + 2e \rightleftharpoons DPNH$$
$$CH_3CH_2OH \rightleftharpoons CH_3CHO + 2H^+ + 2e$$

If these are combined, the electrons produced by one system are used by another, and do not appear in the over-all equation:

$$DPN^+ + CH_3CH_2OH \rightleftharpoons DPNH + CH_3CHO + H^+$$

To determine an oxidation-reduction potential for one half-reaction (or *couple*), the potential for the other couple must be known as well as the equilibrium constant of the over-all reaction. Oxidation-reduction potentials are thus seen to be arbitrary, relative values. The evaluation of potentials is made with reference to an accepted standard; the primary reference half-reaction is:

$$H_2 \rightleftharpoons 2H^+ + 2e$$

According to convention, when hydrogen gas at 1 atmosphere pressure is equilibrated with hydrogen ions at unit activity (essentially 1 N), the oxidation-reduction potential is zero volts. Unfortunately, from this point, there are two conventions in common usage. One, employed by physical chemists, defines as negative the potentials of systems that tend to oxidize the reduced members of other systems, whereas the convention used by most biochemists terms the potentials of reducing systems as negative. Thus, in the biochemical convention, a system in which hydrogen gas at 1 atmosphere is equilibrated with a pH 7.0 buffer has a potential of -0.42 volts. At pH 7.0, hydrogen gas has a greater tendency to give up electrons than at pH zero.

Each half-reaction is characterized by a standard potential, E_0. This is a measure of the tendency of the couple to become more oxidized or reduced when all members of the couple are present at unit activity. For most cases, this same value applies at other concentrations when the oxidized and reduced members are present at equal concentrations. The actual potential of a system, E, is defined thus:

$$E = E_0 + \frac{RT}{n\mathbf{F}} \ln \frac{[ox]}{[red]} = E_0 + \frac{RT}{n\mathbf{F}} \times 2.303 \log \frac{[ox]}{[red]}$$

In this equation R is the so-called gas constant, 1.987 cal./degree; $\mathbf{F}$ is Faraday's constant, 23,063 cal./volt; n is the number of electrons involved, and T is the absolute temperature. For a single-electron transfer, as in the system

$$Fe^{+++} + e \rightleftharpoons Fe^{++}$$

the term $RT/n\mathbf{F} \times 2.303$ approximates 0.06 at ordinary room temperatures, whereas for two-electron systems, as the DPN-DPNH couple, the constant becomes 0.03. In describing a real reaction, the equations for the two half-reactions must be written to include the same number of electrons; that is, the sum of the half-reactions must be a balanced chemical reaction.

When two half-reactions come to equilibrium with each other, the same potential, E, must apply to both. Therefore

$$E_{0_a} + \frac{RT}{n\mathbf{F}} \ln \frac{[ox_a]}{[red_a]} = E_{0_b} + \frac{RT}{n\mathbf{F}} \ln \frac{[ox_b]}{[red_b]}$$

This equation rearranges simply to

$$E_{0_a} - E_{0_b} = \frac{RT}{n\mathbf{F}} \ln \frac{[ox_a]}{[red_a]} - \frac{RT}{n\mathbf{F}} \ln \frac{[ox_b]}{[red_b]}$$

which simplifies to

$$\Delta E = \frac{RT}{n\mathbf{F}} \ln \frac{[ox_a][red_b]}{[red_a][ox_b]}$$

This is equivalent to

$$\Delta E = \frac{RT}{n\mathbf{F}} \ln K_{eq}$$

The relations above indicate that the position of an equilibrium is determined by the difference in potential between the two half-reactions.

A couple is oxidizing or reducing only when referred to another couple. The numerical value assigned to standard potentials measures the tendency to accept electrons from (positive E_0) or donate electrons to (negative E_0) the standard hydrogen system. When systems at condition other than standard are studied, modified E_0 values, designated E_0', are conventionally used (see Table 5). E_0' is meaningful only when the conditions are stated.

If the oxidation-reduction potentials of two systems are known, the equilibrium constant for a reaction between them may be calculated. Such calculations are valid even in cases in which the systems do not react directly with each other and in cases in which direct measurements may be obscured by side reactions. The oxidation of ethanol (1 M) by

DPN (1 M) (in the presence of catalytic amounts of alcohol dehydro-genase) at pH 7.0 may be used to illustrate the calculations. In this example, one half-reaction is the ethanol-acetaldehyde couple, which has an E_0' of -0.200 volts; the other couple, DPN-DPNH, has an E_0' of

$$DPN + ethanol \rightarrow DPNH + acetaldehyde + H^+$$

-0.320 volts. In the reaction the H^+ term may be dropped, since E_0' values are used, and the buffer prevents H^+ from accumulating. ΔE then equals $-0.320 - (-0.200)$. The acetaldehyde-ethanol potential is sub-tracted from that of the DPN system, rather than the reverse (which

TABLE 5

REPRESENTATIVE OXIDATION-REDUCTION POTENTIALS[a]

Reaction	E_0' (pH 7.0, 25°C.) (volts)
Oxygen-water	0.815
Oxygen-hydrogen peroxide	0.682
Ferricyanide-ferrocyanide	0.36
Cytochrome c (Fe^{+++})-cytochrome c (Fe^{++})	0.25
Methylene blue-leucomethylene blue	0.011
Fumarate-succinate	0.000
Oxalacetate-malate	-0.166
Riboflavin-leucoriboflavin	-0.219
Acetaldehyde-ethanol	-0.200
DPN-DPNH	-0.320
H^+ (10^{-7} M)-H_2 (1 atm.)	-0.42
Acetate-acetaldehyde	-0.468

[a] The systems included are only a few examples of the many oxidation-reduction potentials recorded. The range of values encompasses almost all biological systems, but many more powerful oxidizing and reducing systems are known, with potentials extending from $+3.09$ volts ($\frac{3}{2}N_2 + H^+ + e \rightarrow HN_3$) to -3.06 volts ($F_2 + 2H^+ + 2e \rightarrow 2HF$).

would change the sign of ΔE) because the reaction is being considered in the direction in which DPN is the oxidant. Couples considered in the reverse direction, as ethanol-acetaldehyde, have their signs reversed. The assignment of values of standard potentials must be made in accord-ance with convention to obtain results of the proper sign. In the biochemi-cal convention, the potentials are reduction potentials, and the value of the system supplying the oxidant is that of Table 5, while the value for the reducing reagent is substracted. That is, when a system is con-sidered as reacting in the opposite manner to that indicated in Table 5, its E_0 value is reversed in sign.

$$\Delta E = \frac{RT}{n\mathbf{F}} \ln K_{eq}$$

$$-0.120 = \frac{1.987 \times 298}{2 \times 23,063} \ln K_{eq}$$

$$\log K_{eq} = \frac{-0.120 \times 2 \times 23,063}{1.987 \times 298 \times 2.303} = -4.059$$

Thus,

$$K_{eq} = 8.73 \times 10^{-5} = \frac{(DPNH)(acetaldehyde)}{(DPN)(ethanol)}$$

Since equivalent concentrations of DPN and ethanol are being used, (DPNH) = (acetaldehyde) and (DPN) = (ethanol). Then, since DPN + DPNH = 1 M, if DPNH = x, DPN = $1 - x$.

$$\frac{x^2}{(1 - x)^2} = 8.73 \times 10^{-5}$$

x is found to be approximately 0.0093, which means that less than 1 per cent of the DPN would be reduced at equilibrium.

The equations presented are perfectly general, but the values substituted into the equations must apply to the actual system being studied. If H^+ or OH^- participates in the reaction, the complete equation for the reaction must include a pH term. This term may be included in the calculations, or it may be absorbed in a modified standard potential, E_0', characteristic of the couple at the given pH. E_0' of a couple involving two electrons in which 1 hydrogen ion is produced differs from the standard potential, in which the H^+ ion is defined as having unit activity, by -0.03 volts per pH unit, while E_0' for systems in which 2 hydrogen ions are formed changes by -0.06 volts per pH unit at ordinary temperatures. Care must be taken to see that E_0' values are used for comparable conditions when ΔE values are calculated. The use of E_0' and the two conventions of sign in current usage require caution in calculations based on oxidation-reduction potentials.

The effect of pH on oxidation-reduction potentials can be illustrated with the alcohol dehydrogenase system. Instead of pH 7.0, consider the reaction at pH 11.0. The acetaldehyde-ethanol couple will have an E_0' of $-0.200 - (4 \times 0.06) = -0.440$. The DPN-DPNH couple, involving only 1 hydrogen ion, changes from -0.320 to -0.440 volts. ΔE now is zero, and $\log K_{eq}$ is zero. K_{eq}, therefore, is 1.

The relation of ΔE to K_{eq} can be used to relate ΔE to free energy, since $\Delta F_0 = -RT \ln K_{eq}$. Substituting ΔF_0 in the earlier equation gives the equation:

$$\Delta F_0 = -n\Delta E\mathbf{F}$$

ΔF_0 is the standard free energy of the reaction, and must not be confused with the constant $\mathbf{F}$, usually written in bold type to minimize confusion. In the example of the alcohol dehydrogenase reaction at pH 7.0, $\Delta F_0'$ may be evaluated from oxidation-reduction potentials without first calculating the equilibrium constant.

$$\Delta F_0' = -2 \times (-0.120) \times 23{,}063$$

$\Delta F_0' = +5535$ for the reaction in the direction being considered. The same value is obtained, of course, by substituting in the equation

$$\Delta F_0' = -RT \ln K_{eq}$$

That oxidation-reduction potentials are related to equilibrium constants may be used to underscore the fact that they apply to (theoretically) reversible reactions. The oxidized member of any couple will reduce some of the reduced member of any other couple, provided a reaction mechanism exists. The potentials of the two couples permit an estimation only of the possible extent of the reaction; nothing can be predicted about the rate, or indeed, whether the reaction will occur at all. In determining the extent of any reaction, actual concentrations of all reagents must be considered, since the actual equilibrium is important, not the equilibrium of an ideal 1 M solution that might be calculated from ΔE or ΔF_0 values.

The general equation relating ΔE to concentrations requires that each component be included. Therefore, if two moles of one compound participate in a reaction, that compound must be included twice, to give a squared term in the equation. One special case requires additional comment. When two molecules on one side of a half-reaction form one on the other, there is a marked dependence of the oxidizing or reducing ability on absolute concentration. For example, two molecules of reduced glutathione are needed to produce one of oxidized glutathione. This couple has a standard potential at pH 7 of 0.04. At 1 M GSH and GSSG, the potential of a solution would be 0.04 volts. Diluting such a solution ten times, however, changes the potential to 0.07 volts, and as the solution is further diluted it becomes more oxidizing, increasing in potential 0.03 for each factor of 10 in concentration. An equilibrium calculated from standard potentials does not describe the composition of a solution at other concentrations. Of course, this phenomenon is implicit in any formulation of such reactions and does not differ in principle from the effect of concentration on the aldolase reaction (p. 53). In each case the equilibrium constant is a true constant. It must always be remembered, however, that the equilibrium is a statistical property of a system, not an intrinsic function of individual molecules.

E_0 values have been determined for many couples. When used in pairs, oxidation-reduction potentials give the same sort of information as other measures of free energy changes. This information can be expressed as the extent to which a reaction will proceed. This function is useful for determining the answer to such practical questions as: in which direction will a reaction actually proceed? can a given system "pull" another? is a given reaction likely to be reversible? etc.

GENERAL REFERENCES

Johnson, M. J. (1949). *In* "Respiratory Enzymes" (H. A. Lardy, ed.), p. 58. Burgess Publishing, Minneapolis.

Latimer, W. M. (1951). "Oxidation Potentials." Prentice-Hall, New York.

Michaelis, L. (1951). *In* "The Enzymes" (J. B. Sumner and K. Myrbäck, eds.), Vol. II, Part 1, pp. 1–54. Academic Press, New York.

Plaut, G. W. E., and Andersen, L. (1949). *In* "Respiratory Enzymes" (H. A. Lardy, ed.), p. 71. Burgess Publishing, Minneapolis.

Flavoproteins

All known organisms contain oxidative enzymes containing the yellow compound, riboflavin, vitamin B_2. The biosynthetic pathway by which riboflavin is synthesized has been investigated only recently by isotopic methods, which so far have only given slight indications of how the

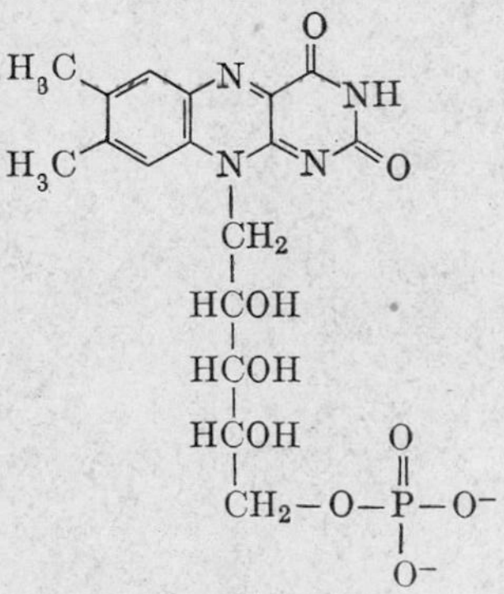

Riboflavin phosphate (FMN)

vitamin is formed. Nothing at all is known of the enzymes concerned. The vitamin is known to exist in two nucleotides that occur as tightly bound components of enzymes. The simpler is called flavin mononucleotide, and has the structure shown above.[1]

[1] R. Kuhn, H. Rudy, and F. Weygand, *Ber.* **69,** 1543 (1936).

Riboflavin differs from all other nucleosides in being a derivative of the sugar alcohol, ribitol. A specific kinase has been found that produces the nucleotide, FMN, from riboflavin and ATP.[2]

The other biologically important flavin is a dinucleotide,[3] made from FMN and ATP in a reaction analogous to the synthesis of DPN,[4] to form flavin adenine dinucleotide (FAD), in which the two nucleotides are joined by a pyrophosphate bond (I).

Enzymes that contain flavin components participate in reactions with many different electron donors and acceptors. In general greater specificity has been found for reductants than for oxidants. The ability to react with certain oxidants has implications as to the physiological functions of the flavoproteins, and therefore has been used as the basis for naming some of the enzymes, such as cytochrome reductases and nitrate reductase. Other flavoproteins have been named for the substrates oxidized, such as xanthine oxidase, glucose oxidase, amino acid oxidases, and aldehyde oxidase. It must be emphasized that the assignment of a name to an enzyme does not limit its abilities to react with other materials, and the flavoproteins present an interesting series of degrees of cross-reaction.

Oxidation of Reduced Pyridine Nucleotides: Old Yellow Enzyme. The most prominent role of the flavoproteins is that of intermediary in the oxidation of reduced pyridine nucleotides. The first yellow enzyme to be purified was obtained from yeast by Warburg and Christian as a factor

$$\text{TPNH} + O_2 + H^+ \rightarrow \text{TPN} + H_2O_2$$

that could link a TPN system (Zwischenferment) with molecular oxygen.[5] When other flavoproteins were purified and found to be different from the first, the name "old yellow enzyme" was given to the original preparation. The enzyme was purified to a high degree by Theorell,[6a] who resolved the enzyme by acid dialysis, and, for the first time, demonstrated reversible cleavage of an enzyme. The yellow, dialyzable prosthetic group was isolated, and chemical analysis of its structure resulted in the establishment of the structure of the first flavin nucleotide, FMN.[1] Recently the old yellow enzyme has been crystallized; the crystalline material is homogeneous according to ultracentrifuge, electrophoresis,

---

[2] E. B. Kearney and S. Englard, *J. Biol. Chem.* **193**, 821 (1951); S. Englard, *Federation Proc.* **4**, 208 (1952).

[3] O. Warburg and W. Christian, *Biochem. Z.* **296**, 294 (1938); E. P. Abraham, *Biochem. J.* **33**, 543 (1939).

[4] A. W. Schrecker and A. Kornberg, *J. Biol. Chem.* **182**, 795 (1950).

[5] O. Warburg and W. Christian, *Biochem. Z.* **257**, 492 (1933).

[6a] H. Theorell, *Biochem. Z.* **272**, 155 (1934); **278**, 263 (1935).

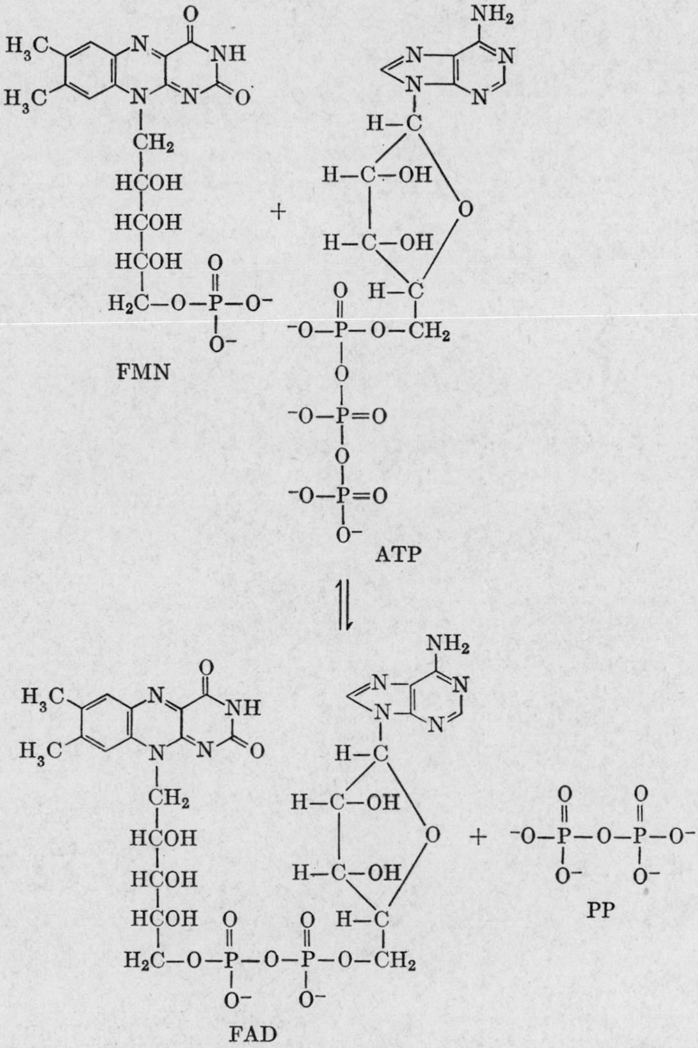

(I)

and solubility test criteria. It has a molecular weight of 104,000 and contains 2 moles of FMN per mole of protein.[6b]

New Yellow Enzyme. Shortly after the isolation of the old yellow enzyme, Haas, in Warburg's laboratory, isolated a second flavoprotein from yeast.[7] This enzyme, the "new yellow enzyme" differs from the

[6b] H. Theorell and Å. Åkeson, *Arch. Biochem. Biophys.* **65,** 439 (1956).
[7] E. Haas, *Biochem. Z.* **298,** 378 (1938).

old yellow enzyme in several respects. Instead of FMN, the prosthetic group of the new yellow enzyme is FAD. Whereas resolved old yellow enzyme can be reactivated with FAD to 70 per cent of its activity with FMN, the new yellow enzyme cannot be reactivated by FMN, but is completely reactivated by FAD. Old yellow enzyme protein + FAD was called "crossed" or "synthetic" yellow enzyme. Old yellow enzyme reacts rapidly with both oxygen and cytochrome c. New yellow enzyme reacts sluggishly with oxygen and does not reduce cytochrome c, but its turnover number with effective electron acceptors, such as methylene blue, is greater than that of the new yellow enzyme.

Cytochrome Reductase of Yeast. A third yeast flavoprotein was isolated by Haas and co-workers.[8] This was called cytochrome reductase because its rate of reaction with cytochrome c is over 150,000 times as great as the rate of cytochrome c reduction by old yellow enzyme. Its rate of reaction with oxygen is only 8 per cent that of old yellow enzyme. The prosthetic group is FMN. Reduction occurs only with TPNH. Another enzyme, DPN-cytochrome reductase, was partially purified in Hogness' laboratory.

Diaphorase and Cytochrome Reductase. Enzymes have been isolated from animal sources that have many of the properties of the various yeast enzymes. The first, and simplest, was liberated from particulate structures by Straub,[9] who employed dilute ethanol and ammonium sulfate at 43°C. The enzyme could then be purified and was named a diaphorase. "Diaphorase" was coined to identify a widespread group of enzymes that transfer electrons from DPNH to dyes. Many of the purified flavoproteins have been found to oxidize pyridine nucleotides, and almost all of the flavoproteins can reduce dyes. Straub's diaphorase reduces methylene blue but not cytochrome c. Slight modification of the isolation procedure was found by workers at the Enzyme Institute of the University of Wisconsin to yield a cytochrome reductase.[10] The relation between these preparations is not known, but it is possible that one is derived from the other. Cytochrome reductase contains 4 atoms of iron for each flavin, whereas Straub's preparation contains little iron.[11] Both proteins have molecular weights around 75,000, and contain 1 equivalent of FAD.

A TPN-cytochrome reductase isolated from liver has essentially the same activity as yeast cytochrome reductase.[12] The liver enzyme appar-

[8] E. Haas, B. L. Horecker, and T. R. Hogness, *J. Biol. Chem.* **136,** 747 (1940).

[9] F. B. Straub, *Biochem. J.* **33,** 787 (1939).

[10] H. Edelhoch, O. Hayaishi, and L. J. Teply, *J. Biol. Chem.* **197,** 97 (1952); H. R. Mahler, N. K. Sarkar, L. P. Vernon, and R. A. Alberty, *ibid.* **199,** 585 (1952).

[11] H. R. Mahler and D. G. Elowe, *J. Biol. Chem.* **210,** 165 (1954).

[12] B. L. Horecker, *J. Biol. Chem.* **183,** 593 (1950).

ently contains FAD as isolated, but is even more active when resolved and combined with FMN. The enzyme is bright yellow, but on reduction the color did not disappear completely. The material responsible for light absorption in addition to that characteristic of flavins has not been identified, but by analogy with other flavoproteins, it may be iron, or some other metal.

A cytochrome reductase from liver microsomes was found by Stritt-matter and Velick to oxidize DPNH and to reduce a microsomal cyto-chrome, but not cytochrome c.[13] This enzyme does not reduce methylene blue, but it does react with ferricyanide.

D-*Amino Acid Oxidase.* Among the flavoproteins that oxidize sub-strates other than reduced pyridine nucleotides, the first to be isolated and one of the most useful is D-amino acid oxidase. This enzyme was detected by Krebs, who distinguished activities specific for D- and L-amino acids.[14] The isolation of the cofactor of sheep kidney D-amino acid oxidase[15] was soon followed by its identification as a flavin dinucleotide containing adenylic acid, and having the structure shown as FAD.[3] Warburg and Christian resolved D-amino acid oxidase by adding HCl to bring the pH to 2.8 to an ammonium sulfate solution of the enzyme. The protein precipitates under these conditions, while FAD remains in solution. The protein can be purified further to a high degree. The iso-lated protein has a very high affinity for FAD, but does not react at all with FMN. These properties have made D-amino acid oxidase measure-ments the assay of choice in determining FAD.

The reaction carried out by D-amino acid oxidase uses molecular oxygen as electron acceptor efficiently. The conventional assay of this

$$\underset{\underset{NH_2}{|}}{\overset{\overset{H}{|}}{R-C-COO^-}} + O_2 + H_2O \longrightarrow \overset{\overset{O}{\parallel}}{R-C-COO^-} + NH_3 + H_2O_2$$

reaction is manometric, although determinations of ammonia or keto acid offer a valid alternative. The apparent oxygen consumption is affected by the fate of the peroxide formed. This may be converted to water and oxygen by catalase (p. 196) or it may be utilized for second-ary oxidations, including the nonenzymatic oxidation of α-keto acids.

Simple Flavoproteins. The specificity of kidney D-amino acid oxidase for D-amino acids is discussed in Chapter VII. Similar flavoproteins have

---

[13] P. Strittmatter and S. Velick, *J. Biol. Chem.* **221,** 277 (1956).

[14] H. A. Krebs, *Biochem. J.* **29,** 1620 (1935).

[15] O. Warburg and W. Christian, *Biochem. Z.* **296,** 294 (1938); **298,** 150 (1938); F. B. Straub, *Nature* **141,** 603 (1938); *Biochem. J.* **33,** 787 (1939).

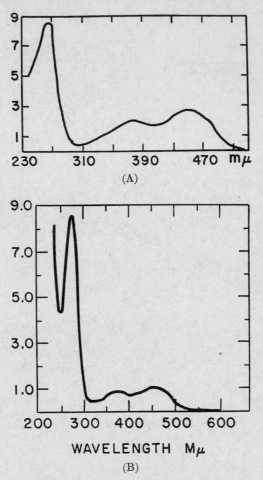

FIG. 22. Absorption spectra of FAD (A)[23] and TPN-cytochrome c reductase (B).[12]

been found in molds.[16] An L-amino acid oxidase from rat kidney uses FMN,[17] while a similar enzyme from snake venom contains FAD.[18] Other enzymes oxidize specific amino acids; glycine[19] and aspartate[20] systems include FAD. Another specific flavoprotein is the glucose oxidase

[16] N. H. Horowitz, *J. Biol. Chem.* **154**, 141 (1944).

[17] M. Blanchard, D. E. Green, V. Nocito, and S. Ratner, *J. Biol. Chem.* **155**, 421 (1944); **161**, 583 (1945).

[18] E. A. Zeller and A. Maritz, *Helv. Chim. Acta* **27**, 1888 (1944); T. P. Singer and E. B. Kearney, *Arch. Biochem.* **27**, 348 (1950); **29**, 190 (1950).

[19] S. Ratner, V. Nocito, and D. E. Green, *J. Biol. Chem.* **152**, 119 (1944).

[20] J. L. Still and E. Sperling, *J. Biol. Chem.* **182**, 585 (1950).

of molds, also called notatin.[21] Recently another FMN-requiring enzyme was found to oxidize glycolic acid.[22] In all of these enzymes, no components are known other than the flavin and amino acids.

Metallo-Flavoproteins. As was mentioned in the case of cytochrome reductase, enzymes are known that contain metal cofactors in addition to flavin. These are called metallo-flavoproteins. The presence of metals introduces complexity into the reaction, since the metals involved, iron, molybdenum, copper, and manganese, all exist in at least two valence states and can participate in oxidation-reduction reactions. The enzymes known to be metallo-flavoproteins include: xanthine oxidase, aldehyde oxidase, nitrate reductase, succinic dehydrogenase, fatty acyl CoA dehydrogenases, hydrogenase, and cytochrome reductases. Before these are discussed in detail some physical properties of flavin will be presented.

Physical Properties of Flavins. The most striking characteristic of riboflavin and its derivatives is the yellow color. This absorption spectrum reveals three major peaks, at 260 mμ, 375 mμ, and 450 mμ.[23] Both the 375 mμ and 450 mμ peaks disappear on reduction. This change in absorption permits the extent of oxidation or reduction to be measured spectrophotometrically or visually; on reduction the yellow color disappears as the leuco form is produced. The reaction is thought to be as indicated in (II). E_0 (pH 0) for simple flavin derivatives is -0.187 volts,

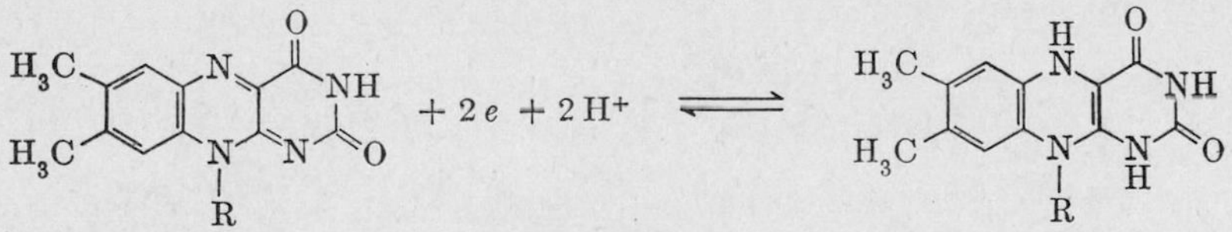

$$+ 2e + 2H^+ \rightleftharpoons$$

(II)

and $E_0{}'$ (pH 7) is about -0.219 volts. These compounds are classical examples of one-electron donors and acceptors. While many oxidations involving two electrons appear to proceed without appreciable time between the transfers of the individual electrons, some compounds exist as stable free radicals, and exhibit altered redox titration curves. The free radical form of riboflavin is seen as a red compound (with an absorption peak at 490 mμ) in strong acid[24] and as a green compound in neutral

[21] D. Muller, *Biochem. Z.* **199,** 136 (1928); C. E. Coulthard, R. Michaelis, W. F. Short, G. Sykes, G. E. H. Shrinshire, A. F. B. Standfast, J. H. Birkinshaw, and H. Raistrick, *Biochem. J.* **39,** 24 (1945); D. Keilin and E. F. Hartree, *ibid.* **42,** 221 (1948).

[22] I. Zelitch and S. Ochoa, *J. Biol. Chem.* **201,** 707 (1953).

[23] O. Warburg and W. Christian, *Biochem. Z.* **298,** 150 (1938).

[24] R. Kuhn and T. Wagner-Jauregg, *Ber.* **67,** 361 (1934).

solution.[25] The free radical form of the nucleotides has been calculated to represent over 10 per cent of the flavin when half reduced in neutral solution.[26] The ability to form stable free radicals may be useful in permitting riboflavin derivatives to transfer electrons efficiently to one-electron systems, such as iron complexes.

Riboflavin and the nucleotides FMN and FAD fluoresce, although at various pH values the relative intensities change sufficiently to permit this property to serve as the basis for a differential analysis of the flavin.[27] Fluorescence disappears as groups dissociating at pH 1.7 and 10.2 are titrated with acid and alkali, respectively. It has been concluded

that the dissociation with the pK of 10.2 is the imide   NH (position 3).[28]

Both the oxidation-reduction potential and the fluorescence of flavin nucleotides are modified profoundly by attachment of the nucleotide to various proteins. Flavin enzymes have been reported to have oxidation-reduction potentials at pH 7 ranging from -0.4 to 0.187. The combination to proteins also results in shifts of the absorption maxima. The 450 mμ band is found at 451 mμ in Straub's diaphorase and at 455 mμ in Haas' yellow enzyme, while the 375 mμ band appears at 359 mμ and 377 mμ in these preparations. Most flavin enzymes do not fluoresce, and it is assumed that the quenching of fluorescence implies binding of the flavin to the enzyme through N-3. Straub's diaphorase, unlike most other flavoproteins, does fluoresce. This may be evidence that this diaphorase is a partially degraded cytochrome reductase.

Xanthine Oxidase. One of the most thoroughly studied flavoproteins is xanthine oxidase. This enzyme has been purified from milk[29] and from liver.[30] Recently the milk enzyme, for many years known as the Schardinger enzyme, has been crystallized in a highly purified state.[31] For many years the absorption spectrum of the purified enzyme was a source of concern, because instead of the two distinct characteristic flavin peaks, an end absorption was always found to obscure the flavin spectrum. This has now been found to be the true spectrum of the enzyme. In addition to FAD, the enzyme contains iron and molybdenum in the ratio, 8 iron:2 FAD:2 Mo per mole of enzyme.[32] Molybdenum does not affect

[25] L. Michaelis, M. P. Schubert, and C. V. Smythe, *J. Biol. Chem.* **116,** 587 (1936).
[26] H. J. Lowe and W. M. Clark, *J. Biol. Chem.* **221,** 983 (1956).
[27] O. A. Bessey, O. H. Lowry, and R. Love, *J. Biol. Chem.* **180,** 755 (1949).
[28] R. Kuhn and P. Boulanger, *Ber.* **69,** 1557 (1936).
[29] B. L. Horecker and L. A. Heppel, *J. Biol. Chem.* **178,** 683 (1949).
[30] H. S. Corran, J. G. Dewan, A. H. Gordon, and D. E. Green, *Biochem. J.* **33,** 1694 (1939).
[31] P. G. Avis, F. Bergel, and R. C. Bray, *J. Chem. Soc.* p. 1100 (1955).
[32] D. A. Richert and W. W. Westerfeld, *J. Biol. Chem.* **209,** 179 (1954).

the spectrum, since the same curve is obtained with Mo-free preparations, and the addition of Mo causes no change in light absorption.

Xanthine oxidase oxidizes a large number of purines; hypoxanthine and xanthine are the substrates usually used, and are converted to uric acid. The spectral changes observed during these oxidations have been

Hypoxanthine   Xanthine   Uric acid

used both to measure the enzyme and as a quantitative assay of the purines.[33] In addition to purines, xanthine oxidase also oxidizes pterines, many different aldehydes, and DPNH. The enzyme is capable of passing electrons to many different acceptors, including molecular oxygen, cytochrome c, nitrate, quinones, and various dyes. The oxidized member of substrate couples, such as uric acid and DPN, can also serve as oxidants. Thus, xanthine oxidase catalyzes a dismutation of xanthine:[34] 2 xanthine $\rightleftharpoons$ hypoxanthine + uric acid.

The organization of xanthine oxidase appears to be quite complex. There is evidence that various substrates are not bound at the same site, and that the primary reaction of different substrates may occur with various ones of the cofactors. The oxidation of purines and aldehydes is inhibited by pteridyl aldehyde[35] and by cyanide,[36] but these reagents do not affect the oxidation of DPNH. It is possible that these inhibitors influence substrate binding sites and primary electron transport, respectively, and that the oxidation of DPNH involves a different binding site and avoids the cyanide-sensitive electron transport mechanism, which may well involve iron. Xanthine oxidase, and probably all flavoproteins, require —SH groups, but a definite function for these groups cannot be ascribed at this time.[37] Similarly, various factors influence the reactions with oxidants differentially. Cyanide inhibits cytochrome reduction, but not the reactions with O_2 or dyes.[36] Reduction of either cytochrome c or nitrate depends upon the presence of molybdenum.[38] These observations

[33] H. M. Kalckar, J. Biol. Chem. 167, 429 (1947).

[34] D. E. Green, Biochem. J. 28, 1550 (1934).

[35] H. M. Kalckar, N. D. Kielgaard, and H. Klenow, J. Biol. Chem. 174, 771 (1948); O. H. Lowry, O. A. Bessey, and E. J. Crawford, ibid. 180, 399 (1949).

[36] B. Mackler, H. R. Mahler, and D. E. Green, J. Biol. Chem. 210, 149 (1954).

[37] J. Harris and L. Hellerman, in Symposium on Inorg. Nitrogen Metabolism, p. 565 (1956).

[38] The interpretation that molybdenum is required for any of the reactions of xanthine oxidase has been questioned by Westerfeld and associates, who consider that the only effect of molybdenum is to stimulate dehydrogenase activity of xanthine oxidase.

and others have been the basis for models of the enzyme in which the role of metals is to permit the transfer of one electron at a time to acceptors, such as cytochrome c, that cannot accept more than one electron. Reactions with dyes and oxygen do not require molybdenum, and may be similar to the reactions of the *non*-metallo-flavoproteins, in which a direct reaction between the reduced flavin and the electron acceptor seems probable. The large number of structural elements that must be included in xanthine oxidase and the variety of reactions catalyzed by this enzyme leave considerable scope for speculation. A peculiar observation that cytochrome c reduction depends on the presence of molecular oxygen[29] is difficult to fit into any scheme. Ordinarily these oxidants compete for electrons. The properties above relate to milk xanthine oxidase. Very similar properties have been found for flavoproteins from mammalian intestine[39] and liver[40] and avian liver,[41] although different ratios of iron to molybdenum and flavin were found. These enzymes differ in the relative rates at which they reduce various oxidants, and the chicken liver enzyme has been called xanthine dehydrogenase, to emphasize the slow rate of the reaction with molecular oxygen.

Aldehyde Oxidase. Besides xanthine oxidase, liver contains a more specific aldehyde oxidase.[42] This enzyme does not oxidize reduced DPN or purines, but does convert a large number of aliphatic and aromatic aldehydes to the corresponding acids. This enzyme contains molybdenum and an iron protoporphyrin in addition to FAD.[43] It is completely reduced by aldehydes, whereas aldehydes bleach xanthine oxidase only partially (compared with hydrosulfite). Aldehyde dehydrogenase oxidizes DPNH at no more than 1 per cent of the rate of aldehyde oxidation. Cytochrome c, oxygen, and dyes all serve as oxidants.

Hydrogenase. Hydrogenase is the name of enzymes that utilize and produce molecular hydrogen. An extensive literature describes the properties of hydrogenase systems in microorganisms, but until recently only cellular or particulate preparations were used. The purification of some bacterial hydrogenases has led to the impression that these enzymes in general are metallo-flavoproteins. When the color of the enzyme is measured, the reduction of the enzyme by hydrogen can be seen. This reaction is reversible, and it has been concluded that the oxidation-

[39] E. C. De Renzo, P. G. Heytler, and S. Stolzenberg, *in Symposium on Inorg. Nitrogen Metabolism*, p. 507 (1956).

[40] R. K. Kielley, *J. Biol. Chem.* **216**, 405 (1955).

[41] C. Raney, D. A. Richert, R. J. Doisy, I. C. Wells, and W. W. Westerfeld, *J. Biol. Chem.* **217**, 293 (1955).

[42] A. H. Gordon, D. E. Green, and V. Subrahmanyan, *Biochem. J.* **34**, 764 (1940).

[43] H. R. Mahler, B. Mackler, D. E. Green, and R. M. Bock, *J. Biol. Chem.* **210**, 465 (1954).

reduction potential of the Clostridial hydrogenase is near that of the hydrogen electrode (-0.42 volts at pH 7).[44] Hydrogenases have been studied in preparations from many organisms, and it remains to be seen whether they are all metallo-flavoproteins. An enzyme from *Micrococcus lactolyticus* appears to require FAD, molybdic trioxide, and ferrous iron.[45] A preparation from *Hydrogenomonas ruhlandii* has been reported to contain manganese and FAD.[46]

Several reactions have been used to assay hydrogenase activity.[47] The most common has been the reduction of dyes, such as methylene blue, with hydrogen (III). This type of assay has the potential defect

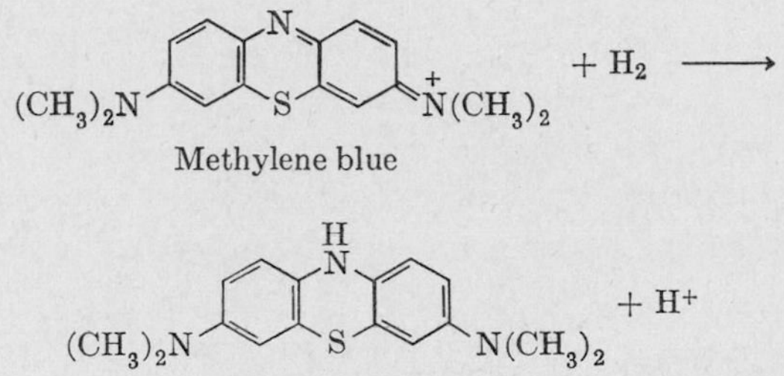

$$(CH_3)_2N \quad \overset{N}{\underset{S}{\bigcirc\bigcirc\bigcirc}} \quad \overset{+}{N}(CH_3)_2 \quad + H_2 \longrightarrow$$

Methylene blue

$$(CH_3)_2N \quad \overset{\overset{H}{N}}{\underset{S}{\bigcirc\bigcirc\bigcirc}} \quad N(CH_3)_2 \quad + H^+$$

Leucomethylene blue

that it may require electron-transferring enzymes in addition to hydrogenase. A similar assay uses the reverse reaction, the evolution of hydrogen from reduced dyes with low oxidation-reduction potentials, such as methyl viologen ($E_0' = -0.446$ volts). This is subject to the same potential dangers as the first method.

Another assay measures the rate of exchange between H_2 and D_2O. The appearance of HD can be followed with a mass spectrometer.

$$H_2 + D_2O \rightleftharpoons HDO + HD$$

A fourth assay measures the rate of interconversion of *ortho* and *para* hydrogen. Since any type of rupture of the hydrogen-hydrogen bond that permits random recombination will be detected by this assay, it was suggested that this is the most specific measurement of hydrogenase. However, the finding that a D_2O medium substitutes the exchange reaction (HD formation) for the *ortho-para* conversion indicates that

[44] A. L. Shug, P. W. Wilson, D. E. Green, and H. R. Mahler, *J. Am. Chem. Soc.* **76**, 3355 (1954).
[45] H. R. Whiteley and E. J. Ordal, *in Symposium on Inorg. Nitrogen Metabolism*, p. 521 (1956).
[46] L. Packer and W. Vishniac, *Biochim. et Biophys. Acta* **17**, 153 (1955).
[47] Various assays are discussed and compared in the article by H. D. Peck, Jr., A. San Pietro, and H. Gest, *Proc. Natl. Acad. Sci. U.S.* **42**, 13 (1956).

both assays require the same reactions; in both one hydrogen must exchange with that of the medium.

The various assays show grossly different relative rates with hydrogenase preparations from different organisms. At this time there is no agreement among workers in the field as to the mechanism of hydrogenase action.

Hydrogenase is apparently inactivated by molecular oxygen. Oxygen can be reduced by hydrogenase in the so-called "Knallgas" reaction. The relation of the reduction of oxygen, which may be slow, to enzyme inhibition is not clear. It has been suggested that an "oxygenated" enzyme may be the inhibited form.

Nitrate Reductase. Nitrate reductase is a name applied to certain flavoproteins containing molybdenum that reduce nitrate to nitrite at the expense of reduced pyridine nucleotides. The enzyme from *Neurospora* requires TPNH,[48] one from soybean leaves uses either TPNH or DPNH, while a preparation from the bacteria of soybean root nodules uses only DPNH.[49] These enzymes require molybdenum and FAD.

$$NO_3^- + TPNH + H^+ \rightarrow NO_2^- + TPN^+ + H_2O$$

Other enzymes catalyze similar reactions in the further reduction of nitrite, hydroxylamine, and organic nitro compounds.

Fatty Acyl CoA Oxidation. The two flavoproteins which have been isolated that oxidize fatty acyl CoA compounds differ in cofactors as well as in substrate specificity. The yellow enzyme, acting on longer fatty acid derivatives, may contain iron in addition to FAD.[50] The green enzyme, acting on 3- to 8-carbon chains, contains 2 atoms of copper per FAD.[51] As in the case of other metal-flavins, the copper appears to be required only for transfer to cytochrome c. The implicit role ascribed to the metal is difficult to reconcile with the recent report of additional flavoproteins with low metal content required to mediate between the fatty acyl CoA oxidases and either dyes or cytochrome c.[52] These enzymes, called "electron-transferring flavoprotein" or ETF, exhibit peaks near 375 and 437 mμ with a shoulder at 460 mμ.

Succinic Dehydrogenase. Soluble succinic dehydrogenase contains extremely tightly bound iron. In general enzymes that contain iron are inhibited by reagents that complex with iron, such as α,α'-dipyridyl or

[48] A. Nason and H. J. Evans, *J. Biol. Chem.* **202,** 655 (1953).
[49] H. J. Evans, *Plant. Physiol.* **29,** 298 (1954).
[50] F. L. Crane, S. Mii, J. G. Hauge, D. E. Green, and H. Beinert, *J. Biol. Chem.* **218,** 701 (1956).
[51] H. R. Mahler, *J. Biol. Chem.* **206,** 13 (1953).
[52] F. L. Crane and H. Beinert, *J. Biol. Chem.* **218,** 717 (1956).

o-phenanthroline. Succinic dehydrogenase merely forms a colored complex and continues to function.[53] The linkage of the flavin of succinic dehydrogenase, FAD, appears to be more secure than the linkages of other flavins. When the flavin is liberated by partial hydrolysis of the enzyme, no free FAD is found, but a series of compounds containing amino acids firmly attached to the flavin.[54]

The mechanism of action of succinic dehydrogenase has been investigated by measuring the exchange of methylene hydrogens with water.[55] The four methylene hydrogens represent two equivalent pairs. Assume that enzyme action requires attachment of one carboxyl group at a given point. Either carboxyl group of succinate may combine, as the two are equivalent, but in each case the adjacent methylene carbon bears "right" and "left" hydrogens, sterically oriented. It is conceivable that the removal of hydrogens may involve only the right or the left position but not both. In this case only two of the hydrogen atoms would be exchangeable. The finding that in fact all four do exchange eliminates all mechanisms that are stereospecific for both hydrogen atoms to be removed. This type of reaction has been called *nonstereospecific*.

The flavoproteins are an extremely important group of enzymes. For years the principal function assigned to them was the mediation of electron transfer between pyridine nucleotides and the cytochrome system. They undoubtedly do carry out this function, by a somewhat more complicated mechanism than previously considered. The possibility exists that during the transfer of electrons the reaction may be coupled with esterification of inorganic phosphate. Besides reducing cytochromes, flavin enzymes react directly with molecular oxygen, forming hydrogen peroxide.

Specific functions for individual flavins include activation of hydrogen among other specific substrates, participation in fatty acid oxidation, and serving as terminal oxidase for some organisms. The great variety of substrates and acceptors used by these enzymes is indicative of the versatility of the flavin structure, which alters its oxidation-reduction potential by several hundred millivolts through combination with specific proteins and cofactors. Teleologically, it is desirable to carry out biological oxidations by a series of reversible steps, in which the free-energy changes are relatively small. This is accomplished by including a series of enzymes that interact with each other and with substrates sequentially, with each couple having a standard potential in the neighborhood of the couples it reacts with. In this way reactions are easily reversed, and can

[53] E. B. Kearney and T. P. Singer, *Biochim. et Biophys. Acta* **17**, 596 (1955).
[54] E. B. Kearney, V. Massey, and T. P. Singer, *Federation Proc.* **15**, 286 (1956).
[55] S. Englard and S. P. Colowick, *J. Biol. Chem.* **221**, 1019 (1956).

be channeled into alternate sequences in response to altered circumstances. This property does not apply to terminal reactions with oxygen, which are irreversible. The reduction of oxygen by flavoproteins results in the formation of H_2O_2, whereas the more active terminal oxidases discussed in the following chapters do not form any detectible peroxide, but reduce oxygen to water.

GENERAL REFERENCES

Theorell, H. (1951). *In* "The Enzymes" (J. B. Sumner and K. Myrbäck, eds.), Vol. II, Part 1, p. 335. Academic Press, New York.

Mahler, H. R. (1956). *Advances in Enzymol.* **17**, 233.

Symposium on Inorg. Nitrogen Metabolism (W. D. McElroy and B. Glass, eds.). Johns Hopkins Press, Baltimore (1956). This volume contains numerous articles on various aspects of flavoprotein enzymes.

Cytochromes

The principal routes for acquiring energy through coupled reactions start with the reactions of carbohydrate and fat metabolism; the efficient utilization of these substrates includes complete oxidation to CO_2, whether by the Krebs cycle or by an alternate pathway. In the oxidation processes electrons are transferred from substrate molecules to pyridine nucleotides or to flavoproteins. Although many flavoproteins are capable of reacting with molecular oxygen, this mechanism for completing electron transfer is not of much quantitative significance. Instead the oxidation of reduced flavins is carried out by the cytochromes, a group of iron porphyrin proteins.

$$\text{substrate} \rightarrow \text{pyridine nucleotides} \rightarrow \text{flavoproteins} \rightarrow \text{cytochromes} \rightarrow O_2$$
$$\uparrow$$
$$\text{substrate}$$

Principal Pathways of Electron Transport

The extended series of electron transport reactions offers the following advantages:

(1) The opportunities for coupling oxidation with other reactions (oxidative phosphorylation) are increased.

(2) The sluggish autooxidation of flavoproteins, which is very sensitive to oxygen tension, is replaced by the rapid reaction of the cytochrome system, which uses oxygen efficiently even at low tensions.

(3) The reduction of oxygen by cytochromes leads directly to water, instead of to the potentially dangerous hydrogen peroxide.

Biosynthesis of Porphyrins, the Prosthetic Groups of Cytochromes. The cytochromes are all proteins conjugated with porphyrins, which means that they are derivatives of the heterocyclic ring system, porphin.

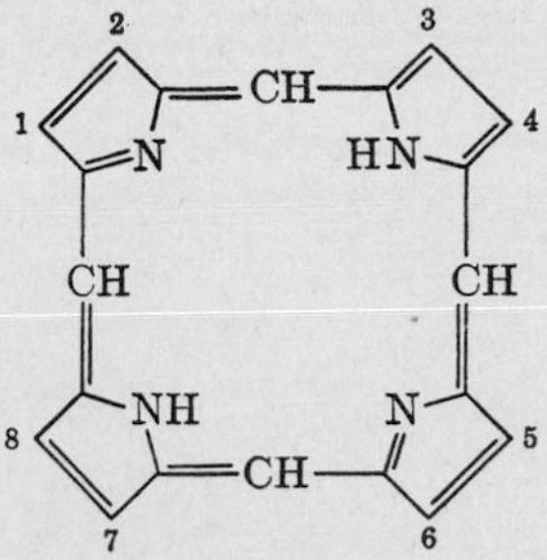

Porphin

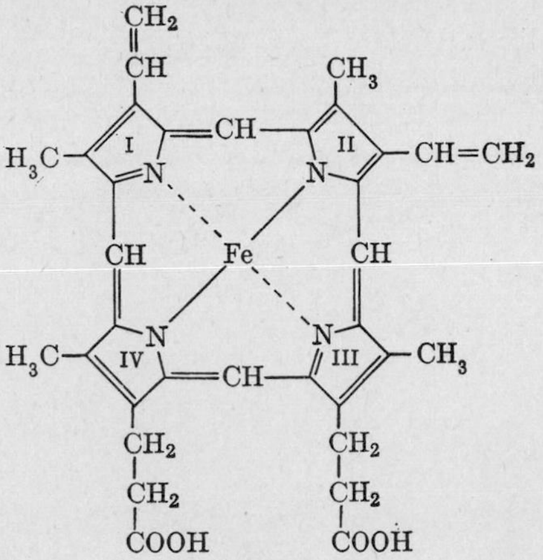

Hemin
(Iron Protoporphyrin IX)

The outlines of the biosynthetic pathway of the complicated porphyrin structure have been established through elegant use of isotopic methods and biochemical genetics and several of the intermediates have been established, but the details of the enzymatic reactions are only now being investigated. The atoms of porphyrins are all derived from "active succinate" and glycine.[1a] These condense to form α-amino-β-ketoadipic acid. The decarboxylation of this β-keto acid results in the formation of

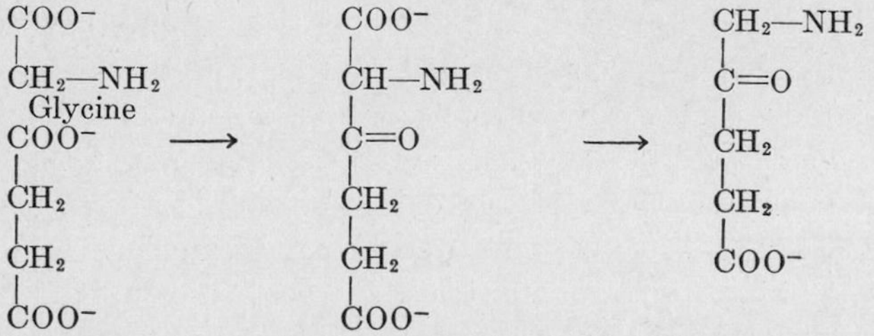

Succinate α-Amino-β-ketoadipate δ-Aminolevulinate
(I)

δ-aminolevulinic acid (I). The condensation of two molecules of δ-amino-levulinic acid establishes the pyrrole ring that is the basic structure of

[1a] Reactions in which a succinate group derived from α-ketoglutarate condenses with glycine to initiate a "succinate glycine cycle" have been reviewed by D. Shemin, *Federation Proc.* **15,** 971 (1956).

porphyrins (II). This condensation, involving two points in each of two substrate molecules, appears to be catalyzed by a single enzyme, δ-levulinic dehydrase amino, which has been studied in preparations of duck erythrocytes and has been purified 270-fold from acetone-dried ox liver.[1b] Many porphyrins are known in which various substituents are found in the eight numbered positions of porphin. These are all derived

Porphobilinogen

(II)

from porphobilinogen; in protoporphyrin, for example, methyl groups are formed by decarboxylation of the acetic acid side chains and vinyl groups come from decarboxylation and oxidation of propionic acid groups. In other porphyrins ethyl, hydroxyethyl, and formyl groups are found. The designation of protoporphyrin indicates the nature of the eight substituents, and IX specifies the order in which they occur, numbered from the pioneer studies of Fisher, who established the structures of porphyrins by synthesis. Protoporphyrin IX is found in many prominent biological compounds, including hemoglobin and the hydroperoxidases (p. 199), in addition to some cytochromes.[2] The physiologically active porphyrins are found as metal complexes. Although any of several metallic ions can be easily inserted into the ring system, the predominant metal associated with porphyrins is iron, and this is the only metal known to be a component of the cytochromes. In hemoglobin, which is not an enzyme,[3] the iron does not undergo oxidation or reduction, but stays in the ferrous state; the iron of the cytochromes functions as part of an electron-transport system, and shuttles between ferrous and ferric states.

Detection and Naming of Cytochromes. The cytochromes, as all compounds containing porphyrins, absorb light markedly. The absorption

---

[1b] K. D. Gibson, A. Neuberger, and J. J. Scott, *Biochem. J.* **61,** 618 (1955).

[2] Studies on the isolation, identification, and chemical synthesis of porphyrins are reviewed in the monograph of Lemberg and Legge (see general references).

[3] The principal function assigned to hemoglobin is oxygen transport. In this process hemoglobin combines stoichiometrically and reversibly with molecular oxygen. Since no catalysis is involved, the process cannot be described as enzymatic.

bands are so pronounced that they are easily detected spectroscopically in intact organisms. Unfortunately most of the cytochromes are firmly bound to cell particles and only recently have these been solubilized. Therefore most of the available information on these important hemoproteins is derived from studies on insoluble, unfractionated preparations. The light-absorbing components of the particles were identified by their absorption bands. As similar but not identical bands were found in various organisms, the nomenclature of the cytochromes became more complicated and less meaningful.

A brief historical résumé is essential for an appreciation of the current status of this field and its terminology. The first demonstration of tissue pigments with absorption bands in the reduced state was made by MacMunn.[4] He found "histohaematins" in a large number of animal species, and associated them with tissue respiration. Because of the ridicule of the more authoritative Hoppe-Seyler, MacMunn's studies were ignored for almost 40 years. Then Keilin[5] confirmed and extended the spectroscopic studies of MacMunn, and found the pigments which he named "cytochromes" in plants and microorganisms as well as animal muscle.

The cytochromes of beef heart have sharp absorption bands in the reduced state; by adjusting the concentration of cell particles, samples can be prepared for spectroscopic examination which show bands at 605 mμ, 564 mμ, 550 mμ, 520 mμ, 445 mμ, 430 mμ, and 415 mμ. These bands represent peaks in the absorption spectra of at least three different pigments.[6] The component with bands at 605 mμ and 445 mμ is called cytochrome a; the bands at 564 mμ, 520 mμ, and 430 mμ are attributed to cytochrome b; cytochrome c is responsible for absorption at 550 mμ, 520 mμ, and 415 mμ. The bands at the longest wavelength (605, 564, and 550 mμ) are the α bands of each cytochrome (a, b, and c). The 520 band is the β band of cytochromes b and c. The bands at the lower wavelengths are called Soret bands or γ bands. The association of the various bands with individual compounds was made by noting the appearances and disappearances on changing from aerobic to anaerobic condition and back, and on the addition of poisons to the preparation.

Very similar patterns have been seen in yeast, bacteria, insects, and many other organisms. As will be seen, the various cytochromes have been assigned certain chemical properties and biological functions. There

[4] C. A. MacMunn, *Phil. Trans. Roy. Soc. London* **B177,** 267 (1886).

[5] D. Keilin, *Proc. Roy. Soc.* **B98,** 312 (1925).

[6] The absorption bands observed at ordinary temperatures are not as sharp as those seen at liquid air temperatures. Cold samples show the presence of additional components, as the diffuse bands seen at higher temperatures are resolved.

has been a tendency to associate these properties with absorption bands, but the recent discovery of cytochromes with similar spectra but different properties has forced a revision of this concept. The role of cytochromes in respiration was developed through studies involving many different organisms, which fortunately have similar cytochromes. The following discussion describes the current ideas about cytochromes as derived from the materials commonly studied, but any aspect may be different in systems from other sources.

The sequence of reactions in which the cytochromes participate is a mechanism for transferring electrons to molecular oxygen via iron complexes that are alternately in ferric and ferrous states. The order of the transfer has been deduced from studies with inhibitors, in which the electron-transport chain is broken so that components below the break are reduced, those above are oxidized; from studies with poised potentials, in which the relative degrees of oxidation and reduction define the oxidation-reduction potentials of the various components; and from rapid kinetic measurements, in which the order of reduction or oxidation can be seen. These methods agree on the following sequence:

electron donor $\rightarrow$ cytochrome b $\rightarrow$ cytochrome c_1 $\rightarrow$ cytochrome c $\rightarrow$
$$\text{cytochrome a} \rightarrow \text{oxygen}$$

Cytochrome c. The one cytochrome that is well studied is cytochrome c. It is a remarkably stable protein, surviving extraction by dilute trichloroacetic acid.[7] The pigment in the extract has been purified by ammonium sulfate and acid precipitation, ammoniacal ammonium sulfate fractionation, electrophoresis, and ion exchange chromatography. The purest material contains 0.46 per cent iron, corresponding to a molecular weight of a little over 12,000 with one atom of iron per molecule.[8] Cytochrome c from ox heart contains a large number of lysine residues, which make the protein very basic, with an isoelectric point over pH 10.[9]

In cytochrome c, and in all the known porphyrin enzymes, the ring contains an iron atom complexed with the 4 nitrogen atoms, forming a hematin. The iron is presumably linked to the protein through imidazole rings of histidine residues. Another linkage binds the porphyrin to the protein; the vinyl side chains are bound to cysteine residues through addition of the sulfhydryl groups to the double bonds, forming thioethers.[10]

[7] D. Keilin and E. F. Hartree, *Proc. Roy. Soc.* **B122,** 298 (1937). Similar methods had been used by H. Theorell, *Biochem. Z.* **285,** 207 (1936).

[8] E. Margoliash, *Biochem. J.* **56,** 529 (1954).

[9] H. Theorell and A. Akeson, *J. Am. Chem. Soc.* **63,** 1809 (1941).

[10] H. Theorell, *Biochem. Z.* **298,** 242 (1938); K.-G. Paul, *Acta Chem. Scand.* **4,** 239 (1950).

Digestion of beef cytochrome c with pepsin or trypsin results in the formation of peptides still conjugated with the porphyrin. Analysis of the amino acids shows the structure of part of cytochrome c to be as indicated below.[11] Similar but not necessarily identical amino acid sequences occur in cytochrome c from other sources.

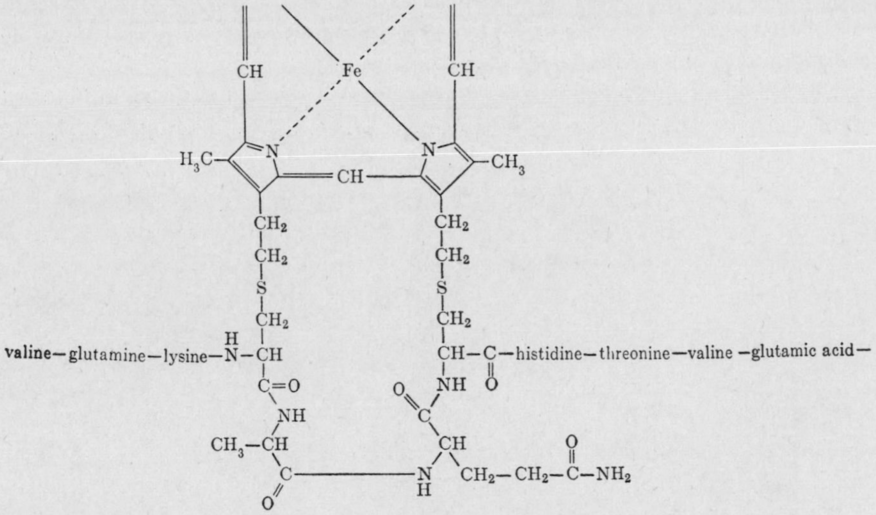

Structure of Part of Cytochrome c

As discussed previously, cytochrome c is reduced by several flavoproteins. On enzymatic reduction or reaction with hydrosulfite, ascorbic acid, or any of several other reducing reagents, the typical spectrum of reduced cytochrome c is produced. The band at 550 mμ is usually used to assay reduced cytochrome c. The reduction causes a change in the iron from the ferric to the ferrous state. The oxidation-reduction potential of the couple, ferricytochrome:ferrocytochrome at neutral pH values (5–8) is +0.256 volts.[12]

Ferricytochrome c reacts very poorly with cyanide and azide to give solutions with modified cytochrome spectra.[13] These complexes have large dissociation constants, and in the concentrations of inhibitors ordinarily used, cytochrome c is largely unbound. Since cytochrome oxidase is inhibited by low concentrations of cyanide and azide, these inhibitors can be used to prevent the oxidation of ferrocytochrome c without interfering with the reduction of ferricytochrome c.

[11] H. Tuppy and G. Bodo, *Monatsh. Chem.* **85**, 807, 1024, 1182 (1954); S. Paleus, A. Ehrenberg, and H. Tuppy, *Acta Chem. Scand.* **9**, 374 (1955).
[12] F. L. Rodkey and E. G. Ball, *J. Biol. Chem.* **182**, 17 (1950).
[13] B. L. Horecker and A. Kornberg, *J. Biol. Chem.* **165**, 11 (1946).

Compounds resembling cytochrome c of heart muscle have been isolated from many sources. The cytochrome c of skeletal muscle of the king penguin has been purified by the same methods used for ox heart cytochrome c, and has been crystallized from "almost saturated" ammonium sulfate.[14] A precipitate of the oxidized cytochrome was obtained that exhibited birefringence but was not crystalline; addition of reducing agent to the ammonium sulfate solution permitted crystals of reduced cytochrome c to form. Measurements of the catalytic activity and extinction coefficients gave values equivalent to those of the ox heart preparation within the limits of error. In a preliminary report, the

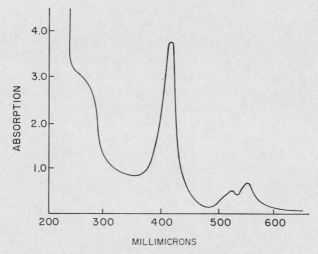

FIG. 23. Absorption spectrum of reduced cytochrome c.[14a]

crystallization of ox heart cytochrome c has been announced.[15] Cytochrome c preparations from the hearts of many other animals also resemble that of ox heart. A cytochrome c from a fungus, *Ustilago*, differs from mammalian cytochrome c in that its isoelectric point is near 7, but spectrally and catalytically the two are very similar.[16] In contrast to the cytochromes of vertebrates, the cytochromes of several bacteria have been found to include compounds with a c-type spectrum but very different reactivities.[17] Using the reduction and oxidation of the c-type cytochromes by bacterial and mammalian enzymes, Kamen and Vernon found that the hemoprotein from each of several photosynthetic bacteria

[14] G. Bodo, *Nature* **176,** 829 (1955).
[14a] D. Keilin, *Ergeb. Enzymforsch.* **2,** 239 (1933).
[15] S. A. Kuby, S. Paleus, K.-G. Paul, and H. Theorell, *Acta Chem. Scand.* **10,** 148 (1956).
[16] J. B. Nielands, *J. Biol. Chem.* **197,** 701 (1952).
[17] M. D. Kamen and L. P. Vernon, *Biochim. et Biophys. Acta* **17,** 10 (1955).

could be differentiated from the others. The c-type cytochrome of *Rhodospirillum rubrum*, for example, was found to have an oxidation-reduction potential near 0.30 volts at pH 8, somewhat dependent on ionic strength. Below pH 7.5 the potential increased 0.03 volts per pH unit, whereas mammalian cytochrome c is not affected by pH changes in this region (5–8). A cytochrome isolated in a high state of purity from *Pseudomonas aeruginosa* is almost identical with mammalian cytochrome c spectroscopically, but is quite different in other physical properties. It fails to react with mammalian enzymes, but reacts with bacterial enzymes to reduce oxides of nitrogen.[18]

Particulate Cytochromes. Most of the information about cytochromes a and b has been obtained with particulate systems. Several investigators have extracted various proteins from washed particles with deoxycholate solutions. Such solubilized fractions can be further fractionated to yield materials that exhibit the spectra attributed to the various cytochromes, but these studies have not yielded materials that can be recombined to reconstruct the efficient electron-transport system of intact particles.

Succinoxidase. Washed particles from heart muscle oxidize succinate to fumarate with oxygen as electron acceptor.[19] This is the so-called succinoxidase preparation. The initial step in the oxidation of succinate has recently been found to be catalyzed by an iron-flavoprotein.[20] When this is isolated from other components of succinoxidase, it fails to react with other enzymes and reduces efficiently only phenazines of the several oxidants that are effective with more complex preparations. Partially fragmented particles have been obtained by differential centrifugation of isobutanol-treated preparations of larger particles from mitochondria or by digestion with trypsin in the presence of cholate. These treatments yield preparations that can reduce cytochrome c.[21]

Cytochrome b. A fundamental difference between soluble succinic dehydrogenase and all of the particles capable of reducing oxidants other than phenazines is the presence in the latter preparations of heme. Reduction of the enzyme by succinate or hydrosulfite causes the appearance of bands near 560 mμ, 525 mμ, and 428 mμ. These are approximately the bands assigned to cytochrome b. When succinoxidase is treated with urethane, the over-all reaction is inhibited and on the addition of substrate the bands of reduced cytochrome b appear, but the other cytochromes remain oxidized.[22] This observation can be duplicated by using

[18] M. D. Kamen and Y. Takeda, *Biochim. et Biophys. Acta* **21**, 518 (1956).
[19] D. Keilin and E. F. Hartree, *Proc. Roy. Soc.* **B129**, 277 (1940).
[20] E. B. Kearney and T. P. Singer, *Biochim. et Biophys. Acta* **17**, 596 (1955).
[21] D. E. Green, P. M. Kohout, and S. Mii, *J. Biol. Chem.* **217**, 551 (1955).
[22] D. Keilin and E. F. Hartree, *Proc. Roy. Soc.* **B127**, 167 (1939).

2,3-dimercaptopropanol (British anti-Lewisite, BAL)[23] or the mold product, antimycin A.[24] The mechanism of the inhibitions is not established; it has been suggested in some cases that the effect of the inhibitor may be caused by disruption of the physical state of the particle instead of the more usual combination with an enzyme. The interpretation of these observations has been that cytochrome b is reduced early in the electron-transport sequence, and the subsequent transfer of electrons to cytochrome c is inhibited. Cytochrome reductases are not inhibited by any of the inhibitors listed, so the effect is not attributed to an interaction with cytochrome c. Preparations that mediate electron transport between succinate and cytochrome c have been described and named SC factors (succinic dehydrogenase-cytochrome c combining factor).[25] At least some of these preparations contain a hemoprotein with an α band at 552 mμ. The extremely close position of this band to the 550 mμ band of cytochrome c led to controversy about the identity of the material. It is now generally accepted that another component, cytochrome c_1 (also called cytochrome e), exists, and may participate in electron transfer from b to c.[26a] The effect of BAL was shown by Slater to interfere with the oxidation of cytochrome b by cytochrome c, but not the oxidation by other oxidants. Therefore he concluded that a "factor" between the two cytochromes is the BAL-sensitive material. The so-called Slater factor has not been identified with any individual component of the electron-transport system.

Many investigators have shown that nonspecific reagents as diverse as calcium phosphate gel, EDTA, histidine, and nonspecific proteins activate succinoxidase preparations in otherwise unfavorable environments. The mechanism of the activation is not established, but it has been repeatedly suggested that the activators in some manner influence the steric orientation of components of the particulate succinoxidase. Another component of electron-transport systems has been implicated by Nason and Lehman. DPNH oxidation by a particulate fraction of rat muscle was found to be decreased by extraction of 10 per cent of the lipid with isooctane; the activity was restored by addition of α-tocopherol (vitamin E) or the lipid extracted from muscle or bovine serum albumin. These lipids are able to reverse the inhibition of cytochrome c reduction caused by antimycin A. It has not been determined whether the tocoph-

[23] E. C. Slater, *Biochem. J.* **45,** 14 (1949).

[24] K. Ahmad, H. G. Schneider, and F. M. Strong, *Arch. Biochem. Biophys.* **28,** 281 (1950).

[25] F. B. Straub, *Z. physiol. Chem.* **272,** 219 (1942); H. A. Neufeld, C. R. Scott, and E. Stotz, *J. Biol. Chem.* **210,** 869 (1954).

[26a] E. Yakushiji and K. Okunuki, *Proc. Imp. Acad. (Tokyo)* **16,** 299 (1940); C. Widmer, H. W. Clark, H. A. Neufeld, and E. Stotz, *J. Biol. Chem.* **210,** 869 (1954).

erol or other lipids participate in electron transport or in maintenance of an optimal spacial orientation of the cytochromes.[26b]

Catalytic properties of cytochrome b have been observed only in particulate preparations. A highly purified preparation of soluble cytochrome b has recently been described as crystalline, but no enzymatic properties have been reported at this time.[27]

Cytochrome Oxidase. Reduced cytochrome c is oxidized in particles by cytochrome a. This is consistent with the difference in oxidation-reduction potentials measured by Ball of $+0.25$ volts for cytochrome c and $+0.29$ volts for cytochrome a.[28] Despite years of intensive investigation, both the chemical nature and physiological functions of cytochrome a remain uncertain. One of the major questions concerns the relation of cytochrome a to the terminal enzyme of the series, cytochrome oxidase, which reacts with molecular oxygen.

Some properties of cytochrome oxidase were discovered before any of the components of the electron-transport system were identified. The catalytic role of the cytochromes was approached indirectly by Warburg, who was impressed with the similarity in certain properties of biological oxidations and model systems in which iron-charcoal catalyzed oxidations. The concept of the "Atmungsferment," the respiratory enzyme, as an iron catalyst that activates molecular oxygen was derived from these studies.[29] Warburg found that the oxidation of biological systems is inhibited by carbon monoxide, which had been shown to combine with ferrous forms of hemoglobin in a light-sensitive complex. In the presence of high concentrations of CO, yeast respiration was found to be greatly decreased, but oxygen consumption was restored by bright light, and the extent of the restoration was dependent on the wavelength as well as intensity of the light. Warburg and Negelein measured the reactivation spectrum of CO-inhibited yeast. The action spectrum showed peaks at 590 mμ, 540 mμ, and 432 mμ, in general resembling the absorption spectrum of a cytochrome.[30] It is possible that the spectrum may not be precise, because of transfer of energy from other pigments to the CO-complex and because of shading by inert pigments; nevertheless, in general it is believed that the spectrum is that of the CO-complex of the terminal enzyme, cytochrome oxidase.

Additional information about the catalytic role of the cytochromes

[26b] A. Nason and I. R. Lehman, *J. Biol. Chem.* **222**, 511 (1956).

[27] I. Sekuzu and K. Okunuki, *J. Biochem.* **43**, 107 (1956).

[28] E. G. Ball, *Biochem. Z.* **295**, 262 (1938).

[29] O. Warburg, *Ber.* **58**, 1001 (1925).

[30] O. Warburg and E. Negelein, *Biochem. Z.* **214**, 64 (1929). Other similar measurements were made with bacteria as well as yeast by F. Kubowitz and E. Haas, *Biochem. Z.* **255**, 247 (1932).

was made through a study of the oxidation of artificial substrates. Ehrlich had found in 1885 that a phenylenediamine and α-naphthol reacted in an oxidative system to form a colored indophenol. Subsequently the enzyme that oxidizes p-phenylenediamine to the compound that couples with α-naphthol was called *indophenol oxidase.*[31] Indophenol oxidase is now believed to be a combined function of cytochrome c and cytochrome oxidase. Cytochrome c is reduced by p-phenylenediamine and is then oxidized by an enzyme that is sensitive to CO in the same manner as the Atmungsferment.

Originally the absorption band at 605 mμ was assigned to a single compound. When the spectra of carbon monoxide- and cyanide-treated preparations from ox heart were studied, it appeared that only part of the cytochrome a reacted with the inhibitors. The reaction was seen spectroscopically as a splitting of the α band in the presence of CO. Spectrophotometrically, a new band was seen at 590 mμ and the 605 mμ absorption decreased in intensity; corresponding changes were seen in the Soret band. The component that retained the original absorption bands was designated cytochrome a, while the part that reacted with the inhibitor was named cytochrome a_3, and has been identified with cytochrome oxidase.[22] It must be emphasized, however, that the only evidence for separate cytochromes a and a_3 is derived from spectral analysis of inhibitor complexes. An alternate explanation of the observations is that the organization of cytochrome oxidase might orient four hemes so that only one CO or HCN molecule could interact with any of the four at one time; in this case all of the four hemes could be identical, all would be cytochrome a, but in the presence of inhibitors two components would be seen. If the inhibitor occupied the space normally used by oxygen, reaction of one heme in four would tie up all of the cytochrome oxidase.[32]

The chemistry of cytochrome oxidase is not well known. From its spectrum the prosthetic group isolated by Negelein[33] is similar to that of a hemin from the snail, *Spirographis.*[34] This compound contains the structure 1,3,5,8-tetramethyl-2-formyl-4-vinylporphin-6,7-propionic acid.[35a] A very similar compound has been crystallized recently from horse heart muscle, but its precise structure is not yet established.[35b] A green protein isolated from bile salt-solubilized particles has been

[31] D. Keilin, *Proc. Roy. Soc.* **B104,** 206 (1929).
[32] E. G. Ball, C. F. Strittmatter, and O. Cooper, *J. Biol. Chem.* **193,** 635 (1951).
[33] E. Negelein, *Biochem. Z.* **266,** 412 (1933).
[34] O. Warburg and E. Negelein, *Biochem. Z.* **244,** 239 (1932).
[35a] H. Fischer and C. von Seeman, *Z. physiol. Chem.* **172,** 594 (1948).
[35b] O. Warburg, H. Gewitz, and W. Völker, *Z. Naturforsch.* **10b,** 541 (1955).

reported to contain copper as well as iron.[36] The metal in native cytochrome oxidase has not been identified with certainty. Cytochrome oxidase forms complexes with HCN and HN_3 that dissociate very slightly ($K_{diss} = 5 \times 10^{-7}$ and 7×10^{-7}, respectively).[37] It has been established that the undissociated acids, not the anions, combine with cytochrome oxidase. These complexes are catalytically inert. Therefore, relatively low concentrations of cyanide or azide eliminate oxygen consumption by cytochromes, since the other cytochromes react very sluggishly, if at all, with oxygen. In the presence of the poisons and substrate, all of the cytochromes appear in their reduced forms.

Soluble Cytochrome a. Compounds with the spectral properties of cytochrome a have been obtained in solution by treatment of particles with bile salts.[25,38] Such preparations appear to require the presence of bile salt to prevent precipitation, but in the presence of the surface-active substance, fractionation can be carried out. The water insolubility is probably a function of a large lipid content. The preparations can be reduced by reduced cytochrome c, and are autooxidizable. The autooxidation results in the formation of water, not H_2O_2. It is interesting to note that this solubilized preparation is rather rapidly reduced by the flavoprotein, aldehyde oxidase.[39]

Structural Integration of Cytochromes. The cytochromes must exist naturally in a structure that permits an easy, rapid, efficient flow of electrons from one to the other, since the molecules are not free to move about, yet they handle the bulk of tissue respiration. The easily extracted cytochrome c may seem to be an exception, but even in this case it has been found that small amounts of undissociated cytochrome c are turned over about 100 times as fast as the soluble pigment.[40]

The succinoxidase system has been described as a particle containing cytochromes b, c, and a in addition to succinic dehydrogenase and any other (unidentified) enzymes needed for oxygen consumption. The oxidation of reduced DPN requires similar factors when particulate preparations are used. The role of cytochrome b in these systems has not been defined to the satisfaction of all. A difficulty in assigning a place to cytochrome b lies in the nature of the kinetic experiments and the assumptions made about the properties of the systems. It is not questioned that this cytochrome is reduced by DPNH or by succinate; the question is

---

[36] See general reference by Green.
[37] B. L. Horecker and J. N. Stannard, *J. Biol. Chem.* **172**, 594 (1948).
[38] E. Yakushiji and K. Okunuki, *Proc. Imp. Acad. (Tokyo)* **17**, 38 (1941); W. W. Wainio, S. J. Cooperstein, S. Kollen, and B. Eichel, *J. Biol. Chem.* **173**, 145 (1948).
[39] J. Hurwitz and S. J. Cooperstein, *J. Biol. Chem.* **212**, 771 (1955).
[40] E. C. Slater, *Biochem. J.* **44**, 305 (1949).

whether it is reduced at a sufficient rate to be an obligatory intermediate in the reduction of cytochrome c.[41]

Microsomal Cytochrome. A cytochrome m was detected in microsomes by C. F. Strittmatter and Ball.[42] This has been identified with the material identified spectrally and named cytochrome b_1.[43] It has also been called cytochrome b_5,[44] which name has been used for a compound with a similar spectrum found in homogenates of a moth larva midgut.[45] This cytochrome from rabbit liver microsomes was purified by P. Strittmatter and Velick,[46] who measured its physical properties. The molecular weight is about 17,000 and the heme appears to be ferriprotoporphyrin IX. The absorption bands of the reduced cytochrome are at 557 mμ, 527 mμ and 423 mμ. The oxidation-reduction potential between pH 5.2 and 6.4 is $+0.03$ volts.[47] Separate enzymes have been detected in microsomes that reduce microsomal cytochrome with DPNH and TPNH. The DPNH enzyme does not reduce cytochrome c, but, since microsomal cytochrome reacts rapidly with cytochrome c, in the presence of small amounts of cytochrome m cytochrome c is reduced rapidly.[48]

Reduction of Oxygen to Water. A problem that has no answer today is the mechanism of oxygen reduction by cytochrome oxidase. The autooxidizable flavoproteins and various model metal-catalyzed reactions all reduce oxygen to hydrogen peroxide. The cytochrome system converts molecular oxygen to water without the intermediate formation of detectable amounts of peroxide. The alternative explanations are: peroxide is formed in amounts too small to detect; four cytochrome chains converge on a single oxygen molecule; or oxygen is incorporated into one of the components of the system, which is reduced by a succession of electrons until it is again at the proper state for reacting with molecular oxygen. It is probable that experiments will not be devised to determine the actual mechanism until more is learned about the nature and organization of the cytochrome oxidase system.

Other Cytochromes. The cytochromes discussed above are representative of porphyrin proteins found in most biological material. There are additional modifications of these compounds known. Certain microorganisms have terminal oxidases that resemble cytochrome a, and have

[41] This question is discussed in the reviews of D. Keilin and E. C. Slater and of B. Chance and G. R. Williams.
[42] C. F. Strittmatter and E. G. Ball, *J. Cellular Comp. Physiol.* **43,** 57 (1954).
[43] H. Yoshikawa, *J. Biochem.* **38,** 1 (1951).
[44] B. Chance and G. R. Williams, *J. Biol. Chem.* **209,** 945 (1954).
[45] B. Chance and A. M. Pappenheimer, Jr., *J. Biol. Chem.* **209,** 931 (1954).
[46] P. Strittmatter and S. Velick, *J. Biol. Chem.* **221,** 253 (1956).
[47] S. Velick and P. Strittmatter, *J. Biol. Chem.* **221,** 265 (1956).
[48] P. Strittmatter and S. Velick, *J. Biol. Chem.* **221,** 277 (1956).

been called a_1 and a_2. In addition, bacterial cytochromes have been labeled b_1, b_2, etc. The functions of these in many cases may resemble the functions of the corresponding mammalian cytochromes. The oxidation-reduction potentials vary considerably, however, and each case must be evaluated separately.[49] A cytochrome b_2 has been identified with yeast lactic dehydrogenase.[50] This enzyme, containing FMN in addition to a

TABLE 6
ABSORPTION BANDS OF SOME CYTOCHROMES

Reduced	α	β	γ
a (heart)	605	—	445
a_1 (*Acetobacter pasteurianum*)	590	—	435–440
b (yeast)	564	525	430
b (*E. coli*)	560	533	432
c (heart)	554	521	415
c (*Pseudomonas*)	550	520	416
c_1 (heart)	554	524	418
c_m (liver)	556	526	423
CO complexes			
a	590	—	430
a_1	590	—	427

porphyrin, has been crystallized.[51] Hill and associates have isolated another iron porphyrin compound, cytochrome f, from green leaves.[52] This material has an E_0' of 0.365 volts. Because of its biological origin, cytochrome f has been implicated in the electron transport of photosynthesis. Still other cytochromes have been found in other organisms, and it is apparent that a large number of variations have been composed about the iron porphyrin theme to accomplish a variety of biochemical functions.

fatty acyl CoA
succinate   } → flavoprotein
α-glycerophosphate
                                               (factor?)
substrates → pyridine nucleotide → flavoproteins → cyt. b → cyt. c_1 → cyt. c →
substrates → pyridine nucleotide → flavoprotein → cyt. m ————————↑

cyt. a → (cyt. a_3?) → O_2
(III)

Integration of Electron Transport. It is not possible to reconcile the conflicting evidence about the nature and functions of the cytochromes

[49] Many bacterial cytochromes are described in the review by L. Smith.
[50] S. J. Bach, M. Dixon, and L. G. Zerfas, *Biochem. J.* **40**, 229 (1946).
[51] C. A. Appleby and R. K. Morton, *Nature* **173**, 749 (1954).
[52] R. Hill and R. Scarisbrick, *New Phytologist* **50**, 98 (1951); H. E. Davenport and R. Hill, *Proc. Roy. Soc.* **B139**, 327 (1952).

at this time. The order of electron transport considered most likely for animal tissues is shown in (III). Other organisms obviously employ variations. It is probable that the flow of electrons is rigidly controlled by the physical structures of cell particles, but it is possible that several parallel pathways occur in individual cells, and that a single mechanism cannot be written for the oxidation of individual electron donors.

GENERAL REFERENCES

Chance, B., and Williams, G. R. (1956). *Advances in Enzymol.* **17**, 65.
Granick, S. (1954). *Chem. Pathways Metabolism* **2**, 287.
Green, D. E. (1956). *In* "Enzymes: Units of Biological Structure and Function" (O. H. Gaebler, ed.), p. 465. Academic Press, New York.
Kamen, M. D. (1956). *In* "Enzymes: Units of Biological Structure and Function" (O. H. Gaebler, ed.), p. 483. Academic Press, New York.
Keilen, D., and Slater, E. C. (1953). *Brit. Med. Bull.* **9**, 87.
Lemberg, R., and Legge, J. W. (1949). "Hematin Compounds and Bile Pigments." Interscience, New York.
Paul, K.-G. (1951). *In* "The Enzymes" (J. B. Sumner and K. Myrbäck, eds.), Vol. II, Part 1, p. 357. Academic Press, New York.
Smith, L. (1954). *Bacteriol. Rev.* **18**, 106.
Stotz, E. H., Morrison, M., and Marinetti, G. (1956). *In* "Enzymes: Units of Biological Structure and Function" (O. H. Gaebler, ed.), p. 401. Academic Press, New York.
Wainio, W. W., and Cooperstein, S. J. (1956). *Advances in Enzymol.* **17**, 329.

The Hydroperoxidases: Catalase and Peroxidase

Hydrogen peroxide is produced in the autooxidation of flavoproteins and many other compounds, and therefore occurs naturally in biological systems. Some organisms, obligate anaerobes, do not contain efficient mechanisms for metabolizing peroxides, which accordingly are toxic to these organisms. Most organisms, however, do contain enzymes that utilize H_2O_2; these enzymes may even be required for essential cellular functions.

Catalase is the name given to enzymes that carry out the reaction:

$$2H_2O_2 \rightarrow O_2 + 2H_2O \text{ (catalatic reaction)}$$

Peroxidases carry out oxidations with various substrates:

$$AH_2 + H_2O_2 \rightarrow A + 2H_2O \text{ (peroxidatic reaction)}$$

Hydrogen peroxide was discovered in 1818 by Thénard,[1] and the ability of biological material as well as inorganic substances to catalyze

[1] L. J. Thénard, *Acad. Sci. Paris* (1818).

the evolution of oxygen from H_2O_2 was documented by him in studies that anticipated the introduction of the term "catalysis" by Berzelius. The activity of biological material in decomposing H_2O_2 has played an important role in the development of enzyme chemistry. Following the demonstration by Jacobson that decomposition of H_2O_2 is *not* a function of all enzymes,[2] Loew coined the name *catalase* for the oxygen-liberating enzyme.[3] Catalase has been crystallized from both animal[4] and bacterial[5] sources, and the pure enzyme has been subjected to intense chemical, physical, and enzymatic investigation. Many interesting properties of catalase have been disclosed, but the nature of the enzyme mechanism remains uncertain.

It is convenient to discuss catalase and the peroxidases separately, but the distinction is arbitrary. Peroxidases are enzymes that catalyze oxidations of various substrates by H_2O_2. Catalase activity may be considered a special case of such an oxidation, in which H_2O_2 serves as both oxidant and reductant. The discovery that catalase also catalyzes the peroxidation of numerous other molecules,[6] albeit slowly, establishes this enzyme as one of the family of *hydroperoxidases*.

CATALASE

Assay Methods. The conventional assay for catalase activity involves determination of the first-order rate constant under specified conditions:[7] 0.01 M H_2O_2 and 0.0067 M phosphate pH 6.3 at 0°C. One milliliter of enzyme is added to 50 ml. of reaction mixture and 5 ml. samples are titrated in H_2SO_4 with $KMnO_4$. The reaction constant is evaluated:

$$K = \frac{1}{t_2 - t_1} \times \log \frac{x_1}{x_2}$$

in which x represents the peroxide concentration (which, of course, may be expressed in terms of the permanganate used in the titration). The rate constant is a measure of the catalase activity in the test solution. The purity or potency of catalase preparations is conventionally expressed as "Katalasefahigkeit," abbreviated Kat.f. Kat.f. = K/grams of enzyme in standard test.

Catalase has been assayed by many other procedures. The most

[2] J. Jacobson, *Z. physiol. Chem.* **16**, 340 (1892).
[3] O. Loew, *U.S. Dept. Agr. Rept.* **68**, 47 (1901).
[4] J. B. Sumner and A. L. Dounce, *J. Biol. Chem.* **121**, 417 (1937).
[5] D. Herbert and J. Pinsent, *Biochem. J.* **43**, 193 (1948).
[6] D. Keilin and E. F. Hartree, *Biochem. J.* **39**, 293 (1945).
[7] H. von Euler and K. Josephson, *Ann.* **452**, 158 (1927).

obvious measurement is the manometric determination of the evolved oxygen, but this is subject to many errors when carried out in conventional manometers.[8] One difficulty encountered is inactivation of the enzyme in the presence of the concentrations of substrate necessary to yield sufficient gas for measurement. Attempts to overcome this difficulty led to the use of peroxide-generating compounds to liberate H_2O_2 slowly, and thus provide adequate amounts of substrate without exposing the enzyme to excessive concentrations.[9] A more serious difficulty encountered with conventional manometers is the limitation on the rate of gas exchange imposed by diffusion at the surface. The apparent rate of reaction is a function of shaking speed under the conditions ordinarily used. The objections to conventional manometry have been overcome, however, by the use of vessels equipped with vigorous stirrers in which pressure changes are recorded automatically from a pressure transducer.[10] A polarographic method[11] has been used successfully, but the most convenient, accurate assay involves determination of H_2O_2 by its ultraviolet absorption in the region of 230 to 250 mμ.[12] This permits rapid rate determinations, even with manually operated instruments.

Crystalline Catalase. Catalase was crystallized from beef liver by Sumner and Dounce in 1937.[4] Liver catalase from many species (human, guinea pig, horse, pig, rat, lamb) crystallizes easily, and very brief, simple procedures have been found to give good yields of active crystals. The enzyme stands shaking with an alcohol–chloroform mixture which denatures hemoglobin. The dark green crystals obtained with 0.45 saturated ammonium sulfate are extremely stable. Bonnichsen crystallized catalase from horse erythrocytes[13] and found no difference between the blood and liver enzyme with respect to amino acid composition, activity, or immunochemical properties. He interpreted his findings to indicate that the enzymes found in both blood and liver cells of the horse to be identical, even though probably synthesized in different tissues. Catalase has also been crystallized from kidney[14] and from a bacterium, *Micrococcus lysodeikticus*.[5]

Prosthetic Group of Catalase. Liver catalases from various animals have molecular weights somewhat over 200,000,[15] and contain 4 iron atoms

[8] R. F. Beers, Jr., and I. W. Sizer, *Science* **117**, 710 (1953).

[9] R. N. Feinstein, *J. Biol. Chem.* **180**, 1197 (1949).

[10] R. E. Greenfield and V. E. Price, *J. Biol. Chem.* **209**, 355 (1954).

[11] B. S. Walker, *Federation Proc.* **1**, 140 (1942).

[12] B. Chance, *J. Biol. Chem.* **179**, 1299 (1949); R. F. Beers, Jr. and I. W. Sizer, *ibid.* **195**, 133 (1952).

[13] R. K. Bonnichsen, *Acta Chem. Scand.* **2**, 561 (1948); *Arch. Biochem.* **12**, 83 (1947).

[14] R. K. Bonnichsen, *Acta Chem. Scand.* **1**, 114 (1947).

[15] J. B. Sumner and N. Gralen, *J. Biol. Chem.* **125**, 33 (1938); K. Agner, *Biochem. J.* **32**, 1702 (1938).

per molecule.[16] The iron is all present as the hematin of protoporphyrin IX.[17] The first crystalline preparations were found to contain in addition to the hematin a blue pigment, identified as biliverdin.[18] This appears to be a degradation product of the original hematin, since the activity varies inversely with the biliverdin content. Crystalline catalases from erythrocytes and from human liver are free from biliverdin,[13] demonstrating that this pigment has no essential role in catalase activity, as once thought.

The appearance of biliverdin in catalase preparations has been suggested as being indicative of the high metabolic turnover of the enzyme in liver, as biliverdin is presumably a degradation product of protoporphyrin. Experiments with radioactive iron show that the iron becomes incorporated into liver catalase very rapidly.[19] By way of contrast, radioactive iron appears much more slowly in erythrocyte catalase. Apparently iron is introduced into both hemoglobin and catalase at the time erythrocytes are formed, and this iron remains firmly bound for the lifetime of the red cell. These observations have been interpreted as showing that liver and erythrocyte catalase are synthesized independently, and one is not formed from the other. Within a few days the specific activity of liver catalase iron approaches that of ferritin (an iron-storing protein), which supports the conclusion that the turnover of catalase in liver is rapid.

Catalase has been studied intensively by chemical and physical methods. The protein component is not unusual in amino acid composition. The iron appears to be present only in the ferric form, and cannot be reduced chemically by any procedure that does not denature the enzyme.[20] There is evidence, however, that during an enzymatic reaction with hydrazoic acid and hydrogen peroxide the iron is reduced to the ferrous state.[21] The iron forms compounds with a large number of molecules.[22] Some of these, such as HCN, HN_3, and H_2S, are inhibitors of the enzyme. The absorption spectrum of catalase shows bands at 403 mμ (Soret), 500 mμ, 540 mμ, and 629 mμ. These shift to give characteristic spectra when HCN, HN_3, or HF combines with enzyme.[23] Magnetic

[16] H. Theorell, *Advances in Enzymol.* 7, 265 (1947).

[17] K. G. Stern, *J. Biol. Chem.* 112, 661 (1936); K. Zeile and H. Hellstrom, *Z. physiol. Chem.* 192, 171 (1930).

[18] R. Lemberg and J. W. Legge, *Biochem. J.* 37, 117 (1943).

[19] H. Theorell, M. Beznak, R. Bonnichsen, K. C. Paul, and A. Akeson, *Acta Chem. Scand.* 5, 445 (1951).

[20] D. Keilin and E. F. Hartree, *Biochem. J.* 39, 293 (1945).

[21] D. Keilin and E. F. Hartree, *Proc. Roy. Soc.* B121, 165 (1936).

[22] K. Agner and H. Theorell, *Arch. Biochem.* 10, 321 (1946).

[23] The absorption peaks of several free and combined hydroperoxidases have been summarized in the review by H. Theorell.

measurements have also been made on catalase and some of its derivatives to learn more about the nature of the iron and its bonds.[24] The unpaired electrons revealed by the magnetic susceptibilities indicate ferric forms for catalase as the free enzyme, with substrate and with inhibitors, except in the case noted above of enzyme + azide + peroxide, which is diamagnetic, and therefore not ferric.

PEROXIDASES

The oxidation of tincture of guaiac by hydrogen peroxide to give a blue color is one of the most sensitive tests for the presence of peroxidase. This reaction was discovered in 1855.[25] In subsequent years the distinctive nature of the enzyme was established and preparations were purified on the basis of oxidation of polyphenols. The assay most commonly used determines the Purpurogallinzahl (PZ), or purpurogallin number.[26] This assay involves incubation of the enzyme under arbitrarily defined conditions with pyrogallol and H_2O_2 (I). At a specified time the reaction is

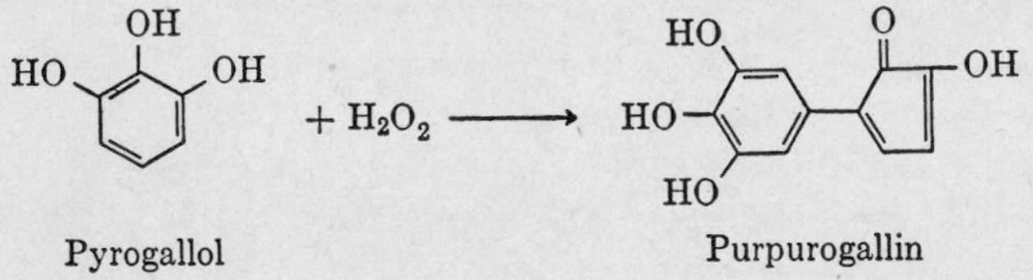

Pyrogallol $+ H_2O_2 \longrightarrow$ Purpurogallin

(I)

stopped with acid and the yellow purpurogallin is extracted from the acidified solution with ether and estimated colorimetrically. PZ is the specific activity; that is, weight of purpurogallin formed in 5 minutes per weight of enzyme. Besides guaiacol and pyrogallol, many other substrates have been used in peroxidase assays, including other phenols (mono- and polyphenols), aromatic amines, leuco dyes, iodide, nitrite, and cytochrome c. Many of these produce colored products on oxidation, so that spectrophotometric methods can be used to follow the reaction. Titrimetric and potentiometric methods have also been used with iodide and hydroquinone, respectively.

The various peroxidases have different affinities for the compounds commonly used as substrates, and the relative affinities of a given enzyme vary with conditions. A yeast peroxidase was isolated on the basis of its reaction with reduced cytochrome c.[27] A similar enzyme has recently been found in bacteria. Cytochrome c peroxidase is not specific, however,

[24] H. Theorell, *Arkiv Kemi Mineral. Geol.* **16A** (3) (1942).
[25] C. F. Schonbein, *Verhandl. naturforsch. Ges. Basel* **I**, 339 (1855).
[26] R. Willstätter and H. Stoll, *Ann.* **416**, 21 (1917).
[27] A. M. Altschul, R. Abrams, and T. R. Hogness, *J. Biol. Chem.* **142**, 303 (1942).

but reacts with other conventional substrates, and other peroxidases also oxidize ferrocytochrome c. An unusual reaction was found with dihydroxyfumaric acid.[28] This compound serves as a peroxidase substrate, but in the presence of manganese, no H_2O_2 is needed. Under these conditions the peroxidase acts as both oxidase and peroxidase. A similar reaction with indoleacetic acid was reported recently.[29] The description of these reactions does not permit a mechanism to be written with any confidence. The participation of H_2O_2 is indicated by the inhibition of the reaction by catalase. Mn^{++} acts catalytically, and may participate in an oxidation cycle. The reduced manganese could form H_2O_2 by reacting with O_2, then return to Mn^{++} by oxidizing an intermediate oxidation form of the substrate. Mn^{++} and substrate do not consume oxygen in the absence of a peroxidase. The oxidation product of indoleacetic acid oxidation has not been identified, but appears to include the atoms of a molecule of oxygen added to the 5-membered ring.

The identification of peroxidase activity with specific enzymes is sometimes difficult because many iron compounds exhibit weak peroxidase activity. The catalytic activities of compounds such as hemoglobin are significant, and the high concentrations of various porphyrins in some preparations make the determination of small amounts of more active, specific peroxidases extremely uncertain. In general, plants are good sources of peroxidase compared with animal tissues. The best known peroxidase is the crystalline enzyme from horseradish root.[30] A *lacto-peroxidase* has been crystallized from milk[31] and highly purified preparations have been obtained from leucocytes (*myeloperoxidase* or *verdo-peroxidase*),[32] fig sap, and yeast.

The various peroxidases are all distinct from each other. One obvious difference is in their absorption spectra, which all show the bands of iron porphyrins, but each of which is slightly different from the other (Fig. 24A).[23] It is not known whether these differences are caused by variations in the porphyrins or in the proteins. Horseradish peroxidase contains iron protoporphyrin IX, which is split from the protein by treatment with acid–acetone.[33] None of the other peroxidases or catalases is split by this procedure. The acid–acetone treatment results in precipitation of a protein that can be reactivated by addition of iron protoporphyrin. Deuterohemin and mesohemin (H and $—CH_2CH_3$ in place of

[28] H. Theorell and B. Swedin, *Nature* **145**, 71 (1940).
[29] R. H. Kenten, *Biochem. J.* **59**, 110 (1955); **61**, 353 (1955).
[30] H. Theorell, *Enzymologia* **10**, 250 (1942).
[31] H. Theorell and K. G. Paul, *Arkiv Kemi Mineral. Geol.* **18A** (12) (1944).
[32] K. Agner, *Acta Physiol. Scand.* **2**, *Suppl.* VIII (1941).
[33] H. Theorell, *Arkiv Kemi Mineral. Geol.* **14B** (20) (1940).

—CH=CH₂) are almost as effective as protohemin in reactivating apoperoxidase but several other porphyrins do not reactivate at all (Fig. 24B).[34] Other metals cannot replace iron in peroxidase.

Peroxidases vary in molecular weight, but appear to contain only a single iron atom per molecule. Reactivation of resolved horseradish

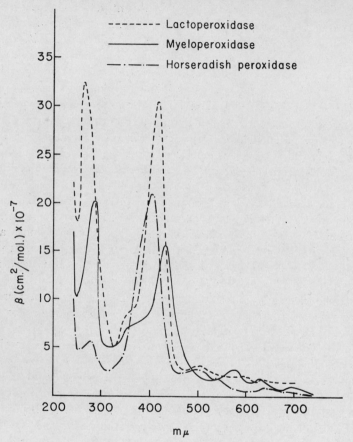

FIG. 24. A: Absorption spectra of peroxidases.[34a]

peroxidase is proportional to added hemin until a maximum value is reached with 1 hemin per protein molecule. The protein of horseradish peroxidase contains an excess of basic amino acids over acidic amino acids but the molecule has an isoelectric point at pH 7.2.[35] The enzyme

[34] H. Theorell, S. Bergstrom, and A. Akeson, *Arkiv Kemi Mineral. Geol.* **16A** (13) (1942).
[34a] Taken from figures collected by H. Theorell, *in* "The Enzymes" from data of Theorell, Agner, and Pedersen (1951). See General References.
[35] H. Theorell and A. Akeson, *Arkiv Kemi Mineral. Geol.* **16A** (8) (1942).

contains 13 per cent polysaccharide, which is believed to contribute acid groups to balance the cationic side chains.

The iron atom of peroxidase is thought to bear an OH group that dissociates with a pK near 5.[22] At pH values near 5 the OH may be replaced by a large variety of anions, which inhibit the reaction with H_2O_2.

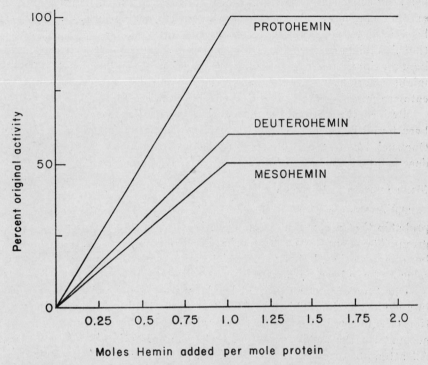

Fig. 24. B: Reconstitution of horseradish peroxidase with iron porphyrins.[34]

The various peroxidases show spectral changes on reduction and on combination with cyanide, azide, fluoride, and carbon monoxide.

CATALASE AS A PEROXIDASE

The distinction between catalase and peroxidase was minimized when Keilin and Hartree found that catalase possesses peroxidase activity.[20] Special circumstances are required to demonstrate peroxidations with catalase, because the catalatic reaction with H_2O_2 is very rapid and efficient. If, however, the peroxide concentration is no greater than that of catalase, all of the peroxide is used in the formation of oxidized (complexed) catalase. This is capable of oxidizing a variety of molecules, including ethanol and formate, as well as more conventional peroxidase substrates. This activity of catalase probably accounts for all of the oxidation of formate by animal preparations. Although the rates of these peroxidations are slow compared with oxygen liberation, large amounts of crystalline catalase are readily available and convenient rates are readily obtained *in vitro*. The peroxidatic activity of catalase has been studed in systems in which H_2O_2 is generated continuously by another enzyme system, such as glucose-glucose oxidase. Catalase has been used to detect H_2O_2 formation by carrying out "coupled oxidations." While any peroxidase would do the same, using conditions favoring peroxidatic reactions of catalase eliminates the catalatic activity that might interfere with other peroxidases, and using large amounts of catalase provides an efficient scavenger for H_2O_2.

Besides the peroxidases already mentioned, similar enzymes have been found in various tissues where they may perform special functions. One such function may be the formation of thyroxine. A special peroxidase will be discussed in connection with tryptophan oxidation.

MECHANISM OF ACTION OF HYDROPEROXIDASES

All of the reactions of the hydroperoxidases are oxidations, and it is probable that a common mechanism is employed by all enzymes of this group. The chemistry of this mechanism has been investigated by a number of approaches, but remains undetermined, although many speculative schemes have been proposed.

One type of mechanism that has been proposed invoked free radicals in a chain reaction. This was attractive in explaining the explosive reaction of catalase, but since conventional reactions between molecules can account for the rate of reaction, such an explanation is not necessary. All attempts to detect free radicals have failed and chain-breakers do not inhibit the reaction. Therefore the free radical hypothesis has been abandoned.

Other catalysts with firmly bound prosthetic groups, including the iron-porphyrin–bearing cytochromes, have been shown to undergo reversible oxidation. Therefore, it would form a consistent pattern if the hydroperoxidases were also to be oxidized and reduced. To explain the action of these enzymes several schemes have been advanced in which the iron shifts from the ferric to the ferrous state and back. All of these schemes have been criticized when applied to catalase because of the inability to detect a ferrous enzyme by magnetic measurements or by inhibition of the reaction with CO. It does not seem that these objections are overwhelming, as the ferrous state may be very short-lived, and escape detection by physical means, and the reduced enzyme may have less affinity for CO than those enzymes that are inhibited by this compound.

Although the chemical nature of intermediate forms of the hydro-peroxidases cannot be determined, certain properties of these enzymes have been ascertained by studying the spectral changes during reactions. Keilin and Mann found that horseradish peroxidase reacts with H_2O_2 to form a red compound.[36] The red compound is an active form of the enzyme, which reverts to its original state on the addition of substrate (donor). Theorell found that a green compound is formed prior to the red compound of Keilin and Hartree.[37] A red compound of catalase was found by Stern when catalase was allowed to react with ethyl hydrogen peroxide (CH_3CH_2—O—OH).[38] This peroxide and certain other organic peroxides are decomposed by catalase at slower rates than H_2O_2.

The properties of the intermediate forms of the hydroperoxidases were investigated in detail by Chance,[39] who developed rapid-flow appa-ratus and spectrophotometry into extremely rapid and sensitive devices. In a rapid-flow apparatus, reagents are mixed rapidly as they are forced from syringes. The mixture passes through a tube whose walls serve as a spectrophotometer cell. By correlating the rate of flow from the syringes with the flow rate through the tube, the time between the formation of a mixture and its appearance at the light beam can be calculated. The solution that remains in the light path after the flow has ceased is used to follow the reaction for longer time periods.

The use of rapid, sensitive methods revealed that catalase and all of the peroxidases studied form a green compound rapidly on the addition of H_2O_2. In the presence of excess peroxide, catalase is converted to the red compound of Stern. This is catalytically inactive. The reactions involved in the decomposition of H_2O_2 by catalase include the formation of the green enzyme and its return to the original brown. Other electron donors that serve as substrates for catalase, such as ethanol, HNO_2, and formic acid, also regenerate free catalase from the green compound. The green forms of all of the hydroperoxidases have been named by Chance complex I forms. The red (inactive) catalase is called complex II. Another inactive complex III also forms on further exposure to H_2O_2.

The reaction of peroxidase complex I with substrates causes the appearance of a red complex II. This conversion was shown by George to be accomplished also by a stoichiometric reduction with ferrocyanide.[40] Further reaction with substrate (or with another equivalent of ferro-cyanide) causes peroxidase to revert to its original free form. The events

[36] D. Keilin and T. Mann, *Proc. Roy. Soc.* **B122,** 119 (1937).
[37] H. Theorell, *Arkiv Kemi Mineral. Geol.* **16A** (2) (1942).
[38] K. G. Stern, *J. Biol. Chem.* **114,** 473 (1936).
[39] See review article by B. Chance in *Advances in Enzymol.* **12.**
[40] P. George, *J. Biol. Chem.* **201,** 413 (1953).

of peroxidase activity thus include two spectrophotometrically discernable forms of the enzyme, complexes I and II. The presence of excess H_2O_2 causes the formation of an enzymatically inactive complex III and alkyl hydrogen peroxide in excess converts the enzyme to another inactive complex, IV.

The term complex was applied to the various forms of the hydroperoxidases to imply an enzyme–substrate combination. It was suggested that complex I is a combination of the enzyme with one peroxide. Various structures have been proposed for this hypothetical compound, in general having H_2O_2 replace a previously bound H_2O, or —OOH displace an —OH from combination with Fe^{+++}. This chemical representation of complex I was subjected to serious doubt when George found that other oxidizing agents, such as hypochlorite, could convert horseradish peroxidase to its complex I. George has suggested that super valence states of iron (Fe^{++++} and Fe^{+++++}) may participate in peroxidations, and it has also been proposed that other parts of the enzyme may be oxidized. At this time the chemical changes of the hydroperoxidases remain obscure. The nature of the one-electron reactions in the transformation of peroxidase complex I to complex II to free peroxidase also presents an interesting problem, since the usual substrates oxidized donate two electrons, and cannot be expected to form stable free radicals.

Although the chemistry of the reactions of hydroperoxidases must be described in terms of unanswered questions, this area is one of the most intriguing of enzyme chemistry, because substrate quantities of enzyme are available for study, and kinetics of interaction between enzyme and various reagents (substrates and inhibitors) can be measured by direct observation. It is to be hoped that further studies with these enzymes will reveal the chemical nature of their reactions.

GENERAL REFERENCES

Chance, B. (1951). *Advances in Enzymol.* **12,** 153.
Chance, B. (1951). *In* "The Mechanism of Enzyme Action" (W. D. McElroy and B. Glass, eds.), p. 389. Johns Hopkins Press, Baltimore, Maryland.
Theorell, H. (1951). *In* "The Enzymes" (J. B. Sumner and K. Myrbäck, eds.), Vol. II, Part 1, p. 397. Academic Press, New York.

Oxidases and Oxygenases

Oxygen is the ultimate electron acceptor in the principal metabolic pathways described in the previous chapters. In most aerobic organisms oxygen is reduced to water by cytochrome oxidase at rates that account for most of the oxygen consumed. Reactions of flavoproteins with oxygen also occur, and produce H_2O_2, which is utilized by the hydroperoxidases. In addition to these reactions, other mechanisms are available for the utilization of molecular oxygen. One group of enzymes, the copper oxidases, have been considered as possible terminal oxidases in place of cytochrome oxidase. The physiological role of the copper oxidases is not certain; even though enzyme series can be constructed that include these enzymes as terminal catalysts, it is not clear how the organism would benefit from such oxidations and it remains to be determined whether their function is concerned with general oxidation or the metabolism of specific substrates.

Copper-Containing Oxidases

Oxidations now known to be catalyzed by copper-containing enzymes were noticed over a century ago, when Schoenbein observed that oxidation of natural substrates resulted in pigment formation in mushrooms.[1] Individual enzymes were gradually identified; laccase by Yoshida in 1883[2] and tyrosinase by Bertrand in 1896.[3] However, it was not until potato polyphenol oxidase was isolated in 1937 by Kubowitz that the role of copper was defined.[4] The family of copper oxidases includes a number of enzymes of both plant and animal origin that may very probably be found to react through similar mechanisms, but which exhibit a number of individual characteristics. The enzymes to be described in this section include potato phenol oxidase, mushroom polyphenol oxidase (tyrosinase), laccase, mammalian and insect tyrosinase, and ascorbic acid oxidase. Each of these differs in certain respects from the others, and undoubtedly other related enzymes will be described from other sources that resemble these, but also display individualities. In these cases, identities in nomenclature must not be extended to imply identities in enzyme structure or activity.

Potato Phenol Oxidase. The studies of Kubowitz on the copper oxidase of potatoes were part of the efforts of Warburg's institute to find enzyme systems that could catalyze oxygen consumption coupled with substrate oxidation. The enzyme was purified on the basis of an assay involving transfer of electrons from pyridine nucleotides to *o*-quinone and from the resulting catechol to oxygen (I). The reduction of catalytic quantities of quinone was probably catalyzed by a flavoprotein present in the Zwischenferment preparation used to reduce TPN. In the presence of

[1] C. F. Schoenbein, *Phil. Mag.* [4] **11**, 137 (1856).
[2] H. Yoshida, *J. Chem. Soc.* **43**, 472 (1883).
[3] G. Bertrand, *Compt. rend.* **122**, 1215 (1896).
[4] F. Kubowitz, *Biochem. Z.* **292**, 221 (1937); **299**, 32 (1939).

adequate amounts of the Zwischenferment system (glucose 6-phosphate, enzyme, and TPN) and catechol, oxygen consumption is proportional to the oxidase. A purification procedure involving repeated acetone and

glucose-6-phosphate $+$ TPN $+$ H_2O $\xrightarrow{\text{Zwischenferment}}$ 6-phosphogluconate $+$ TPNH $+$ 2 H^+

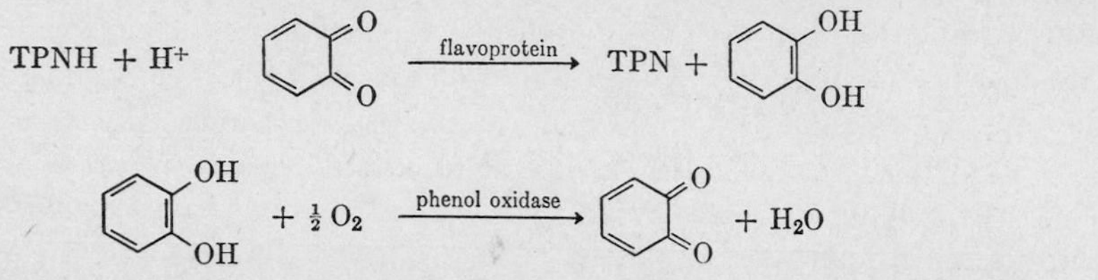

TPNH $+$ H^+ [structure] $\xrightarrow{\text{flavoprotein}}$ TPN $+$ [structure]

[structure] $+$ $\frac{1}{2}$ O_2 $\xrightarrow{\text{phenol oxidase}}$ [structure] $+$ H_2O

_____

Sum: glucose-6-phosphate $+$ $\frac{1}{2}$ O_2 $\longrightarrow$ 6-phosphogluconate $+$ H^+

(I)

ammonium sulfate fractionations and selective denaturation steps resulted in highly purified preparations. The activity of the various fractions was found to be proportional to the copper content, which was about 0.20 per cent in the best preparations. These highly purified preparations display no absorption bands in the visible region, and there is no evidence for any cofactor other than copper.

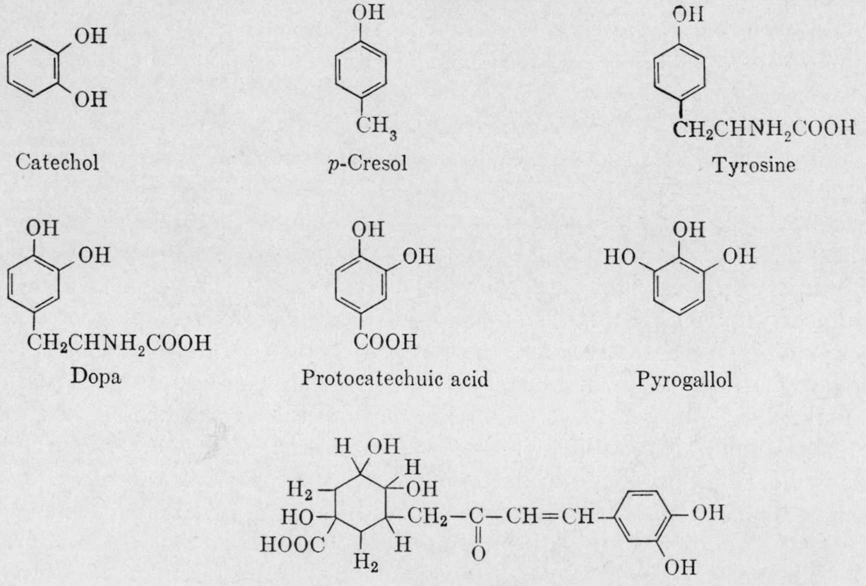

Catechol

p-Cresol

Tyrosine

Dopa

Protocatechuic acid

Pyrogallol

Chlorogenic acid

Substrates for Phenol Oxidases

Potato phenol oxidase is inhibited by cyanide and carbon monoxide, and, more specifically, by salicylaldoxime and diethyldithiocarbamate, which bind fewer metals but react avidly with copper. The activity of inhibited enzyme in each case was restored by the addition of copper, but this alone does not establish the nature of the group in the enzyme that reacts with inhibitors, as the effect of copper could be merely to bind and remove the inhibitor. More definitive experiments involved dialyzing the enzyme against cyanide. In this way the metal is removed from the enzyme as the cyanide complex, which passes through the membrane. The cyanide was then removed and the inactive protein was assayed with various metal additions. Only copper restored activity. Substantially more copper was bound than was originally present, and the original activity was eventually reached. The nonspecific binding of cofactors following resolution of enzymes has been discussed previously, and must be recalled when the metal content of enzymes is assayed.

Besides catechol, the potato enzyme attacks pyrogallol, dopa, adrenaline, tyrosine, and other phenolic compounds. The kinetics of the reactions with various substrates are not easily interpreted. Tyrosine is oxidized at a slow, linear rate and catechol is oxidized much more rapidly. p-Cresol, however, is oxidized rapidly only after a lag period. Other phenolic compounds, including resorcinol and hydroquinone, are not oxidized by this enzyme.

Polyphenol Oxidase or Tyrosinase. An enzyme very similar to the potato enzyme has been isolated from the mushroom, *Psalliota campestris*. Keilin and Mann[5] purified this enzyme using an assay resembling the peroxidase assay, in which pyrogallol is converted to purpurogallin. Mushrooms have also been used for the very extensive studies of Nelson and the other workers of his school.[6] The enzyme called polyphenol oxidase by Keilin and Mann was named tyrosinase by the Nelson group. Keilin and Mann purified mushroom polyphenol oxidase about 2100-fold from extracts of minced mushroom ground with sand. Since the extracts appear to contain large amounts of copper not associated with this enzyme, there is correspondence between copper content and activity only after extensive purification. This enzyme is also inhibited by metal binders, including cyanide, sulfide, azide, and carbon monoxide.

Catechol and pyrogallol are good substrates for the mushroom enzyme. In crude preparations monophenols are also oxidized, but with a lag period. As purification proceeds the ability to oxidize monophenols is lost. The ability to oxidize monophenols was named *cresolase* activity by the Nelson school, as p-cresol was used as a representative substrate.

[5] D. Keilin and T. Mann, *Proc. Roy. Soc.* **B125,** 187 (1938).
[6] J. M. Nelson and C. R. Dawson, *Advances in Enzymol.* **4,** 99 (1944).

The more stable activity toward polyphenols was correspondingly named *catecholase*. These terms, which bear the suffix -*ase* used to denote enzymes, have not been applied to specific catalysts. They are used merely to describe aspects of the activity of phenol oxidases.

The relation between oxidation of monophenols and polyphenols has been investigated extensively by Nelson and his students. The nature of the reactions studied is still not clear, but many interesting aspects have been explored. As in the case of other workers, it was found that there is a tendency for the ability to oxidize monophenols (cresolase) to be lost on purification, while polyphenol oxidation (catecholase) is more stable. The ratio of catecholase to cresolase activity of purified preparations appears to vary with electrophoretic mobility, and it has been suggested that a peptide component is lost from the enzyme, and that the purified preparations represent partially degraded enzyme.[7]

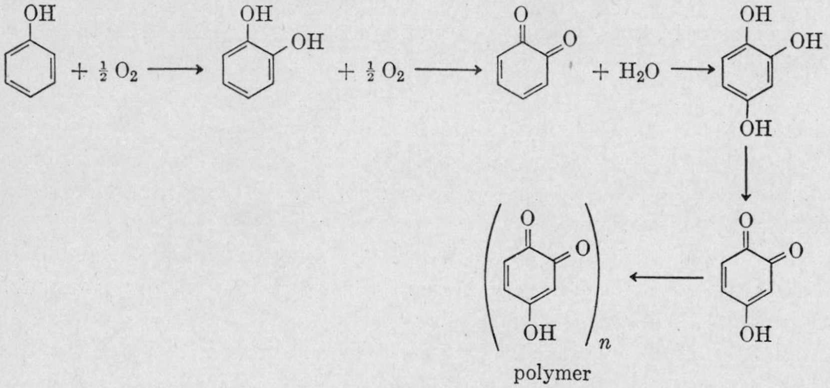

(II)

Many workers have demonstrated the catalytic effect of polyphenols on monophenol oxidation. An early hypothesis suggested that (auto) oxidation of polyphenols could yield H_2O_2 required for oxidation of monophenols. Intensive efforts to detect H_2O_2 in tyrosinase preparations have yielded only negative results and today there is no reason to believe that peroxide is formed or utilized by tyrosinase. Another suggestion considered that quinone produced in polyphenol oxidation might oxidize monophenols to polyphenols. This hypothesis was tested by incubating monophenol, quinone, and enzyme anaerobically; no monophenol was oxidized. On the other hand when ascorbic acid was added to an aerobic system (ascorbic acid reduces quinones to hydroquinones very rapidly), monophenol oxidation proceeded well, showing that an accumulation of quinone is not necessary for phenol oxidation. The current feeling is

[7] M. F. Mallette and C. R. Dawson, *Arch. Biochem.* **23**, 29 (1949).

that in some unspecified manner the act of oxidizing polyphenols activates the enzyme for monophenol oxidation. A scheme for tyrosinase action on phenol has been advanced by Nelson and his associates (II).[8a] When tyrosine is the substrate, dopa is formed, which undergoes the

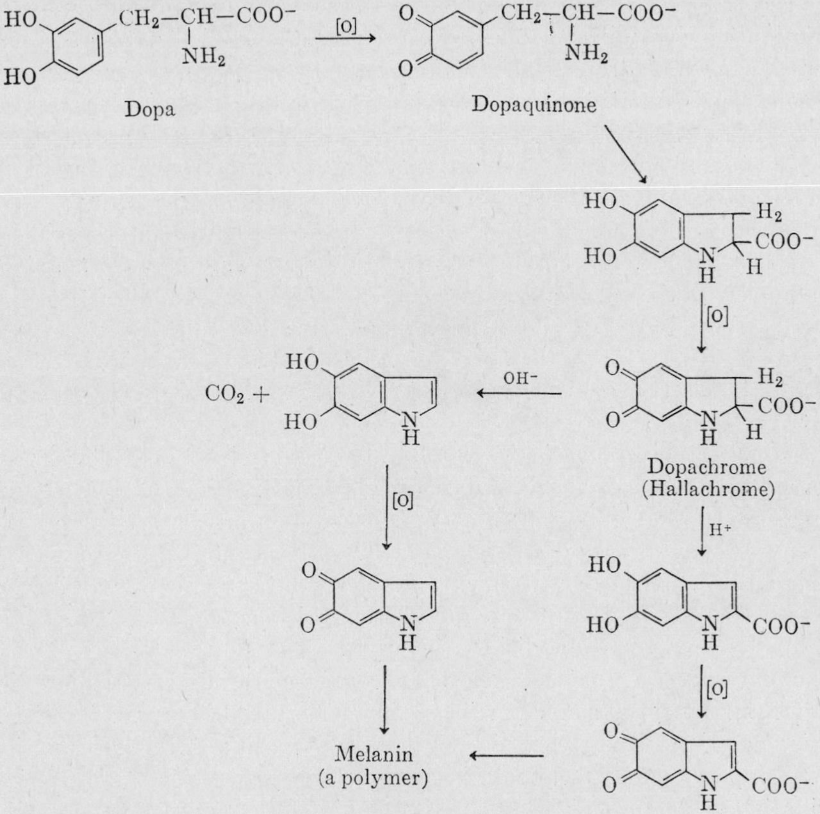

Dopa $\xrightarrow{[O]}$ Dopaquinone

Dopaquinone →

→ $\xrightarrow{[O]}$ Dopachrome (Hallachrome)

CO_2 + ← $\xleftarrow{OH^-}$

Dopachrome (Hallachrome) $\xrightarrow{H^+}$

$\downarrow [O]$ ← $\xleftarrow{}$ Melanin (a polymer)

$\downarrow [O]$

Melanin (a polymer)

(III)

changes shown in (III) in becoming the dark pigment, melanin. Reactions showing oxidation are enzyme catalyzed.[8b]

Studies with mushroom tyrosinase have been hampered by inactivation of purified preparations under assay conditions. Manometric methods have been used with gelatin added to protect the enzyme, but this assay is complicated by several disturbing factors.[9] The so-called chronometric

[8a] M. F. Mallette, *in* "Symposium on Copper Metabolism" (W. D. McElroy and B. Glass, eds.), p. 48. Johns Hopkins Press, Baltimore, Maryland, 1951. *See also* Nelson and Dawson.[6]

[8b] H. S. Mason, *J. Biol. Chem.* **172**, 83 (1948).

[9] M. H. Adams and J. M. Nelson, *J. Am. Chem. Soc.* **60**, 2472 (1938).

method has been used, in which the substrates are catechol and ascorbic acid.[10] The catechol is oxidized to o-quinone, which is immediately reduced by the ascorbic acid. The catechol thus shuttles back and forth until all of the ascorbic acid is consumed. The time required to consume the ascorbic acid is determined by the release of iodine in a starch-iodide solution by the quinone. The iodine reaction is carried out in a second vessel into which drops of the incubation mixture fall through a capillary siphon. Unfortunately, even in this assay, there is rapid inactivation of the enzyme as it works.

Laccase. A polyphenol oxidase has been purified from the sap of the lac tree by Keilin and Mann.[11] *Laccase* differs from the potato and mushroom enzyme in several respects. With regard to substrate specificity, it oxidizes p-phenylenediamine more rapidly than catechol. p-Phenylenediamine is not a substrate for the other polyphenol oxidases described. Laccase apparently is inert with p-cresol. It is not inhibited by carbon monoxide. Unlike the other phenol oxidases, this enzyme is not a pale yellow, but is blue, as is ascorbic acid oxidase (see below). This enzyme, however, is not an ascorbic acid oxidase.

Insect Protyrosinase and Tyrosinase. Tyrosinase has been isolated from grasshopper eggs as a protyrosinase by Allen and Bodine.[12] This precursor already contains copper. Activation occurs when any of a large number of environmental conditions is changed, such as ionic strength, pH, or temperature. Surface-active agents (as aerosols) are effective activators. Cyanide can be used to remove metal from this enzyme also, and activity can be restored by the addition of copper.

Mammalian Tyrosinase. Another copper-containing tyrosinase has been found in cytoplasmic particles of melanocytes of mammalian skin.[13] This enzyme attacks L-tyrosine, but not D-tyrosine, after a lag period. With dopa as substrate, there is no lag. The product is a polymerized oxidation product, a highly colored melanin. The enzyme is more stable than the plant tyrosinases, and can be assayed both manometrically and by estimation of the product. C^{14}-labeled tyrosine has been used to produce labeled, insoluble melanin.

The function of tyrosinase in producing pigments in skin is obvious. The function of the corresponding plant enzymes is less clear. It has been proposed that they function in terminal respiration, but this has not been

[10] W. H. Miller, M. F. Mallette, L. J. Roth, and C. R. Dawson, *J. Am. Chem. Soc.* **66,** 514 (1944).

[11] D. Keilin and T. Mann, *Nature* **145,** 304 (1940).

[12] T. H. Allen and J. H. Bodine, *Science* **94,** 443 (1941).

[13] A. B. Lerner, T. B. Fitzpatrick, and W. H. Summerson, *Federation Proc.* **8,** 218 (1949).

established. The compounds isolated as natural substrates, tyrosine and chlorogenic acid, do not appear to be oxidized in plants as long as cell integrity is maintained. Specific functions for individual oxidases may well be the pattern for this group of enzymes; such a function occurs in the formation of cockroach oothecae, which requires the oxidation of a phenol to a quinone.

Ascorbic Acid Oxidase. Enzymes that oxidize ascorbic acid (IV) have been detected in many types of plants. An enzyme purified from squash

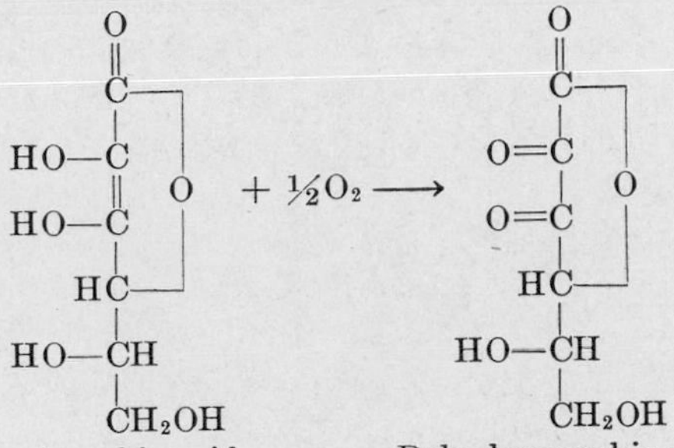

$$
\begin{array}{ccc}
\text{O} & & \text{O} \\
\| & & \| \\
\text{C} & & \text{C} \\
\text{HO—C} & & \text{O=C} \\
\| \quad \text{O} \quad + \tfrac{1}{2}\text{O}_2 \longrightarrow & & \| \quad \text{O} \\
\text{HO—C} & & \text{O=C} \\
\text{HC} & & \text{HC} \\
\text{HO—CH} & & \text{HO—CH} \\
\text{CH}_2\text{OH} & & \text{CH}_2\text{OH} \\
\text{L-Ascorbic acid} & & \text{Dehydroascorbic acid} \\
& \text{(IV)} &
\end{array}
$$

is a blue-green protein containing 0.26 per cent copper and having a molecular weight of about 150,000.[14] Therefore, each molecule contains 6 atoms of copper. The existence of a specific ascorbic acid oxidase has been seriously questioned, since copper alone catalyzes ascorbic acid oxidation, and when bound to inert proteins or calcium phosphate gels behaves like a heat-labile enzyme. However, the models used are much less active than the specific enzyme, and react with a large number of substrates that are not oxidized by ascorbic acid oxidase. A very significant difference between the enzyme and the artificial models is that the enzyme consumes only one atom of oxygen per molecule of ascorbic acid oxidized without any evidence of peroxide formation, whereas the model systems all produce peroxide.

The copper of ascorbic acid oxidase is firmly bound, and is not removed by a cation exchange resin.[15] When radioactive Cu^{64} is used to study exchange of copper, no isotope is incorporated into the enzyme under any conditions except those that permit enzyme activity. Under these conditions there is both exchange of copper and inactivation of the enzyme. It is not known whether these two processes are related. It has been suggested that the exchange reaction implies a reversible reduction of copper during enzyme activity, and that only the cuprous form ex-

14 F. J. Dunn and C. R. Dawson, *J. Biol. Chem.* **189**, 485 (1951).
15 M. Joselow and C. R. Dawson, *J. Biol. Chem.* **191**, 1 (1951).

changes. The mechanism of the exchange is hard to visualize, however, for reduction of the copper would be expected under anerobic conditions in the presence of substrate. There is no exchange when the enzyme is incubated with Cu^{64} and ascorbic acid under nitrogen. In such systems, it has been reported, the color is bleached, but returns when air is admitted. The blue color is probably related to a copper complex, but the structure of such a complex is not known. A suggestion has been made that the colored complex has a square coplanar configuration and includes molecular oxygen. On reduction, the monovalent copper complex may assume a tetrahedral configuration, and this may exchange copper whereas the more stable planar structure does not.[16] There is no known reason to account for the differences in color of the various copper oxidases.

Copper-Free Ascorbic Acid Oxidase. A totally different specific ascorbic acid oxidase has been prepared from fungal spores.[17] This enzyme is not sensitive to copper inhibitors but is inhibited by sulfhydryl reagents. It does not attack ascorbic acid analogs. Like the copper-containing enzyme, fungal ascorbic acid oxidase gives rates proportional to oxygen tension.

OXYGENASES

The flavin oxidases, cytochrome oxidase, and certain of the copper oxidases transfer electrons to oxygen, which is reduced either to H_2O_2 or to H_2O. In each case electrons or hydrogen atoms are removed from the substrate. Recently a group of enzymes has been found to catalyze the oxidation of substrates by the insertion of oxygen atoms. These enzymes have been designated oxygenases. In general, oxygenases attack cyclic compounds, but not all oxidations of cyclic substrates are catalyzed by oxgenases. An enzyme that oxidizes uric acid, uricase, has recently been reported to be a copper oxidase.[18] This enzyme will be discussed in more detail in connection with purine metabolism (Chapter VI). The nature of the reaction of this enzyme with oxygen was investigated with the aid of O^{18} by Bentley and Neuberger,[19] who found that no molecular oxygen was incorporated in the CO_2 produced in the oxidation; instead the oxygen appeared to act merely as an electron acceptor and was found as H_2O_2. This result was consistent with all reaction mechanisms ascribed to biological oxidations at that time.

A fundamental difference between the phenol oxidases and the con-

[16] M. Joselow and C. R. Dawson, *J. Biol. Chem.* **191**, 11 (1951).

[17] G. R. Mandels, *Arch. Biochem. Biophys.* **42**, 164 (1953); **44**, 362 (1953).

[18] H. R. Mahler, G. Hubscher, and H. Baum, *J. Biol. Chem.* **216**, 625 (1955).

[19] R. Bentley and A. Neuberger, *Biochem. J.* **52**, 694 (1952).

ventional terminal oxidases was found by Mason and his collaborators.[20] In the oxidation of 3,4-dimethylphenol to 4,5-dimethylcatechol they found that the oxygen atom inserted into the molecule was derived exclusively from molecular oxygen, not from the solvent (V). This contrasts with such reactions as the formation of phosphoglycerate from phosphoglyceraldehyde, in which a hydrolysis follows the loss of electrons, or the oxidation of fumarate to oxalacetate, which is preceded by

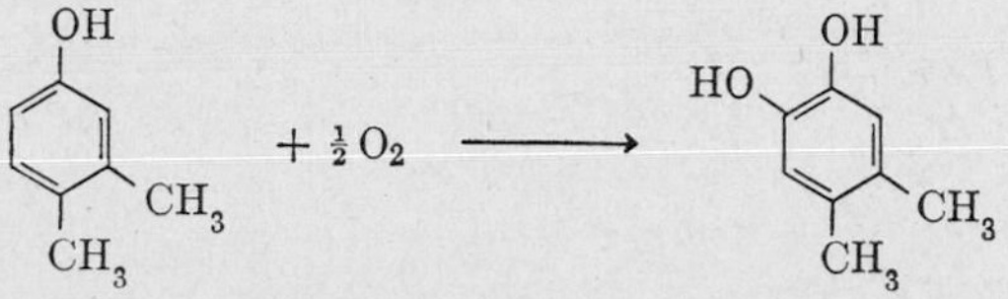

(V)

hydration to malate. In general, it may be concluded that any reaction in which other oxidants can be substituted for oxygen uses oxygen merely as an electron acceptor. The reaction described by Mason is not unique, however, for, since his pioneer studies were announced, a number of reactions have been found to insert atoms from molecular oxygen into organic substrates. In distinction to the oxidases, which transfer electrons to oxygen, the group of enzymes to be described are called oxygenases, to denote their ability to add oxygen to their respective substrates.

The approach adopted by Mason has found application in the study of several specific biological oxidations. In all cases incubations were carried out with O^{18} present either in molecular oxygen or in water. The reaction products were isolated and assayed for O^{18} content. The reactions described below are now classified as oxygenations, as they involve insertion of molecular oxygen into the substrate. Although oxygenation reactions were only recently described, the enzymes themselves were previously known, and have been sufficiently studied to permit organization into three groups: phenolytic, mixed, and peroxidatic oxygenases.

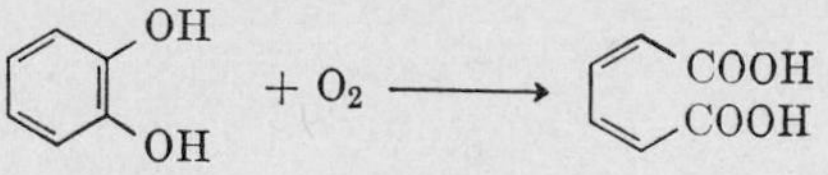

(VI)

Phenolytic Oxygenases. In the oxidation of tryptophan by a *Pseudomonas* an intermediate, catechol, is converted to *cis,cis*-muconic acid by an enzyme named pyrocatechase (VI).[21] This enzyme requires ferrous

[20] H. S. Mason, W. L. Fowlks, and E. Peterson, *J. Am. Chem. Soc.* **77**, 2914 (1955).
[21] M. Suda, H. Hashimoto, H. Matsuoka, and T. Kamahora, *J. Biochem.* **38**, 289 (1951).

ions and sulfhydryl groups for activity. Both atoms of oxygen added are derived exclusively from O_2.[22] The reaction does not seem to require more than one enzyme. Its action is postulated in (VII). A similar reaction appears to occur in the oxidation of 3-hydroxyanthranilic acid (another derivative of tryptophan) by a mammalian liver enzyme.[23]

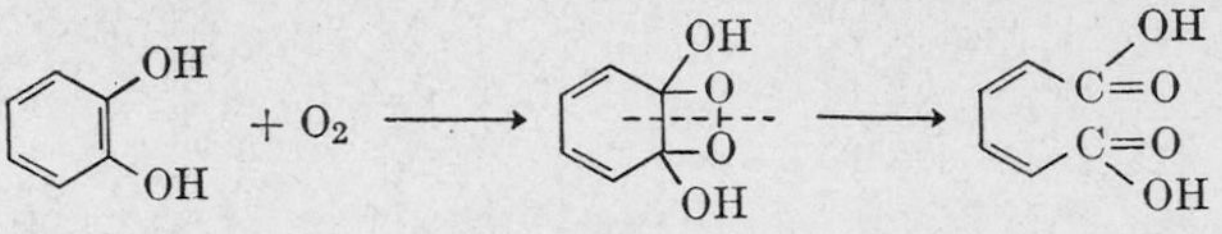

(VII)

In this case the unstable immediate oxidation product has not been isolated, but from several properties it is believed to have the structure shown in (VIII). In separate experiments quinolinic and picolinic acids were isolated and were found to contain one atom of oxygen derived from O_2 and none from water. This oxygenation is classified with pyrocatechase because it seems probable that a similar mechanism is employed. In the formation of the intermediate two atoms of oxygen are

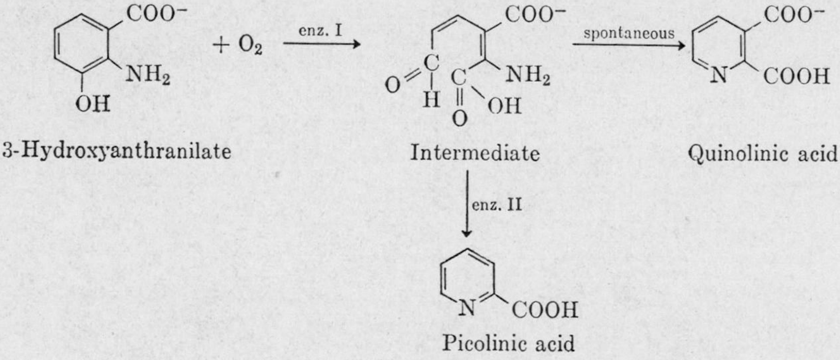

3-Hydroxyanthranilate   Intermediate   Quinolinic acid

Picolinic acid

(VIII)

used, but in the subsequent condensation of the aldehyde to form the pyridine ring, one of these is eliminated as water. This enzyme also appears to require Fe^{++} and sulfhydryl groups. Similar enzymes cleave protocatechuic acid, homogentisic acid, and other phenols, and it is probable that all employ similar mechanisms.[24]

[22] O. Hayaishi, M. Katagiri, and S. Rothberg, J. Am. Chem. Soc. 77, 5450 (1955).
[23] O. Hayaishi, A. H. Mehler, and S. Rothberg, Abstr. 130th Meeting Am. Chem. Soc., Atlantic City, 1956, p. 53C.
[24] D. I. Crandall, in "Amino Acid Metabolism" (W. D. McElroy and B. Glass, eds.), p. 867. Johns Hopkins Press, Baltimore, Maryland, 1955.

Mixed Oxygenases. Several reactions that may be considered to be hydroxylations have been shown to require both molecular oxygen and a reduced pyridine nucleotide. Examples include: hydroxylation of aromatic rings (IX)[25] hydroxylation of steroids (X);[26] and cleavage of an imidazole ring XI.[27]

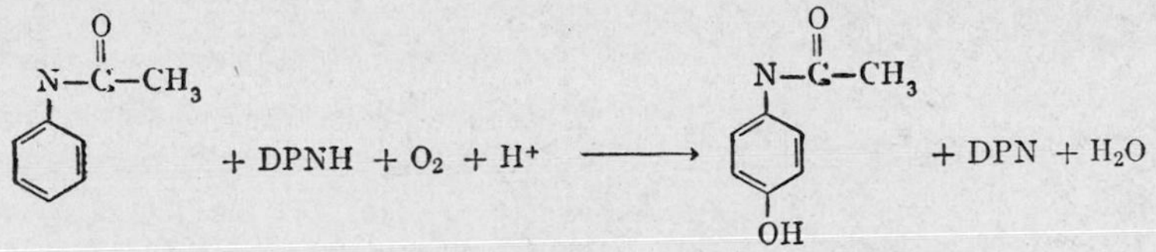

Acetanilide   *p*-Hydroxyacetanilide

Hydroxylation of Aromatic Rings
(IX)

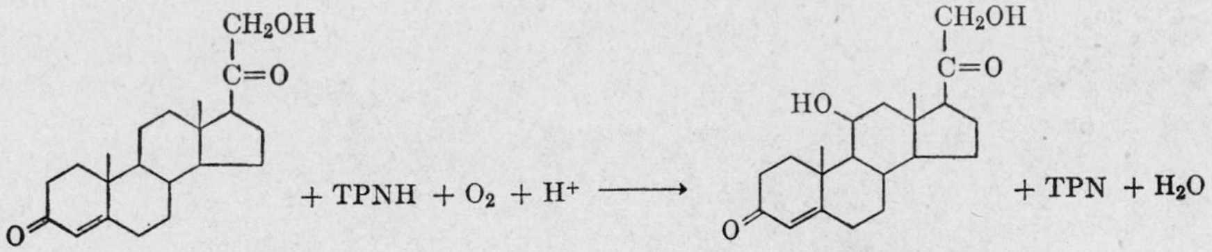

11-Deoxycorticosterone   Corticosterone

Hydroxylation of Steroids
(X)

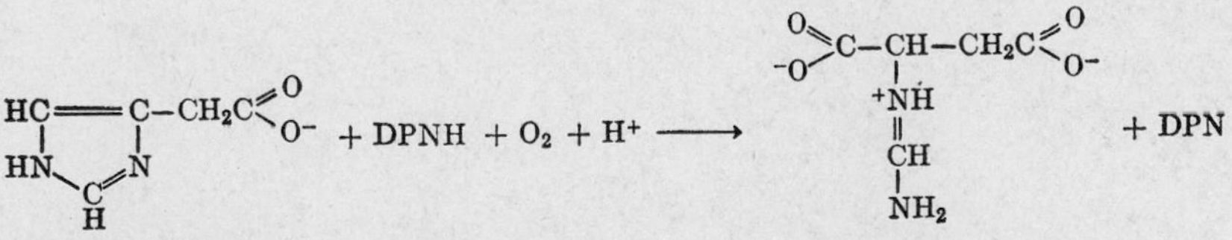

Imidazoleacetate   Formiminoaspartate

Cleavage of an Imidazole Ring
(XI)

It is possible to visualize three mechanisms for these complex reactions: reduction may precede oxidation, it may follow oxidation, or the two processes may occur simultaneously. The first two alternatives are not favored because no partial reactions have been detected and only single enzymes appear to be required. The third alternative is advocated, therefore, but positive evidence in its support is lacking. The hypothetical

[25] H. S. Posner, S. Rothberg, S. Udenfriend, and O. Hayaishi quoted in *Am. Chem. Soc.*[23]

[26] M. Hayano, M. C. Lindberg, R. I. Dorfman, J. E. H. Hancock, and W. von E. Doering, *Arch. Biochem. Biophys.* **59,** 529 (1955).

[27] O. Hayaishi and H. Tabor quoted in *Am. Chem. Soc.*[23]

mechanism would have the enzyme split the oxygen molecule, insert one atom into the substrate, and use the electrons of the reduced pyridine nucleotide to form water from the other atom.

Peroxidase Oxygenations. Two poorly defined reactions are believed to represent a third type of oxygenation. In one, a well-defined enzyme, horseradish peroxidase ($+Mn^{++}$) appears to add molecular oxygen to substrates including dihydroxyfumarate and indoleacetate.[28] The products are not well-defined in this case. Another enzyme, tryptophan peroxidase, has not been purified sufficiently to permit its chemical nature to be ascertained. This enzyme carries out the stoichiometric conversion of tryptophan to formylkynurenine, with the addition of

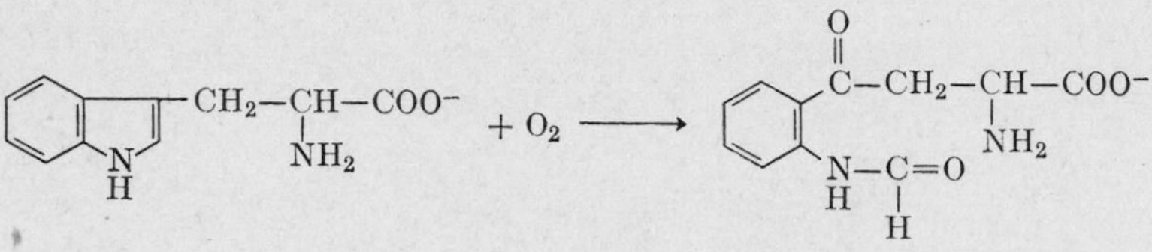

(XII)

two atoms of oxygen derived exclusively from O_2 (XII).[23] Although the net reaction requires merely the addition of oxygen, there is reason to believe that a peroxide is formed as an intermediate, as these reactions are inhibited by catalase. Enzyme-generated peroxide formed by flavoprotein systems supports reactions of horseradish peroxidase in the absence of Mn^{++} and overcomes the inhibition by catalase of mammalian tryptophan peroxidase.

The original reaction of Mason does not fit into the three categories described although further investigation may reveal similarities not now apparent, and still more examples of oxygenation reactions may yet be described. All of the oxygenases seem to be concerned with specific biological transformations rather than with general respiratory mechanisms. The widespread occurrence of oxygenases and the biosynthetic reactions which depend upon them emphasize the importance of direct oxygenation in biological reactions. It is of interest that all of the oxygenations seem to require metal participation, but the metal may be Fe^{++}, Cu^{++}, or iron in a porphyrin. The oxygenases form a series of enzymes that correspond to the various oxidases, and it will be of interest

[28] R. H. Kenten, *Biochem. J.* **61,** 353 (1955). A similar enzyme has been studied in medium from cultures of the mold *Omphalla flavida.* This enzyme, which acts most rapidly near pH 3.5, carries out an oxidation that is followed by several secondary reactions. The properties of the principal product are consistent with the structure of an oxygenated indole [P. M. Ray and K. V. Thimann, *Arch. Biochem. Biophys.* **64,** 175 (1956); P. M. Ray, *Arch. Biochem. Biophys.* **64,** 193 (1956)].

to compare the detailed mechanisms of the two types of oxygen-consuming systems when techniques become available for further analysis.

GENERAL REFERENCES

Dawson, C. R., and Mallette, M. F. (1945). *Advances in Protein Chem.* **2,** 179.

Lerner, A. B. (1953). *Advances in Enzymol.* **14,** 73.

Lerner, A. B., and Fitzpatrick, T. B. (1950). *Physiol. Revs.* **30,** 91.

McElroy, W. D., and Glass, B., eds. (1951). "Symposium on Copper Metabolism." Johns Hopkins Press, Baltimore, Maryland.

Mason, H. S. (1955). *Advances in Enzymol.* **4,** 99.

Chapter V

SUGARS AND SUGAR DERIVATIVES

Polysaccharides

Polysaccharides are widely distributed in nature as food reserves and structural elements. The study of the enzymes involved in the synthesis and breakdown of polysaccharides has developed together with increases in knowledge of the structures of these carbohydrates. Indeed, much of our present knowledge about the detailed structure of the larger polysaccharides has been obtained with the aid of purified specific enzymes.

Polysaccharides are composed of sugars or sugar derivatives linked together through acetal bonds. These bonds are formally represented as the result of elimination of water from a carbonyl group and an alcohol, but only indirect routes of synthesis are known, in which a substituted sugar, a glycoside, participates in a transfer reaction. In a later section reactions involving a specific cofactor, uridine diphosphate, will be considered. This chapter deals with reactions of simple glycosides and sugar phosphates.

Glycosidases

Several enzymes have been known for many years only as catalysts for the hydrolysis of naturally occurring disaccharides (sugars composed of two monosaccharides) and heterosaccharides (glycosides of a sugar with another type of molecule). It has been known for some time that these enzymes are not absolutely specific for disaccharides, but that they attack various analogs in which the glycosyl group of the disaccharide is the common feature. More recently it has been established that in general the enzymes of this group are not restricted to hydrolysis of their respective substrates, but that they catalyze transfers of glycosyl groups. By transfer reactions short-chain polysaccharides (oligosaccharides) are formed.

Galactosidases. Many glycosides of galactose occur naturally. The configuration of the glycoside is α in some and β in other galactosides. The most familiar galactoside is lactose, D-β-galactosyl-4-D-glucose, and the general name β-galactosidase applies to what has been called lactase for many years. The hydrolysis of lactose is catalyzed by β-galactosidases from many sources, including molds, bacteria, snails, plant seeds, animal serum, and animal liver.[1]

[1] For historical introduction, see the general reference of S. Veibel.

Because the two halves of lactose, galactose and glucose, are readily distinguished from each other, the analysis of lactase activity is less complicated than similar studies with polymers of glucose. For many years it was known that lactose could be hydrolyzed into its component sugars.

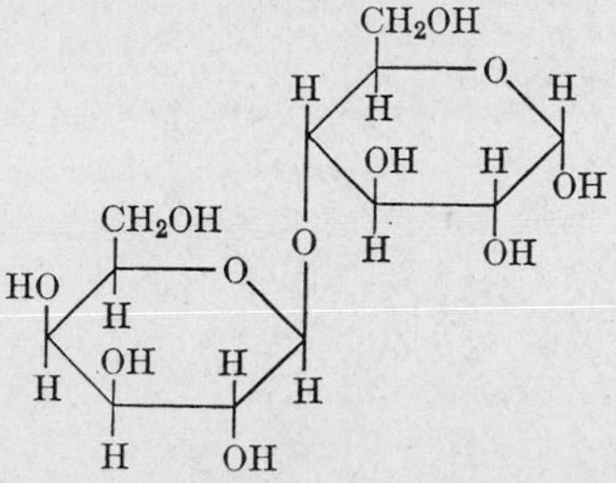

Lactose (4-D-glucose- D-β-galactopyranoside)

For over 40 years synthetic action of β-galactosidases has also been known. Bourquelot and associates[2] showed that β-galactosidase from various sources could synthesize such compounds as ethyl galactoside from the free sugar and ethanol when the substrates were present in high concentrations (I). The equilibrium of the reaction is reached in spite of technical difficulties, such as inactivation of the enzyme and formation of condensation products of galactose with itself.

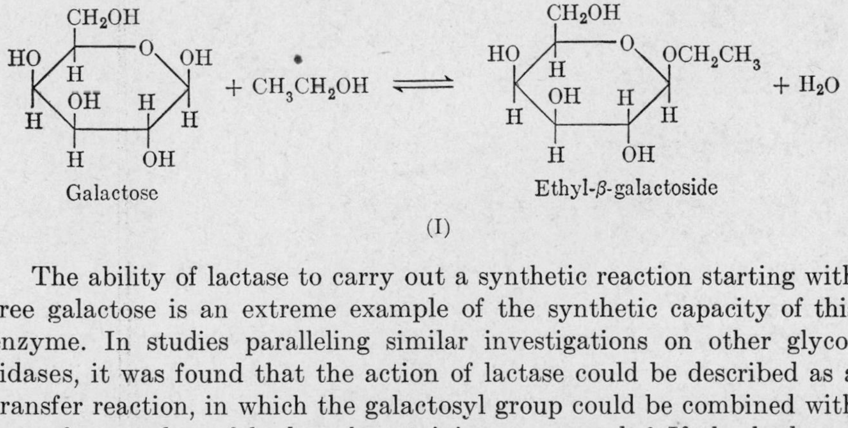

Galactose + CH_3CH_2OH ⇌ Ethyl-β-galactoside + H_2O

(I)

The ability of lactase to carry out a synthetic reaction starting with free galactose is an extreme example of the synthetic capacity of this enzyme. In studies paralleling similar investigations on other glycosidases, it was found that the action of lactase could be described as a transfer reaction, in which the galactosyl group could be combined with any of a number of hydroxyl-containing compounds.[3] If the hydroxyl group were that of water, the product would be free galactose, but other compounds were found to compete with water as acceptors for the galactose of lactose. In particular lactose itself and the hydrolysis products,

[2] E. Bourquelot, J. pharm. chim. 10, 361, 393 (1914).
[3] K. Wallenfels and E. Bernt, Angew. Chem. 64, 28 (1952).

glucose and galactose, are efficient acceptors. The sugars offer several hydroxyl groups as potential sites for condensation, but the primary hydroxyl groups seem to be favored. Therefore, little if any lactose is formed through a reversal of lactase action and the major condensation product formed under conditions that might give lactose is 6-glucose-β-galactoside. The formation of this compound occurs during the hydrolysis of lactose, but more is formed when extra glucose is present. Oligosaccharides composed of galactosyl-1,6-galactose linkages are also formed; in such compounds the reducing end may be either glucose or galactose, but the remainder of the sugar residues are exclusively galactose.

galactose-1-4-glucose + glucose* → galactose-1-6-glucose* + glucose
2 galactose-1-4-glucose → galactose-1-6-galactose-1-4-glucose + glucose
Representative Syntheses of Oligosaccharides by Transgalactosidation

It is obvious that, if an increase in reducing sugar is used to follow lactose hydrolysis, exchange reactions such as lactose (4-glucose-β-galactoside) + glucose → 6-glucose-β-galactoside + glucose will show no net change. Therefore, many studies with β-galactosidase, as well as other glycosidases, were thought to show inhibition by various sugars. Today it is known that these presumed inhibitions do not decrease the rate of splitting of the original glycoside, but merely substitute a sugar for water as the acceptor. A significant difference between the present concept and the former idea of inhibition lies in the deductions that are made. In the older concept mechanisms were postulated to account for the assumed ability of the "inhibitor" to prevent adsorption of the substrate to the enzyme, whereas today it is recognized that activation of the substrate is the same in the presence and absence of the "inhibitor," and that the differences between various "inhibitors" relate to their abilities to accept the glycosyl group and do not furnish any information about the nature of the groups in the substrate that interact with the enzyme.

Consistent with the transfer of galactosyl groups to various acceptors is the hydrolysis of a large series of β-galactosides.[4] The relative rates of hydrolysis of several galactosides differ when enzymes from various sources are used, but this is not unusual for members of a family of enzymes. The distinction of galactosidases from glucosidases poses a more difficult problem. β-Galactosidases have been found to hydrolyze glycosides of several sugars with configurations similar to D-galactose, including L-arabinose and D-galactoaldoheptose. In cases in which crude preparations hydrolyze β-glycosides of both glucose and galactose, no convincing separation of activities has been achieved, but the existence of separate

[4] See general references of W. W. Pigman and S. Veibel.

β-galactosidases and β-glucosidase is strongly suggested by the finding of extracts from several sources that hydrolyze galactosides but not glucosides.[5]

α-Galactosidases are also widely distributed, and often occur together with β-galactosidases. The substrate specificity of α-galactosidases is very similar to that of β-galactosidases, with a variety of different aglycones forming hydrolyzable substrates with galactose, and related sugars replacing galactose also to form substrates.[6] There is no evidence that the galactosidases can split any bond but the terminal glycoside in a polysaccharide. Melibiose, 6-glucose-α-D-galactoside, and raffinose, α-galactosyl sucrose, are typical substrates. Stachyose, a tetrasaccharide in which a

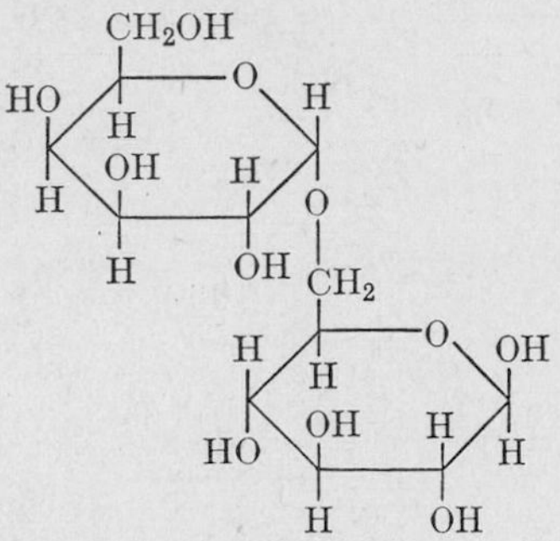

Melibiose (6-glucose-α-D-galactopyranoside)

second α-galactosyl group is added to raffinose, is also split, and the reaction has been shown to cleave the terminal galactose first, leaving raffinose as an intermediate product.[7] The raffinose is then split to a second molecule of galactose and one of sucrose.

α-galactosyl-1–⋮–6-α-galactosyl-1–⋮–4-α-glucosyl-1-2-β-fructofuranoside

first cleavage   second cleavage

α-Glucosidases. Both α- and β-glucosidases are well known, widely distributed enzymes. The ubiquitous maltase is an α-glucosidase. Maltase is named from its ability to hydrolyze maltose, 4-D-glucose-α-D-glucoside.[8] This enzyme exhibits very great specificity for the glucose structure, but is tolerant toward many modifications of the aglycon. Maltases from various sources differ greatly in the relative rates of hydrolysis of different glucosides. Maltase from some sources appears capable of attacking

[5] B. Helferich and R. Griebel, *Ann.* **544**, 191 (1940).
[6] W. W. Pigman, *J. Research Natl. Bur. Standards* **30**, 257 (1943).
[7] J. E. Courtois, C. Anagnastopoulos, and F. Petek, *Compt. rend.* **238**, 2020 (1954).
[8] E. Fisher, *Ber.* **27**, 1429 (1895).

sucrose, but other maltase preparations do not seem to be capable of reaction with this particular α-glucoside.

Maltase and α-glucosidases in general catalyze transfer reactions in the same manner as galactosidases. It was shown many years ago that maltose hydrolysis could be reversed,[9] although the structure of the product was not established conclusively. Experiments with maltose and C^{14}-labeled glucose have shown the incorporation of the isotope into maltose during incubation with an enzyme preparation from intestinal mucosa.[10]

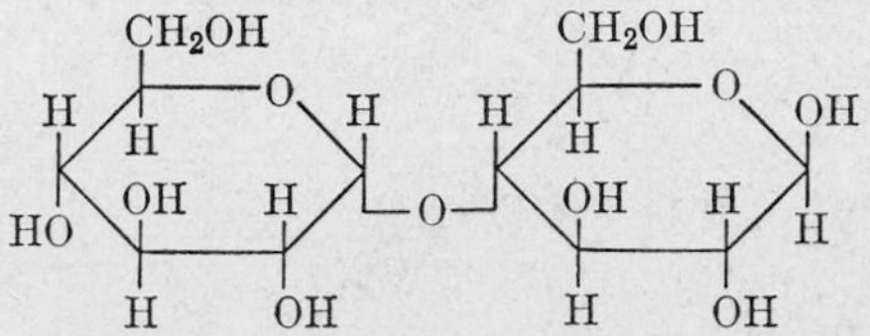

Maltose

This supports the report of reversal of maltase action. Recently the synthesis of isomaltose, in which the primary hydroxyl of carbon 6, rather than the secondary hydroxyl of carbon 4, forms the glycoside, was reported.[11] The thermodynamic and kinetic factors involved in the formation of various disaccharides of glucose have not yet been evaluated. The α-glucosidotransferases of molds have been studied intensively. They do not form 1-4 linkages, but transfer only to the 6-position of glucose. A series of oligosaccharides is formed from maltose by these enzymes. The

2 glucosyl-1-4-glucose → glucosyl-1-6-glucosyl-1-4-glucose + glucose
Maltose   Panose

transfer of glucose to maltose forms panose. At least one more glucosyl-1-6 linkage can be added to panose. Another series of oligosaccharides is formed simultaneously, built on isomaltose, in which all of the glycosides are 1-6 linkages.

Trehalase. Maltase does not attack trehalose, 1-α-glucose-α-glucoside, in which the two glucose residues are equivalent. This sugar is formed by many fungi. The enzyme trehalase, which hydrolyzes the disaccharide,[12] is found in many organisms. It attacks trehalose and certain derivatives in which only one glucose residue bears substitutions.[13a] It has been reported that trehalose occurs in high concentrations in the blood of

[9] C. Hill, *J. Chem. Soc.* **73**, 634 (1898).
[10] E. E. Bacon and J. S. D. Bacon, quoted in review by J. Edelman (see general references).
[11] J. H. Pazur and D. French, *J. Biol. Chem.* **196**, 265 (1952).
[12] E. Bourquelot, *Compt. rend.* **116**, 826 (1893); T. Baba, *Biochem. Z.* **273**, 207 (1934).
[13a] B. Helferich and F. von Stryk, *Ber.* **74**, 1794 (1941).

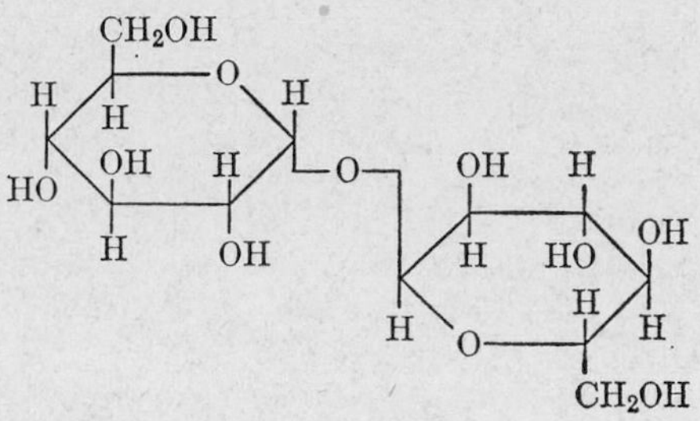

Trehalose

insect larvae and pupae, and that a pathway involving phosphate may participate in its metabolism.[13b]

β-*Glucosidases*. The hydrolysis of β-glucosides is catalyzed by a group of enzymes typified by the β-glucosidase of the sweet almond. The original substrate studied was amygdalin, composed of glucose and mandelic nitrile (benzaldehyde + HCN) (II). The enzyme that hydrolyzed amyg-

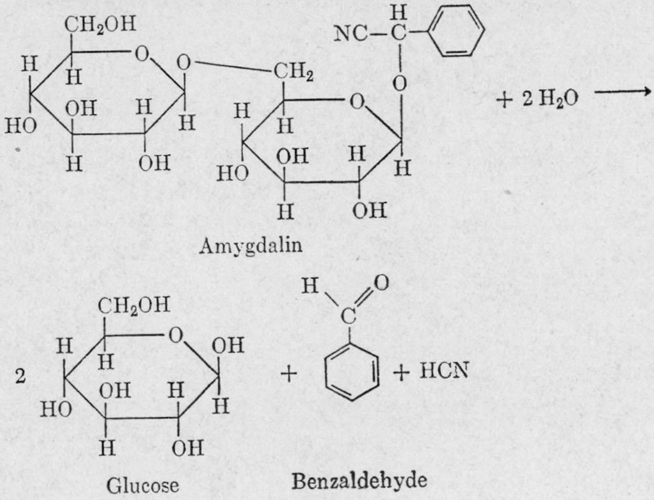

$+ 2 H_2O \longrightarrow$

Amygdalin

Glucose   Benzaldehyde

(II)

dalin was named emulsin. Now that almond extracts are known to contain many glycosidases, the term emulsin has no accepted precise meaning, and the specific designations, such as β-glucosidase, are to be preferred.

Substrate specificity studies with β-glucosidase are not perfectly satisfactory in many cases because of the strong probability of the presence of additional, interfering enzymes. When enzymes of sufficient purity are used, they show absolute specificity for the β-configuration at car-

[13b] G. R. Wyatt and G. F. Kalf, *Federation Proc.* **15**, 388 (1956).

bon 1. Epimerization at carbon 2 forms mannose from glucose; mannosides are split by separate enzymes. 2-Deoxyglycosides are reported to be hydrolyzed by a mannosidase, not by β-glucosidase. Carbon 4 configuration distinguishes glucose from galactose, and, as already discussed, there is reason to believe that separate enzymes attack glycosides of these sugars. Substitutions at carbon 4 do not prevent hydrolysis. The effect of epimerization at carbon 5 to make an L- sugar has not been reported. The substituents on carbon 5 seem to be of secondary importance; the reaction rate is influenced by the presence of a sixth carbon, but pentosides are hydrolyzed, as are glucosides in which carbon 6 bears substitutions.[14]

As in the case of α-glucosidases, the nature of the aglucon affects the rate of the reaction.[15] Indeed, although rates varying by 10^5 have been found, no oxygen linked β-glucoside has been reported to be completely resistant to hydrolysis. The bonds of N- or thioglycosides, on the other hand, appear to be completely resistant to β-glucosidase.

Most of the literature on β-glucosidases deals with hydrolysis of glycosides of compounds other than sugars. The polysaccharide cellulose consists of glucose linked by β-1-4 bonds. On degradation a disaccharide, cellobiose, is formed. Enzymes from molds and snails as well as plants (emulsins) have been found to carry out transfer reactions with cellobiose as glycosyl donor, and several oligosaccharides with β-1-4 and β-1-6 linkages have been detected after prolonged incubation. Thus it appears probable that β-glucosidases resemble the other enzymes that hydrolyze disaccharides in being (relatively) specific only for the glycosyl donor, and in catalyzing transfer reactions in the presence of adequate acceptors.[16]

Synthetic experiments have been carried out with β-glucosidase, and equilibrium constants were determined for systems involving glucose and several simple alcohols.[17] An interesting aspect of this work was the use of acetone as a "neutral" component, to dilute the water concentration to 15 M.

Invertase. The best known of the enzymes that hydrolyze disaccharides is invertase, named for its ability to hydrolyze sucrose to invert sugar, an equimolar mixture of glucose and fructose.[18] Invertase has also been called sucrase and saccharase. Like trehalose, sucrose is composed of sugars joined through their carbonyl groups. In this case either an

[14] W. W. Pigman, *J. Research Natl. Bur. Standards* **26**, 197 (1941).
[15] B. Helferich, *Ergeb. Enzymforsch.* **7**, 83 (1938); S. Peat, W. J. Whelan, and K. A. Hinson, *Nature* **170**, 1056 (1952).
[16] K. V. Giri, V. N. Kigam, and K. S. Srinivasan, *Nature* **173**, 953 (1953).
[17] S. Veibel, *Enzymologia* **1**, 124 (1936).
[18] M. Berthelot, *Compt. rend.* **50**, 980 (1860).

α-glucosidase or a β-fructosidase might split the glycoside. The question as to the mode of action of well-known invertases was thought to have been answered in early work by reference to apparent inhibition by sugars; inhibition by glucose and related compounds was interpreted as indicating that the enzyme being tested was a glucosidase, whereas inhibition by fructose was taken to indicate a fructosidase. These interpretations have no longer any credance, since it has been shown that invertases are transglycosidases, in general β-transfructofuranosidases.

Invertase has been found widely distributed in microorganisms, plants, and animals.[19] The best-studied enzymes are those of yeast and other fungi. In higher animals invertase is produced only in the intestinal mucosa and acts only as a digestive enzyme.

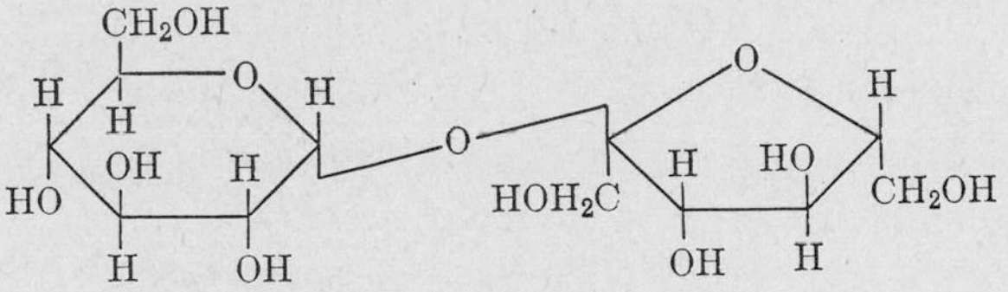

Sucrose

The fructosidase of yeast has been purified 3000-fold, but has not yet been obtained in a pure state.[20] This enzyme transfers fructose from sucrose and other β-fructofuranosides to primary hydroxy groups.[21] Glucose is an acceptor in the transfer reaction, but experiments with C^{14}-labeled glucose show that transfer to the carbonyl group, to reform sucrose does not occur.[22] This is in contrast with the mold enzyme, which incorporates labeled glucose into sucrose as well as into other oligosaccharides. An unexplained observation is the failure to detect sucrose synthesis when fructose was transferred to glucose from methyl-β-fructoside by the mold invertase. Among the major products of invertase action on sucrose are: fructosyl-2-6-glucose, fructosyl-2-1-fructosyl-2-1—glucose (1-kestose), fructosyl-2-6-fructosyl-2-1-glucose (6-kestose), and fructosyl-2-6-glucosyl-1-2-fructose (neokestose).

Inulinase. Inulin is a polysaccharide of the Jerusalem artichoke composed principally of fructosyl-2-1-fructosyl groups. It seems probable that each molecule is built on one sucrose molecule, and it is possible that inulin synthesis is carried out at the expense of sucrose. A fructosidase

[19] See review by C. Neuberg and I. Mandl (general reference).
[20] J. B. Sumner and D. J. O'Kane, *Enzymologia* **12**, 251 (1948).
[21] B. F. Bealing and J. S. D. Bacon, *Biochem. J.* **53**, 277 (1953); W. J. Whelan and D. M. Jones, *ibid.* **54**, xxxiv (1953).
[22] J. Edelman, *Biochem. J.* **57**, 22 (1954).

from Jerusalem artichoke tubers transfers fructose from inulin and intermediate-sized oligosaccharides of the inulin type.[23] The enzyme appears to be quite specific in its acceptor requirements; only the 1-position of fructofuranose, as found in sucrose and inulin, is used.

Glucoinvertases. Besides fructosidases, invertases are known that are α-glucosidases. The distinction between glucoinvertases and maltases has not been established clearly in most cases. Enzymes found in honey are glucotransferases.[24a] Incubation of sucrose with nectar of flowers of *Robinia pseudacacia* resulted in accumulation of free fructose and oligosaccharides containing principally glucose residues.[24a] Transfructosidases have also been found in both plant nectar and insect viscera. A transglucosidase with specificity different from the other enzymes described occurs in honeydew and manna formed by certain insects. This enzyme is produced by the insects, not the plant, and transfers glucosyl groups to the 3-position of the fructose of sucrose.[25]

$$2 \text{ glucosyl-1-2-fructoside} \rightarrow \text{glucosyl-1-3-fructosyl-2-1-glucoside} + \text{glucose}$$
Sucrose   Melezitose

Hydrolysis of Other Oligosaccharides. Enzymes designated α-mannosidases have been found that attack α-glycosides of mannose and related sugars.[26] A β-mannoside has been found in yeast.[27] Certain sulfurcontaining glucosides occur naturally, and the hydrolysis of these compounds by thioglucosidase, present in certain seeds, has been known for many years. The thioglucosidase is accompanied by a sulfatase, which

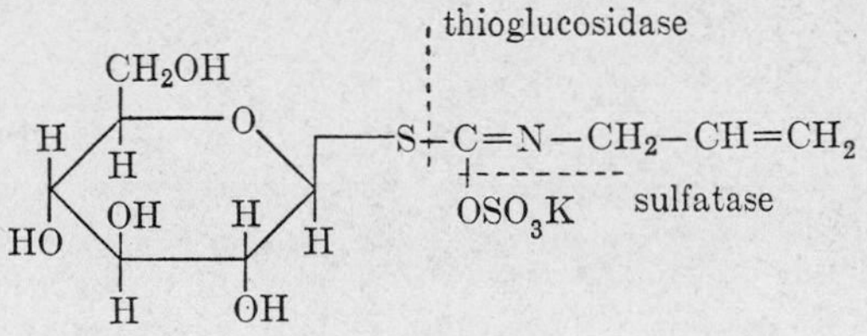

Sinigrin

forms the glucoside of allyl mustard oil (merosinigrin) from sinigrin. Subsequent hydrolysis at the bond indicated liberates glucose and allyl isothiocyanate (allyl mustard oil).[28]

[23] J. Edelman and J. S. D. Bacon, *Biochem. J.* **49**, 446 (1951).
[24a] J. W. White and J. Maker, *Arch. Biochem. Biophys.* **42**, 360 (1953).
[24b] M. Zimmerman, *Experientia* **10**, 145 (1954).
[25] J. S. D. Bacon and B. Dickinson, *Biochem. J.* **61**, xv (1955).
[26] H. Herissey, *Compt. rend.* **172**, 766 (1921); see also reference 6.
[27] M. Adams, N. K. Richtmyer, and C. S. Hudson, *J. Am. Chem. Soc.* **65**, 1369 (1943).
[28] C. Neuberg and O. von Schonbeck, *Biochem. Z.* **265**, 223 (1933).

In addition to sugars, certain modified sugars occur naturally in glycosides. In animals glucuronic acid is found to be conjugated with a large variety of aromatic and terpene compounds. In plants the glucuronides of flavones are known. These compounds all appear to be β-glycosides, and they are hydrolyzed by a widely distributed family of β-glucuronidases.[29a] β-Glucuronidases from animals, plants, and microorganisms all

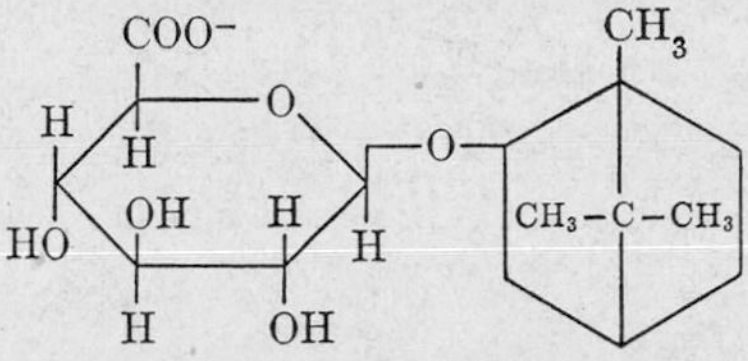

Borneol glucuronide

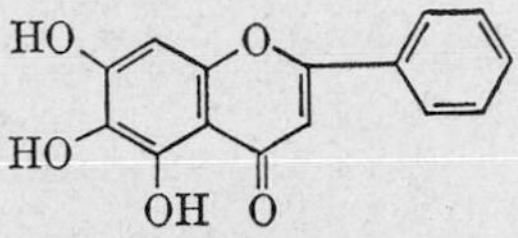

Baicalein
(the flavone aglycone of
the glucuronide, baicalin)

appear to be similar in their abilities to hydrolyze a large variety of β-glucuronides, although the relative rates of hydrolysis, K_m values, and sensitivity to inhibitors differ. The enzymes of animal origin all have pH optima at 4.5 or below, with some having a second optimum at pH 5.0–5.2. Several bacterial enzymes, on the other hand, act most rapidly near pH 6.[29b]

Since the determination of β-glucuronidase is of interest in connection with studies on various physiological states, assay methods are of practical importance.[30] Assays dependent on the liberation of free glucuronic

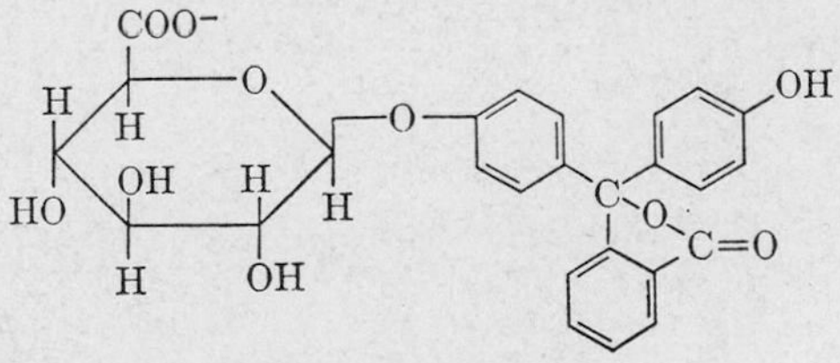

Phenolphthalein glucuronide

acid are subject to many objections both because of the inconvenience in determining glucuronic acid and the possibility of blank reactions and endogenous substrates contributing to the final values. A more convenient assay is the colorimetric determination of phenolphthalein liberated from its glucuronide.

[29a] C. Neuberg and C. Niemann, *Z. physiol. Chem.* **44,** 114 (1905); Y. Sera, *Z. physiol. Chem.* **92,** 261 (1914).

[29b] G. A. Levvy, *Biochem. J.* **58,** 462 (1954).

[30] See general reference by W. H. Fishman (1951).

Synthesis of Larger Polysaccharides

Polysaccharides are high polymers of carbohydrate compounds. Various polysaccharides are polydisperse mixtures characterized by the nature of the monomer(s) and the type(s) of chemical linkage in the polymer. The most prominent and best-studied polysaccharides are composed of glucose; the linkages found are α-1-4 (amylose), α-1-4 and α-1-6 (amylopectin, glycogen), α-1-6 (dextran), and β-1-4 (cellulose). Polymers of fructose also occur naturally; β-2-1 (inulin) and β-2-6 (levan) are the known types of linkage. Other sugars and sugar derivatives including pentoses, galactose, uronic acids, and hexose amines also occur in polysaccharides, some of which contain more than one type of monomer. The known methods for enzymatic synthesis of polysaccharides include transfer from oligosaccharides, transfer from sugar phosphates, and transfer from specific carriers. The degradation of polysaccharides may be the reversal of synthesis, but it also may occur through the activity of hydrolytic enzymes that do not catalyze measureable synthesis.

Dextran Sucrase. Dextrans are produced by the bacteria *Leuconostoc mesenteroides* from sucrose. The reaction is catalyzed by a glucotransferase, which has been purified and which produces little if any free glucose.[31] Similar enzymes appear to be present in other bacteria. The reaction catalyzed by dextran sucrase is:

$$n \text{ glucosyl-1-2-fructoside} \rightarrow (\text{glucose-1-6})_n + n \text{ fructose}$$

The dextrans are not simply straight-chain compounds, however; they contain branches of 1-4 linkages that are created from 1-6 chains by a "branching enzyme." This is similar in its action to a branching enzyme described below in connection with starch and glycogen synthesis. The formation of α-1-6 linkages by dextran sucrase is similar to the reaction already described for α-glucosidases. The difference lies in the affinity for acceptors. While maltase forms oligosaccharides with 3 or 4 glucose residues, dextran sucrase reacts more rapidly with long chains, and the principal products have molecular weights in the thousands (in the presence of branching enzyme, molecular weights may be in the millions); oligosaccharides are also formed, however. The reaction is essentially irreversible.

Amylosucrase. An enzyme from the bacteria *Neisseria perflava* carries out a reaction similar to that of dextran sucrase, but forms α-1-4 linkages instead of α-1-6.[32] The reaction is reversible, but, as in case of dextran

$$n \text{ sucrose} \rightleftharpoons n \text{ fructose} + \text{glucose}_n \text{ (amylose)}$$

[31] E. J. Hehre and J. Y. Sugg, *J. Exptl. Med.* **75,** 339 (1942).
[32] E. J. Hehre and D. M. Hamilton, *J. Biol. Chem.* **166,** 777 (1946).

synthesis, the equilibrium lies far to the right. In nature this enzyme is also accompanied by a branching enzyme, which converts the amylose to a branched amylopectin structure.

Levan Synthesis. Oligosaccharides of the levan type (fructosyl-2-6-fructose) are synthesized by fructosidases. The synthesis of larger polysaccharides is accomplished by special enzymes that appear to have higher affinity for long chains than for short ones. Levan-synthesizing enzymes have been studied in preparations from *Aerobacter levanicum*.[33]

Amylomaltase. Maltose can also serve as a glucosyl donor for the formation of polysaccharides. An enzyme from *E. coli* forms a polymer of α-1,4-glucosides and an equivalent amount of free glucose from maltose.[34] In this case the length of the polymer is found to be dependent on the amount of free glucose present. Apparently glucose acts as an acceptor and starts more chains with a shorter average chain length. This reaction is reversible, and the polysaccharide reacts with glucose to form maltose.

Disproportionating (D) Enzyme. An enzyme from potato tubers catalyzes a reversible transfer of glucose-α-1-4 linkages, but it transfers 2 sugar residues, a maltose unit.[35] The donor of such a unit may be as small as maltotriose (3 glucose residues) and may be a large starch molecule. The transfer builds up linear polymers of glucose (amylose).

Dextran-Dextrinase. A dextran-forming enzyme from *Acetobacter capsularum* converts α-1,4 linkages to α-1,6, but acts only on substrates of intermediate size. This *dextran-dextrinase* catalyzes a reversible reaction in which the terminal residue of either a dextrin (α-1,4) or a dextran (α-1,6) is transferred to the other polysaccharide.[36]

PHOSPHORYLASES

Sugars become available for the major metabolic pathways by being phosphorylated. Polysaccharide formation also requires phosphorylated sugars for *de novo* synthesis of glycosidic bonds. The transglycosidations discussed previously depend upon preformed glycosides. Glucose becomes available for polysaccharide synthesis through the action of hexokinase

$$\text{glucose} + \text{ATP} \xrightarrow{\text{hexokinase}} \text{glucose-6-phosphate} \xrightarrow[\text{glucomutase}]{\text{phospho-}} \text{glucose-1-phosphate}$$

and phosphoglucomutase. Other sugars may be converted indirectly to glucose-1-phosphate, as through the pentose phosphate cycle, or they may

[33] S. Avineri-Shapiro and S. Hestrin, *Biochem. J.* **39**, 167 (1945); see also S. Hestrin and S. Avineri-Shapiro, *ibid.* **38**, 2 (1944); **37**, 450.

[34] J. Monod and A. M. Torriani, *Ann. inst. Pasteur* **78**, 65 (1950); see *Compt. rend.* **227**, 240 (1948).

[35] S. Peat, W. J. Whelan, and W. R. Rees, *Nature* **172**, 158 (1953).

[36] E. J. Hehre, *J. Biol. Chem.* **192**, 161 (1951).

react as glycosyl phosphates. Galactose-1-phosphate is formed directly by phosphorylation. The special reactions of some phosphorylated derivatives of certain sugars will be considered in the next chapter; this section deals with the formation of polysaccharide from glucose-1-phosphate.

Sucrose Phosphorylase. A group of microorganisms, *Leuconostoc mesenteroides, Pseudomonas saccharophila,* and *P. putrefasciens,* contain enzymes that decompose sucrose with the esterification of inorganic phosphate, forming glucose-1-phosphate and fructose.[37] In this reaction phosphate plays the role of water in hydrolysis, and the enzyme is therefore called a phosphorylase; in this case, *sucrose phosphorylase.* Phosphorolysis differs from hydrolysis in the relative ease of reversal. The equilibrium constant for the reaction:

$$K = \text{(sucrose)(phosphate)}/\text{(fructose)(glucose-1-phosphate)}$$

is approximately 0.05 at pH 6.6 and 30°C. Although this reaction does not seem to account for sucrose accumulation in green plants, it is of interest as the first *in vitro* synthesis of sucrose.[38]

Sucrose phosphorylase exhibits a very high degree of specificity for the glucose portion of its substrates, but it reacts with a number of glycosyl acceptors in addition to phosphate and fructose. The enzyme from *P. saccharophila,* but not from *P. putrefaciens,* will react with sorbose,

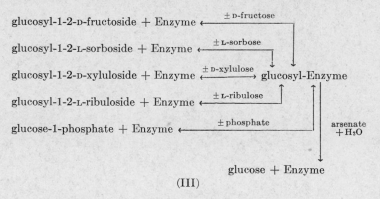

(III)

forming α-D-glucosido-α-L-sorbofuranoside from L-sorbose and glucose-1-phosphate. The same is true for the reaction with D-xylulose. All of the sucrose phosphorylase preparations appear to react with L-arabulose and L-arabinose. Inorganic arsenate enters into the sucrose phosphorylase reaction to cause arsenolysis of any of the substrates. As in the cases of other arsenolyses (triose phosphate dehydrogenase, nucleoside phosphor-

[37] M. Doudoroff, N. Kaplan, and W. Z. Hassid, *J. Biol. Chem.* **148,** 67 (1943).
[38] W. Z. Hassid, M. Doudoroff, and H. M. Barker, *J. Am. Chem. Soc.* **66,** 1416 (1944).

ylase, transacetylase), the explanation for the arsenolysis presumes the formation of an unstable glucose arsenate that hydrolyzes spontaneously.

Sucrose phosphorylase catalyzes a number of exchange reactions, such as the equilibration of glucose-1-phosphate with inorganic phosphate (labeled with P^{32}). The exchange of the group condensed with glucose and the transfer of glucose from one acceptor to another have led to the postulation of an enzyme-glucose compound as an intermediate in all of the reactions (III).[39] The enzyme-glucose compound, according to this concept, reacts with any glucose acceptors that have the proper configuration. Enzyme-substrate compounds have been postulated many times to explain exchange reactions, but the evidence for the existence of such compounds is still incomplete and usually indirect.

Muscle Phosphorylase. The first of the phosphorylases to be discovered was named simply phosphorylase.[40] It was found by Cori and Cori to catalyze the breakdown of glycogen, an animal polysaccharide, with the consumption of inorganic phosphate and the formation of a previously unknown phosphate ester, glucose-1-phosphate.[41] The action of this enzyme together with certain other enzymes has permitted studies that have largely determined the structures of various natural high molecular weight polysaccharides.

Phosphorylase acts reversibly on 1,4 linkages. The enzyme from potato requires a chain of at least 3 glucose units as a "primer" and reacts more rapidly with larger molecules, and the enzyme from muscle requires a relatively complex primer. The reaction continues to increase the chain length until an equilibrium is established[42] in which

$$K_{eq} = \frac{\text{(inorganic P)}}{\text{(glucose-1-phosphate)}}$$

The concentration of glycogen does not appear in the equilibrium constant because the same molecule is both substrate and product, and therefore cancels out of the equation.

$$K_{eq} = \frac{\text{(polysaccharide end groups)(phosphate)}}{\text{(polysaccharide end groups)(glucose-1-phosphate)}}$$

The concentration of primer, while not influencing the amount of glucose polymerized, does determine the average chain length of the product.

Muscle phosphorylase was isolated in two crystalline forms, *a* and

[39] M. Doudoroff, H. A. Barker, and W. Z. Hassid, *J. Biol. Chem.* **168,** 725 (1947).
[40] C. F. Cori and G. T. Cori, *Proc. Soc. Exptl. Biol. Med.* **34,** 702 (1936); **36,** 119 (1937).
[41] C. F. Cori, S. P. Colowick, and G. T. Cori, *J. Biol. Chem.* **121,** 465 (1937); G. T. Cori, S. P. Colowick, and C. F. Cori, *ibid.* **123,** 375 (1938).
[42] G. T. Cori and C. F. Cori, *J. Biol. Chem.* **135,** 733 (1940).

b.[43] The two forms are distinguished enzymatically by the requirement of phosphorylase *b* for adenylic acid. An enzyme (PR) converts phosphorylase *a* to *b*.[44] Since a cofactor, adenylic acid, was required after the action of the enzymatic conversion of phosphorylase *a* to *b*, the enzyme responsible was named "prosthetic group-removing" enzyme. This implication of the nature of PR activity was subjected to reservations when attempts to demonstrate adenylic acid in phosphorylase *a* were unsuccessful, and no adenylic acid was found after PR action. When it was found that the phosphorylase *a* molecule (molecular weight 495,000) is split into halves (molecular weight 242,000) by PR, the initials were retained to represent "phosphorylase-rupturing" enzyme.[45] Similar but not identical fragments with phosphorylase *b* activity were obtained by partial digestion of phosphorylase *a* by trypsin. PR does not seem to have general proteolytic activity, however.

During the conversion of phosphorylase *a* to *b* inorganic phosphate is released. The condensation of phosphorylase *b* to *a* has been found to proceed with utilization of phosphate from ATP.[46] A kinase purified from muscle requires Mn^{++} or Mg^{++} to transfer phosphate to the reconstituted phosphorylase *a*. This phosphate is liberated by PR enzyme (IV). The

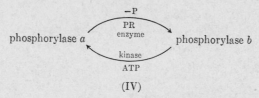

$$-P$$
$$\text{PR enzyme}$$
phosphorylase *a*   phosphorylase *b*
$$\text{kinase}$$
$$\text{ATP}$$

(IV)

occurrence of the kinase is essential for the preparation of crystalline phosphorylase *a*. In the preparation of phosphorylase from rabbit muscle very little phosphorylase *a* is obtained in the extracts. However, during an early step, sufficient metal is extracted from filter paper to activate the kinase, and phosphorylase *a* is formed with the participation of ATP present in the extract.

Liver Phosphorylase. Liver phosphorylase has been highly purified but not crystallized.[47] It has been shown to contain firmly bound phosphate that does not exchange during activity. The enzyme phosphate is lost on incubation with another "inactivating" enzyme, and the resulting "dephosphophosphorylase" is an inactive protein of about the same size as

[43] C. F. Cori, G. T. Cori, and A. A. Green, *J. Biol. Chem.* **151**, 39 (1943); C. F. Cori and G. T. Cori, *ibid.* **158**, 341 (1945).
[44] G. T. Cori and A. A. Green, *J. Biol. Chem.* **151**, 351 (1943).
[45] P. J. Keller and G. T. Cori, *Biochim. et Biophys. Acta* **12**, 235 (1953).
[46] E. H. Fischer and E. G. Krebs, *J. Biol. Chem.* **216**, 121 (1955).
[47] E. W. Sutherland and W. D. Wosilait, *J. Biol. Chem.* **218**, 483 (1956).

the original enzyme. Reactivation of liver phosphorylase is brought about by an ATP, Mg-requiring kinase.[48] The relationship between the kinases for reactivating liver and muscle phosphorylases has not been established.

Phosphorylase and the Structure of Polysaccharides. Substrates for phosphorylase include starch and glycogen. Starch has been fractionated into amylose and amylopectin fractions. Amylose is essentially a linear polymer of α-1-4 linked glucose molecules and produces a blue color with iodine. Amylopectin is a highly branched molecule, containing α-1-6 linkages in addition to α-1-4. Glycogen resembles amylopectin. The branched polysaccharide produce red-brown colors with iodine. The structures of these polysaccharides were inferred from chemical degradation studies, and confirmed by the use of enzymes. The chemical methods that give more information are determination of methyl glucoses following exhaustive methylation and hydrolysis and periodate oxidation. The nonreducing ends yield 2,3,4,6-tetramethylglucose, the bulk of the glucose becomes 2,3,6-trimethyl glucose, and branch points yield 2,3-dimethyl glucose (V). Only nonreducing ends bear three adjacent free hydroxyl

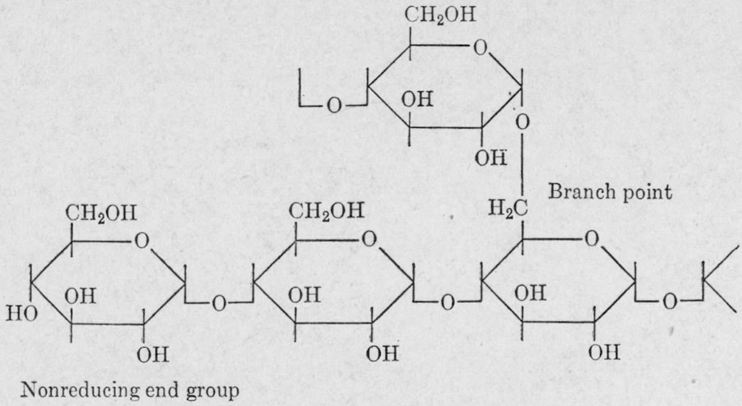

(V)

groups (on carbons 2, 3, and 4). Periodate converts carbon 3 to formic acid; no formic acid is formed from any other glucose residues, so formic acid appearance is a measure of nonreducing end groups. Only one reducing group is found in each molecule of the polysaccharides, and molecular weight determinations by physical methods and determination of reducing groups agree with each other.

The degradation of amylose by phosphorylase can be coupled with reactions for the removal of glucose-1-phosphate, and the reaction proceeds essentially to completion. The degradation of amylopectin or gly-

[48] T. W. Rall, E. W. Sutherland, and W. D. Wosilait, *J. Biol. Chem.* **218**, 483 (1956).

cogen, on the other hand, stops far short of completion, with the accumulation of a polysaccharide resistant to further attack by phosphorylase. This polysaccharide is a limit dextrin. Phosphorylase activity can proceed only from nonreducing ends to branch points; phosphorylase cannot split 1-6 linkages, and, since only single glucose units can be removed by phosphorylase, it cannot by-pass the branch points. The degradation of the branch (which starts with a 1-6 linkage) proceeds until only one sugar residue is left. The other chain (the main stem, with continuous 1-4 linkages) is not split completely to the branch point, but is left as a stub composed of two maltose residues added to the molecule that bears the branch.[49]

A specific 1,6-amyloglucosidase is capable of removing the last glucose residue of a branch hydrolytically.[50] This enzyme is active only with large polysaccharides. Another glucosidase, found in intestine, splits 1,6

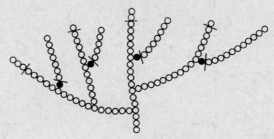

FIG. 25. Portion of glycogen or amylopectin molecule. Lines indicate limits of phosphorylase degradation. Solid·circles represent glucose residues removed by amyloglucosidase. From Larner et al.[52]

linkages of oligosaccharides.[51] The sum of glucosidase + phosphorylase results in complete degradation of glycogen. If the dextrin is isolated after incubation with each enzyme in turn, the chains of glucose residues can be removed in "tiers." Such experiments have established an "arboreal" structure for glycogen (Fig. 25)[52] and have permitted isotopic studies on the rate of turnover of various parts of the polysaccharide molecule.[53] Plant polysaccharides have also been analyzed by enzymatic degradation.

[49] S. Hestrin, J. Biol. Chem. 179, 943 (1949).
[50] G. T. Cori and J. Larner, J. Biol. Chem. 188, 17 (1951).
[51] M. Seiji, J. Biochem. 40, 519 (1953); J. Larner and C. M. McNickle, J. Am. Chem. Soc. 76, 4747 (1954).
[52] K. H. Meyer, P. Bernfeld, R. A. Boissonnas, P. Gurtler, and G. Noelting, J. Phys. & Colloid. Chem. 53, 3M (1949); J. Larner, B. Illingworth, G. T. Cori, and C. F. Cori, J. Biol. Chem. 199, 641 (1952).
[53] M. R. Stetten and D. W. Stetten, Jr., J. Biol. Chem. 207, 331 (1954).

The synthesis of branched polysaccharides involves the prior synthesis of long chains of 1,4 linkages. A "branching enzyme" then transfers a fragment of a chain to the 6 position of a glucose residue within a shortened chain.[54] Animal branching enzymes attack chains with 6 to 11 glucose residues. A very similar enzyme from potato, the Q enzyme, requires a slightly longer chain.

The resultant of phosphorylase and branching enzyme activities is glycogen or amylopectin. No other enzymes are needed. The requirements of branching enzymes for chain lengths seem to result in a fairly regular structure. Formation of a limit dextrin by phosphorylase produces a number of glucose phosphate molecules that can be divided by either the number of nonreducing end groups or twice the number of free glucose molecules liberated from the dextrin by amyloglucosidase to give an average chain length of the so-called outer tier. If the two enzymes are present together, the entire polysaccharide is converted to glucose phosphate and free glucose, and an average chain length between branch points can be determined. By using only one enzyme at a time, the outer chains can be degraded, the exposed branch points removed, the seond tier degraded, the second series of branch points removed, etc. In this way the average chain length of any position of the molecule may be determined and, with isotopes, the rates of turnover of inner and outer tiers can be estimated individually. Average chain lengths between 4 and 9 glucose units have been found for various polysaccharides.[52]

HYDROLYSIS OF STARCH AND GLYCOGEN

The hydrolytic enzymes that attack large polymers of glucose with α-1,4 linkages are called amylases. These are divided into two classes, α and β, not on the basis of the linkage attacked, but on the configuration of the reducing sugar liberated. Both types of amylase liberate maltose, and from the mutarotation of the product, it was concluded that some amylases yield β-maltose while others leave the α-configuration of the polysaccharide.[55] The two classes of enzymes thus established also differ in other major respects.

β-Amylase resembles phosphorylase in that it attacks the nonreducing end of a chain. It removes maltose units as far as a branch point, then stops.[56] α-Amylase is not restricted by branches, but attacks randomly throughout the polysaccharide to form fragments of various sizes,

[54] J. Larner, *J. Biol. Chem.* **202**, 491 (1953); P. N. Hobson, W. J. Whelan, and S. Peat, *J. Chem. Soc.* p. 596 (1951).
[55] R. Kuhn, *Ann.* **443**, 1 (1925); E. Ohlsson, *Z. physiol. Chem.* **189**, 17 (1930).
[56] K. H. Meyer, M. Wertheim, and P. Bernfeld, *Helv. Chim. Acta* **24**, 412 (1941).

which continue to be split.[57] The two types of activity are reflected in different types of changes in the substrate. α-Amylase is often assayed by the decrease in viscosity or change in color on addition of iodine, and is sometimes called the liquefying or dextrinogenic amylase. β-Amylase does not interfere seriously with the assays for α-amylase, but causes a rapid increase in the amount of reducing sugar (maltose). It is therefore sometimes called the saccharogenic amylase, and also is known as exo-amylase, in contrast to endoamylase, which describes the ability of α-amylase to attack the middle of chains.

The amylases and phosphorylase appear to attack randomly, in the so-called "multichain" action.[58a,b] That is, instead of starting an attack on one chain and continuing to digest that chain to completion, each enzyme appears to dissociate completely from the split products after the addition of one molecule of water or phosphate. The subsequent attack is determined by the concentration of various potential substrates and their relative affinities, not their past histories. β-Amylase exhibits similar affinities for both long and short chains (until the chain is reduced to 4 glucose residues).[58b] α-Amylases from various sources require different chain lengths for optional activity; some attack only two linkages from the nonreducing end, while others prefer at least 5 units beyond the sensitive bond.[59] The free aldehyde on the reducing end of the substrate is not required; it may be converted to a carboxyl group and the molecule is still attacked by both α- and β-amylases.[60]

Both α- and β-amylases from several sources have been crystallized. These include α-amylase of human saliva,[61] human[62] and pig[63] pancreas, malt,[64] *Bacillus subtilis*,[65] and a thermophilic bacterium.[66] Human salivary and pancreatic amylases appear identical by all criteria, and it has been suggested that the enzymes from various tissues of a given organism are indeed identical. Crystalline α-amylases from different organisms are easily distinguished by solubility, electrophoresis, substrate affinity, molecular weight, and chemical composition. Crystalline β-amylase has

[57] K. H. Meyer and P. Bernfeld, *Helv. Chim. Acta* **29**, 359E (1941).

[58a] J. Larner, *Federation Proc.* **13**, 247 (1954).

[58b] R. H. Hopkins and B. Jelineck, *Biochem. J.* **56**, 136 (1954).

[59] R. Bird and R. H. Hopkins, *Biochem. J.* **56**, 86 (1954).

[60] K. Svanborg and K. Myrbäck, *Arkiv Kemi* **6**, 113 (1954).

[61] K. H. Meyer, E. H. Fischer, A. Staub, and P. Bernfeld, *Helv. Chim. Acta* **31**, 2158 (1948).

[62] K. H. Meyer, E. H. Fischer, P. Bernfeld, and F. Duckert, *Arch. Biochem.* **18**, 203 (1948).

[63] E. H. Fischer and P. Bernfeld, *Helv. Chim. Acta* **31**, 1839 (1948).

[64] S. Schwimmer and A. K. Balls, *J. Biol. Chem.* **179**, 1063 (1949).

[65] K. H. Meyer, M. Fuld, and P. Bernfeld, *Experientia* **3**, 411 (1947).

[66] L. L. Campbell, *J. Am. Chem. Soc.* **76**, 525 (1954).

been obtained from sweet potatoes[67] and wheat.[68] The animal α-amylases are unusual in their activation by monovalent anions, especially chloride. This effect is greatly dependent on pH.[69]

Amyloglucosidases. Several species of fungi have been found to produce enzymes that split free glucose from polysaccharides without the formation of oligosaccharides. One such enzyme, gluc amylase, has been crystallized from *Rhizopus delemar*.[70]

Levanpolyase. Levans are hydrolyzed by specific bacterial enzymes[71] to oligofructosides that are further metabolized by conventional fructosidases. These latter will not attack intact levans, however.

Cellulase. Cellulose is a polysaccharide composed of β-1,4-glucosides. At this time only a beginning has been made in the separation and identification of various cellulases.[72] No evidence for phosphorolysis of β-glucosides has been presented, but exchange reactions of β-glucosides are known.[73] The inversion of the glucoside bond of maltose has been observed by Fitting and Doudoroff, who obtained a phosphorylase from *Neisseria meningitidis* that forms β-glucose-1-phosphate.[74] The relation between β-glucose-1-phosphate and cellulose remains speculative at this time. Cellulases analogous to amylases seem likely to be found, as both glucose and cellobiose (glucosyl-β-1-4-glucose) are found as degradation products of cellulose by fungal preparations.

Hyaluronidase. Hyaluronic acid is a polymer composed of alternating N-acetylglucosamine and glucuronic acid residues. The polymer is degraded by an enzyme, hyaluronidase, which occurs widely distributed in animal tissues, and is especially abundant in umbilical cord and leeches. A good source of hyaluronidase is testis. Testicular hyaluronidase splits oligosaccharides in multiples of the repeating unit, glucuronidoacetylglucosamine, including the single unit, N-acetylhyalobiuronic acid.[75] Bacterial hyaluronidases apparently attack the same linkage as animal enzymes, but merely split the linkage without the addition of water. A

[67] A. K. Balls, R. R. Thompson, and M. K. Walden, *J. Biol. Chem.* **173**, 9 (1948).
[68] K. H. Meyer, P. F. Spahr, and E. H. Fischer, *Helv. Chim. Acta* **36**, 1924 (1953).
[69] P. Bernfeld, A. Staub, and E. H. Fischer, *Helv. Chim. Acta* **31**, 2165 (1948).
[70] J. Fukumoto, Y. Sakazaka, and K. Minamii, *Symposia on Enzyme Chem. (Japan)* **9**, 94 (1954).
[71] S. Hestrin and J. Goldblum, *Nature* **172**, 1646 (1953).
[72] See general reference by Pigman (1951).
[73] P. Knoiman, P. A. Roelofsen, and S. Sweeris, *Enzymologia* **16**, 237 (1953); S. A. Barker, E. J. Bourne, and M. Stacey, *Chem. & Ind. (London)* p. 1287 (1953); K. V. Giri, V. N. Nigam, and K. S. Srinivasan, *Nature* **173**, 953 (1954); H. W. Buston and A. Jabbar, *Chem. & Ind. (London)* p. 48 (1954).
[74] C. Fitting and M. Doudoroff, *J. Biol. Chem.* **199**, 153 (1954).
[75] B. Weissmann, K. Meyer, P. Sampson, and A. Linker, *J. Biol. Chem.* **208**, 417 (1954).

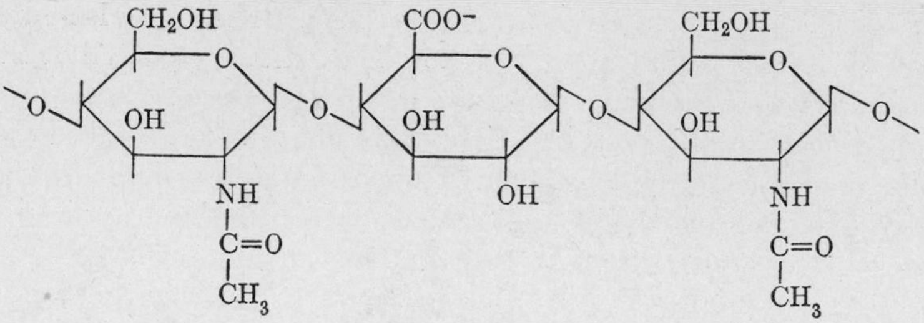

Segment of Hyaluronic Acid

double bond thus is introduced into the glucuronic acid component.[76] Testicular hyaluronidase has been found to catalyze exchange reactions as well as hydrolyses.[77] The disaccharide *N*-acetylhyalobiuronic acid is transferred to form oligosaccharides indistinguishable from those formed

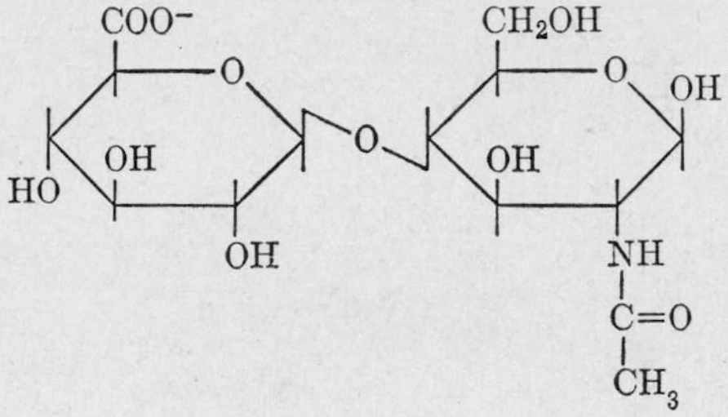

N-Acetylhyalobiuronic acid

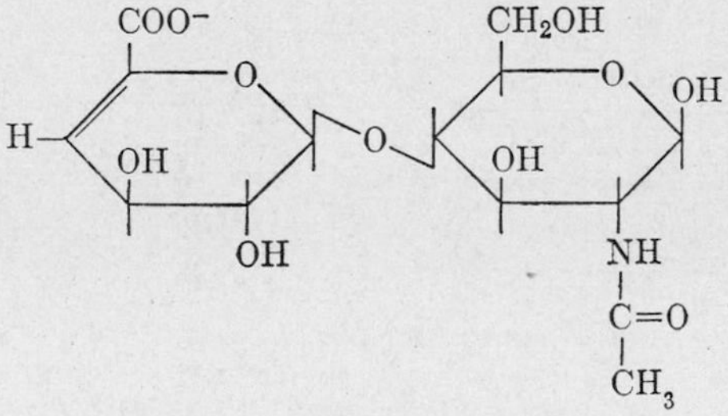

Product of Bacterial Hyaluronidase

during hydrolysis of hyaluronic acid. Further degradation of the oligosaccharides derived from hyaluronic acid is catalyzed by a β-glucosaminidase found in liver and testis. Successive hydrolyses by the β-glucosaminidase and β-glucuronidase convert the oligosaccharides to monosaccharides.[78]

[76] A. Linker, K. Meyer, and P. Hoffman, *J. Biol. Chem.* **219**, 13 (1956).
[77] B. Weissmann, *J. Biol. Chem.* **216**, 783 (1955).
[78] A. Linker, K. Meyer, and B. Weissmann, *J. Biol. Chem.* **213**, 237 (1955).

Certain other polysaccharides are also known to be hydrolyzed enzymatically. Polygalacturonic acid is the basic structure of pectins. This is hydrolyzed by enzymes that require free carboxyl groups, and form oligosaccharides as small as the disaccharide 4-(α-D-galacturonopyranosido)-D-galacturonopyranose. The complete mechanisms of the synthesis and degradation of these compounds and the several other polysaccharides, arabans, mannans, galactans, alginic acid, chitin, etc., are still far from known.

General References

Cori, C. F. (1946). *Harvey Lectures Ser.* **41**, 253.

Edelman, J. (1956). *Advances in Enzymol.* **17**, 189.

Fishman, W. H. (1955). *Advances in Enzymol.* **16**, 361.

Fishman, W. H. (1951). *In* "The Enzymes" (J. B. Sumner and K. Myrbäck, eds.), Vol. I, Part 1, p. 635; Part 2, p. 769. Academic Press, New York.

Gottschalk, A. (1951). *In* "The Enzymes" (J. B. Sumner and K. Myrbäck, eds.), Vol. I, Part 1, p. 551. Academic Press, New York.

Hassid, W. Z., Doudoroff, M., and Barker, H. A. (1951). *In* "The Enzymes" (J. B. Sumner and K. Myrbäck, eds.), Vol. I, Part 2, p. 1014. Academic Press, New York.

Hehre, E. J. (1951). *Advances in Enzymol.* **11**, 267.

Kertesz, Z. I. (1951). *In* "The Enzymes" (J. B. Sumner and K. Myrbäck, eds.), Vol. I, Part 2, p. 745. Academic Press, New York.

Lineweaver, H., and Jansen, E. F. (1951). *Advances in Enzymol.* **11**, 267.

Neuberg, C., and Mandl, I. (1951). *In* "The Enzymes" (J. B. Sumner and K. Myrbäck, eds.), Vol. I, Part 1, p. 527. Academic Press, New York.

Peat, S. (1951). *Advances in Enzymol.* **11**, 339.

Pigman, W. W. (1944). *Advances in Enzymol.* **4**, 51.

Pigman, W. W. (1951). *In* "The Enzymes" (J. B. Sumner and K. Myrbäck, eds.), Vol. I, Part 2, p. 725. Academic Press, New York.

Stacey, M. (1954). *Advances in Enzymol.* **15**, 301.

Veibel, S. (1951). *In* "The Enzymes" (J. B. Sumner and K. Myrbäck, eds.), Vol. I, Part 1, p. 583. Academic Press, New York.

Reactions of Uridine Nucleotides

The biological occurrence of complex nucleotides of uracil was found in two simultaneous unrelated studies concerned with the effect of penicillin on bacteria and with the interconversion of glucose and galactose. When penicillin was used to inhibit the growth of *Staphylococcus*, acid-soluble, ultraviolet-absorbing compounds were found to accumulate in the cells.[1] Purification of the material responsible for the ultraviolet absorption resulted in the separation of three compounds. All of these

[1] J. T. Park, *Phosphorus Metabolism Symposium* **1**, 93 (1951).

contained the structure uracil-ribose-P-P-X. In the simplest compound X appears to be an acidic derivative of an acetylated amino sugar which is linked to the pyrophosphate by a glycosidic bond. The other compounds contain amino acids added to this sugar derivative.

Glucose-4-Epimerase. When galactose is incubated with the yeast, *Saccharomyces fragilis*, it is first phosphorylated to give galactose-1-phosphate.[2] Subsequent metabolism involves the conversion to the corresponding glucose derivative; the conversion involves an inversion of the substituents on carbon 4.[3]

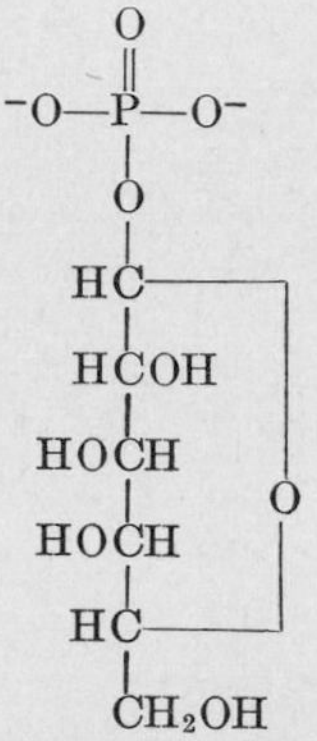

$$
\begin{array}{c}
\text{O} \\
\| \\
{}^{-}\text{O}-\text{P}-\text{O}^{-} \\
| \\
\text{O} \\
| \\
\text{HC}-\! \\
| \\
\text{HCOH} \\
| \\
\text{HOCH} \quad\quad \text{O} \\
| \\
\text{HOCH} \\
| \\
\text{HC}-\! \\
| \\
\text{CH}_2\text{OH}
\end{array}
$$

Galactose-1-phosphate 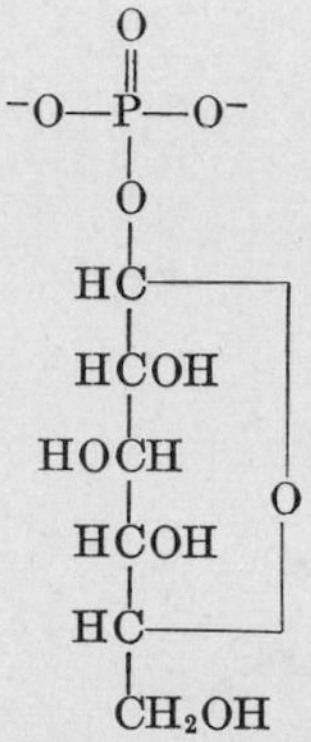   Glucose-1-phosphate

Inversions of this type by attack from the opposite side of a substituted carbon were studied by Walden, and are known as Walden inversions. The system that carries out the interconversion of galactose and glucose phosphates was called Waldenase; later, when this conversion was found to require several enzymes, the enzyme that catalyzes the inversion step was named Waldenase. All inversions are not Walden inversions, however, and since there is no evidence to favor a Walden mechanism, the enzyme has been renamed an epimerase (glucose and galactose are epimers, since they differ only in the configuration of one carbon atom). It was apparent during the early studies on this system that a soluble cofactor, cowaldenase, was required for activity. This was isolated and identified as uracil-ribose-phosphate-phosphate-glucose. The cofactor is called uridine diphosphoglucose, and abbreviated UDPG.[4]

Structure of UDPG. The structure of UDPG was determined by analysis. Mild acid hydrolysis liberates an equivalent of glucose, and stronger

[2] H. W. Kosterlitz, *Biochem. J.* **37**, 322 (1943).

[3] R. Caputto, L. F. Leloir, C. E. Cardini, and A. C. Paladini, *J. Biol. Chem.* **184**, 333 (1950).

[4] C. E. Cardini, A. C. Paladini, R. Caputto, and L. F. Leloir, *Nature* **165**, 191 (1950).

acid hydrolysis yields an equivalent of inorganic phosphate. The residue after acid hydrolysis was crystallized as a barium salt and identified chromatographically and crystallographically as uridine-5'-phosphate. The pyrophosphate linkage was suspected from titration studies, and its existence was confirmed by hydrolysis with potato pyrophosphatase. The pyrophosphatase liberated glucose-1-phosphate, thus establishing the nature of the glycosidic bond.

Pyrophosphate Uridyl Transferase. The over-all epimerase reaction is composed of three individual reactions. The sugar phosphates are not converted directly into each other, but are first incorporated into uridine nucleotides. Epimerization occurs in the hexose component of the UDP-sugar, which then is cleaved to yield the epimerized sugar phosphate. The synthesis of both UDPG and the corresponding galactose derivative, UDPGal, are catalyzed by enzymes called uridyl transferases. A pyrophosphate (PP) uridyl transferase[5] has been studied in yeast and in several mammalian tissues. It catalyzes the reaction

$$\text{UDPG} + \text{PP} \rightleftharpoons \text{UTP} + \text{glucose-1-phosphate}$$

The equilibrium constant of this reaction is near 1. This reaction is conveniently followed by measuring the appearance of glucose phosphate through coupled reactions. Glucose-1-phosphate is converted by phosphoglucomutase to glucose-6-phosphate, and the latter is oxidized by the Zwischenferment system, which forms TPNH as a measure of glucose phosphate split from the uridine nucleotide. UTP and UDPG can be separated by paper chromatography, and P^{32} has been used to determine glucose 1-phosphate incorporation into the nucleotide. This underscores the reaction mechanism, which involves transfer of a uridine-5'-phosphate group.

$$\text{URP} \vert \text{PP} + \text{G-P*} \rightleftharpoons \text{URP} \vert \text{P*G} + \text{PP}$$

This transferase has shown no activity when other nucleotides were substituted for UTP, and glucose-1-phosphate is the only carbohydrate compound known to react.

Mechanism of 4-Epimerase Activity. The mechanism of the conversion of UDPG to the galactose derivative, UDPGal, is not established.[6] Several types of reaction have been suggested. The mechanism considered most probable involves oxidation of the carbinol group to a carbonyl, followed by nonstereospecific reduction back to a carbinol. Support

[5] A. Munch-Petersen, H. M. Kalckar, E. Cutolo, and E. E. B. Smith, *Nature* **172**, 1036 (1953).

[6] L. F. Leloir, *Arch. Biochem. Biophys.* **33**, 186 (1951).

for this description of epimerization comes from the finding that the epimerase requires DPN for activity.[7a] DPNH cannot replace DPN. Since the only known function for DPN is to accept electrons, reaction (I) is suggested. No direct evidence has been obtained for the reduction of DPN or the formation of a 4-keto sugar. Therefore other mechanisms which have been suggested must still be considered as possibilities. If the sugar were to split between carbons 3 and 4, on recombination a mixture of

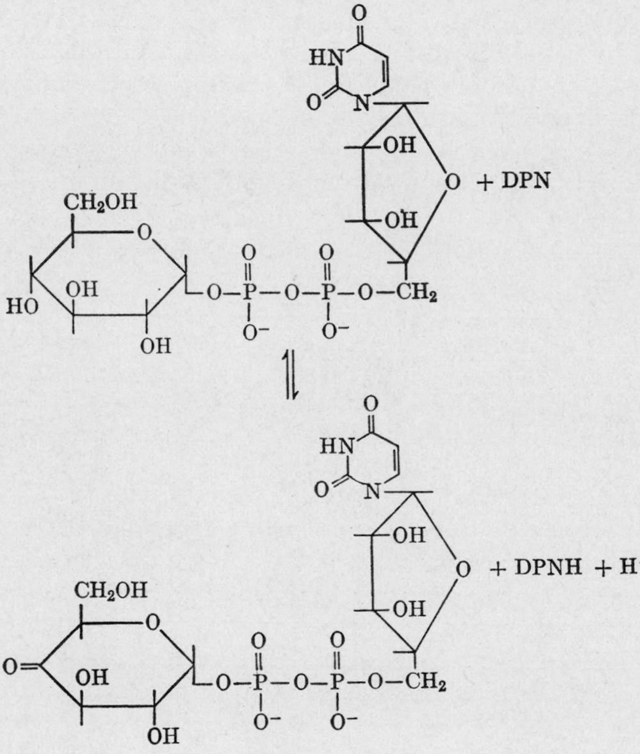

(I)

diastereoisomers could be obtained. Similarly, dehydration to form a double bond between carbons 3 and 4, or 4 and 5, could be followed by hydration with the hydroxyl located on either side of carbon 4.

The mechanism of the reaction has been investigated with water labeled with tritium and with O^{18}. Neither isotope was found to enter the sugar.[7b] These results eliminate all mechanisms that require the

[7a] E. S. Maxwell, *J. Am. Chem. Soc.* **78**, 1074 (1956).
[7b] L. Anderson, A. M. Landel, and O. F. Diedrich, *Biochim. et Biophys. Acta* **22**, 573 (1956); H. M. Kalckar and E. S. Maxwell, *ibid.* **22**, 588 (1956); A. Kowalsky and D. E. Koshland, Jr., *ibid.* **22**, 575 (1956).

participation of water molecules or hydrogen or hydroxyl ions. It is possible, however, to write various mechanisms, including oxidation and reduction, cleavage of the carbon chain, and other rearrangements, such as that proposed for aconitase in which a hydride ion shift was suggested, which are consistent with the isotope results.

Epimerase Equilibrium. When substrate quantities of the cofactor are incubated with yeast epimerase, an equilibrium mixture of 3 parts UDPG and 1 part UDPGal is formed. This was determined by isolating the UDP compounds and determining the sugar released on mild acid hydrolysis by paper chromatography. Essentially the same equilibrium is found when the complete Waldenase reaction is carried out with catalytic amounts of cofactor.[8] The equilibrium concentrations can be determined by measuring glucose-1-phosphate in the enzymatic assay described above and measuring galactose-1-phosphate as the additional material reacting in the assay in the presence of the epimerase system. An alternate assay involves determining the amounts of the free sugars formed on hydrolysis; glucose and galactose can be separated by paper chromatography.

GP Uridyl Transferase. UDPGal participates in a nonpyrophosphate-utilizing reaction to form UDPG when incubated with glucose-1-phosphate and an enzyme formed in galactose-adapted yeast.[9] This is a transfer reaction which is freely reversible.

$$\text{UDPGal} + \text{glucose-1-phosphate} \rightleftharpoons \text{UDPG} + \text{galactose-1-phosphate}$$

Since the substrates include glycosyl phosphates, the enzyme has been called GP uridyl transferase.[10a] It is the only known mechanism for introducing galactose into a uridine compound. A direct reaction between galactose-1-phosphate and UTP, as occurs with glucose-1-phosphate, has not been observed with a purified enzyme. Crude extracts have given the impression that a pyrophosphorylase-type reaction can occur with galactose as well as with glucose, but this could easily be an artifact. The distinction between the apparent direct reaction and indirect reaction sequences is a familiar and challenging aspect of enzyme chemistry. In crude extracts small amounts of glucose-1-phosphate may be present, and may initiate the following pair of reactions:

$$\text{glucose-1-phosphate} + \text{UTP} \rightarrow \text{UDPG} + \text{PP}$$
$$\underline{\text{UDPG} + \text{galactose-1-phosphate} \rightarrow \text{UDPGal} + \text{glucose-1-phosphate}}$$
$$\text{sum: galactose-1-phosphate} + \text{UTP} \rightarrow \text{UDPGal} + \text{PP}$$

[8] L. F. Leloir, C. E. Cardini, and E. Cabib, *Anales asoc. quim. arg.* **40,** 228 (1952); quoted in review by Leloir (1953).

[9] H. M. Kalckar, B. Braganca, and A. Munch-Petersen, *Nature* **172,** 1038 (1953).

[10a] E. S. Maxwell, H. M. Kalckar, and R. M. Burton, *Biochim. et Biophys. Acta* **18,** 444 (1955).

Thus the apparent existence of a UDPGal pyrophosphorylase can be explained by the action of two other enzymes, and the occurence of uridine pyrophosphorylases other than that specific for UDPG remains uncertain. The three reactions discussed above provide a mechanism for

(a) obtaining an active co-Waldenase: PP uridyl transferase;
(b) reversibly combining sugars with the cofactor: GP uridyl transferase;
(c) interconverting glucose and galactose: epimerase.

Similar reactions have recently been found in plants to be involved in pentose metabolism.[10b] D-Xylose-1-phosphate was found to react with UTP to form UDPXy. A second enzyme, which may be different from the epimerase of glucose-galactose interconversions, epimerizes carbon 4 of the pentose to produce the UDP derivative of L-arabinose. The number of specific pyrophosphorylases for sugar phosphates in plant tissue is not known at this time.

Enzyme Defect in Galactosemia. The hereditary disease, galactosemia, is characterized by high blood levels of galactose. The accumulation of this sugar appears to be responsible for the manifestations of the disease. Individuals suffering from the disease have been found to be lacking in the enzyme GP uridyl transferase. Other enzymes, PP uridyl transferase and epimerase, are present in normal amounts.[11a] Therefore, it may be concluded that galactose metabolism requires the epimerase series of reactions, and that galactose is not metabolized significantly in the absence of one of the enzymes of this series. Deficiencies of individual enzymes are well known as consequences of mutations in simple organisms, but very few examples of genetic relation to enzyme formation are known for more complicated organisms, such as humans. Other enzymes have been implicated in congenital diseases of carbohydrate metabolism in humans. Glucose-6-phosphatase has been found to be essentially absent from livers of certain cases of glycogen storage disease; in two other types of glycogen storage disease it has been suggested that either the branching or the debranching enzymes, respectively, may be lacking.[11b]

Nucleotide Kinases. To generate UTP biologically a uridine nucleotide must first be formed. The synthesis of UMP will be discussed later. A kinase partially purified from calf liver catalyzes the phosphorylation of

[10b] V. Ginsburg, E. F. Neufeld, and W. Z. Hassid, *Proc. Natl. Acad. Sci. U.S.* **42,** 333 (1956).

[11a] H. M. Kalckar, E. P. Anderson, and K. J. Isselbacher, *Biochim. et Biophys. Acta* **20,** 262 (1956).

[11b] B. Illingworth and G. T. Cori, *J. Biol. Chem.* **199,** 653 (1952); G. T. Cori and C. F. Cori, *ibid.* **199,** 661 (1952).

UMP in a reaction that greatly resembles that of adenylate kinase.[12] The reaction requires ATP and results in the formation of ADP and UDP. The complete purification of kinases of this type has not been achieved, but the partial separations already accomplished indicate that at least three, and possibly more enzymes with varying substrate specificities exist. These enzymes, studied in calf liver extracts, all use ATP as the phosphate donor, and show at least relative specificity for adenylic acid, uridylic acid, and cytidylic acid as phosphate acceptors.

$$ATP + XMP \rightleftharpoons ADP + XDP$$
Nucleoside Monophosphate Kinase Reaction

$$XTP + YDP \rightleftharpoons XDP + YTP$$
Nucleoside Diphosphate Kinase Reaction

The addition of a third phosphate to any of the nucleoside pyrophosphates, such as UDP, is catalyzed by an enzyme found in yeast and several animal tissues. Since the reaction catalyzed by this enzyme appears to be nonspecific with respect to the base of the nucleotides, it has been named nucleoside diphosphate kinase (nudiki).[13] The phosphate donor in this reaction may be any of the nucleoside triphosphates. As in the case of the nucleoside monophosphate kinases, there are no significant differences in the free energies of hydrolysis of the various nucleoside triphosphates, so all of the reactions are freely reversible with equilibrium constants near 1. Since the phosphorylation of nucleoside diphosphates is reversible, it is necessary that each of the corresponding triphosphates serve as phosphate donor. The phosphorylation of the monophosphates is similarly reversible, but in this case one of the sites on the enzyme appears to react only with adenine, and involves the conversion of ATP to ADP. The complementary reaction involves $XMP \rightleftharpoons XDP$. Thus the nonadenine nucleotides are never equivalent to ATP, and the reaction is limited to the phosphorylation of a specific nucleoside monophosphate by ATP.

The existence of nudiki complicates studies in which the specificity of enzymes for nucleotides is examined. However, in some cases, kinases have been shown to react with other than adenine nucleotides. In particular, pyruvate kinase will transfer phosphate to several of the nucleoside diphosphates, including UDP.[14]

Synthesis of Nonreducing Disaccharides. UDPG serves as a glycosyl donor in a reaction catalyzed by an enzyme in yeast in which glucose-6-

[12] J. L. Strominger, L. A. Heppel, and E. S. Maxwell, *Arch. Biochem. Biophys.* **52,** 488 (1954).

[13] P. Berg and W. K. Joklik, *J. Biol. Chem.* **210,** 657 (1954).

[14] J. L. Strominger, *Biochim. et Biophys. Acta* **16,** 616 (1955).

phosphate serves as the glycosyl acceptor.[15] The reaction results in the formation of a nonreducing disaccharide phosphate ester. The carbohydrate component was identified as trehalose, and it was shown that UDP is formed in corresponding quantities. The reaction is therefore:

$$UDPG + G\text{-}6\text{-}P \rightarrow UDP + trehalose\text{-}P \ (G\text{-}1\text{-}1'\text{-}G\text{-}6\text{-}P)$$

A very similar reaction was found in extracts of wheat germ and other plant material. In this case the acceptor molecule is free fructose,[16] and the reaction is

$$UDPG + fructose \rightarrow UDP + sucrose$$

This enzyme is very specific; it does not react with sorbose, fructose phosphates, glucose, or any other compound tested in place of fructose, and glucose-1-phosphate cannot replace UDPG. Unlike the case of sucrose phosphorylase, the equilibrium of this reversible reaction favors sucrose synthesis, and the presence of this enzyme in the cells of higher plants implicates this reaction in sucrose synthesis.

An even more favorable mechanism for sucrose synthesis is suggested by the finding of another enzyme in wheat germ.[17] This forms sucrose phosphate in a reaction almost identical to that described above.

$$UDPG + fructose\text{-}6\text{-}phosphate \rightarrow UDP + glucosyl\text{-}1\text{-}2\text{-}fructoside\text{-}6\text{-}phosphate$$

Sucrose phosphorylated on position 6 of the fructose moiety has been isolated from sugar beet leaves.[18] The hydrolysis of the phosphate ester would produce sucrose in an essentially irreversible reaction, and could account well for the large accumulation of sucrose in many plants. The synthesis of sucrose from phosphorylated sugars (glucose phosphate via UDPG and fructose-6-phosphate) could explain the results of isotope experiments in which it was found that free fructose did not become labeled while both halves of the sucrose formed were equally labeled.[19] Experiments with leaf homogenates indicate that the formation of sucrose phosphate from UDPG and fructose-6-phosphate is indeed carried out by sugar beets.[20]

UDPG Dehydrogenase. UDPG has been found to serve as an oxidizable substrate for a DPN-requiring dehydrogenase.[21] This enzyme carries out

[15] L. F. Leloir and E. Cabib, *J. Am. Chem. Soc.* **75**, 5445 (1953).
[16] C. E. Cardini, L. F. Leloir, and J. Chiriboga, *J. Biol. Chem.* **114**, 149 (1955).
[17] L. F. Leloir and C. E. Cardini, *J. Biol. Chem.* **114**, 157 (1955).
[18] J. G. Buchanan, *Arch. Biochem. Biophys.* **44**, 190 (1953).
[19] P. V. Vittorio, G. Krotkov, and G. B. Reed, *Can. J. Botany* **32**, 369 (1954).
[20] D. P. Burma and D. C. Mortimer, *Arch. Biochem. Biophys.* **62**, 16 (1956).
[21] E. S. Maxwell, H. M. Kalckar, and J. L. Strominger, *Arch. Biochem. Biophys.* **65**, 2 (1956).

a two-step reaction, in which UDPG is oxidized to the corresponding glucuronic acid derivative, while two equivalents of DPN are reduced. No evidence could be found for the separation of the reaction into two steps, and no aldehyde intermediate could be detected.

Glucuronide Formation. UDP glucuronic acid (UDPGA) had previously been isolated as a factor necessary for glucuronide formation in liver.[22] An enzyme had been found that could form glucuronides. The reaction requires the stoichiometric participation of a factor isolated from boiled extracts of liver, identified as UDPGA. UDPGA was called a cofactor for glucuronide synthesis, even though no catalytic activity could be found. The formation of the glucuronide of *o*-aminophenol proceeded to completion, limited by the amount of "cofactor" added. No direct activation of glucuronic acid or its 1-phosphates (α and β) could be found; that is, glucuronic acid and its simple derivatives are inert in glucuronide formation. The finding of a system for synthesizing UDPGA from UDPG, of course, obviates the necessity for finding a direct activation of glucuronic acid, and explains why glucose is used while glucuronic acid is not for glucuronide synthesis *in vivo*. The reaction sequence most probably followed is shown in (II).

$$\text{glucose} + \text{ATP} \rightarrow \text{glucose-6-phosphate} + \text{ADP}$$
$$\text{glucose-6-phosphate} \rightarrow \text{glucose-1-phosphate}$$
$$\text{glucose-1-phosphate} + \text{UTP} \rightarrow \text{UDPG} + \text{PP}$$
$$\text{UDPG} + 2 \text{ DPN} \rightarrow \text{UDPGA} + 2\text{DPNH}$$
$$\text{UDPGA} + \text{ROH} \rightarrow \text{ROGA} + \text{UDP}$$

(II)

N-Acetylglucosamine. Uridine nucleotides are involved in the metabolism of another glucose derivative, glucosamine. The UDP glycoside of *N*-acetylated glucosamine (UDPAG) has been isolated from natural sources.[23] An enzyme found in liver nuclei seems to carry out a pyrophosphorylase reaction with UTP and glucosamine but not with acetylglucosamine.[24] The formation of UDPAG has been observed when *N*-acetylglucosamine-1-phosphate was incubated with a soluble liver enzyme and UTP.[25]

The sequence of reactions involved in the synthesis of UDPAG is believed to proceed as shown in (III). The subsequent reactions of UDPAG are still conjectural. By analogy with other reactions of UDP glycosides, it is suggested that glycosides of both *N*-acetylglucosamine and glucuronic acid can be formed, and if these compounds react with each other alter-

[22] I. D. E. Storey and G. J. Dutton, *Biochem. J.* **59**, 279 (1955).
[23] E. Cabib, L. F. Leloir, and C. E. Cardini, *J. Biol. Chem.* **203**, 1055 (1953).
[24] F. Maley, G. F. Maley, and H. A. Lardy, *J. Am. Chem. Soc.* **78**, 5303 (1956).
[25] F. Maley and H. A. Lardy, *Science* **129**, 1207 (1956).

nately, the structure of hyaluronic acid can be formed. Similar polysaccharides (chitin) form structural elements in molds, arthropods, and other organisms. The existence of various enzymes of the series above in *Neurospora* (which synthesizes chitin) lends support to the scheme as being concerned with polysaccharide formation, but direct evidence is still lacking.

$$\text{glucose-6-phosphate} + \text{glutamine} \rightarrow \text{glucosamine-6-phosphate} + \text{glutamic acid}$$

$$\text{glucosamine-6-phosphate} + \text{acetyl CoA} \rightarrow N\text{-acetylglucosamine-6-phosphate} + \text{CoA}$$

$$N\text{-acetylglucosamine-6-phosphate} \rightarrow N\text{-acetylglucosamine-1-phosphate}$$

$$N\text{-acetylglucosamine-1-phosphate} + \text{UTP} \rightarrow \text{UDPAG} + \text{PP}$$

(III)

Additional Complex Nucleotides. Several nucleotides isolated from various organisms have been identified recently. These include guanosine diphosphate mannose, which can be formed in a pyrophosphorylase reaction:[26]

$$\text{GTP} + \text{mannose-1-phosphate} \rightarrow \text{GDP mannose} + \text{PP}$$

Uridine diphosphate also is found in combination with galactosamine (in liver)[27] and with glucosamine phosphate and glucosamine sulfate (in oviduct).[28] The metabolic functions of these nucleotides has not been ascertained, but there is an obvious tendency to suspect that they are intermediates in the biosyntheses of chondroitin sulfate and other polysaccharides.

The occurrence of amino acids in the uridine nucleotides originally isolated by Park suggests that these compounds are involved in the activation or condensation of amino acids, but no enzymatic reactions of these compounds are known. A specific function in the formation of bacterial cell walls has been suggested following the finding of similar ratios of D-glutamic acid, D- and L-alanine, lysine, and amino sugar (1:3:1:1) in cell walls and a uridine nucleotide, both isolated from *Staphylococcus aureus*.[29]

GENERAL REFERENCES

Kalckar, H. M. (1954). *In* "The Mechanism of Enzyme Action" (W. D. McElroy and B. Glass, eds.), p. 675. Johns Hopkins Press, Baltimore.
Leloir, L. F. (1953). *Advances in Enzymol.* **14**, 193.
Leloir, L. F. (1956). *Proc. 3rd Intern. Congr. Biochem., Brussels, 1955* p. 154.

[26] A. Munch-Petersen, *Acta Chem. Scand.* **10**, 928 (1956).
[27] H. G. Pontis, *J. Biol. Chem.* **216**, 195 (1955).
[28] J. L. Strominger, *Biochim. et Biophys. Acta* **17**, 283 (1955).
[29] J. T. Park and J. L. Strominger, *Science* **125**, 99 (1957).

Chapter VI

POLYNUCLEOTIDES AND THEIR COMPONENTS

Nucleic Acids

Nucleic acids are high molecular weight polymers of nucleotides. Two types are known, identified by the sugar characteristic of each; ribose and 2-deoxyribose. There may be a metabolic relationship between these types but the chemistry and enzymology of the two are sufficiently distinct to require separate treatments. The original isolation of nucleic acids was made by Miescher[1] from nuclei of pus cells, and he subsequently found a good source in the heads of salmon sperm. From these materials and from thymus gland nucleic acid, the bases adenine, guanine, thymine, and cytosine were isolated,[2] together with phosphoric acid and a sugar, identified by Levene as deoxyribose.[3] A similar material isolated from yeast was found to have uracil in place of thymine[4] and ribose instead of the deoxy sugar.[5] The finding of material in plants similar to the nucleic acid of yeast led to an erroneous designation of animal and plant nucleic acids. The so-called thymus nucleic acid is now referred to as deoxyribonucleic acid (DNA) and occurs in all organisms, in general in a fixed amount per cell; the amount is characteristic of the species. DNA is thought to be the material that transmits inherited characteristics to each cell. The constant amount of DNA per cell thus reflects the uniform genetic composition of the cells of a complex organism.[6] Germ cells of diploid organisms contain half the amount of DNA of somatic cells,[7] and polyploid cells of certain plants contain multiple quantities of DNA.[8] The nucleic acid of the type originally found in yeast occurs in variable amounts in all organisms, and is called pentose or ribose nucleic acid (PNA or RNA). The name nucleic acid implies a nuclear source, which is justified in the case of DNA, but is not true of RNA. Most RNA exists

[1] F. Miescher, *Hoppe-Seyler's Med. chem. Unters.* **1871**, 441, 502.

[2] J. Piccard, *Ber.* **7**, 1714 (1874); A. Kossel, *Z. physiol. Chem.* **8**, 404 (1884); **12**, 241 (1888); A. Kossel and A. Newmann, *Ber.* **27**, 2215 (1894); A. Kossel and H. Steudel, *Z. physiol. Chem.* **37**, 177 (1902–03); P. A. Levine, *ibid.* **37**, 402 (1902–03).

[3] P. A. Levine, L. A. Mikeska, and T. Mori, *J. Biol. Chem.* **85**, 785 (1930).

[4] A. A. Scoli, *Z. physiol. Chem.* **31**, 161 (1900–1901).

[5] P. A. Levine and W. A. Jacobs, *Ber.* **42**, 2102, 2469, 2474, 2703 (1909).

[6] A. Boivin, R. Vendrely, and C. Vendrely, *Compt. rend.* **226**, 1061 (1948).

[7] A. E. Mirsky and H. Ris, *Nature* **163**, 666 (1949).

[8] H. Ris, and A. E. Mirsky, *J. Gen. Physiol.* **33**, 125 (1949).

in cytoplasmic microsomes,[9] and only a small fraction is associated with nuclei. The cellular concentration of RNA is variable, indicative of a metabolic, rather than a genetic, function.

The study of nucleic acids is concerned with the elucidation of the structures of these compounds as well as their metabolism. At the present time the detailed structures are not known, but recently the types of linkages involved have been established. As in the cases of proteins and polysaccharides, information about the nature of the enzymes that attack nucleic acids and knowledge about the structures of their substrates have been acquired together as interdependent developments.

Ribonucleic Acid

Ribonucleic acid occurs in all organisms, apparently associated with proteins. The nucleoproteins have been purified by physical methods, but the best preparations are undoubtedly not homogeneous except in the case of certain viruses. The relationship between the two components of even the viruses is still under investigation. On denaturing the protein with physical changes, salts, alcohol, etc., the nucleic acids are liberated. Salts, acid, and organic solvents are used in the purification of protein-free RNA.

The purified preparations from most sources are not homogeneous; even the RNA of tobacco mosaic virus is polydisperse, although this may be a reflection of association and disassociation of similar particles. Molecular weights of RNA are estimated between 10,000 and 600,000.[10] RNA preparations from various sources contain variable ratios of bases. The relative amounts of purines usually exceed the pyrimidines by a small factor and guanine and cytosine tend to exceed adenine and uracil, respectively.[11] These differences are well established, and force the abandonment of the old concept of a regular tetranucleotide structure. It is not known how many specific types of RNA molecules may be formed in a given cell, but it appears that RNA can be fractionated into components with different ratios of bases, and there is no reason to believe that further separations will not be accomplished.

The structure of RNA appears to be very regular with respect to the type of bonds used. Many other linkages have been suggested, but all of the available evidence points to a relatively simple pattern. All of the

[9] G. H. Hogeboom, W. C. Schneider, and G. E. Palade, *J. Biol. Chem.* **172**, 619 (1948).

[10] See review by D. O. Jordan *in* "The Nucleic Acids" (E. Chargaff and J. N. Davidson, eds.), Vol. I, p. 447. Academic Press, New York (1955).

[11] E. Chargaff, B. Magasanik, E. Vischer, C. Green, R. Doniger, and D. Elson, *J. Biol. Chem.* **186**, 51 (1950); G. W. Crosbie, R. M. S. Smellie, and J. N. Davidson, *Biochem. J.* **54**, 287 (1953); E. Volkin and C. E. Carter, *J. Am. Chem. Soc.* **73**, 1516 (1951).

bases are present as ribosides, probably in the β configuration.[12] The ribosides are joined in a linear polymer by phosphate groups. These are esterified doubly; one bond to carbon 3' of one nucleoside, the other to the 5' position of another nucleoside (I).[13]

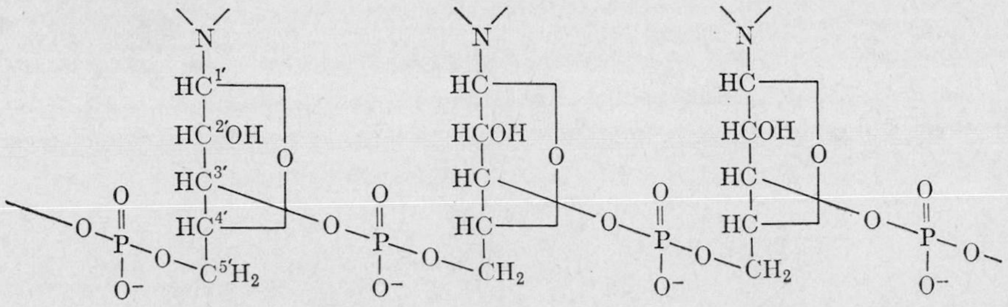

(I)

RNA is easily hydrolyzed in both acid and alkaline solutions. The reaction products from alkaline hydrolysis have been shown to be mixtures of 2' and 3' nucleotides.[14] The reaction mechanism by which these are formed involves the splitting of the phosphate bond with the primary alcohol (C-5') and the simultaneous formation of a 2' bond. This produces a cyclic phosphate diester (II).

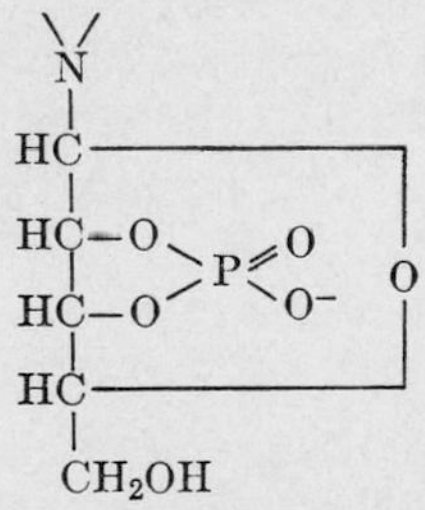

(II)

Cyclic diesters are easily hydrolyzed to give essentially random mixtures of the corresponding 2' and 3' monoesters.[15]

Ribonuclease. The classical enzyme for the hydrolysis of RNA is pancreatic ribonuclease. The presence of a nucleic acid destroying activity in pancreas was noted in 1903. The enzyme was found to be heat

[12] J. Davoll, B. Lythgoe, and A. R. Todd, *J. Chem. Soc.* p. 833 (1946).
[13] L. A. Heppel, R. Markham, and R. J. Hilmoe, *Nature* **171,** 1152 (1953).
[14] C. E. Carter and W. E. Cohn, *Federation Proc.* **8,** 190 (1949); D. M. Brown, G. D. Fasman, D. I. Magrath, A. R. Todd, W. Cochran, and M. M. Woolfson, *Nature* **172,** 1184 (1953).
[15] D. M. Brown, D. I. Magrath, and A. R. Todd, *J. Chem. Soc.* p. 2708 (1952).

stable by Jones, who consequently doubted its enzymatic nature.[16] Eventually the enzyme from beef pancreas was crystallized by Kunitz.[17] The enzyme splits RNA optimally near pH 7 and is stimulated by cations. It is a small symmetrical molecule (M.W. around 15,000). There is no evidence for any constituents but amino acids. The amino acid sequence has been almost completely determined,[18] and it seems likely that a complete structure for this enzyme will be known in the near future. There is also the prospect that studies with partially degraded enzyme preparations will reveal the structural requirements for enzyme activity.[19]

Assay of Ribonuclease. Several kinds of assay have been found useful for studying ribonuclease (RNAase). A spectrophotometric assay is based on the empirical, unexplained observation that there is a decrease in optical density at 300 mμ during hydrolysis of RNA.[20] RNA can be precipitated from solutions of its degradation products by uranyl acetate in trichloroacetic acid; the soluble nucleotides can be measured spectrophotometrically or chemically.[17] A manometric assay determines the formation of secondary acid groups through release of CO_2 from a bicarbonate buffer;[21] acid groups have also been determined by titration.[22] Recently cyclic nucleotides have been used as substrates in place of nucleic acid.[23]

Reactions of Ribonuclease. The degradation of ribonucleic acid by RNAase was found to result in the accumulation of a mixture of pyrimidine mononucleotides and a so-called "core." The core is not an individual structure; it is a mixture of oligonucleotides in which the bases are predominantly purines.[24] These compounds indicate the random nature of the ribonucleic acid structure. Ribonuclease hydrolyzes esters of doubly esterified phosphate in which one of the substituents is the hydroxyl group in position 3' of a pyrimidine nucleotide; the other substituent, which in ribonucleic acid is a 5' hydroxyl group, is removed.[25a] Since groups substituted on phosphates esterified with 3' positions of purine

[16] W. Jones and M. E. Perkins, *Am. J. Physiol.* **55**, 557 (1923).
[17] M. Kunitz, *J. Gen. Physiol.* **24**, 15 (1940).
[18] C. H. W. Hirs, W. H. Stein, and S. Moore, *J. Biol. Chem.* **221**, 151 (1956); R. R. Redfield and C. B. Anfinsen, *J. Biol. Chem.* **221**, 385 (1956).
[19] G. Kalnitsky and E. E. Anderson, *Biochim. et Biophys. Acta* **16**, 302 (1955); F. Rogers, *Compt. rend. trav. lab. Carlsberg Sér. chim.* **29**, 329 (1955).
[20] M. Kunitz, *J. Biol. Chem.* **164**, 563 (1946).
[21] J. A. Bain and H. P. Rusch, *J. Biol. Chem.* **153**, 659 (1944).
[22] M. Kunitz, *J. Gen. Physiol.* **24**, 15 (1940).
[23] C. B. Anfinsen, W. F. Harrington, A. Hvidt, K. Linderstrøm-Lang, M. Ottesen, and J. Schellman, *Biochim. et Biophys. Acta* **17**, 141 (1955).
[24] R. Markham and J. D. Smith, *Biochem. J.* **52**, 565 (1952).
[25a] G. Schmidt, R. Cubiles, N. Zollner, L. Hecht, N. Strickler, K. Seraidarian, M. Seraidarian, and S. J. Thannhauser, *J. Biol. Chem.* **192**, 715 (1951).

nucleosides are not hydrolyzed, oligonucleotides containing purine nucleosides linked in the 3′ position to the 5′ position of another nucleoside remain as end products of ribonuclease digestion. If the purine nucleoside is linked to a pyrimidine nucleoside, the product is a dinucleotide; larger

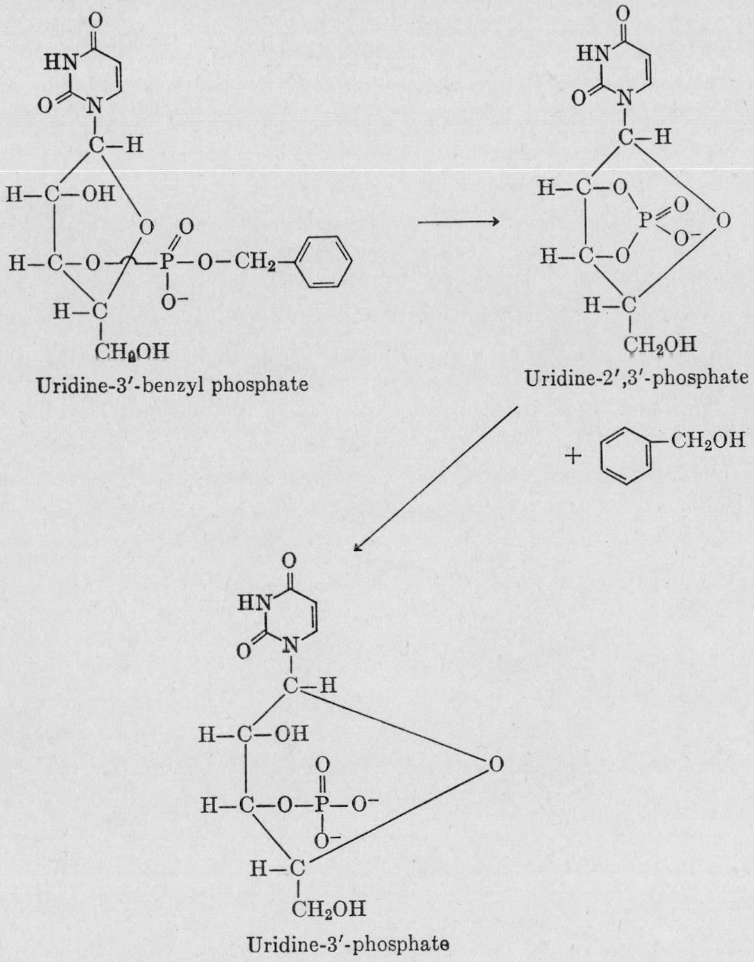

Uridine-3′-benzyl phosphate

Uridine-2′,3′-phosphate

Uridine-3′-phosphate

(III)

oligonucleotides occur as products when the substrate contains purine nucleosides linked together. The isolation of such products, polynucleotides consisting of a series of purine nucleotides terminated by a pyrimidine nucleoside 3′-phosphate, is evidence for sequences of varying numbers of consecutive purine nucleotides in the nucleic acid substrate.

Similarly, the isolation of pyrimidine mononucleotides indicates the occurrence of sequences of at least two consecutive pyrimidine nucleotides.

The specificity of ribonuclease has been studied with small synthetic substrates. When diesters of various nucleotides are subjected to RNAase, only those compounds that are derivatives of pyrimidine nucleoside 3′-phosphate are hydrolyzed; purine nucleotides and pyrimidine 2′- or 5′-phosphates are resistant to this enzyme.[25b,26] With both model substrates and ribonucleic acid, the action of RNAase has been shown to include the intermediate formation of cyclic nucleotides (III).[24]

Synthesis by Transfer. Ribonuclease catalyzes transfer reactions in addition to hydrolyses. Both cyclic nucleotides and diesters of 3′-nucleotides serve as substrates for the transfer. Acceptors for the phosphate include the primary alcohol groups of nucleosides, nucleoside cyclic phosphates, and simple alcohols. Simple nucleotides do not serve as acceptors. The products of transfer reactions, including dinucleoside monophosphates and the so-called cyclic dinucleotides, can also serve as acceptors, and by repeated transfer reactions of cyclic nucleotides small polymers have been built up (IV). It has been suggested that this reaction may play a role in nucleic acid biosynthesis.[27]

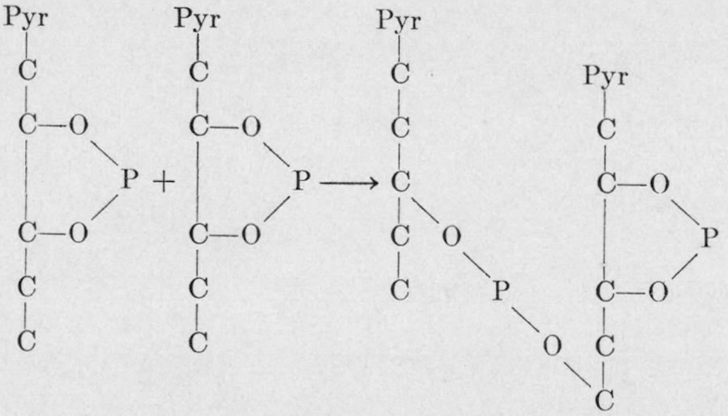

Representative Transfer Reaction Catalyzed by Ribonuclease
(IV)

An enzyme isolated from spleen shows the same specificity as pancreatic ribonuclease.[28] This enzyme differs only in certain physical properties from the pancreatic enzyme. It has been used to show the intermediate formation of cyclic-ended oligonucleotides. Previous studies with pancreatic ribonuclease had been carried out under conditions that could

[25b] D. M. Brown and A. R. Todd, *J. Chem. Soc.* p. 2040 (1953).

[26] R. B. Merrifield and D. W. Woolley, *J. Biol. Chem.* **197**, 521 (1952).

[27] L. A. Heppel and P. R. Whitfeld, *Biochem. J.* **60**, 1 (1955); L. A. Heppel, P. R. Whitfeld, and R. Markham, *ibid.* **60**, 8 (1955).

[28] H. S. Kaplan and L. A. Heppel, *J. Biol. Chem.* **222**, 907 (1956).

permit the back reaction, but with the spleen enzyme, cyclic dinucleotides were formed before sufficient cyclic mononucleotides accumulated to allow measurable recombination.

Venom Phosphodiesterase. A phosphodiesterase from the venom of several species of snakes exhibits extreme specificity with regard to nucleotides. It hydrolyzes only components from a phosphate esterified at a 5′ position.[29] Thus, 5′ nucleotides are liberated from RNA. The venom diesterase attacks both purine and pyrimidine nucleotides (Fig. 26). This activity was useful in establishing that 3′-5′ linkages, rather than 2′-3′ linkages, occur in nucleic acids. A similar enzyme occurs in intestines.

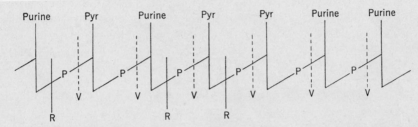

Fig. 26. Cleavage of ribonucleic acid by ribonuclease (vertical lines labeled R) and snake venom 5′-phosphodiesterase (broken lines labeled V). P represents phosphate linking 3′ and 5′ positions of adjacent nucleotides.

Snake venom phosphodiesterase does not attack dinucleotides in which a free phosphomonoester occurs on a 3′ position. Several phosphomonoesterases exist that can remove monoester phosphates from these and, indeed, from RNA.

Other Ribonucleases. A diesterase more specific than that of venom was found in spleen.[30] This enzyme attacks substituted nucleotide 3′-phosphates. Both purine and pyrimidine nucleotides serve as substrates. This enzyme neither forms nor attacks cyclic diesters, and was thus useful in establishing that the 3′, not the 2′, position is esterified in purine as well as pyrimidine nucleoside components of RNA. Synthetic 2′-nucleotide phosphate diesters are not split by this enzyme.

A peculiar specificity of action was found with an enzyme from guinea pig liver nuclei.[31] This enzyme splits ordinary dinucleotide bonds to form 3′-phosphate esters. When it is given cyclic nucleotides as substrates, on the other hand, it liberates exclusively 2′ esters. It is not certain, however,

[29] J. M. Gulland and E. M. Jackson, *Biochem. J.* **32**, 597 (1938); W. E. Cohn and E. Volkin, *J. Biol. Chem.* **203**, 319 (1953).

[30] R. J. Hilmoe and L. A. Heppel, *Federation Proc.* **12**, 217 (1953); D. M. Brown, L. A. Heppel, and R. J. Hilmoe, *J. Chem. Soc.* p. 40 (1956).

[31] L. A. Heppel, P. Ortiz, and S. Ochoa, *Science* **123**, 415 (1956).

that a single enzyme catalyzes the two activities. An interesting limitation of the nuclear diesterase is that it does not attack polymers with terminal 5′-phosphate monoesters. Two other preparations have also been found to open cyclic nucleotides to form 2′ esters. These are fractions from pancreas and spleen.

Synthesis of Polyribonucleotides. The synthesis of large polymers of ribose nucleotides is catalyzed by an enzyme named polynucleotide phosphorylase. This enzyme was first found in extracts of *Azotobacter vinelandii*,[32] and similar preparations have been obtained in other bacteria.[33] This type of enzyme has not yet been identified in animal tissues.

The reaction catalyzed by polynucleotide phosphorylase is the reversible formation of polymer from nucleoside 5′-pyrophosphates. Mg^{++} is required. The diphosphates of adenosine, inosine, guanosine, uridine, and cytidine all react, but the reaction with GDP is poor compared with the other nucleotides. The polymer contains 3′-5′ linkages, and when

$$n \ \mathrm{XRPP} \rightleftharpoons (\mathrm{XRP})_n + n \ \mathrm{P}$$

pyrimidine polymers are formed, they are susceptible to ribonuclease digestion.[34] One end of the polymer bears a 5′-monophosphate, and the other end is a nucleoside 5′ diester with unsubstituted 2′ and 3′ positions.

The affinity of the *Azotobacter* enzyme for nucleotides is not high; K_m values near 10^{-2} M have been found. When mixtures of nucleoside diphosphates are used as substrates, the enzyme synthesizes mixed polymers, with various bases interspersed more or less randomly in the product. The distribution of bases was studied by analyzing the oligonucleotides formed by complete digestion with ribonuclease, which produces the same sort of mixture of pyrimidine nucleotides and purine-rich oligonucleotides from the synthetic mixed polymers as from RNA. The synthetic mixed polymers tend to have the size of some naturally occurring RNA molecules, about 30 nucleotide residues per molecule. Polymers of single nucleotides have been made with molecular weights as high as 2×10^6.

It is not known whether primers are needed for the start of a polynucleotide chain. Terminal pyrophosphate groups, which would be expected from the use of a nucleoside diphosphate as the primary acceptor, have not been detected. Also, nucleoside monophosphates have not been incorporated into polynucleotides. It is possible that the enzyme preparations contain a primer. The length of the chain formed probably depends

[32] M. Grunberg-Manago and S. Ochoa, *J. Am. Chem. Soc.* **77**, 3165 (1955).

[33] U. Littauer, *Federation Proc.* **15**, 302 (1956); R. F. Beers, Jr., *Federation Proc.* **15**, 13 (1956).

[34] L. A. Heppel, J. D. Smith, P. Ortiz, and S. Ochoa, *Federation Proc.* **15**, 273 (1956).

in part upon the amount of primary acceptor and in part on the concentrations of substrate and inorganic phosphate. As in the case of polysaccharide phosphorylase, the concentration of polymer does not seem to enter the equilibrium calculation. The equilibrium appears to be reached when 60–80 per cent of the acid-labile phosphate is released as inorganic phosphate.[35]

Polynucleotide phosphorylase can be assayed in several ways. The liberation of inorganic phosphate can be used to follow the reaction. Both disappearance of acid-soluble nucleotides and formation of acid-insoluble nucleotides have been measured. The reverse reaction in the presence of P^{32}-labeled phosphate allows an exchange reaction, the incorporation of P^{32} into nucleotides, to be used as an assay. The exchange reaction was the first reaction of this enzyme to be discovered, and studies on the responsible enzyme led to the finding of polynucleotide synthesis. Another assay based on the reverse reaction has been devised for rapid spectrophotometric determinations. A polymer of adenylic acid is incubated with the phosphorylase in the presence of phosphate, phosphopyruvate, pyruvate kinase, DPNH, and lactic dehydrogenase. The formation of ADP is thus coupled with the formation of pyruvate which reacts stoichiometrically with DPNH, so that the entire reaction can be followed at 340 mμ.

Polynucleotide phosphorylase is the only known enzyme that forms high polymers of nucleotides, and therefore it is believed to have an important function in the synthesis of ribonucleic acids. Several points remain to be determined about RNA synthesis, however. The relatively poor reaction with the guanine nucleotide is hard to reconcile with the large amounts of guanine found in nucleic acids. Another question relates to the mechanism of control of the synthesis to form specific nucleic acids. While it is possible to devise hypothetical control mechanisms, such as the participation of protein as a template, continued investigation is required to answer the question.

DEOXYRIBONUCLEIC ACIDS

Studies on the structure of DNA are much more dependent on physical and chemical data than on enzymatic. Specific biological functions have been found for DNA from various sources, so it is concluded that specific structures of individual DNA molecules exist, perhaps each structure related to a genetic function. The bases in DNA appear to occur in pairs; adenine + thymine and guanine + cytosine.[36] The struc-

---

[35] S. Ochoa and L. A. Heppel, in "Chemical Basis of Heredity" (W. D. McElroy and B. Glass, eds.), p. 615. Johns Hopkins Press, Baltimore (1957).

[36] E. Chargaff, Federation Proc. 10, 654 (1951); E. Chargaff, S. Zamenhof, G. Brawerman, and L. Kerin, J. Am. Chem. Soc. 72, 3825 (1950).

ture of the best preparations of DNA appears from physical measurements to be a very long molecule composed of two helical strands.[37]

The four bases shown in Fig. 27 account for essentially all of the nitrogen in most preparations of DNA. In certain bacteriophage 5-hydroxymethylcytosine occurs in place of cytosine,[38] and 5-methylcytosine has been found in many DNA preparations.[39] The occurrence of modified bases does not seem to change the 1:1 ratio of purines to pyrimidines. Early studies were interpreted as showing the presence of equivalent quantities of all of the bases, and it was suggested that the unit of DNA structure is a tetranucleotide in which each base is represented. The current interpretation of more precise analytical data is that in a strand of DNA the bases may be arranged in any order (that is, randomly distributed according to analysis, although presumably carefully organized to serve a biological function). The regularity in purine:pyrimidine ratios is derived from a pairing of adenine with thymine and guanine with cytosine on associated strands. The number of different molecular species of DNA in a given preparation is not known, but the finding that many specific biological properties are transmitted among bacteria via DNA indicates that the number of molecular species from a given organism may be very large. Models indicate that such a structure can be held together by hydrogen bonds between the paired purines and pyrimidines from adjacent chains. The absence of the hydroxyl group from position 2' leaves only the 3'-5' phosphate diester to hold the mononucleotides together.

Deoxyribonucleases. Enzymes that hydrolyze DNA are called deoxyribonucleases (DNAases). In addition to specific enzymes found in various extracts,[40] phosphodiesterases such as occur in snake venom hydrolyze DNA. Deoxyribonuclease has been isolated from pancreas.[41] The crystalline enzyme is a protein with a molecular weight around 60,000 and an isoelectric point near pH 5. It requires divalent cations for activity. Like many previously discussed hydrolytic enzymes, it is reversibly denatured by heat.

Assay of Deoxyribonuclease. Hydrolysis of DNA is followed by methods similar to those used with RNA. Another unexpected optical shift occurs, an increase in density at 260 mμ occurs on hydrolysis;[41] this may be caused by rupture of hydrogen bonds involving the bases.

[37] J. D. Watson and F. H. C. Crick, *Nature* **171**, 737, 964 (1953).
[38] G. R. Wyatt and S. S. Cohen, *Biochem. J.* **55**, 774 (1953).
[39] G. R. Wyatt, *Biochem. J.* **48**, 581, 584 (1951).
[40] T. Araki, *Z. physiol. Chem.* **38**, 84 (1903); J. P. Greenstein and W. V. Jenrette, *J. Natl. Cancer Inst.* **2**, 301 (1942).
[41] M. Kunitz, *J. Gen. Physiol.* **33**, 349 (1950).

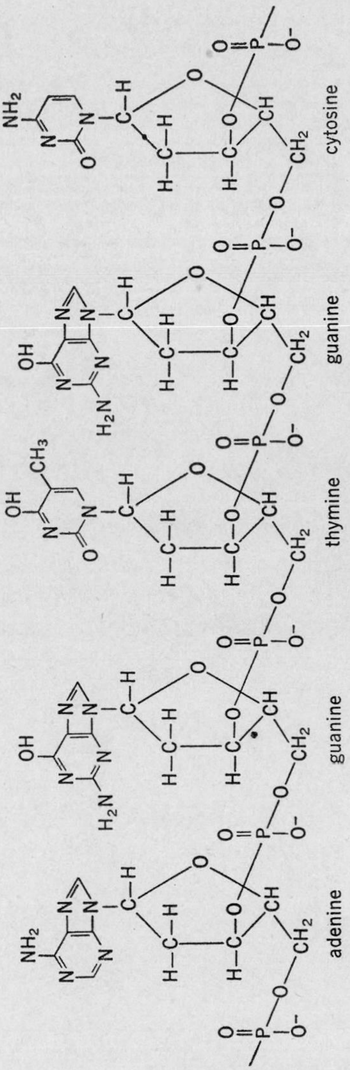

Fig. 27. Representative segment of DNA.

For undetermined reasons, the optical density of DNA is less than the sum of the densities of the component nucleotides.[42] Since DNA is a very long polymer, with a molecular weight of approximately 1,000,000, there is a decrease in viscosity on hydrolysis.[43] The binding of basic dyes is characteristic only of large polymers, and has been used to assay hydrolysis.[44] The most widely used assays, as with ribonuclease, measure either acid produced by hydrolysis of phosphate diesters or the formation of acid-soluble nucleotides.[45]

Reaction of Deoxyribonuclease. DNAase acts on polymers of various lengths and forms a mixture of oligonucleotides with very little mononucleotide. Dinucleotides containing purines, pyrimidines, and both have been isolated from digests.[46] The substrate specificity is obviously not so restricted as that of pancreatic RNAase, but has yet to be defined completely. The mechanism of action is also obscure. The linkage of phosphate to the 3′ position is split by pancreatic DNAase; the monoesters found in the products are all 5′-phosphates.[47] From analogy with ribonucleic acid metabolism, it may be anticipated that DNAase with different specificities will be found, and that some will be found to split the 5′ ester bonds.

Synthesis of DNA. An enzymatic synthesis of polymers of deoxyribonucleotides has been described recently.[48] Enzyme preparations from *E. coli* form acid-insoluble products that are digestible by DNAase. Study of the reaction was complicated by multiple requirements, but extensive purification has permitted the determination of some of the properties of the system. The enzyme is relatively inactive when a single nucleotide is used as a substrate, and appears to have a requirement for the simultaneous presence of four deoxynucleoside triphosphates, deoxy ATP, deoxy GTP, deoxy CTP, and deoxy TTP. Only the triphosphates, not the corresponding disphosphates, are active. In addition to nucleotides, a "primer" is required. DNA from various sources serves as primer; Mg^{++} is also required. The products of the reaction are polymer and inorganic pyrophosphate. Reversibility has been indicated by experiments in which labeled inorganic pyrophosphate was incorporated into the nucleotides.

[42] B. Magasanik and E. Chargaff, *Biochim. et Biophys. Acta* 7, 396 (1951).
[43] G. Schmidt, *Enzymologia* 1, 135 (1936).
[44] N. B. Kurnick, *Arch. Biochem. Biophys.* 43, 97 (1953).
[45] V. Allfrey and A. E. Mirsky, *J. Gen. Physiol.* 36, 227 (1952).
[46] R. L. Sinsheimer, *J. Biol. Chem.* 208, 445 (1954).
[47] J. D. Smith and R. Markham, *Biochim. et Biophys. Acta* 8, 350 (1952).
[48] A. Kornberg, I. R. Lehman, M. J. Bessman, and E. S. Simms, *Biochim. et Biophys. Acta* 21, 197 (1956); M. J. Bessman, I. R. Lehman, E. S. Simms, and A. Kornberg, *Federation Proc.* 66, 153 (1957).

$$n(\text{deoxy ATP} + \text{GTP} + \text{CTP} + \text{TTP}) + \text{DNA} \rightleftharpoons \text{DNA-(AP-GP-CP-TP)}_n + 4n\,\text{PP}$$

The enzyme fraction that forms deoxyribonucleotides may contain one or more enzymes that participate in the polymerization. It remains to be determined whether the polymer is formed in a specific or random manner, and, in either case, what mechanism is used to form biologically specific DNA.

General References

Brown, G. B. (1956). *Federation Proc.* **15**, 823.

Chargaff, E., and Davidson, J. N., eds. (1955). "The Nucleic Acids," Vols. I and II. Academic Press, New York.

Laskowski, M. (1951). *In* "The Enzymes" (J. B. Sumner and K. Myrbäck, eds.), Vol. I, Part 2, p. 956. Academic Press, New York.

McElroy, W. D., and Glass, B., eds. (1957). "Chemical Basis of Heredity." Johns Hopkins Press, Baltimore, Maryland.

Markham, R. (1956). *Proc. 3rd Intern. Congr. Biochem., Brussels 1955* p. 144.

Ochoa, S. (1956). *Federation Proc.* **15**, 832.

Symposium on Biochemistry of Nucleic Acids (1951). *J. Cellular Comp. Physiol.* **38**, *Suppl.* 1.

Symposium on Genetic Recombination (1955). *J. Cellular Comp. Physiol.* **45**, *Suppl.* 2.

Symposium on Nucleoproteins (1953). *Can. J. Med. Sci.* **31**, 222.

Purines and Pyrimidines

Within the last few years enzymatic processes have been found for the synthesis and degradation of both purines and pyrimidines. These bases are essential components of all cells, and most cells seem to be capable of synthesizing their own, although certain fastidious microorganisms and several mutant strains of other organisms require one or more purines or pyrimidines for growth. Both purines and pyrimidines occur as components of nucleotides. The only known biosynthetic formation of purine rings is one in which the intermediates are all derivatives of ribose-5-phosphate, and the product is an intact nucleotide. In pyrimidine biosynthesis, nucleotide formation does not take place until the ring is formed. It is possible that undiscovered pathways of synthesis of purines and pyrimidines do not require nucleotide intermediates. The existence of enzymes that form ribotides from preformed bases requires that alternate mechanisms of synthesis be considered. Organisms that require purines and pyrimidines depend upon the formation of nucleotides from the bases, and it remains to be determined whether this is a remnant of a biosynthetic pathway or whether it is a device to take advantage of the bases synthesized by other organisms. Certain modifications of the bases may take place with intact nucleotides, but more extensive degradation pathways have been described for the free bases.

PURINE BIOSYNTHESIS

Pigeon liver has been used as the system of choice in studying purine biosynthesis, originally because birds were known to excrete large amounts of purine (uric acid), and later because of the knowledge gained with this favorable system. It was found that the first purine compound to accumulate is inosinic acid.[1] Studies with labeled precursors showed that the purine skeleton of this compound was built from the units shown in (I).[2]

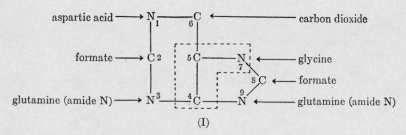

(I)

Enzymatic Formation of Purines. Some of the steps in the formation of this structure have not been isolated, but the outline of the synthesis shown in Fig. 28 is now well established. The structure is built upon ribose-5-phosphate, with the first substitution following the mechanism described for the synthesis of nucleotides with preformed bases.[3]

ribose-5-phosphate + ATP → 5-phosphoribose-1-pyrophosphate (PRPP) + AMP

PRPP reacts with glutamine to yield the 1-amino derivative of ribose-5-phosphate, phosphoribosylamine (1).[4] In the presence of glycine and ATP, an enzyme fraction of pigeon liver catalyzes amide formation, to form glycinamide ribotide (2).[5] This fraction contains the enzymes for formation of glycinamide ribotide from PRPP, glutamine, glycine, and ATP, but phosphoribosylamine was found to substitute for PRPP and glutamine.

[1] G. R. Greenberg, *J. Biol. Chem.* **190,** 611 (1951).
[2] M. P. Schulman, J. C. Sonne, and J. M. Buchanan, *J. Biol. Chem.* **196,** 499 (1952); J. C. Sonne, I. Lin, and J. M. Buchanan, *ibid.* **220,** 369 (1956); B. Levenberg, S. C. Hartman, and J. M. Buchanan, *ibid.* **220,** 379 (1956).
[3] A. Kornberg, I. Lieberman, and E. S. Simms, *J. Biol. Chem.* **215,** 389 (1955).
[4] D. A. Goldthwait, G. R. Greenberg, and R. A. Peabody, *Biochim. et Biophys. Acta* **18,** 148 (1955).
[5] D. A. Goldthwait, R. A. Peabody, and G. R. Greenberg, *J. Biol. Chem.* **221,** 569 (1956); S. C. Hartman, B. Levenberg, and J. M. Buchanan, *ibid.* **221,** 1057 (1956).

The skeleton of glycinamide contributes atoms 7, 5, 4, and 9 of the purine. The 5-membered ring skeleton is completed with the addition of an "active formate" to the amino nitrogen (3).[5] The transfer of "active formate" requires a cofactor derived from folic acid. The chemistry of

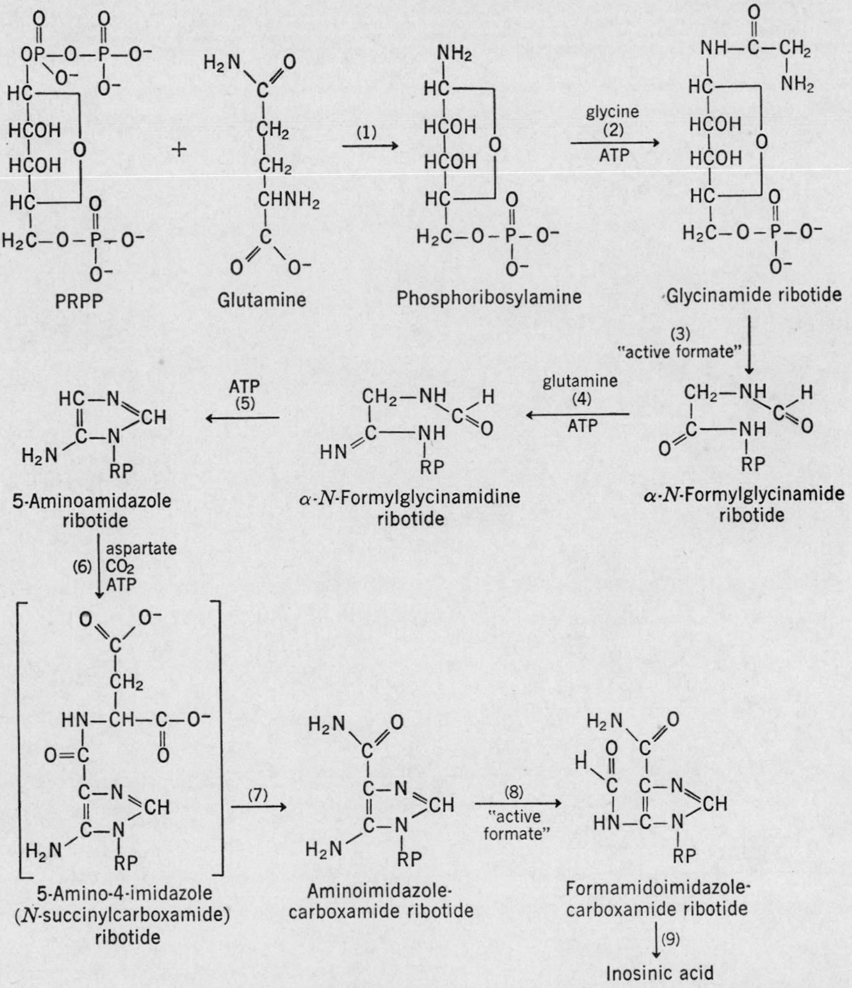

FIG. 28. Biosynthesis of purines.

the cofactor will be considered later. Before the ring is closed, the substituted amide group is converted to a substituted amidine in a reaction that requires glutamine and ATP (4).[6] The enzyme that catalyzes this reaction was separated from the catalyst of the next step by ammonium

[6] B. Levenberg and J. M. Buchanan, *J. Am. Chem. Soc.* **78**, 504 (1956).

sulfate fractionation. In the next step, the formyl group and the amidine nitrogen that participates in the glycoside bond are condensed in a reaction that requires ATP (5). This condensation results in closure of a 5-membered ring, an imidazole, designated 5-aminoimidazole ribotide.[6]

The detailed mechanism of the next steps are obscure. Both CO_2 and aspartic acid contribute to the next product in a reaction that also requires ATP. A preliminary report indicates that an intermediate may be formed by the addition of both of these compounds, CO_2 adding to position 4 of aminoimidazole ribotide and forming an amide with the amino group of aspartic acid (6). This tentative intermediate would release fumarate in becoming 5-aminoimidazole-4-carboxamide ribotide (7).[7] Another hypothetical intermediate would result from the addition of a second "active formate" (8). The enzymatic addition product has not

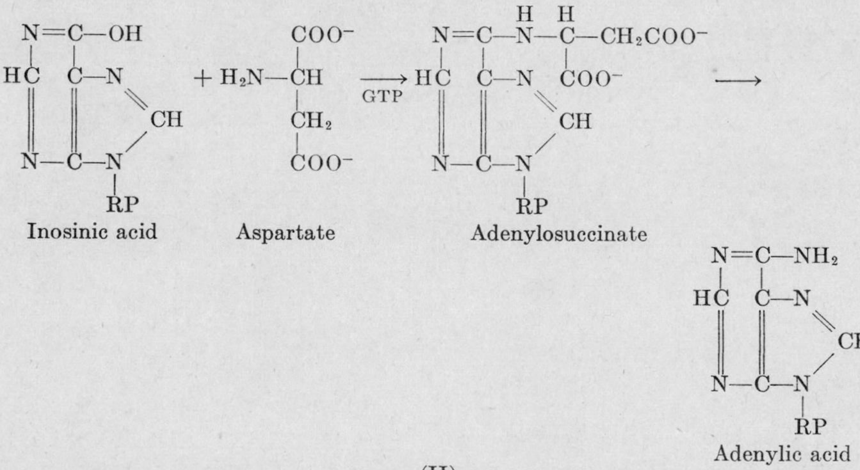

Inosinic acid   Aspartate   Adenylosuccinate

Adenylic acid

(II)

been isolated, but a synthetic compound in which the formyl group is added to the 5-amino group is active in the next, and final, step: cyclization of the 6-membered ring (9).[8] Therefore, it is probable that the formyl group is added enzymatically to the amino, rather than to the amide nitrogen. The gaps in our knowledge of purine biosynthesis are being filled in rapidly, and it is to be anticipated that in the near future the scheme outlined in Fig. 26 will require no brackets to indicate uncertainties.

Interconversion of Purines. Inosinic acid is converted to the nucleotides characteristic of nucleic acids by changes in the purine substituents

[7] L. N. Lukens and J. M. Buchanan, *Federation Proc.* **15**, 305 (1956).
[8] L. Warren and J. G. Flaks, *Federation Proc.* **15**, 379 (1956).

at the nucleotide level. Adenylic acid is formed from inosinic acid through the intermediate formation of adenylosuccinate.[9] This condensation product splits to yield adenylic acid and fumarate.[10] In the initial condensation reaction, GTP is required (II).

Guanine formation occurs following oxidation in position 2, to form xanthylic acid.[11] This oxidation has been demonstrated with a DPN-requiring enzyme from bone marrow. Subsequent amination occurs through different mechanisms in animal and bacterial systems. In bone marrow extracts the amide of glutamine is transferred,[12] while in bacterial extracts ammonia is utilized more readily than glutamine.[13] The utilization of ammonia by the bacterial system is accompanied by the hydrolysis of ATP to AMP and PP. In addition to these reactions, it is known that adenine and guanine can be interconverted, but the enzymatic mechanisms are not known.

Variation in Purine Synthesis. An important development in purine biochemistry came from studies on sulfonamide inhibition of bacteria. A compound was found to accumulate in cultures of *E. coli* incubated with sulfadiazine.[14] This was identified as 4-aminoimidazole-5-carboxamide.[15] In the present concept, this compound represents a degradation product of the corresponding ribotide, but it can also be used as a purine precursor. The free imidazole reacts with ribose-1-phosphate in the presence of nucleoside phosphorylase to form a nucleoside[16] which can be phosphorylated by a kinase.[17] Thus, variations of the basic scheme can and do occur.

The reactions involved in purine biosynthesis illustrate the unpredictable routes taken in complex biological syntheses. It will be of great interest to learn more of the properties of the individual enzymes as they are further separated.

PYRIMIDINE BIOSYNTHESIS

Isotope studies as well as *de novo* synthesis in the presence of precursors implicate CO_2, NH_3, and aspartic acid in the biosynthesis of pyrimi-

[9] I. Lieberman, *J. Biol. Chem.* **223**, 327 (1956).
[10] C. E. Carter and L. H. Cohn, *J. Am. Chem. Soc.* **77**, 499 (1955).
[11] R. Abrams and M. Bentley, *J. Am. Chem. Soc.* **77**, 4179 (1955).
[12] M. Bentley and R. Abrams, *Federation Proc.* **15**, 218 (1956).
[13] H. S. Moyed and B. Magasanik, *Federation Proc.* **15**, 318 (1956).
[14] M. R. Stetten and C. R. Fox, Jr., *J. Biol. Chem.* **163**, 333 (1945).
[15] W. Shive, W. W. Ackermann, M. Gordon, M. E. Getzendaner, and R. E. Eakin, *J. Am. Chem. Soc.* **69**, 725 (1947).
[16] E. D. Korn, F. C. Charalampous, and J. M. Buchanan, *J. Am. Chem. Soc.* **75**, 3610 (1953).
[17] G. R. Greenberg, *Federation Proc.* **12**, 211 (1953).

dines.[18] In other systems it can be shown that carbamyl phosphate is the donor of the carbamyl group.[19] These condense to form ureidosuccinate. In some systems citrulline appears to be a carbamyl donor;[20] this is supported by the stimulation of pyrimidine synthesis by glutamic acid derivatives,[21] which will be shown later to support citrulline formation.

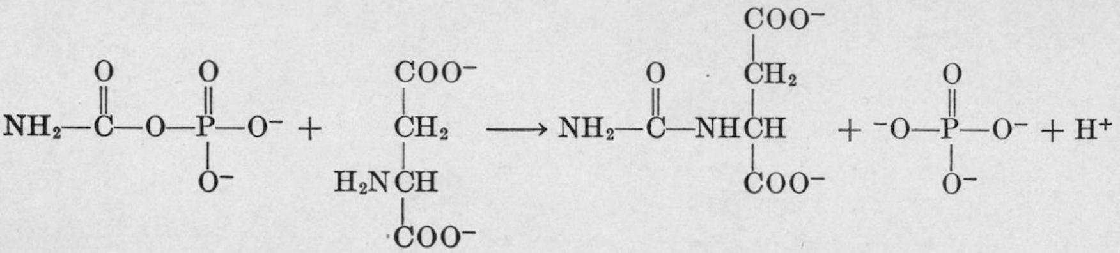

$$NH_2-\overset{O}{\overset{\|}{C}}-O-\overset{O}{\underset{O^-}{\overset{\|}{P}}}-O^- \;+\; \underset{\underset{\cdot COO^-}{H_2N\overset{|}{C}H}}{\overset{\overset{COO^-}{|}}{C}H_2} \;\longrightarrow\; NH_2-\overset{O}{\overset{\|}{C}}-NH\underset{COO^-}{\overset{\overset{COO^-}{|}}{C}H} \;+\; {}^-O-\overset{O}{\underset{O^-}{\overset{\|}{P}}}-O^- + H^+$$

Hydrolysis of ureidosuccinate to aspartate, CO_2, and NH_3 is catalyzed by a bacterial enzyme,[22] but the existence of such an enzyme does not carry any implications for the synthetic mechanism.

Ureidosuccinate can be cyclized to form either a 5-membered or a 6-membered ring (III), and enzymes that carry out both of these reactions have been separated from an anaerobic bacterium grown on orotic acid.[23]

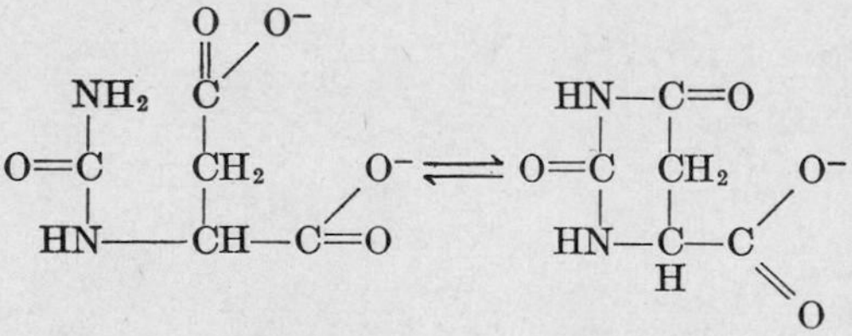

Ureidosuccinate   Dihydroorotate

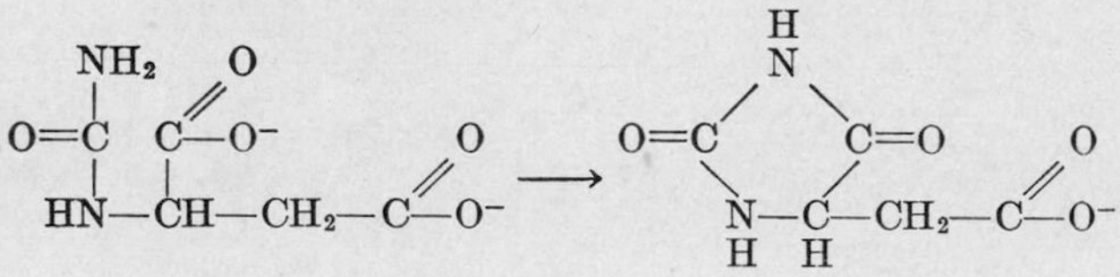

Ureidosuccinate   Hydantoin acetate

(III)

Dihydroorotic acid is oxidized by a DPN-requiring dehydrogenase to the pyrimidine, orotic acid.[24] Orotic acid is an efficient precursor of other pyrimidines and is required by certain microorganisms. The conversion

[18] U. Lagerkvist, P. Reichard, and G. Ehrensvard, *Acta Chem. Scand.* **5**, 1212 (1951).
[19] M. E. Jones, L. Spector, and F. Lipmann, *J. Am. Chem. Soc.* **77**, 819 (1955).
[20] L. H. Smith and D. Stetten, *J. Am. Chem. Soc.* **76**, 3864 (1954).
[21] P. Reichard, *Acta Chem. Scand.* **8**, 795 (1954).
[22] I. Lieberman and A. Kornberg, *J. Biol. Chem.* **212**, 909 (1955).
[23] I. Lieberman and A. Kornberg, *J. Biol. Chem.* **207**, 911 (1954).
[24] I. Lieberman and A. Kornberg, *Biochim. et Biophys. Acta* **12**, 223 (1953).

of orotic acid to other pyrimidines requires condensation to form a nucleotide. This is catalyzed by an orotic-specific pyrophosphorylase, which substitutes orotic acid for the pyrophosphate of PRPP. After this nucleotide is formed, orotidylic acid is decarboxylated to yield uridylic acid (IV).[25]

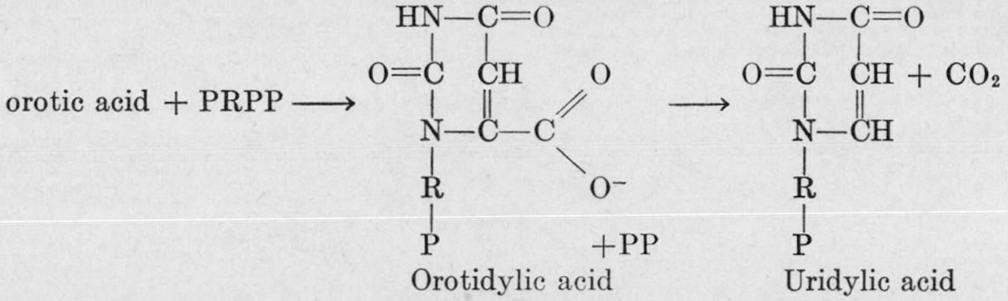

orotic acid + PRPP ⟶

$$
\begin{array}{ccc}
\text{HN—C=O} & & \text{HN—C=O} \\
\text{O=C} \ \ \text{CH} \ \ \text{O} & \longrightarrow & \text{O=C} \ \ \text{CH} + CO_2 \\
\text{N—C—C} & & \text{N—CH} \\
\text{R} \ \ \ \ \ \text{O}^- & & \text{R} \\
\text{P} \ \ \ +PP & & \text{P}
\end{array}
$$

Orotidylic acid   Uridylic acid
(IV)

Interconversion of Pyrimidines. Uridylic acid is converted to cytidylic acid by an enzyme isolated from *E. coli* that catalyzes the reaction shown in (V).[26]

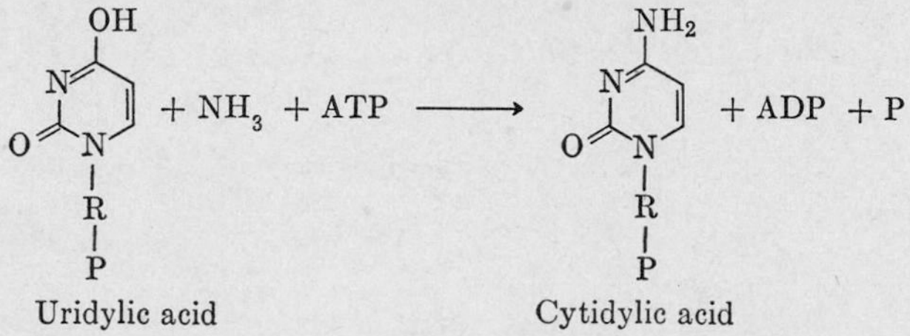

Uridylic acid + NH_3 + ATP ⟶ Cytidylic acid + ADP + P

Uridylic acid   Cytidylic acid
(V)

The mechanism of formation of thymine and hydroxymethylcytosine are not yet known, but the addition of a carbon to position 5 appears to involve a transfer from a folic acid derivative. As in the case of the purines, the relation between ribose and deoxyribose nucleotides is obscure. Deoxyribonucleosides can be formed from deoxyribose-1-phosphate by a nucleoside phosphorylase reaction.

DEGRADATION OF PURINES

Purine nucleotides are modified by a variety of enzymes, including those that attack the phosphate group and those that split the glycosidic bond. In addition to these, there are changes in the bases in which the amino groups of adenylic acid and guanylic acid are removed by specific deaminases.[27] With these cases excepted (i.e., the formation of inosinic and xanthylic acids), degradation of purines occurs after hydrolysis to yield the free base.

[25] I. Lieberman, A. Kornberg, and E. S. Simms, *J. Biol. Chem.* **215,** 403 (1955).
[26] I. Lieberman, *J. Biol. Chem.* **219,** 307 (1956).
[27] G. Schmidt, *Z. physiol. Chem.* **179,** 243 (1928); **208,** 185 (1932).

Uric Acid Formation. In vertebrates purines are oxidized to uric acid. This reaction is catalyzed by xanthine oxidase (or dehydrogenase), which attacks both hypoxanthine and xanthine.[28] Since adenine and guanine nucleotides can give rise to the hydroxylated purines either as the nucleotide, nucleoside, or free base, all of the naturally occurring purines of animals can be converted to uric acid. Adenine may also be oxidized to 2,8-dihydroxy-4-aminopurine,[29] which is excreted in the urine. The formation of uric acid from any of its precursors is followed conveniently spectrophotometrically (Fig. 29).[30]

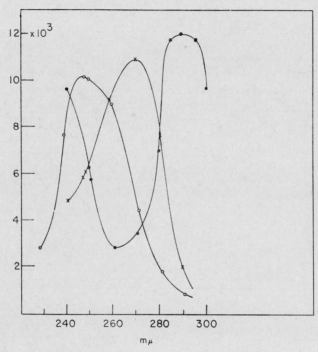

FIG. 29. Absorption spectra of representative purines. Molecular extinction of hypoxanthine (O), xanthine (×), and uric acid (●).[30]

Uricase. Many organisms, including humans, excrete uric acid as the end-product of purine metabolism. Most other mammals, however, oxidize uric acid further. The enzyme uricase has been known for over 50 years, but the nature of the reaction catalyzed is still uncertain. The enzyme has been purified in many laboratories, but until recently was

[28] E. G. Ball, *J. Biol. Chem.* **128,** 51 (1939).
[29] H. Klenow, *Biochem. J.* **50,** 404 (1952).
[30] H. M. Kalckar, *J. Biol. Chem.* **167,** 429 (1947).

always associated with insoluble particles. Extraction of acetone-dried liver mitochondria with alkaline buffers yielded a soluble preparation, which was purified extensively to give a colorless protein, homogeneous by ultracentrifugal and electrophoretic criteria. Inhibition studies and microanalysis indicate that firmly bound copper is a component of the enzyme.[31] A soluble uricase has also been obtained from *Neurospora*.[32] The yield of mold uricase is increased by growing the *Neurospora* in the presence of uric acid. Inhibition studies with *Neurospora* uricase give results consistent with a copper-protein nature for this enzyme also, but this has not been established. Both enzymes are absolutely specific for uric acid, and are inhibited by other purines.

The action of uricase involves consumption of oxygen, and under some conditions a comparable amount of CO_2 is evolved. A good yield of allantoin is formed under these conditions. The formation of allantoin is not the first step, however; spectrophotometric observations show an intermediate that absorbs at higher wavelengths than uric acid.[33] Low pH or the presence of borate also minimizes the accumulation of allantoin.[34] The allantoin that does form has been shown to contain nitrogen atoms provided by the 6-membered and 5-membered rings equally distributed between its ureido group and hydantoin ring.[35] This was interpreted as indicating a symmetrical intermediate, but it has been suggested that tautomerization of allantoin itself might have caused the randomization of labeled atoms.[36] There is no evidence that any of the reactions following the initial oxidation is enzymatic.

The structures shown on page 272 have been considered as products of uricase action, but the mechanisms by which they are formed are still hypothetical. Evidence for the presence of additional compounds has been found.

A large number of products, including some of the products of uricase action, are produced from uric acid by peroxidases + enzymatically generated H_2O_2.[37] The nonspecific action of peroxidases emphasizes the part played by nonenzymatic reactions of the unstable products of uric acid oxidation.

Purine Degradation by Anaerobic Bacteria. A totally different degradation of purines carried out by certain anaerobic bacteria, *Clostridium acidiurici* and *C. cylindrosporum* has been largely elucidated by Rabinowitz.

[31] H. R. Mahler, G. Hubscher, and H. Baum, *J. Biol. Chem.* **216**, 625 (1955).
[32] R. C. Greene and H. K. Mitchell, *Arch. Biochem. and Biophys.* **70**, 603 (1957).
[33] E. Praetorius, *Biochim. et Biophys. Acta* **2**, 602 (1948).
[34] F. W. Klemperer, *J. Biol. Chem.* **160**, 111 (1945).
[35] G. B. Brown, P. M. Roll, and L. F. Cavalieri, *J. Biol. Chem.* **171**, 835 (1947).
[36] E. S. Canellakis and P. P. Cohen, *J. Biol. Chem.* **213**, 379 (1955).
[37] E. S. Canellakis, A. L. Tuttle, and P. P. Cohen, *J. Biol. Chem.* **213**, 397 (1955).

These organisms convert all of the purines they attack (uric acid, hypoxanthine, guanine, and the nucleosides inosine and guanosine) to xanthine.[38] Xanthine is hydrolyzed to 4-ureido-5-imidazolecarboxylic acid.[39] A second enzyme that requires divalent cations splits the ureide to leave 4-amino-5-imidazolecarboxylic acid.[40] The second step is inhibited by metal-chelating agents; these were used to permit the product of the first enzyme to be accumulated and identified. Aminoimidazolecarboxylic acid is decarboxylated by a third bacterial enzyme, leaving 4-aminoimidazole.

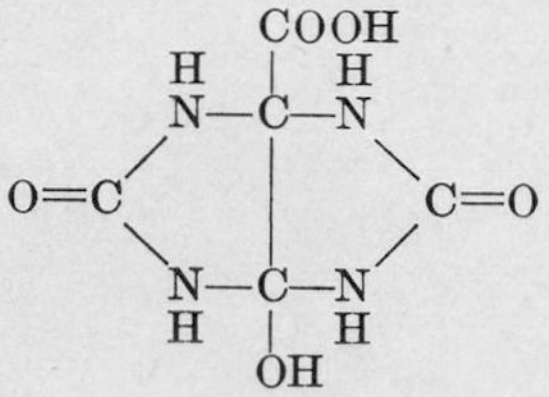

HDC
Hydroxyacetylenediureinecarboxylic acid

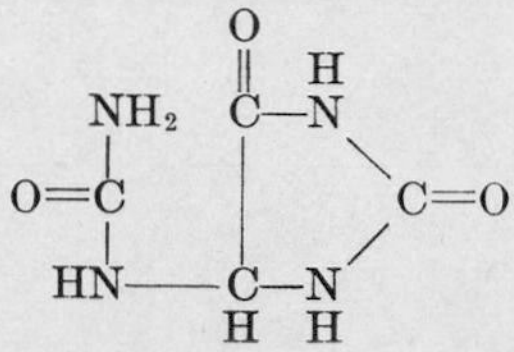

Allantoin

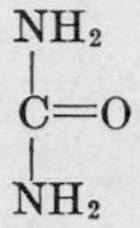

Urea

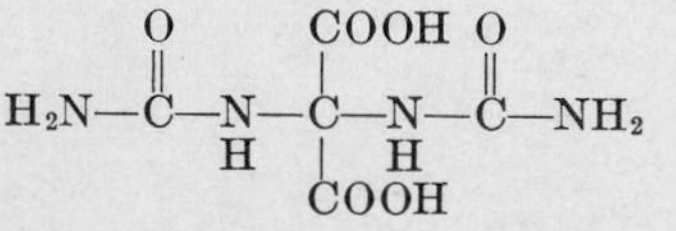

2,2-Diureidomalonic acid

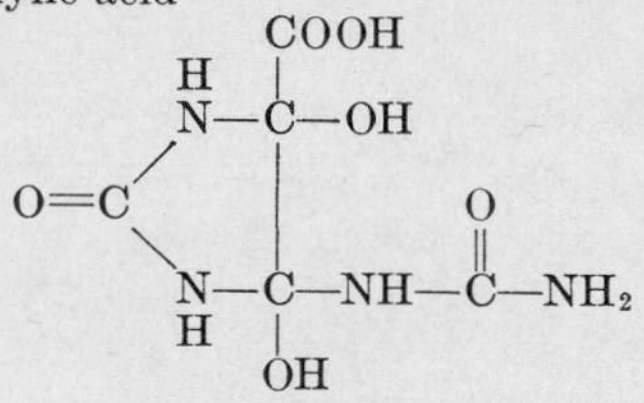

UIDC
5-Ureido-2-imidazolidone-4,5-diol-4-carboxylic acid

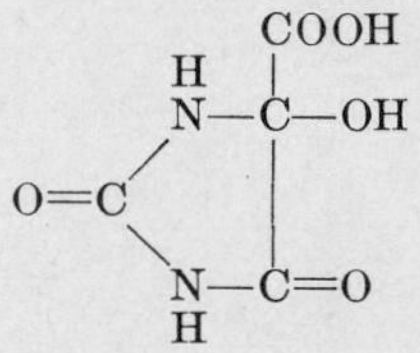

Alloxanic acid

The next steps have not been resolved, but include opening the ring between N-3 and C-4 and hydrolysis of the amino group to form formiminoglycine (Fig. 28).[41] Formiminoglycine is converted to glycine, formate, and ammonia, but, unlike the earlier hydrolytic steps, this sequence requires a cofactor of the folic group and also requires ADP and inorganic phosphate.[42] ATP is formed in amounts equivalent to formiminoglycine used and equimolar quantities of formate, ammonia, and glycine are produced.

Folic Acid in the Degradation of Purines. The conversion of formiminoglycine to glycine is accomplished by a transfer of the formimino group to tetrahydrofolic acid (THF) (VI).[43] The formimino group is assigned to

[38] J. C. Rabinowitz and H. A. Barker, *J. Biol. Chem.* **218**, 161 (1956).

[39] J. C. Rabinowitz and W. E. Pricer, Jr., *J. Biol. Chem.* **218**, 189 (1956).

[40] J. C. Rabinowitz, *J. Biol. Chem.* **218**, 175 (1956).

[41] J. C. Rabinowitz and W. E. Pricer, Jr., *J. Biol. Chem.* **222**, 537 (1956).

[42] R. D. Sagers, J. V. Beck, W. Gruber, and I. C. Gunsalus, *J. Am. Chem. Soc.* **78**, 694 (1956); J. C. Rabinowitz and W. E. Pricer, Jr., *ibid.* **78**, 1513 (1956).

[43] J. C. Rabinowitz and W. E. Pricer, Jr., *J. Am. Chem. Soc.* **78**, 5702 (1956).

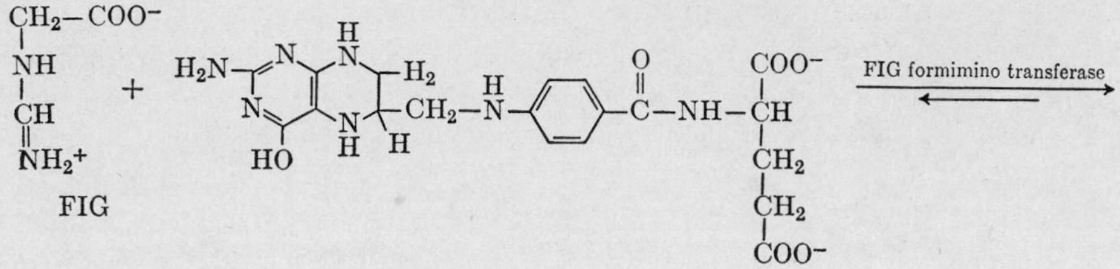

Tetrahydrofolic acid

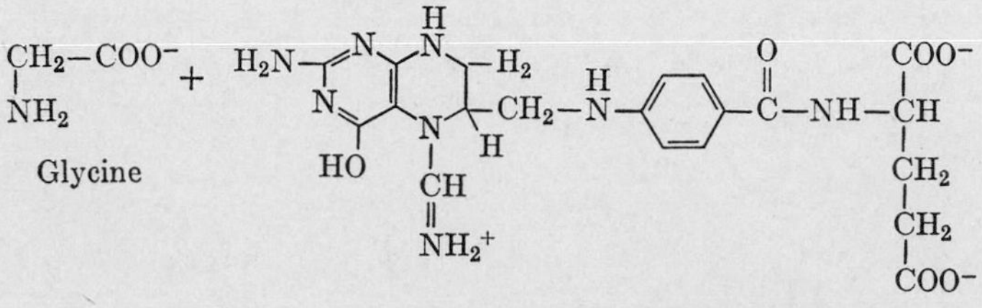

5-Formiminotetrahydrofolic acid
(VI)

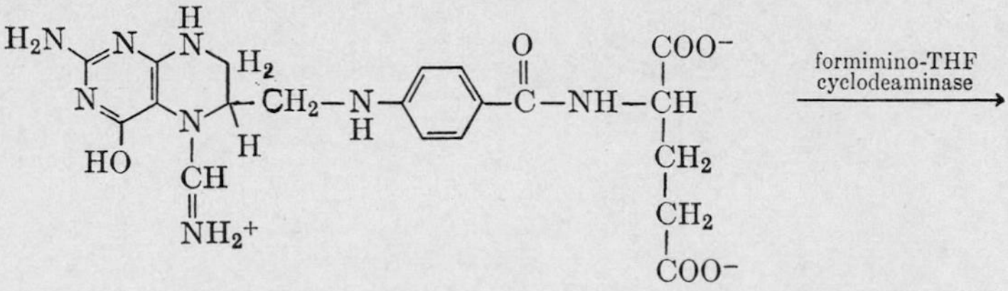

5-Formiminotetrahydrofolic acid

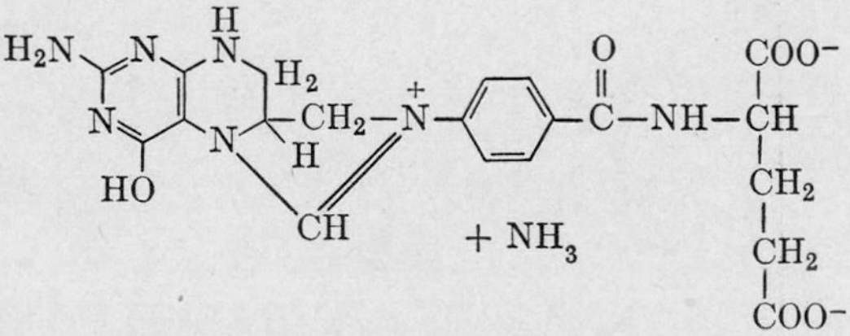

5,10-Anhydroformyltetrahydrofolic acid
(VII)

position 5, since this derivative does not decompose rapidly in oxygen, and 5-substitutions are known to stabilize the reduced ring in the presence of oxygen, whereas tetrahydrofolic acid itself or with substitutions in other positions is very labile to oxygen.

Subsequent steps in the liberation of the final products all involve tetrahydrofolic acid derivatives, and may be used conveniently to illus-

trate the properties of these compounds. Solutions of formiminotetrahydrofolic acid show an absorption peak near 285 mμ, and essentially no
absorption above 320 mμ. On exposure to dilute acid formiminotetrahydrofolic acid is converted at room temperatures to a compound that
has an absorption maximum in acid at 350 mμ and in neutral solutions

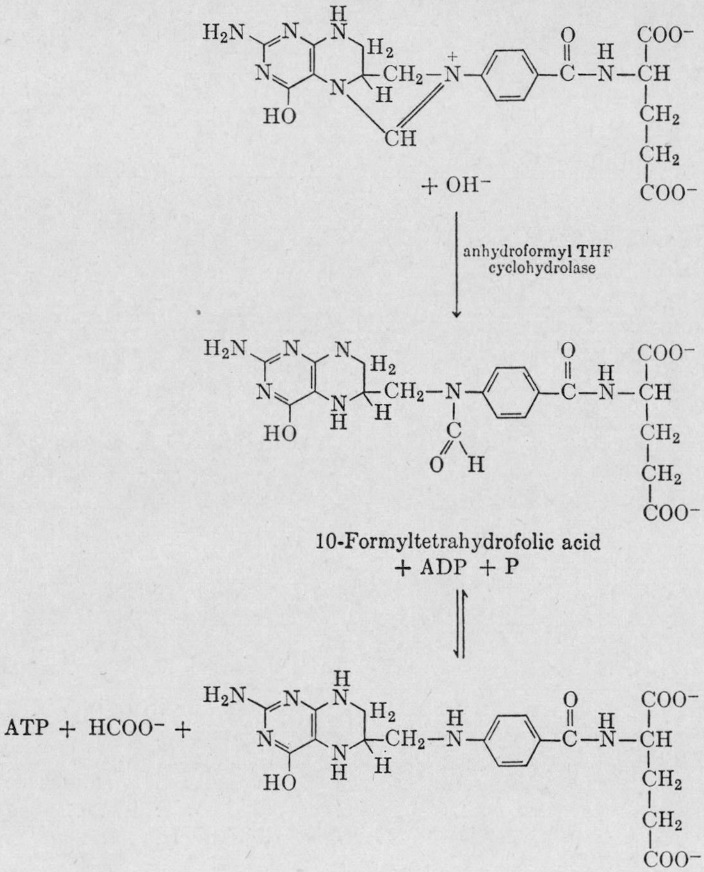

$+$ OH$^-$

anhydroformyl THF
cyclohydrolase

10-Formyltetrahydrofolic acid
$+$ ADP $+$ P

ATP $+$ HCOO$^-$ $+$

(VIII)

at 356 mμ. An enzyme in extracts of clostridia converts formiminotetrahydrofolic acid to the same product as formed by acid, and simultaneously
releases one equivalent of ammonia. The product has been identified as
5,10-anhydroformyltetrahydrofolic acid (VII). The formation of the
cyclic anhydro compound has been used to measure the formation of
formiminotetrahydrofolic acid. In one assay system the reaction has been
carried out in the absence of the second enzyme, and aliquots have been

treated with acid at time intervals, and the absorption read at 350 mμ or 356 mμ. Another assay system measures the increase in optical density at 356 mμ when formiminotetrahydrofolic acid is produced in the presence of an excess of the second enzyme.

The imidazolinium ring is opened by an enzyme, cyclohydrolase, to form N^{10}-formyltetrahydrofolic acid. This product does not absorb much light above 300 mμ, but has an absorption peak near 260 mμ. Both the

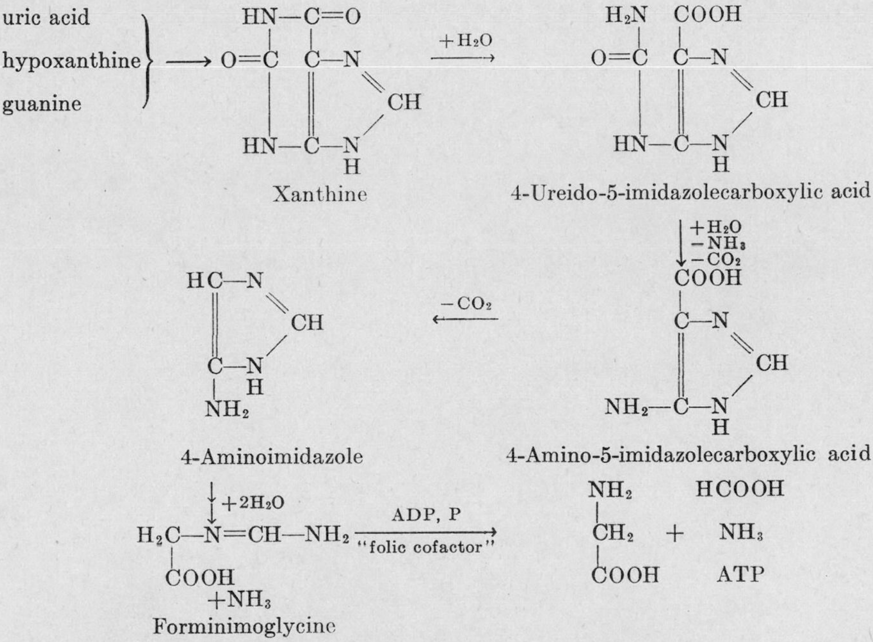

FIG. 30. Bacterial degradation of purines.

formation and subsequent hydrolysis of the anhydro compound can therefore be followed spectrophotometrically by measuring the appearance and disappearance of the 356 mμ peak. N^{10}-formyltetrahydrofolic is distinguished from N^5-derivatives by both lability in oxygen and a more rapid cyclization in acid.

N^{10}-formyltetrahydrofolic acid has been known to be formed from tetrahydrofolic acid and formate in a reaction coupled with hydrolysis of ATP to ADP and inorganic phosphate (VIII). The reaction that finally releases the formate derived from purines is the reverse of the activation reaction.[44]

[44] G. R. Greenberg, L. Jaenicke, and M. Silverman, *Biochim. et Biophys. Acta* **17**, 589 (1955).

DEGRADATION OF PYRIMIDINES

In animals the conversion of thymine and uracil to the corresponding ureides, β-ureidoisobutyric and β-ureidopropionic acids, has been demonstrated. The ureides are degraded to the free β-amino acids, which are excreted. The reactions that catalyze these degradations have been found in liver.[45] The first step is a reduction of the pyrimidine in which TPNH is the electron donor (IX).[46] This reaction has not been studied thermo-

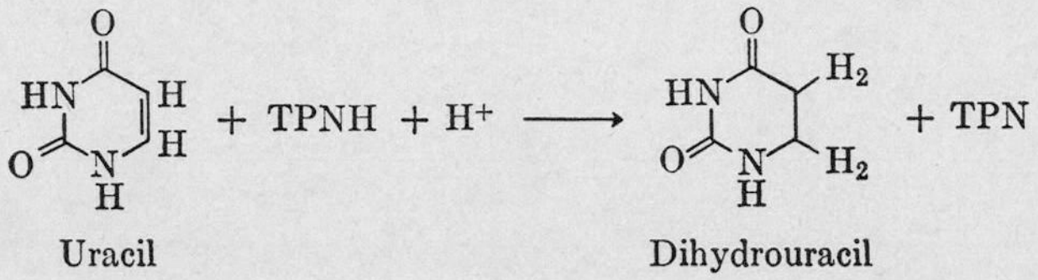

Uracil + TPNH + H$^+$ $\longrightarrow$ Dihydrouracil + TPN

(IX)

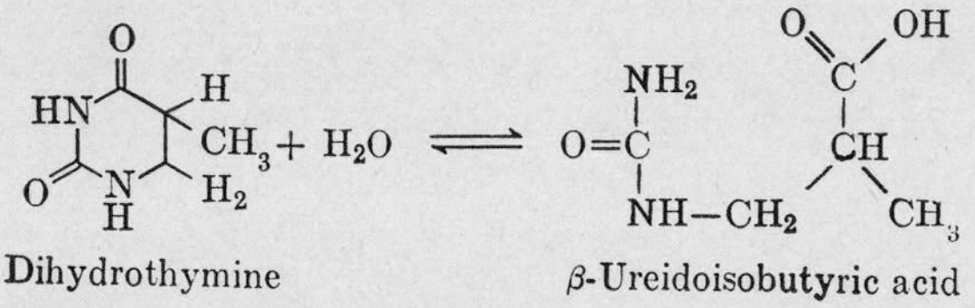

Dihydrothymine + H$_2$O $\rightleftharpoons$ β-Ureidoisobutyric acid

(X)

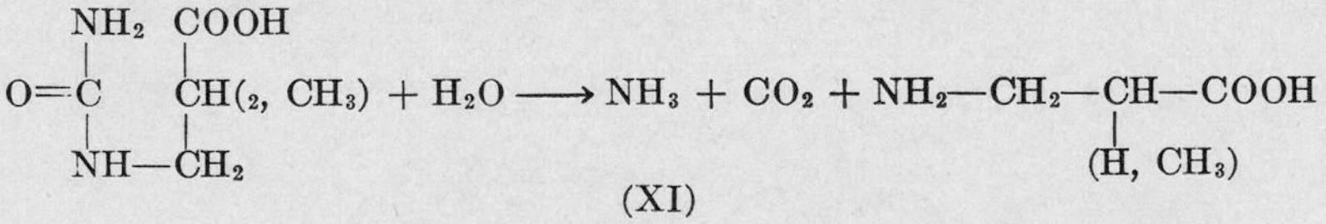

$$O=\overset{\overset{\displaystyle NH_2}{|}}{C}\quad \overset{\overset{\displaystyle COOH}{|}}{CH}(_2, CH_3) + H_2O \longrightarrow NH_3 + CO_2 + NH_2-CH_2-CH-COOH$$
$$\underset{NH-CH_2}{\qquad} \qquad\qquad\qquad\qquad\qquad (H, CH_3)$$

(XI)

Barbituric acid $\longrightarrow$ Urea + Malonate

5-Methylbarbituric acid

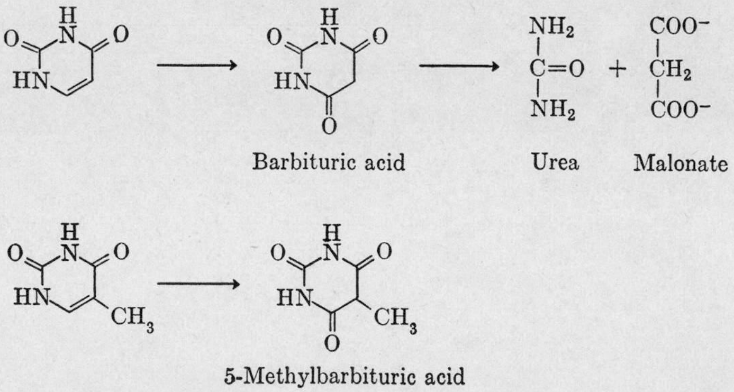

(XII)

[45] K. Fink, *J. Biol. Chem.* **218**, 9 (1956); R. M. Fink, C. McGaughey, R. E. Cline, and K. Fink, *J. Biol. Chem.* **218**, 1 (1956).
[46] E. S. Canellakis, *J. Biol. Chem.* **221**, 315 (1956).

dynamically, but presumably is reversible with an equilibrium similar to that of the orotic–dihydroorotic acid system.

The cleavage of the reduced ring is a reversible reaction, yielding β-ureidopropionic acid in the case of dihydrouracil and β-ureidoisobutyric acid in the case of dihydrothymine (X).[47]

Degradation of the ureido compounds is irreversible; hydrolysis by crude extracts of acetone-dried liver yields CO_2, NH_3, and a β-amino acid, β-alanine or β-aminoisobutyric acid (XI).[46,47]

Bacterial cells that grow on pyrimidines contain adaptive enzymes that carry out the following reactions. Cytosine is deaminated to give uracil and 5-methylcytosine is similarly converted to thymine. The two dihydroxypyrimides are oxidized by an enzyme that reacts with methylene blue, but not with oxygen. The products are barbituric acid and 5-methylbarbituric acid from uracil and thymine, respectively (XII).[48] Although intact bacteria oxidize thymine completely to CO_2 and NH_3, cell-free extracts have not been found to carry the oxidation beyond 5-methylbarbituric acid. Barbituric acid, on the other hand, is hydrolyzed to urea and malonic acid. The urea is further hydrolyzed and the malonic acid is oxidized as its CoA derivative to form acetyl CoA and CO_2.[49]

GENERAL REFERENCES

Buchanan, J. M., and Wilson, D. W. (1953). *Federation Proc.* **12,** 646.
Greenberg, G. R. (1953). *Federation Proc.* **12,** 651.
Schmidt, G. (1955). *In* "The Nucleic Acids" (E. Chargaff and J. N. Davidson, eds.), Vol. I, p. 555. Academic Press, New York.

[47] S. Grisolia and D. P. Wallach, *Biochim. et Biophys. Acta* **18,** 449 (1955).
[48] O. Hayaishi and A. Kornberg, *J. Biol. Chem.* **197,** 717 (1952).
[49] O. Hayaishi, *J. Biol. Chem.* **215,** 125 (1955).

Chapter VII

AMINO ACIDS

Amino Acid Decarboxylases

Amino acids are the units from which proteins are constructed, and thus are essential components of all living organisms. The syntheses of the various amino acids are reactions of prime importance, and most organisms have the capacity to generate some or all of the amino acids required for their proteins. In addition to the obviously important reactions leading to protein synthesis, many reactions of amino acids lead to the formation of compounds of biological importance, including vitamins, hormones, purines, pyrimidines, porphyrins, and other compounds with specific functions. The various amino acids are metabolized in several ways. In the following sections, several of the specific reactions of amino acids will be discussed. Certain of these reactions are similar for many amino acids, such as decarboxylation and transamination, and some are quite nonspecific, as certain amino acid oxidases. These will be discussed first, then various special reactions of individual amino acids will be considered.

The formation of amines corresponding to certain amino acids has been known for a century,[1] and for over 50 years it has been known that certain amino acids can be decarboxylated by animal and microbial preparations.[2] Hanke and Koessler found that various strains of *E. coli* contained decarboxylases for different amino acids, and that the ability to decarboxylate individual amino acids was so distributed among strains as to indicate that specific enzymes existed for the decarboxylation of each of the susceptible amino acids.[3] The reactions of amino acid decarboxylases are illustrated in Fig. 31.

Gale studied the formation of bacterial decarboxylases for arginine, lysine, ornithine, histidine, tyrosine, and glutamic acid.[4] One result of his studies was the demonstration that these enzymes occur in variable amounts in the cells and that environmental conditions, such as pH of the medium, influence the amount of enzyme formed. The failure to

[1] A. Muller, *J. prakt. Chem.* **70**, 65 (1857).
[2] A. Ellinger, *Z. physiol. Chem.* **29**, 334 (1900).
[3] M. T. Hanke and K. K. Koessler, *J. Biol. Chem.* **50**, 131 (1922).
[4] E. F. Gale, *Advances in Enzymol.* **6**, 1 (1946).

detect amino acid decarboxylases in various strains of bacteria implies that these enzymes are not essential for the growth of the organisms; it is possible, however, that with enzymes whose concentrations vary as these do, small amounts may be present at all times to form essential amines. Another important development of Gale's work was the resolution of bacterial decarboxylases into apoenzyme and coenzyme by alkaline ammonium sulfate precipitation of the protein.[5]

Pyridoxal Phosphate, Codecarboxylase. An independent approach to the nature of the amino acid decarboxylases was made by Gunsalus, Umbreit, and collaborators. They found that the production of tyrosine decarboxylase by *Streptococcus faecalis* depended on the vitamin, pyridoxine.[6] In the absence of pyridoxine the cells grew but had little decarboxylase. However, addition of the vitamin permitted deficient cells to decarboxylate tyrosine, and dried cells exhibited active enzyme in the presence of pyridoxal (a derivative of pyridoxine) and ATP, implying the formation of an active cofactor from these substances.[7] Pyridoxal is a more active growth factor for a strain of *Streptococcus faecalis* than pyridoxine; both synthetic pyridoxal and pyridoxamine exhibit 5000 to 9000 times the activity of the hydroxy compound.[8]

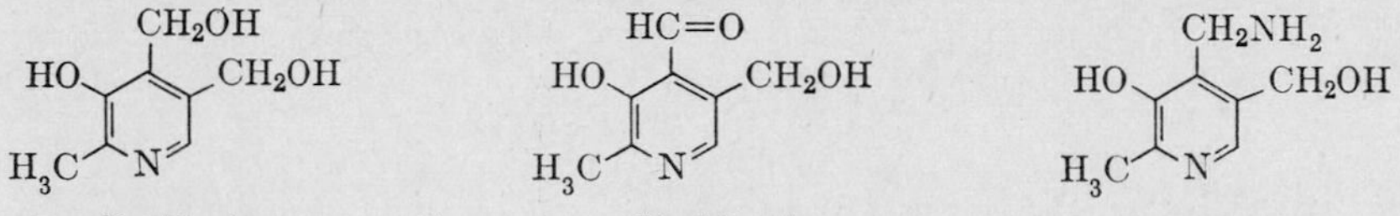

Pyridoxine   Pyridoxal   Pyridoxamine

Interesting differences of opinion enlivened efforts to elucidate the structure of the active cofactor. Eventually it was established by Umbreit and Gunsalus that codecarboxylase is pyridoxal phosphorylated on the hydroxymethyl group at position 5.[9]

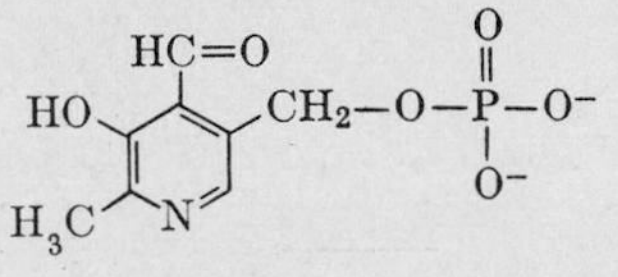

Pyridoxal phosphate

The mechanism of chemical reactions involving this cofactor will be

[5] E. F. Gale and H. M. R. Epps, *Biochem. J.* **38,** 232 (1944).
[6] W. D. Bellamy and I. C. Gunsalus, *J. Bacteriol.* **48,** 191 (1944).
[7] I. C. Gunsalus, W. D. Bellamy, and W. W. Umbreit, *J. Biol. Chem.* **155,** 685 (1944).
[8] E. E. Snell and A. N. Rannefeld, *J. Biol. Chem.* **157,** 475 (1945).
[9] W. W. Umbreit and I. C. Gunsalus, *J. Biol. Chem.* **179,** 279 (1949); J. Baddiley and A. P. Mathias, *J. Chem. Soc.* 2583 (1952).

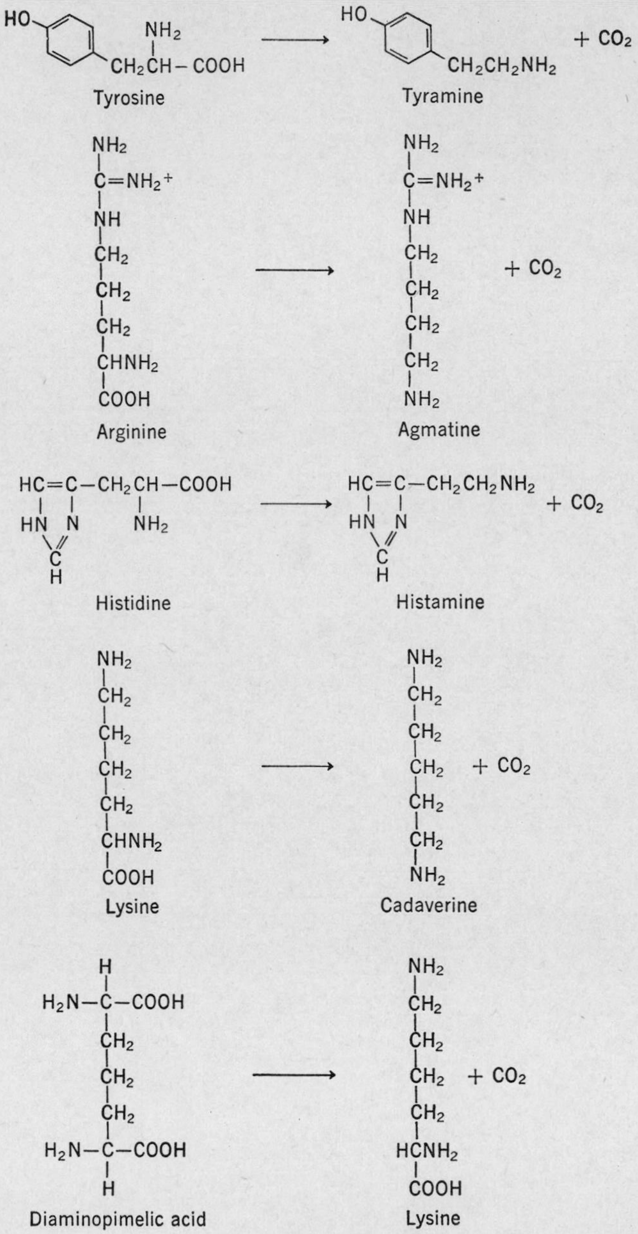

$$\text{Tyrosine} \longrightarrow \text{Tyramine} + CO_2$$

$$\text{Arginine} \longrightarrow \text{Agmatine} + CO_2$$

$$\text{Histidine} \longrightarrow \text{Histamine} + CO_2$$

$$\text{Lysine} \longrightarrow \text{Cadaverine} + CO_2$$

$$\text{Diaminopimelic acid} \longrightarrow \text{Lysine} + CO_2$$

FIG. 31. Reactions of amino acid decarboxylation (*continued on opposite page*).

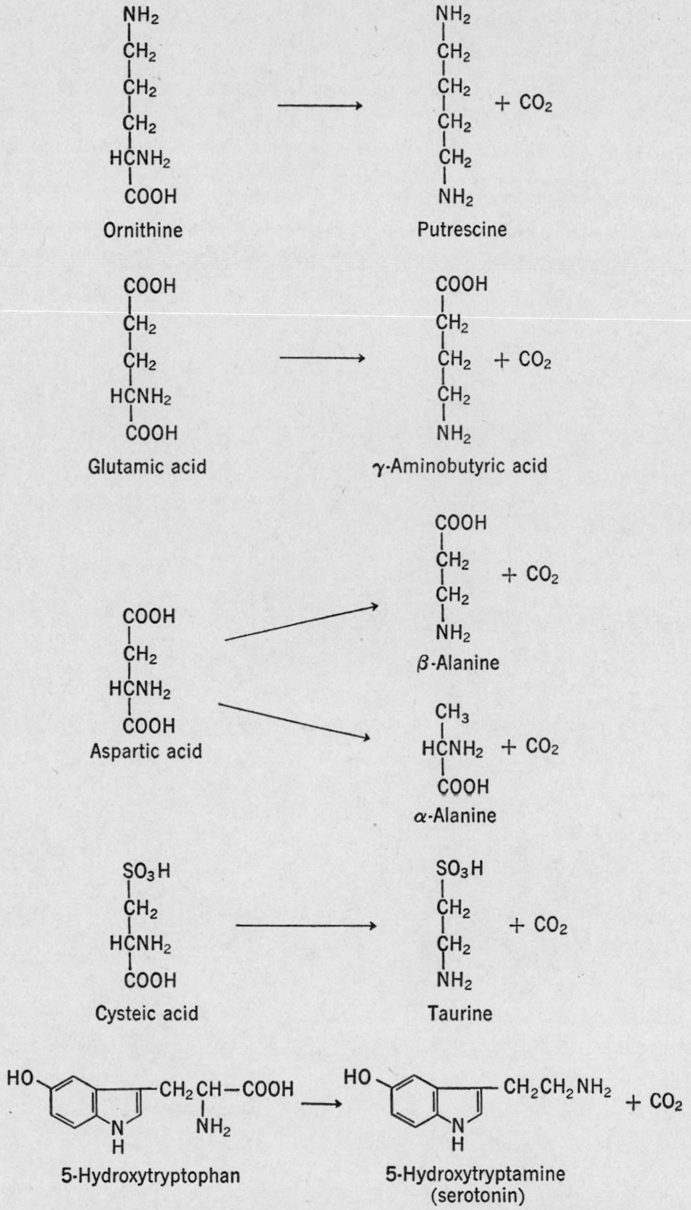

$$\begin{array}{c}
NH_2 \\
| \\
CH_2 \\
| \\
CH_2 \\
| \\
CH_2 \\
| \\
HCNH_2 \\
| \\
COOH
\end{array}
\qquad \longrightarrow \qquad
\begin{array}{c}
NH_2 \\
| \\
CH_2 \\
| \\
CH_2 \\
| \\
CH_2 \\
| \\
CH_2 \\
| \\
NH_2
\end{array} + CO_2$$

Ornithine   Putrescine

$$\begin{array}{c}
COOH \\
| \\
CH_2 \\
| \\
CH_2 \\
| \\
HCNH_2 \\
| \\
COOH
\end{array}
\qquad \longrightarrow \qquad
\begin{array}{c}
COOH \\
| \\
CH_2 \\
| \\
CH_2 \\
| \\
CH_2 \\
| \\
NH_2
\end{array} + CO_2$$

Glutamic acid   γ-Aminobutyric acid

$$\begin{array}{c}
COOH \\
| \\
CH_2 \\
| \\
HCNH_2 \\
| \\
COOH
\end{array}$$

Aspartic acid

$$\begin{array}{c}
COOH \\
| \\
CH_2 \\
| \\
CH_2 \\
| \\
NH_2
\end{array} + CO_2$$

β-Alanine

$$\begin{array}{c}
CH_3 \\
| \\
HCNH_2 \\
| \\
COOH
\end{array} + CO_2$$

α-Alanine

$$\begin{array}{c}
SO_3H \\
| \\
CH_2 \\
| \\
HCNH_2 \\
| \\
COOH
\end{array}
\qquad \longrightarrow \qquad
\begin{array}{c}
SO_3H \\
| \\
CH_2 \\
| \\
CH_2 \\
| \\
NH_2
\end{array} + CO_2$$

Cysteic acid   Taurine

HO—[indole ring]—CH₂CH—COOH → HO—[indole ring]—CH₂CH₂NH₂ + CO_2 with NH₂

5-Hydroxytryptophan   5-Hydroxytryptamine
                             (serotonin)

FIG. 31. Reactions of amino acid decarboxylation (*continued*).

discussed in a later section (p. 358). The biosynthesis of the vitamin is completely unknown. Hurwitz has recently shown that liver contains a system for oxidizing pyridoxine to pyridoxal, and has purified the kinase that phosphorylates pyridoxal.[10] This enzyme phosphorylates many analogs in addition to pyridoxal. It appears to be absolutely specific for ATP, and requires a divalent cation, either Mg^{++}, Fe^{++}, Co^{++}, Ni^{++}, or Zn^{++}. Other ions Fe^{+++}, Al^{+++}, Mn^{++}, and Ca^{++} inhibit the kinase.

Bacterial Amino Acid Decarboxylases. Tyrosine decarboxylase was used extensively in studies on the nature of the amino acid decarboxylases. This activity is found in several microorganisms and the enzyme was purified over 100-fold from *S. faecalis.*[11] It is easily resolved, and the apoenzyme has been used to assay pyridoxal phosphate (PALPO). Final concentrations of PALPO between 10^{-5} M and 10^{-4} M are conveniently measured by the rate of CO_2 liberation from tyrosine.[12] Phenylalanine is decarboxylated by this or a similar enzyme.[13] Tyrosine decarboxylase and all of the other amino acid decarboxylases described in this section act only on the L-isomers of their respective substrates.

The formation of agmatine from arginine is catalyzed by an enzyme obtained from acetone-dried *E. coli.*[14] This enzyme is resolved by ammonium sulfate precipitation. It has a pH optimum at 5.2 and a K_m of 7.5 $\times 10^{-4}$.

Histidine decarboxylase from *E. coli* has a pH optimum near 4.0,[15] while a similar enzyme assayed in *Clostridium welchii* has a pH optimum near 2.5.[16] Extracts of acetone- or dioxane-treated *C. welchii*, however, have maximal rates of histidine decarboxylation at pH 4.5. The reason for the difference in pH optima of intact cells and extracts is conventionally attributed to pH effects on cell permeability, but other factors may be responsible. It is noteworthy in this connection that the K_m of histidine is the same for both types of preparation. For years no cofactor could be found for this enzyme, but recently it was found that both pyridoxal phosphate and Fe^{+++} or Al^{+++} are required.[17]

Lysine is decarboxylated by a specific enzyme from *E. coli.*[5] In the cell, a pH optimum of 4.5 was found, but cell-free extracts showed an optimum at pH 6.0. This enzyme is resolved by ammoniacal ammonium sulfate. With $C^{14}O_2$, Hanke and Siddiqui demonstrated the reversibility

[10] J. Hurwitz, *J. Biol. Chem.* **205,** 935 (1953).
[11] H. M. R. Epps, *Biochem. J.* **38,** 242 (1944).
[12] W. W. Umbreit, W. D. Bellamy, and I. C. Gunsalus, *Arch. Biochem.* **7,** 185 (1945).
[13] R. W. McGilvery and P. P. Cohen, *J. Biol. Chem.* **174,** 814 (1948).
[14] E. S. Taylor and E. F. Gale, *Biochem. J.* **39,** 52 (1945).
[15] E. F. Gale, *Biochem. J.* **34,** 392 (1940).
[16] H. M. R. Epps, *Biochem. J.* **39,** 42 (1945).
[17] B. M. Guirard and E. E. Snell, *J. Am. Chem. Soc.* **76,** 4745 (1954).

of the decarboxylation,[18] which, like all the amino acid decarboxylations, proceeds essentially to completion.

L-Lysine is produced in some organisms by decarboxylation of *meso*-diaminopimelic acid.[19] This enzyme has a pH optimum over 7, in contrast with the acidic pH optima of the other bacterial amino acid decarboxylases. It is extremely specific; it does not attack higher or lower homologs, or compounds in which the methylene carbons bear a methyl or hydroxyl group. This has also been shown to be a PALPO-requiring enzyme.

Ornithine decarboxylation has been studied with preparations from *E. coli*[5] and *C. septicum*.[14] Since the latter organism does not attack any other amino acids, the specificity of this enzyme appears to be very great. Partical resolution of enzyme and coenzyme occurs in extracts of *C. septicum* on standing.

Bacterial glutamic carboxylase has been used extensively for the assay of glutamic acid. A remarkably stable enzyme was purified from *E. coli* by Najjar and Fisher.[20] Lyophilized *C. welchii* has been a very convenient source of this activity.[21] The action of this PALPO-requiring enzyme is to remove the α-carboxyl group, leaving γ-aminobutyric acid. Evidence has been presented to show that the decarboxylation of β-hydroxyglutamic acid by several bacteria is catalyzed by an enzyme different from glutamic carboxylase.[22] γ-Methylene glutamic acid is decarboxylated by several plant extracts.[23] Glutamic decarboxylase is widely distributed in plants, but other amino acid decarboxylases have not been obtained from plants.

Aspartic acid undergoes two different reactions; in some bacteria it loses its α-carboxyl group to form β-alanine in a reaction similar to those previously described for other amino acids.[24] In *C. welchii*, a β-decarboxylation occurs to form α-alanine.[25] This reaction can be inhibited by the cationic detergent, cetyl trimethylammonium bromide; since glutamic decarboxylase is not inhibited by the cationic detergent, this inhibitor is useful in obtaining a specific system for determining glutamic acid in the presence of aspartic. Aspartic acid, of course, can be determined as the difference in CO_2 produced by the uninhibited and inhibited preparations.

[18] M. E. Hanke and M. S. H. Siddiqui, *Federation Proc.* **9**, 181 (1950).
[19] D. L. Dewey, D. S. Hoare, and E. Work, *Biochem. J.* **58**, 523 (1954); D. S. Hoare, *ibid.* **59**, xxii (1955).
[20] V. Najjar and J. Fisher, *J. Biol. Chem.* **206**, 215 (1954).
[21] A. Meister, H. Sober, and S. V. Tice, *J. Biol. Chem.* **189**, 591 (1951).
[22] W. W. Umbreit and P. Heneage, *J. Biol. Chem.* **201**, 15 (1953).
[23] L. Fowden and J. Done, *Biochem. J.* **55**, 548 (1953).
[24] D. Billen and H. C. Lichstein, *J. Bacteriol.* **58**, 215 (1949).
[25] A. Meister, H. Sober, and S. V. Tice, *J. Biol. Chem.* **189**, 577 (1951).

Mammalian Amino Acid Decarboxylases. Amino acid decarboxylases occur in many animal tissues. Histidine decarboxylase of kidney has a pH optimum near 9.[26] The decarboxylation product, histamine, has very powerful pharmacological effects in animals, and its formation in minute quantities has been measured pharmacologically. This nonspecific type of assay has also been used to detect histidine decarboxylase in other tissues. More specific assays have recently shown this enzyme to occur in mast cells,[27] in which histamine accumulates; the stored histamine is released by rupture of these fragile cells.

A tyrosine decarboxylase of animal tissues was separated from other decarboxylases by kaolin adsorption.[28] This treatment separates tyrosine decarboxylase from a very active dopa decarboxylase,[29] which occurs in many tissues. This latter enzyme also attacks certain other hydroxylated phenylalanines. It has been resolved and used for assay of phosphorylated pyridoxal. Crude preparations are sufficiently active to permit the reaction to be followed with conventional manometric apparatus for the determination of liberated CO_2.

Pharmacological evidence was obtained several years ago that indicated that tryptophan is decarboxylated to tryptamine by both animal and bacterial enzymes. More recent studies have failed to detect this reaction, but instead have shown decarboxylation to occur only after oxidation of the indole nucleus to yield 5-hydroxytryptophan.[30] Decarboxylation of 5-hydroxytryptophan produces 5-hydroxytryptamine, serotonin, which has important, though incompletely defined functions in animal physiology. In some animal livers there is an enzyme that decarboxylates cysteic acid to taurine.[31] Glutamic decarboxylase has been found in animal brain,[32] where it is responsible for the formation of γ-aminobutyric acid. This product has been implicated in nervous function as an inhibitor of synaptic transmission.[33]

The enzymes listed above are a family of highly specific decarboxylases. Although several preparations have not shown any cofactor requirements, the drastic measures required to resolve some of these enzymes leave the burden of proof upon those who would have an amino acid decarboxylase act without pyridoxal phosphate.

[26] E. Werle and K. Heitzer, *Biochem. Z.* **299,** 420 (1938).
[27] R. W. Schayer, *Am. J. Physiol.* **186,** 199 (1956).
[28] P. Holtz, K. Credner, and H. Walter, *Z. physiol. Chem.* **262,** 111 (1939).
[29] P. Holtz, R. Heise, and K. Ludtke, *Arch. exptl. Pathol. Pharmakol. Naunyn-Schmiedeberg's* **191,** 87 (1938).
[30] C. T. Clark, H. Weissbach, and S. Udenfriend, *J. Biol. Chem.* **210,** 139 (1954).
[31] H. Blaschko, *Biochem. J.* **36,** 571 (1942); W. J. Wingo and J. Awapara, *J. Biol. Chem.* **187,** 267 (1950).
[32] E. Roberts and S. Frenkel, *J. Biol. Chem.* **188,** 789 (1951).
[33] A. Bazemore, K. A. C. Elliott, and E. Florey, *Nature* **178,** 1052 (1956).

GENERAL REFERENCES

Blaschko, H. (1945). *Advances in Enzymol.* **5**, 67.
Gale, E. F. (1946). *Advances in Enzymol.* **6**, 1.
Gunsalus, I. C. (1950). *Federation Proc.* **9**, 556.
Meister, A. (1955). "Symposium on Amino Acid Metabolism" (W. D. McElroy and
    B. Glass, eds.), p. 3. Johns Hopkins Press, Baltimore.
Schales, O. (1951). *In* "The Enzymes" (J. B. Sumner and K. Myrbäck, eds.), Vol. II,
    Part 1, p. 216. Academic Press, New York.
Snell, E. E. (1953). *Physiol. Revs.* **33**, 509.

Transamination

The reversible transfer of an amino group from an amino acid to a keto acid was first described by Braunstein and Kritsman,[1] who named this type of enzyme *aminopherase*.[2] This term has been superseded by the generally accepted *transaminase*.[3] Early work established the existence in heart muscle of two transaminases, catalyzing the reactions shown in (I). Claims for a third enzyme, an aspartic-alanine transaminase, were

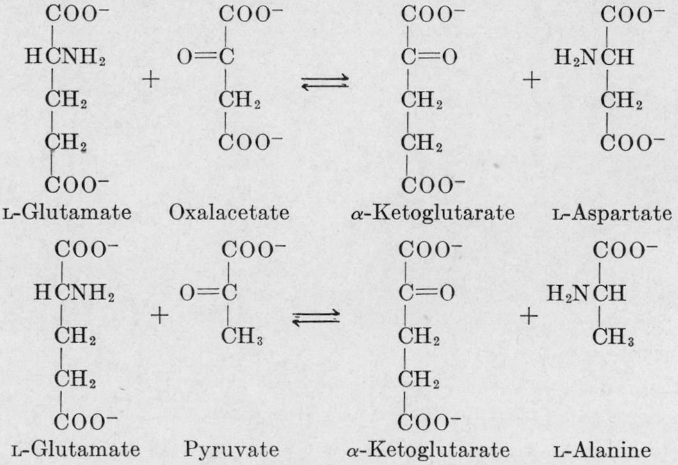

$$
\begin{array}{cccc}
\text{COO}^- & \text{COO}^- & \text{COO}^- & \text{COO}^- \\
| & | & | & | \\
\text{HCNH}_2 & \text{O}{=}\text{C} & \text{C}{=}\text{O} & \text{H}_2\text{NCH} \\
| \quad + & | \quad \rightleftharpoons & | \quad + & | \\
\text{CH}_2 & \text{CH}_2 & \text{CH}_2 & \text{CH}_2 \\
| & | & | & | \\
\text{CH}_2 & \text{COO}^- & \text{CH}_2 & \text{COO}^- \\
| & & | & \\
\text{COO}^- & & \text{COO}^- & \\
\text{L-Glutamate} & \text{Oxalacetate} & \alpha\text{-Ketoglutarate} & \text{L-Aspartate}
\end{array}
$$

$$
\begin{array}{cccc}
\text{COO}^- & \text{COO}^- & \text{COO}^- & \text{COO}^- \\
| & | & | & | \\
\text{HCNH}_2 & \text{O}{=}\text{C} & \text{C}{=}\text{O} & \text{H}_2\text{NCH} \\
| \quad + & | \quad \rightleftharpoons & | \quad + & | \\
\text{CH}_2 & \text{CH}_3 & \text{CH}_2 & \text{CH}_3 \\
| & & | & \\
\text{CH}_2 & & \text{CH}_2 & \\
| & & | & \\
\text{COO}^- & & \text{COO}^- & \\
\text{L-Glutamate} & \text{Pyruvate} & \alpha\text{-Ketoglutarate} & \text{L-Alanine}
\end{array}
$$

(I)

interpreted as indicating the sum of the two reactions above in the presence of catalytic amounts of glutamic acid. After several years, during which the nature of the transamination reaction as a pyridoxal phosphate system was established, the existence of many additional transamination reactions was established.

[1] A. E. Braunstein and M. G. Kritsman, *Enzymologia* **2**, 129 (1937).
[2] A. E. Braunstein, *Enzymologia* **7**, 25 (1939).
[3] P. P. Cohen, *J. Biol. Chem.* **136**, 565 (1940).

Assay of Transamination. Since the sum of keto acids and amino acids does not change in a transamination, specific reactions are required to assay the reaction products. Some of the methods used are: oxidation of α-ketoglutarate to succinate and determination of succinate with succinic dehydrogenase; decarboxylation of oxalacetate with aniline citrate; decarboxylation with specific amino acid decarboxylases; separation of products on paper chromatograms; and spectrophotometric determination of those keto acids that exhibit specific absorption.

Number of Transaminases. Feldman and Gunsalus found a large number of amino acids to transaminate with α-ketoglutarate in the presence

$$
\underset{\text{Histidinol phosphate}}{
\begin{array}{c}
\text{HC}=\!\!\!=\text{C}-\text{CH}_2-\text{CH}-\text{CH}_2-\text{O}-\overset{\overset{\text{O}}{\|}}{\underset{\underset{\text{O}^-}{|}}{\text{P}}}-\text{O}^- \\
\text{HN}\quad\text{N}\qquad\ \ \text{NH}_2 \\
\ \ \ \ \diagdown\ \diagup \\
\ \ \ \ \text{C} \\
\ \ \ \ \text{H}
\end{array}}
\ +\
\begin{array}{c}
\text{COO}^- \\
|\\
\text{C}=\text{O} \\
|\\
\text{CH}_2 \\
|\\
\text{CH}_2 \\
|\\
\text{COO}^-
\end{array}
\ \rightleftharpoons
$$

$$
\underset{\substack{\text{Imidazole acetol phosphate}\\ \text{(II)}}}{
\begin{array}{c}
\text{HC}=\!\!\!=\text{C}-\text{CH}_2-\overset{\overset{\text{O}}{\|}}{\text{C}}-\text{CH}_2-\text{O}-\overset{\overset{\text{O}}{\|}}{\underset{\underset{\text{O}^-}{|}}{\text{P}}}-\text{O}^- \\
\text{HN}\quad\text{N} \\
\ \ \ \ \diagdown\ \diagup \\
\ \ \ \ \text{C} \\
\ \ \ \ \text{H}
\end{array}}
\ +\
\begin{array}{c}
\text{COO}^- \\
|\\
\text{CHNH}_2 \\
|\\
\text{CH}_2 \\
|\\
\text{COO}^-
\end{array}
$$

of dried bacteria.[4] Similarly, Cammarata and Cohen found glutamate produced when animal tissue extracts were incubated with α-ketoglutarate and any of many amino acids.[5] The varying ratios of activities in various preparations led to the suggestion that many specific transaminases occur in animals.

At least three distinct transaminases with interesting substrate specificities occur in *E. coli*.[6] Rudman and Meister separated one fraction that used the following amino and corresponding keto acids: glutamic acid, aspartic acid, tryptophan, phenylalanine, and tyrosine. Another fraction uses glutamic acid and amino acids with aliphatic side chains. A third uses only valine, alanine or α-aminobutyric acid, and the corresponding keto acids. The third enzyme is unusual in its failure to react with glutamic acid, which is an active reagent with most transaminases, and is

[4] L. I. Feldman and I. C. Gunsalus, *J. Biol. Chem.* **187,** 821 (1950).
[5] P. S. Cammarata and P. P. Cohen, *J. Biol. Chem.* **187,** 439 (1950).
[6] D. Rudman and A. Meister, *J. Biol. Chem.* **200,** 591 (1953).

commonly used in exploring for the existence of transaminases. A preparation from *Brucella abortus* was also found to contain a transaminase that reacts with alanine and leucine but not with glutamic acid.[7]

A transaminase isolated from *Neurospora* reacts with certain substrates which do not contain carboxyl groups, but have phosphate esterified to the carbon adjacent to that bearing the amino group (II).[8] Histidinol phosphate, histidine, arginine, glutamic acid and their keto analogs react with this enzyme.

Transamination with Glutamine and Asparagine. Meister and collaborators have described two types of transaminase in which the amino donor is glutamine or asparagine. Transamination from glutamine to any of more than 30 α-keto acids leads to the formation of α-ketoglutaramic acid, which is hydrolyzed by a specific amidase (III).[9] The reaction

$$
\begin{array}{c}
\underset{\text{Glutamine}}{
\begin{array}{c}
\text{O} \\
\parallel \\
\text{C--NH}_2 \\
\mid \\
\text{CH}_2 \\
\mid \\
\text{CH}_2 \\
\mid \\
\text{CHNH}_2 \\
\mid \\
\text{C=O} \\
\mid \\
\text{O}^-
\end{array}}
\;+\;
\begin{array}{c}
\text{R} \\
\mid \\
\text{C=O} \\
\mid \\
\text{C=O} \\
\mid \\
\text{O}^-
\end{array}
\longrightarrow
\begin{array}{c}
\text{R} \\
\mid \\
\text{CHNH}_2 \\
\mid \\
\text{C=O} \\
\mid \\
\text{O}^-
\end{array}
\;+\;
\underset{\text{Ketoglutaramic acid}}{
\begin{array}{c}
\text{O} \\
\parallel \\
\text{C--NH}_2 \\
\mid \\
\text{CH}_2 \\
\mid \\
\text{CH}_2 \\
\mid \\
\text{C=O} \\
\mid \\
\text{C=O} \\
\mid \\
\text{O}^-
\end{array}}
\;\underset{+\text{H}_2\text{O}}{\longrightarrow}\;
\begin{array}{c}
\text{O} \\
\parallel \\
\text{C--O}^- \\
\mid \\
\text{CH}_2 \\
\mid \\
\text{CH}_2 \\
\mid \\
\text{C=O} \\
\mid \\
\text{C=O} \\
\mid \\
\text{O}^-
\end{array}
\;+\;\text{NH}_4^+
\end{array}
$$

(III)

mechanism in the case of asparagine is similar.[10] Using high concentrations of α-ketosuccinamic acid, the reverse reaction leading to asparagine synthesis has been observed (IV).[11]

$$
\begin{array}{c}
\underset{\alpha\text{-Ketosuccinamate}}{
\begin{array}{c}
\text{O} \\
\parallel \\
\text{C--NH}_2 \\
\mid \\
\text{CH}_2 \\
\mid \\
\text{C=O} \\
\mid \\
\text{C=O} \\
\mid \\
\text{O}^-
\end{array}}
\;+\;
\begin{array}{c}
\text{R} \\
\mid \\
\text{CHNH}_2 \\
\mid \\
\text{C=O} \\
\mid \\
\text{O}^-
\end{array}
\;\rightleftharpoons\;
\underset{\text{Asparagine}}{
\begin{array}{c}
\text{O} \\
\parallel \\
\text{C--NH}_2 \\
\mid \\
\text{CH}_2 \\
\mid \\
\text{CHNH}_2 \\
\mid \\
\text{C=O} \\
\mid \\
\text{O}^-
\end{array}}
\;+\;
\begin{array}{c}
\text{R} \\
\mid \\
\text{C=O} \\
\mid \\
\text{C=O} \\
\mid \\
\text{O}^-
\end{array}
\end{array}
$$

(IV)

[7] R. A. Altenborn and R. D. Housewright, *J. Biol. Chem.* **204,** 159 (1953).
[8] B. N. Ames and B. L. Horecker, *J. Biol. Chem.* **220,** 113 (1956).
[9] A. Meister, *J. Biol. Chem.* **200,** 571 (1953).
[10] A. Meister, H. A. Sober, S. V. Tice, and P. E. Fraser, *J. Biol. Chem.* **197,** 319 (1952).
[11] A. Meister and P. E. Fraser, *J. Biol. Chem.* **210,** 37 (1954).

ω-Transamination. Another variation of the transamination pattern is transfer of ω-amino groups. Enzymes from *Neurospora*[12] and liver[13] convert ornithine to glutamic-γ-semialdehyde in the presence of various α-keto acids. Enzymes from brain, liver, and microorganisms have been found to form glutamate from α-ketoglutarate and either γ-amino butyrate or β-alanine.[14]

Specific Transaminases. Transamination reactions are important in the metabolism of tyrosine, kynurenine, proline, and lysine. These reactions will be discussed later. Transamination in the synthesis of glucosamine has already been mentioned.

Pyridoxamine Phosphate in Transamination. In general the transaminases are easily shown to use PALPO as a cofactor. The notable exception at this time is the glutamine system, which has no known cofactor. In considering probable mechanisms of action of this cofactor, it seemed reasonable to postulate the formation of pyridoxamine phosphate (PAMPO). The early attempts to show a cofactor function for PAMPO were discouragingly negative when the classical pig heart glutamic-aspartic transaminase was used, although the cruder liver and bacterial enzymes used both PALPO and PAMPO. With crystalline PAMPO, it was shown that a period of preincubation is required to form an active enzyme, and that the same ultimate activity can be reached when either PALPO or PAMPO is added to pig heart glutamic-aspartic transaminase; the saturation curves for the two can be superimposed.[15]

D-*Transaminases.* All of the reactions described above are specific for the L-isomers of the amino acids. Certain microoganisms have been found to contain transaminases specific for D-amino acids. A specific D-transaminase was purified from *Bacillus subtilis*.[16] This enzyme has been shown to react with a dicarboxylic acid and any of several amino acids. It requires PALPO.

General Properties of Transaminases. Transaminases appear to require the complete amino acid structure. Peptides have not been found to participate in transaminations. The equilibria of transaminations in general indicate very little standard free-energy change. The one conspicuous exception is the formation of glycine from glyoxylate, which proceeds almost to completion, and is reversed with great difficulty.

[12] J. R. S. Fincham, *Biochem. J.* **53**, 313 (1953).
[13] A. Meister, *J. Biol. Chem.* **206**, 587 (1954).
[14] S. P. Bessman, J. Rossen, and E. C. Layne, *J. Biol. Chem.* **201**, 385 (1953); E. Roberts, P. Ayengar, and I. Posner, *ibid.* **203**, 195 (1953).
[15] A. Meister, H. A. Sober, and E. A. Peterson, *J. Biol. Chem.* **206**, 89 (1954).
[16] C. B. Thorne, *Symposium on Amino Acid Metabolism, Baltimore, 1954* p. 41 (1955).

GENERAL REFERENCES

Cohen, P. P. (1951). *In* "The Enzymes" (J. B. Sumner and K. Myrbäck, eds.), Vol. I, Part 2, p. 1040. Academic Press, New York.

Meister, A. (1955). *Advances in Enzymol.* **16**, 185.

Meister, A. (1955). "Symposium on Amino Acid Metabolism" (W. D. McElroy and B. Glass, eds.), p. 3. Johns Hopkins Press, Baltimore.

Oxidation of Amino Acids

Amino acids are readily oxidized by many organisms. The mechanisms used may be complex, involving transamination, deamination, or other alterations before the oxidative step. There are in addition reactions in which amino acids serve as the immediate substrates for oxidative enzymes. A widely distributed group of enzymes is composed of relatively nonspecific oxidases. These will be considered before the specific oxidases and dehydrogenases.

The nonspecific oxidases were first detected by Krebs in animal liver and kidney.[1] Both D- and L-amino acids are oxidized by crude preparations. Two types of activity were distinguished: L-amino acid oxidation, which is sensitive to HCN and to octanol, and D-amino acid oxidation, which is not inhibited by these reagents. The former is associated with tissue particles, while the latter is soluble.

D-*Amino Acid Oxidase.* The D-amino acid oxidizing activity is the more active and, under ordinary handling, the more stable. Purification studies indicate that a single enzyme, D-amino acid oxidase, is responsible for essentially all oxidation of D-amino acids in animal preparations, although specific enzymes appear to attack D-aspartic and D-glutamic acids. D-Amino acid oxidase was the first enzyme to be resolved into a protein and flavin adenine dinucleotide (FAD).[2] The enzyme, precipitated with acid ammonium sulfate, is active only in the presence of FAD, and is used for both qualitative and quantitative determination of this cofactor.

The reaction catalyzed by D-amino acid oxidase is:

$$RCHCOO^- + O_2 + H_2O \rightarrow RCOCOO^- + R'NH_2 + H_2O_2$$
$$\underset{\underset{R'}{|}}{\overset{|}{NH}}$$

[1] H. A. Krebs, *Biochem. J.* **29**, 1620 (1935).

[2] O. Warburg and W. Christian, *Biochem. Z.* **295**, 261 (1938); **296**, 294 (1938); **297**, 447 (1938); **298**, 150 (1938).

Obviously a number of assay methods could be used to measure disappearance of amino acid, appearance of keto acid, appearance of ammonia, or appearance of H_2O_2. The traditional method, manometric determination of oxygen consumption, is most frequently employed. This reaction is useful for the determination of D-amino acids, although the determination is not very sensitive as the K_m is over 10^{-3}. The reaction has a pH optimum near 9.

The available information on relative rates of reaction is very difficult to interpret in terms of the nature of enzyme–substrate interaction. The substrate oxidized most rapidly is tyrosine, but many other amino acids with cyclic side chains are also attacked. The aromatic groups may be those of naturally occurring compounds, but synthetic compounds, such as quinolyl-, furyl-, and pyridylalanine, also serve as substrates.[3] The rates are markedly affected by substitutions in the rings; tyrosine is the most rapidly oxidized, whereas phenylalanine is only attacked at moderate rates. Polybasic amino acids are attacked slowly and polyacidic compounds, including glutamic acid, are very poor substrates if they react at all, unless the acidic groups are well separated in a molecule, such as 1-aminohendecane-1,11-dicarboxylic acid. Small-chain aliphatic compounds are also oxidized efficiently; in this series the rate is highest with alanine and decreases with increasing chain length (see Table 7).

The reaction mechanism most often proposed is a transfer of hydrogen atoms from the amino group and the α-carbon to the flavin, leaving an imino acid, which is thought to hydrolyze spontaneously. While there is no direct evidence for the intermediate formation of an imino acid, support for this hypothesis is found in the requirement for a hydrogen atom on both the nitrogen and α-carbon atoms. N-Monomethyl amino acids and proline are substrates, but neither N-dimethyl compounds nor compounds such as α-aminoisobutyric acid are attacked. Acylated amino acids, including peptides, are not substrates.[4]

Like most flavoproteins, D-amino acid oxidase is inhibited by sulfhydryl reagents sch as p-chloromercuribenzoate.[5] The phenylcarboxylic acids form a group of very efficient competitive inhibitors with respect to substrate.[6] A group of heterocylcic compounds, including quinine and atabrine, appear to compete with the coenzyme.[7]

D-Amino acid oxidases have been found in extracts of mycelia of many

[3] A. E. Bender and H. A. Krebs, *Biochem. J.* **46,** 210 (1950).
[4] Evidence from several laboratories has been summarized in the review by Krebs. (See General References.)
[5] T. P. Singer and E. S. G. Barron, *J. Biol. Chem.* **157,** 241 (1945).
[6] G. R. Bartlett, *J. Am. Chem. Soc.* **70,** 1010 (1948).
[7] L. Hellerman, H. Lindsay, and M. R. Bovarnick, *J. Biol. Chem.* **163,** 553 (1946).

TABLE 7

RELATIVE RATES OF OXIDATION OF D-AMINO ACIDS BY SHEEP KIDNEY
D-AMINO ACID OXIDASES[a,b]

Amino acid	O_2 absorbed (μl./mg. dry matter of enzyme prep./hr.)
STRAIGHT-CHAIN α-AMINOMONOCARBOXYLIC ACIDS	
Alanine	64
α-Aminobutyric	31
α-Aminovaleric	20
α-Aminocaproic	36
α-Aminocaprylic	1.9
α-Aminohendecanoic	0
α-Aminolauric	0
α-Aminostearic	0
BRANCHED-CHAIN ALIPHATIC α-AMINOMONOCARBOXYLIC ACIDS	
Valine	35
Leucine	13.9
Isoleucine	22
ALIPHATIC α-AMINODICARBOXYLIC ACIDS	
Aspartic	1.4
Glutamic	0
α-Aminoadipic	0
α-Aminopimelic	0
1-Aminohendecane-1,11-dicarboxylic	41
α-AMINO-ACIDS WITH CYCLIC SUBSTITUENTS	
Phenylalanine	26
Tyrosine	190
p-Aminophenylalanine	65
p-Dimethylaminophenylalanine	32
p-(β-Aminoethyl)phenylalanine	16
p-Guanidinomethylphenylalanine	33
β-Quinolyl(4)alanine	8.3
β-Quinolyl(2)alanine	59
β-6-Methoxyquinolyl(4)alanine	27
β-Pyridyl(4)alanine	95
β-Furyl(2)alanine	11.9
ϵ-N-Sulfanilyllysine	0
Histidine	6.2
Tryptophan	37
DIAMINOMONOCARBOXYLIC ACIDS	
Lysine	0.6
Ornithine	3.1
UNCLASSIFIED α-AMINO ACIDS	
Serine	42
Threonine	2.1
Proline	148
Piperidine-α-carboxylic	2.6
Methionine	80
Cystine	1.9

[a] From H. A. Krebs, "The Enzymes: Chemistry and Mechanism of Action" (J. B. Sumner and K. Myrbäck, eds.), Vol. II, Part 1, p. 506. Academic Press, New York, 1951.
[b] The values were obtained with arbitrary concentrations of substrates, and do not indicate variations in K_m. The relative rates are none-the-less valuable for practical use of the enzyme, and probably will parallel k_3 values, when this rate constant is measured for the various substrates.

molds.[8] While the mold enzymes appear to resemble the animal enzyme in general, there are many details in which they differ. The relative activity of the *Neurospora* enzyme for the various amino acids is quite different from the values given by the kidney enzyme. The former does not attack polybasic amino acids, tryptophan, or quinolylalanine, but, on the other hand, oxidizes D-glutamic acid. Proline also is not a substrate for the *Neurospora* enzyme. Benzoate does not inhibit this preparation, and the pH optimum is a little lower than that of the kidney enzyme, between pH 8.0 and 8.5. The relative rates reported for *Neurospora* also differ somewhat from the values reported for other molds. It should be noted, however, that the mold enzymes have all been studied in quite crude preparations. D-Amino acid oxidase associated with insoluble particles of the bacterium *Proteus morganii* has also been reported.[9]

Although D-amino acids have recently been found in a number of naturally occurring materials, the total amounts are modest, and the physiological role of D-amino acid oxidase remains obscure.

L-*Amino Acid Oxidase.* In 1944 L-amino acid oxidases were described in 3 sources: animal kidney,[10] snake venom,[11] and bacteria.[12] The enzyme was also detected in animal liver. The purification from rat kidney required large amounts of tissue because of the low activity. A 200-fold purification gave a preparation believed to be essentially pure. This protein has a molecular weight somewhat over 120,000, and contains 2 flavin mononucleotides per mole. If the protein isolated is indeed the enzyme, it has one of the lowest turnover numbers determined; only 6 molecules of amino acid are oxidized per molecule of protein per minute, whereas D-amino acid oxidase has a turnover number of more than 1000.

L-Amino acid oxidase of rat kidney appears to catalyze the same type of reaction as D-amino acid oxidase. It oxidizes the monoamino, monocarboxylic acids, but does not attack either dicarboxylic or polybasic compounds. Proline and *N*-methylamino acids are oxidized. A unique property of this enzyme is its ability to oxidize hydroxy acids.[13] Only L-hydroxy acids are attacked. The ratio of hydroxy acid oxidation rate to amino acid rate is constant throughout purification. In general, hydroxy acid oxidation proceeds somewhat more rapidly than amino acid oxidation, and the hydroxy acids corresponding to the basic amino acids

[8] N. H. Horowitz, *J. Biol. Chem.* **154**, 141 (1944); R. L. Emerson, M. Puziss, and S. G. Knight, *Arch. Biochem.* **25**, 299 (1950).

[9] P. K. Stumpf and D. E. Green, *Federation Proc.* **5**, 157 (1946).

[10] M. Blanchard, D. E. Green, V. Nocito, and S. Ratner, *J. Biol. Chem.* **155**, 421 (1944).

[11] E. A. Zeller and A. Maritz, *Helv. Chim. Acta* **27**, 1888 (1944).

[12] P. K. Stumpf and D. E. Green, *J. Biol. Chem.* **153**, 387 (1944).

[13] M. Blanchard, D. E. Green, V. Nocito, and S. Ratner, *J. Biol. Chem.* **161**, 583 (1945).

are oxidized slowly, whereas the amino acids themselves are not substrates. The requirement for a hydrogen atom on the α-carbon is characteristic of this enzyme also; α-hydroxyisobutyric acid is not oxidized.

L-Amino acid oxidase activity has been found in the venom of many species of snakes, and also in various other organs of these animals. The enzyme has been purified 16-fold from dried water moccasin venom.[14] The purified enzyme was homogeneous by several physical criteria, and appears to be a protein of 62,000 molecular weight containing one FAD per mole. The specificities of enzymes from various species are not identical. In general only monocarboxylic, monobasic compounds are oxidized, and, in contrast to findings with the animal enzyme, proline is not a substrate.

L-Amino acid oxidation is catalyzed by preparations from various molds, but little is known about most of these. *Neurospora* secretes an enzyme into the medium under certain conditions.[15] If only one enzyme is present, it has a broader range of activity than any of the other L-amino acid oxidases, since it attacks all of the naturally occurring amino acids except proline.

The ability to oxidize L-amino acids is widespread among bacteria, but the enzymes responsible have not been well characterized. *Proteus vulgaris* is able to oxidize a large series of L-amino acids, but on storage at 0°C. loses the ability to oxidize 9 out of 20 amino acids tested. The relative rates of oxidation of the other 11 amino acids do not remain constant. An active cell-free preparation was obtained, but it is not known how many amino acid oxidases may exist in these organisms.[12]

Glycine Oxidase. Glycine, with two hydrogen atoms on the α-carbon, is optically inactive, and therefore is neither D nor L. L-Amino acid oxidase does not attack glycine although it does oxidize the corresponding hydroxy compound, glycolic acid. Glycine has been reported to be oxidized by a specific oxidase that accompanies D-amino acid oxidase throughout its purification.[16] The evidence for the existence of a glycine oxidase different from D-amino acid oxidase depends upon differential inactivation during removal of the flavin.

GENERAL REFERENCES

Krebs, H. A. (1951). *In* "The Enzymes" (J. B. Sumner and K. Myrbäck, eds.), Vol. II, Part 1, p. 499. Academic Press, New York.

Meister, A. (1955). "Symposium on Amino Acid Metabolism" (W. D. McElroy and B. Glass, eds.), p. 3. Johns Hopkins Press, Baltimore.

[14] T. P. Singer and E. B. Kearney, *Arch. Biochem. Biophys.* **27**, 348 (1950); **29**, 190 (1950).

[15] A. E. Bender, H. A. Krebs, and N. H. Horowitz, *Biochem. J.* **45**, xxi (1949).

[16] S. Ratner, V. Nocito, and D. E. Green, *J. Biol. Chem.* **152**, 119 (1944).

Glutamic Acid and Amino Acids Derived from It

GLUTAMIC DEHYDROGENASE

Glutamic dehydrogenase occupies a special position among the reactions of amino acids because of the importance of L-glutamic acid in metabolism. Although transaminases can produce amino acids without the participation of glutamic acid, enzymes that require this compound are more widespread, and appear to contribute most of the transamination *in vivo*. This is attested by the rapid equilibration of glutamic acid nitrogen with the nitrogen of labeled amino acids. Therefore the production of glutamic acid is an important step in the formation of other amino acids, and in the converse situation, in which amino acids are being degraded, glutamic acid formation is often a required preliminary step. The reaction catalyzed by glutamic dehydrogenase is thus of importance in the introduction and removal of amino groups of amino acids. α-Ketoglutarate links glutamic acid metabolism with the Krebs cycle (I).

$$
\begin{array}{ccc}
\text{COO}^- & & \text{COO}^- \\
| & & | \\
\text{CHNH}_3{}^+ & & \text{C}{=}\text{O} \\
| & & | \\
\text{CH}_2 & + \text{OH}^- + \text{DPN}^+ \rightleftharpoons & \text{CH}_2 + \text{NH}_4{}^+ + \text{DPNH} \\
| & & | \\
\text{CH}_2 & & \text{CH}_2 \\
| & & | \\
\text{COO}^- & & \text{COO}^- \\
\text{Glutamate} & & \alpha\text{-Ketoglutarate}
\end{array}
$$

$$(\text{I})$$

Glutamic dehydrogenase has been detected in bacteria, yeast, plants, and animal tissues.[1] The enzyme has been purified extensively only from liver, and the properties of the ezyme from other sources are not known very precisely. It has been reported that glutamic dehydrogenase of plants requires DPN, while the enzyme of yeast and *E. coli* requires TPN. The mammalian enzyme uses both coenzymes.

Liver glutamic dehydrogenase was crystallized simultaneously by Olson and Anfinsen[2] and by Strecker,[3] using entirely different procedures. The crystalline enzyme appears to be homogeneous and to have a molec-

---

[1] E. Adler, N. B. Das, H. von Euler, and U. Heyman, *Compt. rend. trav. lab. Carlsberg Sér. Chim.* **22,** 15 (1938); H. von Euler, E. Adler, and T. S. Eriksen, *Z. physiol. Chem.* **248,** 227 (1937); H. von Euler, E. Adler, G. Günther, and N. B. Das, *ibid.* **254,** 61 (1938).

[2] J. Olson and C. Anfinsen, *J. Biol. Chem.* **197,** 67 (1952); **202,** 841 (1953).

[3] H. J. Strecker, *Arch. Biochem. Biophys.* **46,** 128 (1953).

ular weight of 1,000,000. It is a simple protein with an isoelectric point between pH 4 and 5. There is no evidence for any bound nucleotide.

The reaction depends upon all of the components shown in equation (I). It has been proposed that an imino acid is an intermediate, but no evidence for the existence of such a compound has been reported. If a nonenzymatic reaction between ammonia and α-ketoglutarate were to form the true substrate, the apparent requirements for enzyme saturation should be equivalent for both compounds and the apparent requirement for each should depend on the concentration of the other, because the hypothetical imino acid would be formed in a second-order reaction of ammonia and the keto acid. Instead, a K_m of 1.2×10^{-4} was found for α-ketoglutarate and 5.7×10^{-2} for NH_4^+. Other K_m values for this enzyme are 2×10^{-3} for glutamate, 2.5×10^{-5} for DPN, and 1.8×10^{-5} for DPNH. These values are influenced by ions in the medium and pH, and are not fixed properties of the enzyme. As with other enzymes that use both DPN and TPN, glutamic dehydrogenase also acts with desamino DPN and acetylpyridine DPN.

Equilibrium measurements give an average value of 1.9×10^{-15} for the constant:

$$K = (\alpha\text{-ketoglutarate})(NH_4^+)(DPNH)(H^+)/(\text{glutamate})(H_2O)(DPN)$$

This extreme figure does not imply irreversibility, because the removal of the H^+ and H_2O terms leaves a figure near 10^{-5} at neutral pH, and the presence of an extra component in the numerator contributes to this deviation from unity. Actually, with high concentrations of ammonia the reaction does proceed essentially quantitatively in the direction of glutamate formation, but with low ammonia concentrations an extensive reaction can be measured starting with glutamate and DPN. The equilibria measured with TPN and desamino DPN differ slightly from that found with DPN, and indicate standard potentials of these nucleotides 0.005 volts more negative than that of the DPN:DPNH couple. The relative rates of reaction under optimal conditions with DPN, TPN, and desamino DPN were measured as 100, 60, and 35, respectively. The enzyme is absolutely specific for L-glutamic acid; change of chain length or substitution on any of the functional groups causes complete loss of activity.

Glutamic dehydrogenase participates in coupled reactions with other enzymes that occur naturally in the same tissues. Hunter has studied a dismutation of α-ketoglutarate in which one molecule is oxidized to succinate $+CO_2$ while the other is reduced to glutamate.[4] Bessman and collaborators have emphasized the metabolic significance of glutamic decar-

[4] F. E. Hunter and W. S. Hixon, *J. Biol. Chem.* **181**, 67 (1949).

boxylase in brain by demonstrating a transaminase that acts on two products of glutamate metabolism, γ-aminobutyric acid and α-ketoglutarate, to form succinic semialdehyde and a new molecule of glutamate.[5a]

Although most organisms depend heavily upon glutamic dehydrogenase for the incorporation of ammonia into amino acids, this enzyme is not the sole representative of its kind. A strain of *B. subtilis* that contains no glutamic dehydrogenase was found to contain an L-alanine dehydrogenase.[5b] The deamination catalyzed by alanine dehydrogenase is reversible; the enzyme is specific for DPN.

$$\begin{array}{c} \text{CH}_3 \\ | \\ \text{HCNH}_2 \\ | \\ \text{COO}^- \end{array} + \text{DPN}^+ + \text{H}_2\text{O} \rightleftharpoons \begin{array}{c} \text{CH}_3 \\ | \\ \text{C}=\text{O} \\ | \\ \text{COO}^- \end{array} + \text{NH}_3 + \text{DPNH} + \text{H}^+$$

INTERRELATIONS OF GLUTAMIC ACID, PROLINE, AND ORNITHINE

Nutritional and isotopic studies have shown that the 5-carbon amino acids, glutamic acid, proline, and ornithine, are capable of interconversion and that in different groups of organisms these amino acids do indeed replace each other.[6] In animals the reactions of these compounds

$$\begin{array}{ccc} \begin{array}{c} \text{COOH} \\ | \\ \text{CH}_2 \\ | \\ \text{CH}_2 \\ | \\ \text{CHNH}_2 \\ | \\ \text{COOH} \end{array} & \begin{array}{c} \text{CH}_2\text{---CH}_2 \\ | \qquad | \\ \text{CH}_2 \quad \text{CH---COOH} \\ \diagdown \ \diagup \\ \text{N} \\ \text{H} \end{array} & \begin{array}{c} \text{CH}_2\text{NH}_2 \\ | \\ \text{CH}_2 \\ | \\ \text{CH}_2 \\ | \\ \text{CHNH}_2 \\ | \\ \text{COOH} \end{array} \\ \text{Glutamic acid} & \text{Proline} & \text{Ornithine} \end{array}$$

have been inferred almost completely from the results of N^{15} and deuterium studies. The formation of proline from glutamic acid requires cyclization and reduction. Glutamic acid cyclizes readily to become pyrrolidonecarboxylic acid, but this compound is quite inert metabolically except in those organisms that hydrolyze it back to glutamic acid (II).[7] An analog of glutamic acid reduced to the proline level, α-amino-δ-hydroxyvaleric acid, also fails to act on a metabolite in several systems that use glutamic acid and proline, although oxidation to an active material is indicated by results obtained with *Neurospora*.[8]

[5a] S. P. Bessman, J. Rossen, and E. C. Layne, *J. Biol. Chem.* **201**, 385 (1953).

[5b] J. M. Wiame and A. Pierard, *Nature* **176**, 1073 (1956).

[6] See reviews by M. R. Stetten and H. J. Vogel (General References).

[7] E. Abderhalden, *Z. physiol. Chem.* **68**, 487 (1910); H. Weil-Malherbe and H. A. Krebs, *Biochem. J.* **29**, 2077 (1935).

[8] M. Neber, *Z. physiol. Chem.* **240**, 70 (1936); H. M. Srb, J. R. S. Fincham, and D. Bonner, *Am. J. Botany* **37**, 533 (1950).

$$\begin{array}{ccc}
\text{COOH} & & \text{CH}_2\!\!-\!\!\text{CH}_2 \\
| & & \\
\text{CH}_2 & \mp \text{H}_2\text{O} & \text{O}\!\!=\!\!\text{C}\quad\text{CH}\!\!-\!\!\text{COOH} \\
| & \rightleftharpoons & \\
\text{CH}_2 & & \text{N} \\
| & & \text{H} \\
\text{CHNH}_2 & & \\
| & & \\
\text{COOH} & &
\end{array}$$

Glutamic acid   Pyrrolidonecarboxylic acid
                              (II)

An active intermediate in the oxidation of proline by particles from rabbit kidney appears to be glutamic-γ-semialdehyde (III).[9] The iso-

$$\begin{array}{cccc}
\text{HC}\!\!=\!\!\text{O} & & \text{CH}_2\!\!-\!\!\text{CH}_2 & \text{CH}_2\!\!-\!\!\text{CH}_2 \\
| & & & \\
\text{CH}_2 & \rightleftharpoons & \text{HC}\quad\text{CH}\!\!-\!\!\text{COOH} & \text{H}_2\text{C}\quad\text{C}\!\!-\!\!\text{COOH} \\
| & & & \\
\text{CH}_2 & & \text{N} & \text{N} \\
| & & & \\
\text{CHNH}_2 & & & \\
| & & & \\
\text{COOH} & & &
\end{array}$$

Glutamic semialdehyde   Pyrroline-5-carboxylic acid   Pyrroline-1-carboxylic
                                                              acid
                              (III)

meric α-keto-δ-aminovaleric acid does seem to be formed in the oxidation of proline or ornithine, but does not fulfill the role of an intermediate in amino acid interconversions.[10] Both of these compounds cyclize readily to form pyrrolinecarboxylic acids, differing only in the position of the unsaturated bond. N^{15} experiments have indicated that *in vivo* these compounds do not equilibrate, so that interconversion at this level through tautomerism does not occur.[11]

Proline Biosynthesis. Pathways of interconversion have been established in detail for microorganisms. A series of *E. coli* mutants was found to support the following sequence:[9] glutamate → glutamic-γ-semi-aldehyde → proline. Glutamic semialdehyde has been shown to exist largely as the cyclized Δ^1-pyrroline-5-carboxylate under physiological conditions. Therefore the only reactions that require defining are the oxidation-reduction steps. The mechanism of the reduction of the γ-carboxyl group of glutamic acid is not known. From analogy with other systems, it seems likely that the large positive standard free energy of carboxyl reductions precludes a direct reduction and that an acyl derivative is the true substrate. The enzyme has not been studied yet, and no requirements for this reaction are known. The reduction of the pyrroline ring has been

[9] H. J. Vogel and B. D. Davis, *J. Am. Chem. Soc.* **74,** 109 (1952).
[10] H. A. Krebs, *Enzymologia* **7,** 53 (1939); J. V. Taggart and R. B. Krakauer, *J. Biol. Chem.* **177,** 641 (1949).
[11] M. R. Stetten, *J. Biol. Chem.* **189,** 499 (1951).

studied with preparations from *Neurospora*[12] and from rat liver.[13] The *Neurospora* enzyme uses either DPNH or TPNH as a source of electrons whereas the animal enzyme uses only DPNH.

Ornithine Biosynthesis. Two reaction sequences leading to ornithine synthesis from glutamic acid have been described. In *Neurospora*[14] and the yeast *Torulopsis utilis*[15] the sequence involves the simple transamination of glutamic semialdehyde (IV). The reaction proceeds much

$$
\begin{array}{cccc}
HC{=}O & COO^- & CH_2NH_3{}^+ & COO^- \\
| & | & | & | \\
CH_2 & CH_2 & CH_2 & CH_2 \\
| & +\quad | & \rightleftharpoons \quad | & +\quad | \\
CH_2 & CH_2 & CH_2 & CH_2 \\
| & | & | & | \\
CHNH_3{}^+ & CHNH_3{}^+ & CHNH_3{}^+ & C{=}O \\
| & | & | & | \\
COO^- & COO^- & COO^- & COO^- \\
\text{Glutamic semialdehyde} & \text{Glutamate} & \text{Ornithine} & \text{α-Ketoglutarate}
\end{array}
$$

(IV)

more favorably in the direction: ornithine $+$ α-ketoglutarate $\rightarrow$ glutamic semialdehyde $+$ glutamate, but this is a consequence of the cyclization of the semialdehyde.

A more elaborate pathway has been described for *E. coli.*[16] In this organism an enzyme that uses acetyl CoA can form the N-acetyl derivative of glutamic acid.[17] Only crude extracts were used in the experiments reported, and these also acetylate many other amino acids; aspartic acid is acetylated to a greater extent than glutamic. Mutant studies show that acetylglutamic acid is reduced to acetylglutamic semialdehyde, but the mechanism of this reaction is as obscure as the reduction of the free amino acid. N-Acetylglutamic semialdehyde participates in a transamination reaction with glutamic acid to yield α-N-acetylornithine and α-ketoglutarate. The enzyme responsible has been named acetylornithine δ-transaminase, and has been shown to require pyridoxal phosphate.[16]

$$
\begin{array}{ll}
COO^- & CH_2{-}NH_3{}^+ \\
| & | \\
CH_2 & CH_2 \\
| & | \\
CH_2 \quad\;\; O & CH_2 \quad\;\; O \\
|\quad H \;\; \| & |\quad H \;\; \| \\
CH{-}N{-}C{-}CH_3 & CH{-}N{-}C{-}CH_3 \\
| & | \\
COO^- & COO^- \\
\text{Acetylglutamate} & \text{α-N-Acetylornithine}
\end{array}
$$

[12] T. Yura and H. J. Vogel, *Biochim. et Biophys. Acta* **12**, 582 (1955).
[13] M. E. Smith and D. M. Greenberg, *Nature* **177**, 1130 (1956).
[14] J. R. S. Fincham, *Biochem. J.* **53**, 313 (1953).
[15] P. H. Abelson and H. J. Vogel, *J. Biol. Chem.* **213**, 355 (1955).
[16] H. J. Vogel, *Proc. Natl. Acad. Sci.* **39**, 578 (1953).
[17] W. K. Mass, G. D. Novelli, and F. Lipmann, *Proc. Natl. Acad. Sci.* **39**, 1004 (1953).

Acetylornithine is hydrolyzed by an enzyme named acetylornithinase.[16] This activity is stimulated specifically by cobaltous ions.

METABOLISM OF GLUTAMINE

Speck first demonstrated the synthesis of glutamine from glutamic acid and ammonia with extracts of pigeon liver and showed that ATP is required.[18] Elliott studied this reaction in peas, and purified the enzyme involved 2000-fold without finding any evidence for more than a single catalyst.[19] In addition to the synthetic reaction, this enzyme also catalyzes transfer reactions, in which the γ-glutamyl group of glutamine is transferred to hydroxylamine, hydrazine, or ammonia (an exchange reaction studied with N^{15}).[20,21] The two activities have been called glutamine synthetase and glutamyl transferase, but both are manifestations of a single protein (V). A very similar enzyme from *Proteus vulgaris* also catalyzes both the synthetase and transferase reactions, but does not require cofactors for the transferase.[22] This activity is stimulated by cupric ions.

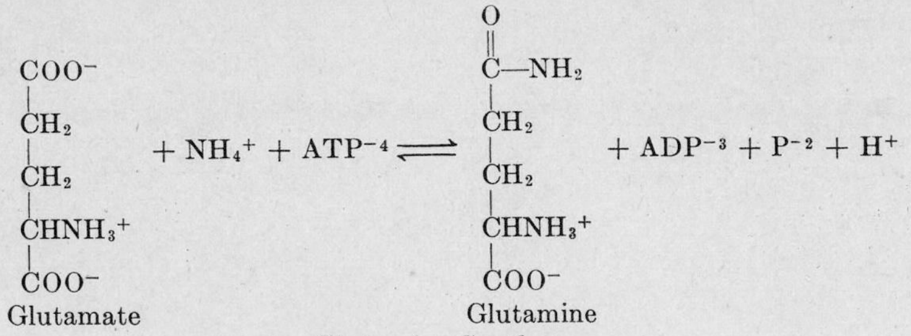

Glutamine Synthetase

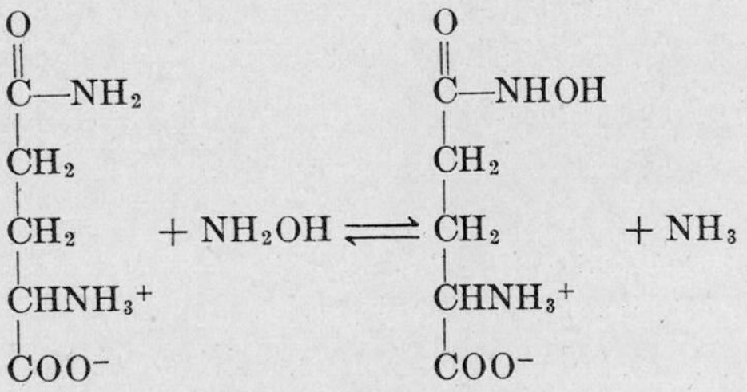

Glutamyl Transferase
(V)

[18] J. F. Speck, *J. Biol. Chem.* **168**, 403 (1947); **179**, 1387 (1949).

[19] W. H. Elliott, *J. Biol. Chem.* **201**, 661 (1953).

[20] P. K. Stumpf and W. D. Loomis, *Arch. Biochem.* **25**, 451 (1950); M. Schon, N. Grossowicz, A. Lajtha, and H. Waelsch, *Nature* **167**, 891 (1951).

[21] L. Levintow and A. Meister, *J. Biol. Chem.* **209**, 265 (1954).

[22] H. Waelsch, *Phosphorus Metabolism II*, 109 (1952).

The equilibrium constant of the synthetase reaction is 1.2×10^3 at pH 7.[21] Divalent cations are required for this reaction; Mg^{++} is more effective than Mn^{++}, and Co^{++} gives less active preparations than the other ions. The metal used affects the relative rates with different substrates. L-Glutamate is always as active as any substrate. D-Glutamate is also a substrate, and under some conditions reacts at rates comparable with those found with the L-isomer but under other conditions the L-isomer forms glutamine more rapidly. Hydroxylamine can replace ammonia, and with this more reactive compound the rates of reaction of D- and L-glutamate are equal. Both D- and L-α-aminoadipic acid, the next higher homolog of glutamic acid, react with hydroxylamine in the synthetase reaction, but these compounds are inert with ammonia.

The transferase reaction requires catalytic amounts of ADP and inorganic phosphate as well as a metal. In this reaction Mn^{++} is more effective than Mg^{++}. Cysteine or another —SH compound activates the transferase as well as the synthetase reaction. L-Glutamine is much more effective in the exchange reaction than the D-isomer, and analogs of glutamine have low or no activity.

The same enzyme catalyzes an arsenolysis of glutamine. This reaction also requires ADP and Mn^{++} or $Mg.^{++}$ The arsenolysis of D-glutamine and of the hydrazide or hydroxamate of L-glutamate proceeds more slowly, and homologous amides are not hydrolyzed by this system.

Glutamine synthetase catalyzes an exchange reaction in which the terminal P of ATP is equilibrated with inorganic P. This reaction requires the presence of glutamate. It is possible that the exchange reaction actually occurs as a result of both forward and reverse reactions, and does not represent a partial reaction. The participation of small amounts of glutamine in the reverse reaction may explain the failure of compounds other than L-glutamate to support the phosphate exchange reaction, since the analogs of L-glutamic react slowly if at all in reverse or transfer reactions.

All of these reactions appear to be a function of a single enzyme. Since some substrates participate in one reaction but not another, it seems necessary to postulate partial reactions with different substrate requirements. At this time there is no evidence for any dissociable intermediates. Synthetic amidophosphate and γ-glutamyl phosphate are not attacked by the enzyme.[23] It seems likely, therefore, that two separate steps are involved in the enzyme activity, and that these have different substrate requirements. All of the reactions require adenine nucleotides, but it is not known how the nucleotides react. It has been shown that in the synthetase reaction (VI) an oxygen atom is transferred from glutamate to inorganic P, indicating a direct reaction between glutamate

[23] L. Levintow and A. Meister, *Federation Proc.* **15**, 299 (1956).

and ATP.[24] If there is indeed a common intermediate in the various reactions of L-glutamate, the second step is the one that shows the greater specificity. The exchange reactions and arsenolysis presumably require a thermodynamically unfavorable reaction for initiation, and if the affinity of the enzyme (in the second step) for compounds other than L-glutamate is low, only L-glutamate and its derivatives will react at measurable rates.

$$\text{glutamate} + \text{ATP} + \text{Enz} \rightleftharpoons \text{intermediate} \begin{array}{c} \text{NH}_3 \nearrow \quad \text{glutamine} + \text{ADP} + \text{P} + \text{Enz} \\ \searrow \text{NH}_2\text{OH} \\ \quad \text{glutamic hydroxamic acid} \\ \quad +\text{ADP} + \text{P} + \text{Enz} \end{array}$$

Hypothetical Reaction Outline for Glutamic Synthetase

(VI)

The intermediate is presumed to contain all of the components of glutamate and ATP together with the enzyme, linked to permit transfer of O from the γ-carboxyl to the terminal P. The substitution of As for P in the back reaction results in the formation of glutamate and ammonia without the formation of ATP.

Glutamine is hydrolyzed by enzymes from many animal, plant and microbial sources. An anion-stimulated enzyme was named glutaminase I.[25] The name glutaminase II was given to an enzyme that requires the presence of a keto acid for glutamine-splitting activity, and was later found to be a combination of two enzymes: a transaminase that forms α-ketoglutaramate and an amidase that attacks this intermediate product. Similar enzymes attack asparagine.

Other reactions of glutamine include participation in purine biosynthesis and transpeptidation. This compound seems to be important in transport of ammonia, and much of the glutamate of proteins bears a γ-amide. These considerations and the transaminations of glutamine make the synthesis and reactions of this compound of considerable biological interest.

General References

Cohen, P. P. (1951). *In* "The Enzymes" (J. B. Sumner and K. Myrbäck, eds.), Vol. II, Part 2, p. 897. Academic Press, New York.

Meister, A. (1956). *Physiol. Revs.* **36**, 103.

Stetten, M. R. (1955). "Symposium on Amino Acid Metabolism" (W. D. McElroy and B. Glass, eds.), p. 277. Johns Hopkins Press, Baltimore.

Vogel, H. J. (1955). "Symposium on Amino Acid Metabolism" (W. D. McElroy and B. Glass, eds.), p. 335. Johns Hopkins Press, Baltimore.

[24] A. Kowalsky, C. Wyttenback, L. Langer, and D. Koshland, *J. Biol. Chem.* **219**, 719 (1956).

[25] J. P. Greenstein and V. E. Price, *J. Biol. Chem.* **178**, 695 (1949).

Aspartic Acid, Threonine, and Lysine

Aspartase. The first of the simple deaminases to be described was aspartase, the enzyme that converts L-aspartate to fumarate and ammonia (I). This enzyme has been found in microorganisms and plants,

$$
\begin{array}{ccc}
\mathrm{COO^-} & & \mathrm{H} \qquad \mathrm{COO^-} \\
| & & \diagdown \;\; / \\
\mathrm{HCNH_2} & & \mathrm{C} \\
| & \rightleftharpoons & \| \qquad\qquad + \mathrm{NH_3} \\
\mathrm{HCH} & & \mathrm{C} \\
| & & / \;\; \diagdown \\
\mathrm{COO^-} & & \mathrm{^-OOC} \qquad \mathrm{H} \\
\text{Aspartate} & & \text{Fumarate}
\end{array}
$$

(I)

not in animals.[1,2] Quastel and Woolf found the reaction to proceed anaerobically with toluene-treated *E. coli*.[1] Gale studied the influence of many factors on washed cells, and later obtained active extracts.[3] A toluene-sensitive fraction was precipitated with low concentrations of ammonium sulfate. This was reported to be stimulated 5 times by adenosine. This observation has not been explained. The fraction precipitating at higher ammonium sulfate concentrations was found to contain a more stable aspartase, insensitive to toluene or adenosine. Crude preparations of aspartase contain fumarase, so that malate is formed in addition to fumarate. Woolf used cyclohexanol to inactivate fumarase, and was then able to demonstrate that malate does not participate in the reaction.[4] The aspartase reaction is freely reversible.[1]

D-*Aspartic Oxidase.* Aspartase and transaminases account for a major part of the metabolism of L-aspartic acid. D-Aspartic acid is oxidized by an enzyme present in liver and kidney.[5] This is an oxidase that converts aspartate to oxalacetate and ammonia while reducing oxygen to hydrogen peroxide. The oxidase was resolved by ammonium sulfate precipitation and dialysis to a protein that could be reactivated by FAD but not by FMN.[6] The enzyme differs from D-amino acid oxidase in its insensitivity to benzoate. The only other known substrate for the partially purified D-aspartic oxidase is D-glutamate, but since the relative rates of oxidation of the two amino acids vary during the preparation of the enzyme, it is

[1] J. H. Quastel and B. Woolf, *Biochem. J.* **20**, 545 (1934).
[2] A. I. Virtanen and J. Tarvanen, *Biochem. Z.* **250**, 193 (1932).
[3] E. F. Gale, *Biochem. J.* **32**, 1583 (1938).
[4] B. Woolf, *Biochem. J.* **23**, 472 (1929).
[5] J. L. Still, M. V. Buell, W. E. Knox, and D. E. Green, *J. Biol. Chem.* **179**, 831 (1949).
[6] J. L. Still and E. Sperling, *J. Biol. Chem.* **182**, 585 (1950).

probable that two separate enzymes are present. The ratio of rates with glutamate and aspartate also varies with species.

FORMATION OF HOMOSERINE AND THREONINE FROM ASPARTATE

Evidence from several lines of investigation indicated a relationship between L-aspartic acid and L-threonine. Studies with isotopically labeled acetate in yeast and bacteria showed that the distribution of label in the 4 carbon atoms of aspartate was the same as found in threonine.[7] Both aspartate and homoserine were found to suppress incorporation of labeled CO_2 into threonine,[8] and the label of aspartate was found to appear in a corresponding position in threonine.[9] Mutants of *Neurospora*[10] and *E. coli*[11] were found to use homoserine to form threonine; other mutants accumulated homoserine, and in *E. coli* it was found that aspartate was converted to homoserine.[12] In Lactobacilli threonine was found to minimize an aspartic acid requirement.[13] All of these findings support a scheme of reversible reactions:

$$\text{aspartate} \rightleftharpoons \text{homoserine} \rightleftharpoons \text{threonine}$$

The steps between aspartate and homoserine have been described in detail by Black and Wright, who separated three enzymes from baker's yeast.[14] The first step is the formation of β-aspartyl phosphate (BAP)

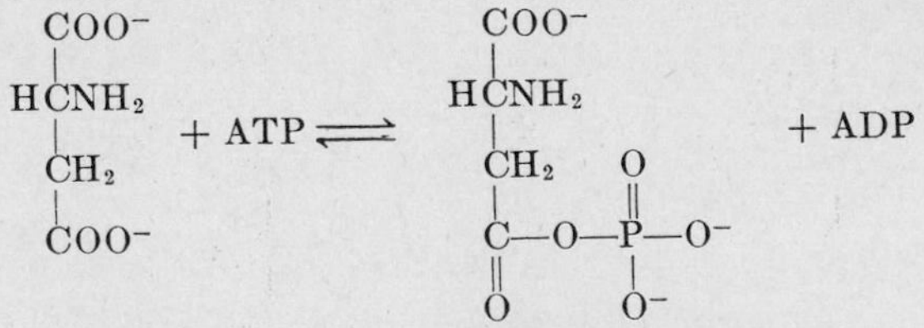

Aspartate β-Aspartyl phosphate
(II)

from aspartate and ATP (II). β-Aspartokinase is a typical kinase, requiring Mg^{++} and forming ADP. The reaction is reversible, with an equilibrium constant about 3.5×10^{-4}. The enzyme is specific for L-aspartate; there is no reaction with D-aspartate or L-glutamate. The reaction rate is

[7] C. Cutinelli, G. Ehrensvard, L. Reio, E. Saluste, and R. Stjernholm, *Acta Chem. Scand.* **5**, 351 (1951).

[8] P. H. Abelson, *J. Biol. Chem.* **206**, 335 (1954).

[9] A. M. Delluva, *Arch. Biochem. Biophys.* **45**, 443 (1953).

[10] H. J. Teas, N. H. Horowitz, and M. Fling, *J. Biol. Chem.* **172**, 651 (1948).

[11] G. N. Cohen and M. L. Hirsch, *J. Bacteriol.* **67**, 182 (1954).

[12] M. L. Hirsch and G. N. Cohen, *Compt. rend.* **238**, 1342 (1954).

[13] J. M. Ravel, L. Woods, B. Felsing, and W. Shive, *J. Biol. Chem.* **206**, 391 (1954).

[14] S. Black and N. G. Wright, *J. Biol. Chem.* **213**, 27, 39, 51 (1955).

maximal from pH 5 to 9. The product, BAP, is very unstable. It was identified by conversion to the stable hydroxamic acid.

β-Aspartyl phosphate is reduced to aspartic-β-semialdehyde (ASA) by aspartic semialdehyde dehydrogenase. The reaction resembles that catalyzed by triose phosphate dehydrogenase, but specifically requires TPN (III). The equilibrium constant is about 3×10^6, so that at neutral

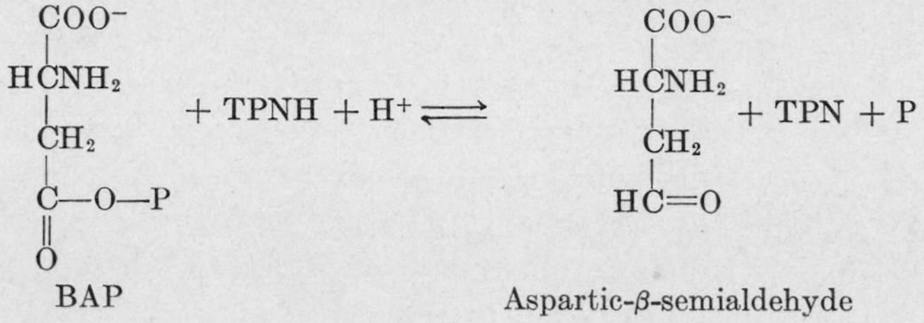

$$
\begin{array}{c}
\text{COO}^- \\
| \\
\text{HCNH}_2 \\
| \\
\text{CH}_2 \\
| \\
\text{C---O---P} \\
\| \\
\text{O}
\end{array}
+ \text{TPNH} + \text{H}^+ \rightleftharpoons
\begin{array}{c}
\text{COO}^- \\
| \\
\text{HCNH}_2 \\
| \\
\text{CH}_2 \\
| \\
\text{HC}{=}\text{O}
\end{array}
+ \text{TPN} + \text{P}
$$

BAP   Aspartic-β-semialdehyde

(III)

pH with concentrations of reagents ordinarily used the reaction proceeds quantitatively in the direction of TPNH oxidation; at higher pH values and with high concentrations of inorganic phosphate the reverse reaction is favored.

L-Homoserine is formed by a second reduction reaction, in which L-aspartic semialdehyde and DPNH participate (IV). This reaction has

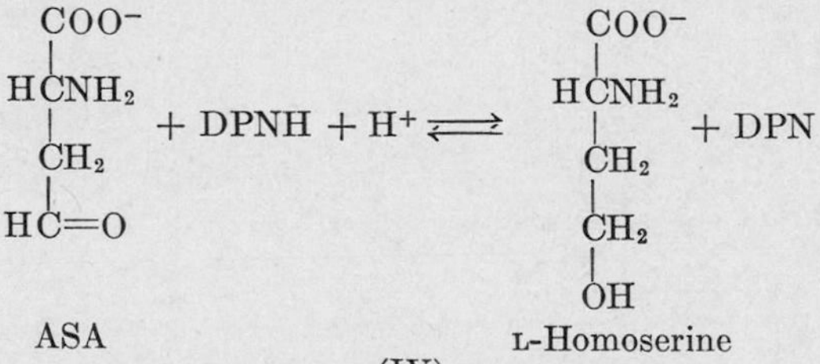

$$
\begin{array}{c}
\text{COO}^- \\
| \\
\text{HCNH}_2 \\
| \\
\text{CH}_2 \\
| \\
\text{HC}{=}\text{O}
\end{array}
+ \text{DPNH} + \text{H}^+ \rightleftharpoons
\begin{array}{c}
\text{COO}^- \\
| \\
\text{HCNH}_2 \\
| \\
\text{CH}_2 \\
| \\
\text{CH}_2 \\
| \\
\text{OH}
\end{array}
+ \text{DPN}
$$

ASA   L-Homoserine

(IV)

$$
\begin{array}{c}
\text{CH}_2\text{OH} \\
| \\
\text{CH}_2 \\
| \\
\text{CHNH}_2 \\
| \\
\text{COO}^-
\end{array}
+ \text{ATP} \xrightarrow[\text{I}]{\text{Mg}^{++}}
\begin{array}{c}
\text{O} \\
\| \\
\text{CH}_2\text{---O---P---O}^- \\
| \qquad\quad | \\
\text{CH}_2 \qquad \text{O}^- \\
| \\
\text{CHNH}_2 \\
| \\
\text{COO}^-
\end{array}
\xrightarrow{\text{II}}
\begin{array}{c}
\text{CH}_3 \\
| \\
\text{CHOH} \\
| \\
\text{CHNH}_2 \\
| \\
\text{COO}^-
\end{array}
$$

L-Homoserine   Phosphohomoserine   L-Threonine

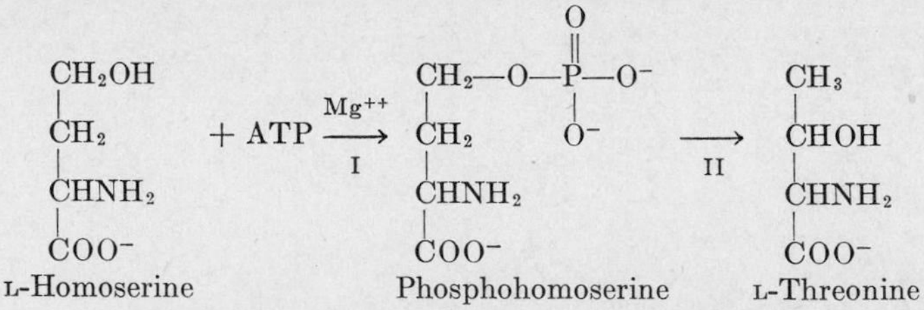

(V)

an equilibrium constant near 10^{11}. TPNH reacts at about $\frac{1}{3}$ the rate of DPNH.

The detailed reactions by which homoserine is converted to threonine have not yet been described. The process requires at least two enzymes, which have been separated from yeast by Watanabe and Shimura.[15] The first enzyme is a kinase that uses ATP to form O-phosphohomoserine. This product is converted to threonine by a second enzyme (V).

LYSINE

Lysine is an essential amino acid for animals. Not only is this compound not synthesized, but it does not contribute significant quantities of either nitrogen or carbon to other amino acids.[16] The biosynthesis of lysine occurs in bacteria and fungi, and presumably in higher plants. *Neurospora* and *E. coli* have been found to use different pathways. The pathway in *Neurospora* involves the synthesis of α-aminoadipic acid, which appears to be reduced to the semialdehyde, which then, in a transamination, yields lysine.[17] The corresponding α-amino-ϵ-hydroxycaproic

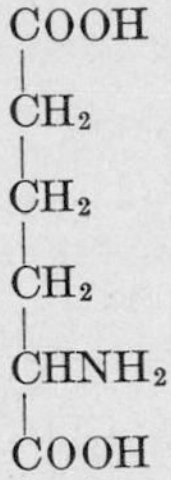

$$
\begin{array}{cc}
\text{COOH} & \text{CH}_2\text{NH}_2 \\
| & | \\
\text{CH}_2 & \text{CH}_2 \\
| & | \\
\text{CH}_2 & \text{CH}_2 \\
| & | \\
\text{CH}_2 & \text{CH}_2 \\
| & | \\
\text{CHNH}_2 & \text{CHNH}_2 \\
| & | \\
\text{COOH} & \text{COOH} \\
\alpha\text{-Aminoadipic acid} & \text{Lysine}
\end{array}
$$

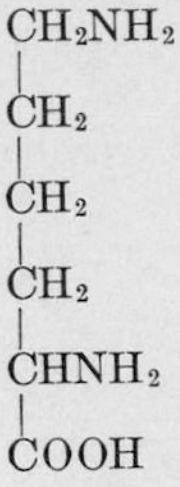

acid also can be used for lysine synthesis, but is less efficient. The enzymes that participate in this sequence are not yet described. In *E. coli* lysine is formed from diaminopimelic acid.[18] This precursor is formed by cell-free extracts, which use aspartic acid preferentially as a nitrogen source.[19] Pyruvate and glutamate also stimulate diaminopimelic synthesis, as do ATP, DPN, TPN, and Mg^{++}. Synthetic β-hydroxydiaminopimelic acid can be used by a mutant blocked in the synthesis of the intermediate.

Diaminopimelic acid is decarboxylated by a specific enzyme that requires pyridoxal phosphate.[20] Only the *meso* isomer is attacked, and the product is L-lysine. L,L-Diaminopimelic has been found to yield lysine in some systems; this observation has been explained as caused by a racemase that forms the *meso* compound.[21]

[15] Y. Watanabe and K. Shimura, *J. Biochem.* **43**, 283 (1956).
[16] N. Weissman and R. Schoenheimer, *J. Biol. Chem.* **140**, 779 (1941).
[17] E. Windsor, *J. Biol. Chem.* **192**, 607 (1951).
[18] B. D. Davis, *Nature* **169**, 534 (1952); D. L. Dewey and E. Work, *ibid.* **169**, 533 (1952).
[19] C. Gilvarg, *Federation Proc.* **15**, 261 (1956).
[20] D. L. Dewey, D. S. Hoare, and E. Work, *Biochem. J.* **58**, 523 (1954).
[21] D. S. Hoare, *Biochem. J.* **59**, xxii (1955).

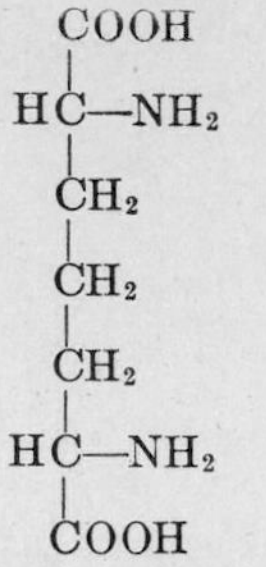

$$\begin{array}{ccc}
\text{COOH} & & \text{COOH} \\
| & & | \\
\text{HC—NH}_2 & & \text{H}_2\text{N—CH} \\
| & & | \\
\text{CH}_2 & & \text{CH}_2 \\
| & & | \\
\text{CH}_2 & & \text{CH}_2 \\
| & & | \\
\text{CH}_2 & & \text{CH}_2 \\
| & & | \\
\text{HC—NH}_2 & & \text{HC—NH}_2 \\
| & & | \\
\text{COOH} & & \text{COOH}
\end{array}$$

meso-Diaminopimelic acid 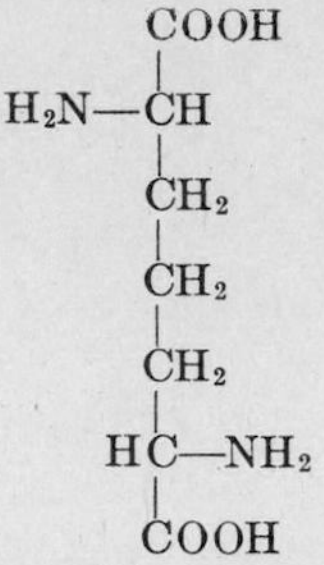   L,L-Diaminopimelic acid

Lysine degradation proceeds by several routes in various organisms, but, in general, the individual reactions have not been isolated. Isotopic studies *in vivo* with plants,[22] animals,[23] and *Neurospora*[24] show conversion of lysine to pipecolic acid (VI). Since the α-amino nitrogen is lost in this

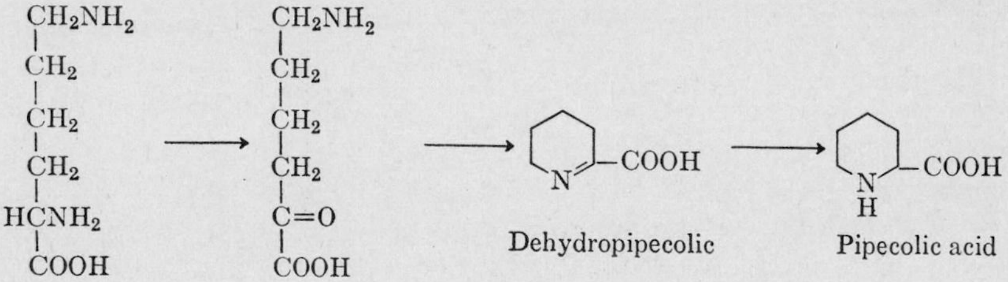

(VI)

reaction, while the ϵ-amino group is retained, the intermediate formation of an α-keto derivative is postulated, followed by cyclization to dehydropipecolic acid. This must be reduced to pipecolic acid. An oxidation of pipecolic acid in animals leads to the formation of α-aminoadipic acid.[25] This is converted to the corresponding α-keto acid, then to glutaric acid, and eventually to α-ketoglutaric acid. The enzymes responsible for these reactions are not known. In *Neurospora* similar reactions occur.[26] An additional compound formed from lysine is this organism is ϵ-N-acetyl-α-hydroxycaproic acid.

Lysine can serve as a substrate for bacterial fermentations. Isotopic experiments with dried cells of a *Clostridium* show the methyl group of acetate and the α-carbon of butyrate to come from the 2-carbon of

[22] P. Lane, *Arch. Biochem. Biophys.* **47,** 228 (1953); N. Grobbelaar and F. Steward, *J. Am. Chem. Soc.* **75,** 4341 (1953).

[23] M. Rothstein and L. Miller, *J. Am. Chem. Soc.* **75,** 4321 (1953).

[24] R. Schweet, J. Holden, and P. Lowery, *J. Biol. Chem.* **211,** 517 (1954).

[25] M. Rothstein and L. Miller, *J. Am. Chem. Soc.* **76,** 1459 (1954).

[26] R. Schweet, J. Holden, and P. H. Lowery, *Symposium on Amino Acid Metabolism, Baltimore, 1954* p. 496 (1955).

lysine, while carbon 6 of lysine becomes mainly the methyl carbon of butyrate.[27] α-Aminoadipate is inert in this system. Other bacterial systems decarboxylate lysine to cadaverine, which can be oxidized by amine oxidases, to be discussed later.

GENERAL REFERENCES

Black, S., and Wright, N. G. (1955). "Symposium on Amino Acid Metabolism" (W. D. McElroy and B. Glass, eds.), p. 591. Johns Hopkins Press, Baltimore.

Erkama, J., and Virtanen, A. I. (1951). *In* "The Enzymes" (J. B. Sumner and K. Myrbäck, eds.), Vol. I, Part 2, p. 1244. Academic Press, New York.

Work, E. (1955). "Symposium on Amino Acid Metabolism" (W. D. McElroy and B. Glass, eds.), p. 462. Johns Hopkins Press, Baltimore.

The Urea Cycle

The formation of urea is quantitatively the most important process in the catabolism of nitrogen compounds by humans and other species that excrete nitrogen as urea. The synthesis of this small molecule is a complicated procedure, involving many amino acids and requiring the consumption of ATP. The outline of the reaction sequence that produces urea was revealed by Krebs and Henseleit,[1] who found several amino acids, ornithine, citrulline, and arginine, to stimulate urea synthesis catalytically. Their interpretation of their observations was a cyclic process, which today is firmly established through the work of several investigators, who have described the entire urea cycle in terms of specific enzyme reactions. The gaps in our knowledge today are left by difficulties in studying the detailed properties of some of the enzymes.

Urea synthesis is a truly cyclic process (I), in which ornithine serves as a base on which the structure of urea is constructed, then released together with free ornithine that renews the cycle.

Citrulline Synthesis. The first syntheses of citrulline and arginine by the addition of NH_3 and CO_2 to ornithine were achieved in the laboratory of Cohen, where soluble enzymes of liver supplemented with cell particles were found to carry out these syntheses.[2] The soluble extract was found to contain the several activities involved in the reactions of the urea cycle, while the particles were found to function solely as a source of ATP. The synthesis of citrulline was found to require ornithine, CO_2, NH_3, and

[27] T. C. Stadtman, *Symposium on Amino Acid Metabolism, Baltimore, 1954* p. 493 (1955).

[1] H. A. Krebs and K. Henseleit, *Z. physiol. Chem.* **210**, 33 (1932).
[2] P. P. Cohen and M. Hayano, *J. Biol. Chem.* **166**, 239, 251 (1946); **172**, 405 (1948).

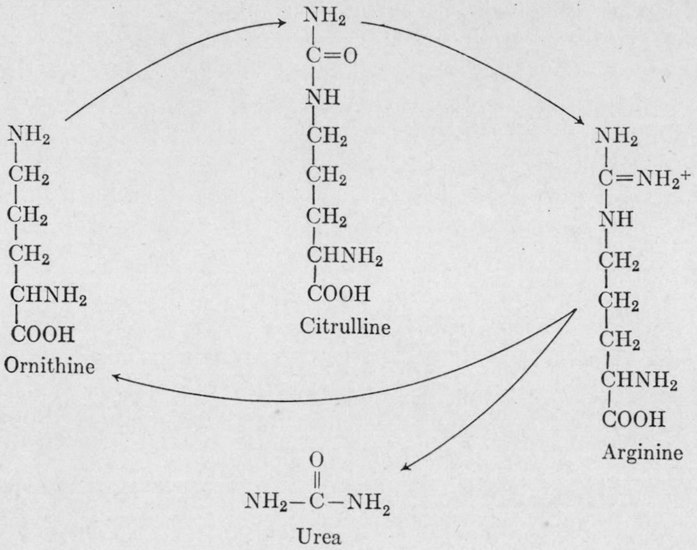

$$
\begin{array}{ccc}
& NH_2 & \\
& | & \\
& C=O & \\
& | & \\
& NH & \\
NH_2 & | & NH_2 \\
| & CH_2 & | \\
CH_2 & | & C=NH_2{}^+ \\
| & CH_2 & | \\
CH_2 & | & NH \\
| & CH_2 & | \\
CH_2 & | & CH_2 \\
| & CHNH_2 & | \\
CHNH_2 & | & CH_2 \\
| & COOH & | \\
COOH & Citrulline & CH_2 \\
Ornithine & & | \\
& & CHNH_2 \\
& & | \\
& & COOH \\
& & Arginine
\end{array}
$$

$$
\begin{array}{c}
O \\
\parallel \\
NH_2-C-NH_2 \\
Urea
\end{array}
$$

(I)

also glutamic acid, ATP, and Mg^{++}.[3] Since citrulline is ornithine plus a carbamyl group, the ureido derivative of glutamic acid, carbamylglutamic acid, was tested as a potential intermediate. The first experiments reported indicated an important role for carbamylglutamic acid and it was concluded that the synthesis of citrulline involved a transfer of the carbamyl group.[4] This concept of the reaction mechanism gained support

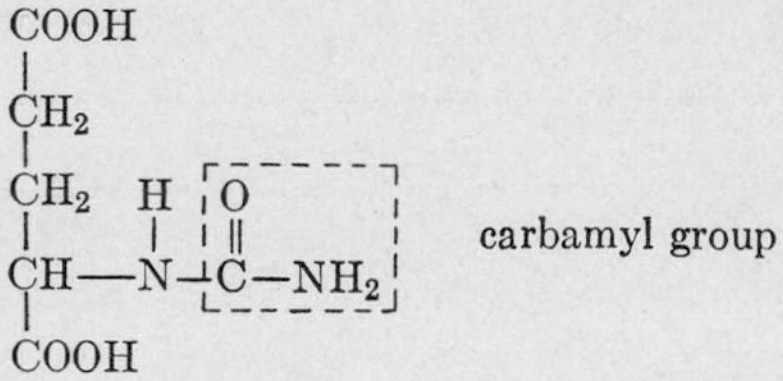

$$
\begin{array}{l}
COOH \\
| \\
CH_2 \\
| \\
CH_2 \quad H \; \lceil O \; \rceil \\
| \qquad | \; | \parallel \; | \\
CH-N\!-\!C-NH_2 \quad \text{carbamyl group} \\
| \qquad \lfloor \_\_\_\_ \rfloor \\
COOH
\end{array}
$$

Carbamylglutamic acid

from the finding that a compound resembling carbamyglutamic acid in its stimulation of citrulline synthesis was formed by liver enzymes from glutamic acid, ammonia, CO_2, and ATP. This simple concept was somewhat complicated by the findings of ATP and NH_3 requirements in the reaction of carbamylglutamate with ornithine. Finally, isotopic experiments forced a radical revision of the scheme.[5] It was found that neither

[3] P. P. Cohen and S. Grisolia, *J. Biol. Chem.* **174**, 389 (1948).
[4] P. P. Cohen and S. Grisolia, *J. Biol. Chem.* **182**, 747 (1950).
[5] S. Grisolia, R. Burris, and P. P. Cohen, *J. Biol. Chem.* **191**, 203 (1951).

the nitrogen nor the carbon of carbamylglutamic acid appeared in citrulline, which was made from ornithine, CO_2, and NH_3.

The function of carbamylglutamate has not been elucidated. Grisolia and collaborators[6] have purified a material formed by incubation of carbamylglutamate with liver preparations, NH_3, CO_2, and ATP in the absence of ornithine. This so-called compound X was reported to react enzymatically with ornithine to form citrulline. The compound was found to contain glutamate and to be very labile, easily losing NH_3, CO_2, and inorganic phosphate, but not at equal rates. The structure of this compound is not established, but it has been suggested that a family of similar compounds can serve in the synthesis of citrulline. Although no other amino acids were found to substitute for glutamic acid, several acyl derivatives of this compound were found to react in place of the carbamyl derivative. Active compounds include acetyl-, chloroacetyl-, propionyl-, and formylglutamate.

Although the nature of compound X remains obscure, it has been found that another intermediate is formed from it, and that this intermediate, carbamyl phosphate, contains the atoms that are added to

$$
\begin{array}{ccc}
& O & O \\
& \parallel & \parallel \\
H_2N & -C-O-P-O^- \\
& & | \\
& & O^-
\end{array}
$$

Carbamyl phosphate

ornithine to form citrulline. Carbamyl phosphate was originally found as a product of a reaction catalyzed by bacterial extracts, in which citrulline is degraded to ornithine. This system will be discussed later.

Arginine Synthesis. The conversion of citrulline to arginine was studied by Ratner and collaborators. It requires only the replacement of the ureido oxygen with an imino group ($=NH$) to form a guanidine group. This exchange is carried out very indirectly. Originally glutamic acid was implicated in this step as well as in the synthesis of citrulline. The usefulness of glutamic acid is now known to be a consequence of its oxidation, which leads to ATP formation, and, through the Krebs cycle, to oxalacetate, which by transamination becomes aspartate, the immediate nitrogen donor in arginine synthesis. The nitrogen of glutamic acid thus enters urea indirectly via aspartic acid and arginine.

The transfer of nitrogen from aspartate to citrulline occurs via a condensation to form an intermediate, argininosuccinate, ASA. The condensation requires the participation of ATP, which is split to AMP and inorganic pyrophosphate during the reaction. The condensation is

[6] See general references by S. Grisolia.

reversible, and the equilibrium appears to favor splitting of ASA and synthesis of ATP (II). The condensing enzyme has been purified from

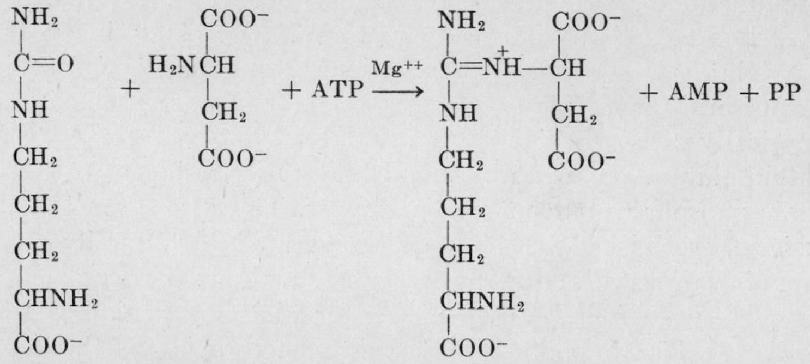

$$
\begin{array}{llll}
NH_2 & COO^- & & NH_2 & COO^- \\
| & | & & | & | \\
C{=}O & H_2NCH & \xrightarrow{Mg^{++}} & C{=}\overset{+}{N}H{-}CH & \\
| & | & +ATP \longrightarrow & | & | & +AMP+PP \\
NH & CH_2 & & NH & CH_2 \\
| & | & & | & | \\
CH_2 & COO^- & & CH_2 & COO^- \\
| & & & | \\
CH_2 & & & CH_2 \\
| & & & | \\
CH_2 & & & CH_2 \\
| & & & | \\
CHNH_2 & & & CHNH_2 \\
| & & & | \\
COO^- & & & COO^- \\
\text{Citrulline} & \text{Aspartate} & & \text{Argininosuccinate}
\end{array}
$$

(II)

liver and a second enzyme fraction from either liver or yeast has been found to stimulate the condensation. The second enzyme has been identified as inorganic pyrophosphatase, which "pulls" the reaction by removing one of the products. There is no measurable reaction in systems from which any of the reagents is omitted; that is, there is no evidence for activation of either citrulline or aspartate prior to condensation.[7] Although argininosuccinate is a reactive molecule that readily forms 5- and 6-membered rings, it has been isolated and characterized.[8] Another liver enzyme cleaves argininosuccinate to arginine and fumarate (III).[9] The nature of the splitting was not easily determined; since the enzyme

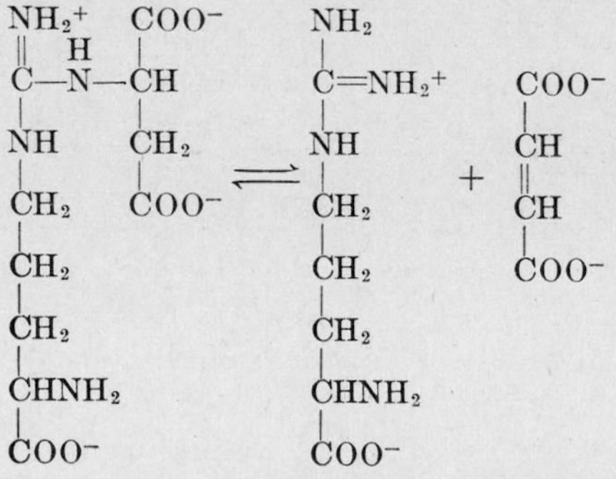

$$
\begin{array}{llll}
NH_2{}^+ & COO^- & NH_2 \\
\| \quad H & | & \| \\
C{-}N{-}CH & & C{=}NH_2{}^+ & COO^- \\
| & | & | & | \\
NH & CH_2 & NH & CH \\
| & | & \rightleftharpoons & | & \| \\
CH_2 & COO^- & CH_2 & + & CH \\
| & & | & | \\
CH_2 & & CH_2 & COO^- \\
| & & | \\
CH_2 & & CH_2 \\
| & & | \\
CHNH_2 & & CHNH_2 \\
| & & | \\
COO^- & & COO^- \\
& & \text{Arginine} & \text{Fumarate}
\end{array}
$$

(III)

[7] S. Ratner and B. Petrack, *J. Biol. Chem.* **200,** 161 (1953); *Arch. Biochem. Biophys.* **65,** 582 (1956).

[8] S. Ratner, B. Petrack, and O. Rochanovsky, *J. Biol. Chem.* **204,** 95 (1953).

[9] J. B. Walker and J. Myers, *J. Biol. Chem.* **203,** 143 (1953); S. Ratner and B. Petrack, *ibid.* **191,** 693 (1951).

fumarase contaminated preparations of the splitting enzyme, it was difficult to distinguish between a hydrolytic reaction, forming malate, and
an elimination reaction, forming fumarate. The fumarase equilibrium
favors malate formation, and malate was the major product found until
enzyme preparations free from fumarase were obtained. Then it could be
shown that fumarate is the primary product and malate is produced secondarily by fumarase.

The reactions of argininosuccinic acid were elucidated largely through
studies with plant preparations. Davison and Elliott found extracts of
pea meal to catalyze a reaction between arginine and fumarate to form
a compound identified by its migration on paper chromatograms.[10] Similar
enzymes were found in other plants, animal tissues, bacteria, and yeast.
Walker and Myers studied this activity in extracts of *Chlorella* and jackbean meal.[9] Other unsaturated acids were not able to replace fumarate,
and creatine, glycocyamine, guanidine, and methylguanidine were all
inactive when substituted for arginine. The formation of argininosuccinate was shown to be reversible. The equilibrium constant for the reaction

$$\text{argininosuccinate} \rightleftharpoons \text{arginine} + \text{fumarate}$$

was determined to be about 1.1×10^{-2} at 38°C., pH 7.5.[11] Besides arginine, the analog canavanine, also reacts with the enzyme (IV).

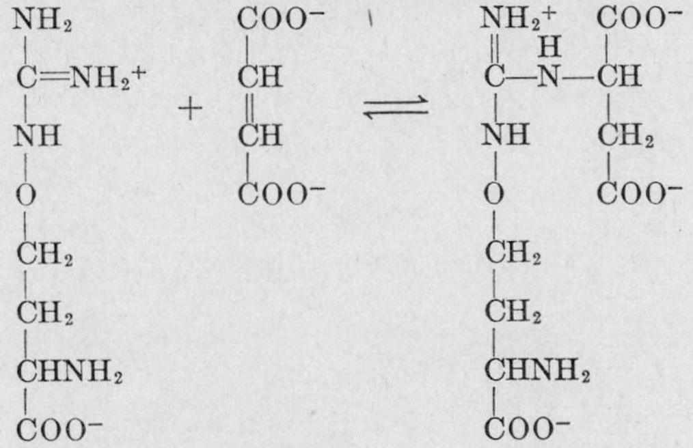

Canavanine + Fumarate ⇌ Canavaninosuccinate
(IV)

Arginase. Urea is formed by the hydrolysis of arginine catalyzed by
the enzyme arginase, which was demonstrated in liver in 1904 by Kossel
and Dakin (V).[12] A 400-fold purification from horse liver gave a preparation that appears to be essentially homogeneous.[13] Arginase requires a

[10] D. C. Davison and W. H. Elliott, *Nature* **169**, 313 (1952).
[11] S. Ratner, W. P. Anslow, Jr., and B. Petrack, *J. Biol. Chem.* **204**, 115 (1953).
[12] A. Kossel and H. D. Dakin, *Z. physiol. Chem.* **41**, 321 (1904); **42**, 181 (1904).
[13] D. M. Greenberg, A. B. Bagot, and D. A. Roholt, Jr., *Arch. Biochem. Biophys.* **62**, 446 (1956).

divalent cation, Co^{++}, Mn^{++}, and Ni^{++}. Cd^{++} has no effect and Zn^{++} inhibits. The metals combine very slowly with the enzyme; several hours are required for complete activation. The properties of the enzyme are greatly affected by the metal constituent. Mn-arginase has a pH optimum at 10.2 whereas Co-arginase is most active at pH 6.5–7.0, where Mn-arginase has essentially no activity.[14] A mixture of these metals added to

$$
\begin{array}{l}
NH_2 \\
| \\
C{=}NH_2{}^+ \qquad\qquad\qquad O \qquad\quad NH_2 \\
| \qquad\qquad\qquad\qquad\qquad\quad\| \qquad\quad | \\
NH \qquad + H_2O \longrightarrow NH_2{-}C{-}NH_2 + CH_2 \qquad + H^+ \\
| \qquad\qquad\qquad\qquad\qquad\qquad\qquad\quad | \\
CH_2 \qquad\qquad\qquad\qquad\qquad\qquad\qquad CH_2 \\
| \qquad\qquad\qquad\qquad\qquad\qquad\qquad\qquad | \\
CH_2 \qquad\qquad\qquad\qquad\qquad\qquad\qquad CH_2 \\
| \qquad\qquad\qquad\qquad\qquad\qquad\qquad\qquad | \\
CH_2 \qquad\qquad\qquad\qquad\qquad\qquad\qquad CHNH_2 \\
| \qquad\qquad\qquad\qquad\qquad\qquad\qquad\qquad | \\
CHNH_2 \qquad\qquad\qquad\qquad\qquad\qquad COO^- \\
| \\
COO^-
\end{array}
$$

(V)

arginase gives a preparation with a broad pH curve covering both op-tima. Since the ratio of metal concentrations determines the relative activity at various pH values, it is probable that only a single protein is involved instead of separate Mn^{++} and Co^{++} activated enzymes. The heat stability of the enzyme is also determined by the metal; Mn^{++} gives greater protection than Co^{++}, which protects better than Ni^{++}.

The enzyme purified from beef liver attacks not only arginine, but also a number of related structures. Chain length is important; canavanine which has essentially the same chain length as arginine (with an —O— in place of —CH_2—) is attacked,[15] but the higher and lower homologs of arginine (4- and 6-carbon chains) are not hydrolyzed.[16] The carboxyl group is not essential, as decarboxylated arginine, agmatine, is split,[17] but esters of arginine are not.[18] The α-amino group is not essential; it may be absent as in δ-guanidovaleric acid,[19] replaced by a hydroxyl group (argininic acid),[20] or substituted as in octopine,[21] α-N-benzylarginine,[22]

[14] D. A. Roholt, Jr. and D. M. Greenberg, Arch. Biochem. Biophys. **62**, 446 (1956).

[15] M. Damadaran and K. G. A. Narayanan, Biochem. J. **34**, 1449 (1940).

[16] K. Thomas, J. Kapfhammer, and B. Flaschentrager, Z. physiol. Chem. **129**, 75 (1922).

[17] M. M. Richards and L. Hellerman, J. Biol. Chem. **134**, 237 (1940).

[18] S. Edlbacher and P. Bonem, Z. physiol. Chem. **145**, 69 (1925).

[19] K. Thomas, Z. physiol. Chem. **88**, 465 (1913).

[20] A. Hunter and H. E. Woodward, Biochem. J. **35**, 1298 (1941).

[21] S. Akasi, J. Biochem. **26**, 129 (1937).

[22] K. Felix, H. Muller, and K. Dirr, Z. physiol. Chem. **178**, 192 (1928).

or carbamylarginine.[23] When an α-substitution is present, however, the group must be in the L-configuration; D-arginine is not a substrate. Phosphoarginine is also inert with arginase.

Arginine Desimidase. An alternate mechanism for arginine degradation occurs in several species of bacteria. The over-all reaction was called arginine dihydrolase by Hills,[24] who found the products to be ornithine, CO_2, and $2NH_3$. Working with whole cells, he did not find a reaction with citrulline. Horn[25] found citrulline to be produced from arginine by *Pseudomonas aeruginosa* (*B. pyocyaneus*), and named the enzyme responsible arginine desimidase. Akamatsu found that arginine dihydrolase could be resolved into metarginase which formed citrulline, and citrullinase which split citrulline to ornithine.[26] Slade and collaborators,[27] Korzenovsky and Werkman,[28] Oginsky and Gehrig,[29] and Knivett[30] independently studied the nature of these reactions in *Streptococcus faecalis* and other bacteria. The hydrolysis of arginine to citrulline has not been shown to require cofactors. Arginine desimidase is inhibited by arsenite, hydroxylamine, semicarbazide, and certain heavy metals, but this information has not led to an understanding of the nature of the reaction.

Citrullinase. The bacterial enzyme that degrades citrulline has been called citrullinase, citrulline ureidase and citrulline phosphorylase. The reaction requires inorganic phosphate, Mg^{++}, and ADP, and ATP is formed together with NH_3, CO_2, and ornithine. Arsenate supports the breakdown of citrulline in the absence of phosphate and adenine nucleotides. The mechanism of the phosphorolysis was shown by Jones *et al.*[31] to be straightforward; the first products are ornithine and carbamyl phosphate, which had previously been considered to be too unstable to exist free. Carbamyl phosphate transfers its phosphate to ADP in a reversible kinase reaction, and the carbamyl group also can react with ornithine to form citrulline (VI). These reversible reactions explain the requirement for stoichiometric amounts of adenine nucleotides for a coupled reaction to remove the labile phosphate. Carbamic acid may equilibrate non-

[23] A. Hunter, *Biochem. J.* **32**, 826 (1938).

[24] G. M. Hills, *Biochem. J.* **34**, 1057 (1940).

[25] F. Horn, *Z. physiol. Chem.* **216**, 244 (1953).

[26] S. Akamatsu and T. Sekine, *J. Biochem.* **38**, 349 (1951); H. D. Slade and W. C. Slamp, *J. Bacteriol.* **64**, 455 (1952).

[27] H. D. Slade, C. C. Daughty, and W. C. Slamp, *Arch. Biochem. Biophys.* **48**, 338 (1954).

[28] M. Korzenovsky and C. H. Werkman, *Arch. Biochem. Biophys.* **41**, 233 (1952); **46**, 174 (1953).

[29] E. L. Oginsky and R. F. Gehrig, *J. Biol. Chem.* **198**, 791 (1952); **204**, 721 (1953).

[30] V. A. Knivett, *Biochem. J.* **50**, xxx (1952); **56**, 602, 606 (1954).

[31] M. E. Jones, L. Spector, and F. Lipmann, *J. Am. Chem. Soc.* **77**, 819 (1955).

enzymatically with NH_3 and CO_2, but the equilibration is greatly accelerated by carbonic anhydrase, which catalyzes the equilibration of CO_2 and H_2CO_3. Sufficient carbamic acid exists in neutral solutions to permit the kinase to form carbamyl phosphate, which can react with various carbamyl acceptors. This system appears adequate to account for synthesis of citrulline by bacterial preparations. Mammalian liver does not

$$
\underset{\text{Carbamate}}{H_2N-\overset{\displaystyle O}{\overset{\|}{C}}-O^-} + ATP \rightleftharpoons ADP + \underset{}{H_2N-\overset{\displaystyle O}{\overset{\|}{C}}-O-\overset{\displaystyle O}{\overset{\|}{\underset{|}{P}}}-O^-} \quad \text{Carbamyl phosphate}
$$

$$+ \quad O^-$$

$$\text{Ornithine } NH_2CH_2CH_2CH_2\overset{NH_2}{\overset{|}{C}H}COO^-$$

$$NH_2-\overset{\displaystyle O}{\overset{\|}{C}}-NHCH_2CH_2CH_2\overset{NH_2}{\overset{|}{C}H}COO^- \qquad + HO-\overset{\displaystyle O}{\underset{\displaystyle O}{\overset{|}{P}}}-O^-$$

$$\underset{\text{Citrulline}}{}$$

$$(VI)$$

seem to contain carbamyl phosphate kinase; therefore the more complex mechanism including compound X is retained to lead indirectly to carbamyl phosphate.[32]

Transamidinase. The guanidine group of arginine may be split by a transamidination reaction, formally very similar to arginase action. In this case an amine acts as acceptor for the amidine group to form a new guanidine group, whereas in arginase action water is the acceptor, forming urea. Fuld described a reaction in kidney in which arginine reacts with glycine to form ornithine and glycocyamine (a precursor of creatine) (VII).[33] Both Fuld and Walker[34a] have purified the enzyme, which cat-

$$
\begin{array}{ccccc}
\overset{NH_2}{\underset{|}{NH-CH=NH_2^+}} & & \overset{NH_2}{\underset{|}{CH_2}} & & \\
\overset{|}{CH_2} & & \overset{|}{CH_2} & & \\
\overset{|}{CH_2} & + \overset{NH_2}{\underset{|}{CH_2}} & \rightleftharpoons \overset{|}{CH_2} & + \overset{NH_2}{\underset{|}{NH-C=NH_2^+}} & \\
\overset{|}{CH_2} & \overset{|}{COO^-} & \overset{|}{CH_2} & \overset{|}{CH_2} & \\
\overset{|}{CHNH_2} & & \overset{|}{CHNH_2} & \overset{|}{COO^-} & \\
\overset{|}{COO^-} & & \overset{|}{COO^-} & & \\
\text{Arginine} & \text{Glycine} & \text{Ornithine} & \text{Glycocyamine}
\end{array}
$$

$$(VII)$$

[32] S. Grisolia, H. J. Grady, and D. P. Wallach, *Biochim. et Biophys. Acta* **17**, 277 (1955); R. D. Marshall, L. M. Hall, and P. P. Cohen, *ibid.* **17**, 279 (1955).

[33] M. Fuld, *Federation Proc.* **15**, 257 (1956).

[34a] J. B. Walker, *J. Biol. Chem.* **218**, 555 (1956); **221**, 771 (1956).

alyzes a reversible reaction in which glycine, ornithine, canaline, lysine, and possibly other diamino acids accept the amidine group. Purification of the enzyme and kinetic studies on competition between various substrates have been described by Ratner and Rochovansky.[34b]

Urease. Urea is an inert end-product of metabolism of many organisms, including man. Soluble enzymes that decompose urea have been

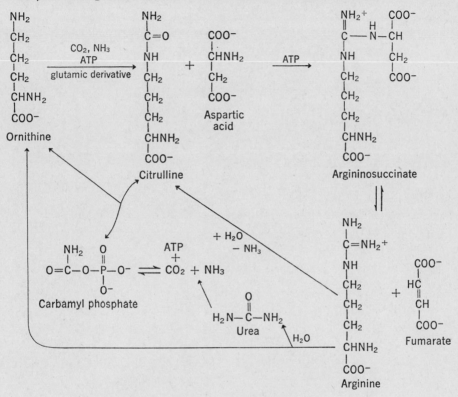

FIG. 32. Urea cycle and variations.

known for many years,[35] and it has been indicated that they occur in animal tissues as well as in microorganisms and plants. More recently, however, it has been established that the only decomposition of urea in animals is catalyzed by bacteria.[36]

The formation of NH_3 and CO_2 from urea is catalyzed by urease. This enzyme was crystallized from jackbean meal by Sumner in 1926.[37] This

[34b] S. Ratner and O. Rochovansky, *Arch. Biochem. Biophys.* **63**, 277, 296 (1956).

[35] K. Shibata, *Beitr. chem. Physiol. Pathol.* **5**, 384 (1903–1904).

[36] R. Z. Dintzis and A. B. Hastings, *Proc. Natl. Acad. Sci.* **39**, 571 (1953).

[37] J. B. Sumner, *J. Biol. Chem.* **69**, 435 (1926).

first crystallization of an enzyme was of great value in establishing the protein nature of enzymes and in permitting modern physical and chemical analyses of biological catalysts.

Urease is a simple protein with a molecular weight of 473,000 and an isoelectric point of 5.0–5.1. It contains 31 sulfhydryl groups per mole, not all of which are required for activity. Urease is sensitive, however, to both heavy metals and oxidizing agents.

The reaction carried out by urease is extremely specific. The mechanism of the reaction is not understood. No intermediates have been established and the enzyme requires no cofactors. Carbamate has been suggested as an intermediate, but it is possible that this is a secondary product. Very active urease has been obtained from bacteria.[38] Although the properties of the proteins differ, the catalytic activities of urease from various sources appear to be very similar.

General References

Greenberg, D. M. (1951). *In* "The Enzymes" (J. B. Sumner and K. Myrbäck, eds.), Vol. I, Part 2, p. 893. Academic Press, New York.

Grisolia, S. (1951). "Phosphorus Metabolism Symposium" (W. D. McElroy and B. Glass, eds.), p. 619. Johns Hopkins Press, Baltimore.

Grisolia, S., and Marshall, R. O. (1955). "Symposium on Amino Acid Metabolism" (W. D. McElroy and B. Glass, eds.), p. 258. Johns Hopkins Press, Baltimore.

Krebs, H. A. (1951). *In* "The Enzymes" (J. B. Sumner and K. Myrbäck, eds.), Vol. II, Part 2, p. 866. Academic Press, New York.

Ratner, S. (1954). *Advances in Enzymol.* **15**, 319.

Ratner, S. (1955). "Symposium on Amino Acid Metabolism" (W. D. McElroy and B. Glass, eds.), p. 231. Johns Hopkins Press, Baltimore.

Sumner, J. B. (1951). *In* "The Enzymes" (J. B. Sumner and K. Myrbäck, eds.), Vol. I, Part 2, p. 873. Academic Press, New York.

β-Hydroxy-α-amino Acids

Serine-Glycine Interconversion. Feeding experiments in animals showed the formation of glycine from serine and threonine.[1] An interconversion of serine and glycine was indicated by microbial nutritional experiments[2] and confirmed by isotope experiments with both microbial and animal preparations.[3] The carboxyl and α-carbon of serine are converted to gly-

---

[38] A. D. Larsen and R. E. Kallio, *J. Bacteriol.* **68**, 67 (1954).

[1] F. Knoop, *Z. physiol. Chem.* **89**, 151 (1914).
[2] R. R. Roepke, R. L. Libby, and M. H. Small, *J. Bacteriol.* **48**, 409 (1944).
[3] D. Shemin, *J. Biol. Chem.* **162**, 297 (1946); G. Ehrensvard, E. Sperber, E. Saluste, L. Reio, and R. Stjernholm, *ibid.* **169**, 759 (1947).

cine, and glycine can supply all of the skeleton of serine, with the α-carbon contributing to both the α- and β-carbons of serine.

The conversion of serine to glycine has been studied in extracts of liver[4] and bacteria.[5] The reverse reaction has also been studied with liver enzymes.

$$\text{L-serine} \rightleftharpoons \text{glycine} + \text{formaldehyde}$$

Model reactions of this type have been studied in which the catalyst is pyridoxal plus a metal.[6] The enzymatic reactions all appear to use pyridoxal phosphate as a cofactor, and in the case of a bacterial system, Mn^{++} is also required. A major difference between the enzymatic and the model reactions is the requirement for a folic acid cofactor in the former. The formation of glycine and acetaldehyde from L-threonine and L-allo-threonine has been described by Lin and Greenberg. Their partially purified enzyme, threonine aldolase, was not shown to require any co-factors, and the reaction was not reversed.[7] This is in contrast to the results of nonenzymatic experiments in which pyridoxal and a metal catalyze the reversible cleavage of threonine.

Some reactions of folic acid derivatives have been discussed as components of systems that synthesize and degrade purines. In the case of serine metabolism the function of folic acid is basically the same; namely, transfer of a one-carbon moiety. Whereas the carbon of previously discussed one-carbon transfers has been at the oxidation level of formate, the β-carbon of serine is at the oxidation level of formaldehyde. Reactions involving formaldehyde derivatives (hydroxymethyl compounds) of folic acid have recently been reported.

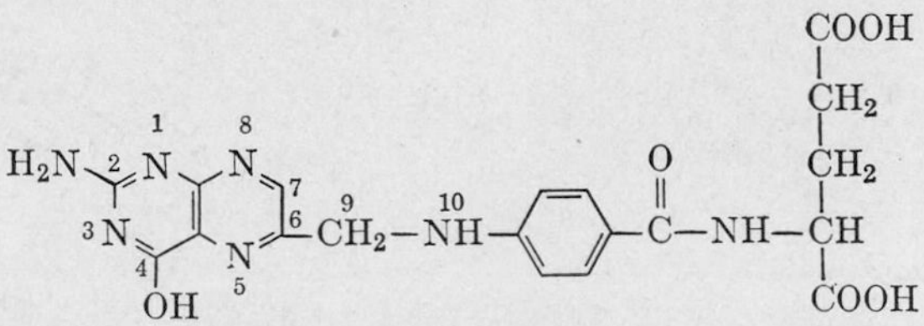

Folic acid

Folic acid is the parent compound of a large series of pteridine derivatives. Folic acid itself has vitamin activity, but for activity in one-carbon transfer reactions it must first be reduced to a tetrahydro derivative

[4] N. Alexander and D. M. Greenberg, *J. Biol. Chem.* **214,** 821 (1955).
[5] B. E. Wright and T. C. Stadtman, *J. Biol. Chem.* **219,** 863 (1956).
[6] D. E. Metzler, J. B. Longnecker, and E. E. Snell, *J. Am. Chem. Soc.* **76,** 639 (1954).
[7] S. C. Lin and D. M. Greenberg, *J. Gen. Physiol.* **38,** 181 (1954).

(hydrogens added to positions 5, 6, 7, and 8).[8] This reduction occurs enzymatically with DPNH as the reducing agent. Dihydrofolic acid is also reduced enzymatically, and may be an intermediate in the formation of tetrahydrofolic acid. The reactions to be discussed below are presented as occurring with tetrahydrofolic acid, but it is possible that in many systems analogs with additional glutamic acid residues linked in a peptide chain may be the active cofactors. The conversion of serine to glycine in *Clostridium* H. F. does not use tetrahydrofolic acid efficiently, but requires the so-called conjugates.[9]

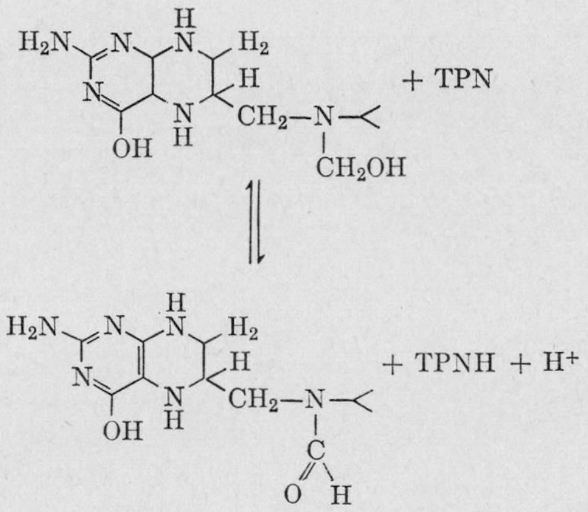

(I)

The one-carbon unit from serine is transferred to the N^{10} of tetrahydrofolic acid. This reaction requires pyridoxal phosphate and Mn^{++}.[10] There is evidence that the first step in this reaction is a condensation of serine with tetrahydrofolic acid in a reaction that does not require additional cofactors. The condensation product was identified by paper chromatography as a colored compound with a characteristic fluorescence. In a second step pyridoxal phosphate and Mn^{++} are required, and glycine

[8] R. L. Blakely, *Nature* **173**, 729 (1954); L. Jaenicke, *Biochim. et Biophys. Acta* **17**, 588 (1955); G. R. Greenberg, L. Jaenicke, and M. Silverman, *ibid.* **17**, 589 (1955).

[9] B. E. Wright, *J. Biol. Chem.* **219**, 873 (1956). It is probable that the natural cofactors in animal systems also are conjugates of tetrahydrofolic acid. Assays of fresh extracts for folic acid vitamin activity reveal the presence of very little active material in most tissues, but the amounts increase upon autolysis. Both oxidized and reduced pteridines occur as folic acid derivatives containing one or more additional glutamic acid residues. Enzymes that hydrolyze peptide bonds to liberate folic acid activity have been found in many organisms and have been named conjugases [E. L. R. Stokstad, *Vitamins* **3**, 163 (1954)].

[10] L. Jaenicke, *Federation Proc.* **15**, 281 (1956).

and hydroxymethyltetrahydrofolic acid are formed. N^{10}-Hydroxymethyltetrahydrofolic acid is formed nonenzymatically from formaldehyde and tetrahydrofolic acid. Formaldehyde does not form a stable compound with folic acid. Hydroxymethyltetrahydrofolic acid is a rather stable compound, and does not exchange significantly with free formaldehyde.[11]

Hydroxymethyltetrahydrofolic acid specifically reacts back to form serine from glycine. The enzyme from pigeon liver requires Mn^{++},[10] but no metal requirement has been established for the enzyme from mammalian liver.[4] In the presence of substrate quantities of tetrahydrofolic acid the equilibrium mixture contains slightly more serine than glycine, but until accurate assays for free formaldehyde and hydroxymethyltetrahydrofolic acid are obtained, equilibrium constants cannot be determined with accuracy. The conversion of serine to glycine is pulled to completion by the presence of pyridine nucleotides and an enzyme present in liver and bacteria. The liver enzyme acts much more rapidly with TPN than with DPN, and oxidizes hydroxymethyltetrahydrofolic acid to N^{10}-formyltetrahydrofolic acid (I).[10] The reversal of this series of reactions appears to be required to introduce formate into the β-position of serine. Formyltetrahydrofolic acid may be formed by a transfer reaction from various one-carbon donors (e.g., purines, formimino amino acids) or through a coupled reaction with ATP.[8b]

tetrahydrofolic acid + formate + ATP $\rightleftharpoons$ formyltetrahydrofolic acid + ADP + P

Serine and Threonine Dehydration. The reaction that is now spoken of as dehydration of α-amino, β-hydroxy acids was originally considered as a deamination. The reaction observed is:

$$RCHOHCHNH_2COOH \rightarrow RCH_2COCOOH + NH_3$$

This activity was measured anaerobically by Gale and Stephenson[12] as release of ammonia by bacterial cells. The system was not stable, but was protected by low concentrations of phosphate and reducing agents. At 0°C. inactivation occurred that was reversed by a boiled bacterial extract plus phosphate and a reducing agent. The reaction was better defined by Chargaff and Sprinson,[13] who also studied the reaction in liver. Several bacterial species were found to deaminate both D- and L-serine and threonine. Phosphate-, methyl-, and ethyl-substituted hydroxyl groups prevented the reaction.

The cofactor requirements for enzymes of this group are somewhat

[11] R. L. Kisliuk, *Federation Proc.* **15**, 289 (1956).
[12] E. F. Gale and M. Stephenson, *Biochem. J.* **32**, 392 (1938).
[13] E. Chargaff and D. B. Sprinson, *J. Biol. Chem.* **151**, 273 (1943).

uncertain. Lichstein and collaborators have reported effects of biotin and biotin + adenylic acid in increasing the rate of deamination by bacterial preparations.[14a,b] Since some preparations were stimulated more by yeast extract than by known additions, it was proposed that a cofactor form of biotin occurred in the yeast. Wood and Gunsalus purified an enzyme from *E. coli* that removes NH_3 from L-serine and L-threonine.[15] This enzyme has an almost absolute requirement for AMP and glutathione. Other adenine nucleotides and reducing agents were found to be ineffective. The D-isomers of the substrates inhibit the enzyme.

Metzler and Snell used the same strain of *coli* that Wood and Gunsalus had used, and by modifying the growth conditions obtained cells that yielded little activity toward L-serine, but were active with D-serine. The D-enzyme was shown to require pyridoxal phosphate.[16]

While this work was proceeding, Yanofsky and Reissig independently found the L-serine deaminase of *Neurospora*.[17] This enzyme is activated by pyridoxal phosphate, and is not stimulated by ATP, biotin, or glutathione. The *Neurospora* enzyme is somewhat more active with threonine than with serine, and acts only on the L-isomers.

Sayre and Greenberg have separated two enzymes from sheep liver.[18] One is a specific dehydrase for L-serine; no related compounds were attacked. The other enzyme is quite specific for L-threonine; allothreonine, D-threonine, and other analogs are inert but about 3 per cent of the rate given by L-threonine was found with L-serine. The cofactors of these enzymes have not been satisfactorily established. Serine dehydrase preparations have been found to be stimulated by pyridoxal phosphate, AMP, or glutathione. Combinations of these do not give increased activation, and there is significant activity in the absence of added cofactors. Threonine dehydrase is not influenced by any cofactors when the enzyme is freshly prepared. Aging or lyophilizing, which appear to cause extensive inactivation, permit demonstration of small amounts of activation by pyridoxal phosphate, AMP, and glutathione, but again combinations do not give increased activation.

GENERAL REFERENCE

Sakami, W. (1955). "Symposium on Amino Acid Metabolism" (W. D. McElroy and B. Glass, eds.), p. 658. Johns Hopkins Press, Baltimore.

[14a] H. C. Lichstein and W. W. Umbreit, *J. Biol. Chem.* **170**, 729, 423 (1947).
[14b] H. C. Lichstein, *J. Biol. Chem.* **177**, 125 (1949).
[15] W. A. Wood and I. C. Gunsalus, *J. Biol. Chem.* **181**, 171 (1949).
[16] D. E. Metzler and E. E. Snell, *J. Biol. Chem.* **198**, 363 (1952).
[17] C. Yanofsky and J. L. Reissig, *J. Biol. Chem.* **202**, 567 (1953).
[18] F. W. Sayre and D. M. Greenberg, *J. Biol. Chem.* **220**, 787 (1956).

Metabolism of Sulfur-Containing Amino Acids

The sulfur-containing amino acids, cysteine and methionine, are essential components of proteins and also participate in special reactions. In animals methionine is of particular importance because of its role in the transfer of methyl groups. Cysteine is a component of glutathione, γ-glutamyl-cysteinyl-glycine. The synthesis of these important amino acids occurs in plants and microorganisms by as yet undefined pathways. Certain interconversions at the amino acid level have been studied in animals, and some of the enzymes concerned have been identified. More is known about the degradation of the amino acids.

Sulfhydryl-Disulfide Interconversion. Cysteine is readily oxidized to the disulfide, cystine. This oxidation is catalyzed very efficiently by traces

$$
\begin{array}{ccccc}
H_2C-SH & HS-CH_2 & & H_2C-S-S-CH_2 \\
| & | & & | \quad\quad | \\
HCNH_2 + & HCNH_2 + X \longrightarrow & HCNH_2 & HCNH_2 + XH_2 \\
| & | & & | \quad\quad | \\
COO^- & COO^- & & COO^- \quad COO^- \\
\text{Cysteine} & & & \text{Cystine}
\end{array}
$$

of metals, especially iron, copper, and manganese, with oxygen as the ultimate electron acceptor. It, therefore, is difficult to study the formation of cystine in biological systems, but evidence has been found that cytochrome c in animal preparations[1] and dehydroascorbic acid in plants[2] can serve as electron acceptors from cysteine and glutathione. Cystine is reduced to cysteine by a DPNH-requiring dehydrogenase found in yeast and plant preparations.[3] These reactions are very similar to those described for glutathione (GSH).[4] GSH is also oxidized to the disulfide (GSSG) by most crude extracts with a variety of electron acceptors. Reduction of the disulfide is catalyzed in plants by TPNH-specific dehydrogenases. Partially purified preparations have been named gluta-

$$G-S-S-G + TPNH + H^+ \rightarrow 2GSH + TPN$$

thione reductase. Similar enzymes have been detected in animal liver and yeast.[5] These have been reported to use DPN, but less efficiently than TPN. Glutathione reductase of *E. coli* has been found to contain FAD.[6]

[1] D. Keilin, *Proc. Roy. Soc.* **B106,** 418 (1930).
[2] F. G. Hopkins and E. J. Morgan, *Biochem. J.* **30,** 1446 (1936).
[3] W. J. Nickerson and A. H. Romano, *Science* **115,** 676 (1952).
[4] E. E. Conn and B. Vennesland, *J. Biol. Chem.* **192,** 17 (1951); L. W. Mapson and D. R. Goddard, *Biochem. J.* **49,** 592 (1951).
[5] E. Racker, *J. Biol. Chem.* **217,** 855 (1955).
[6] R. E. Agnis, *J. Biol. Chem.* **213,** 217 (1955).

Cysteine Desulfhydrase. Cysteine undergoes a desulfuration that is believed to be analogous to the dehydration of serine. The enzyme cysteine desulfhydrase requires pyridoxal phosphate and forms pyruvate, H_2S, and NH_3.[7] This enzyme has been found in animal liver and presumably occurs in microorganisms that release H_2S from cysteine. Evidence has been obtained with H_2S^{35} that indicates some reversibility of the reaction, but no thermodynamic data are available. A similar enzyme has been reported to form α-ketobutyrate, NH_3, and H_2S from homocysteine.[8]

Cysteinesulfinic Acid. Cysteine is oxidized by enzyme systems present in bacteria and in liver to the corresponding sulfinic acid.[9] It has been suggested that the unstable sulfenic acid is an intermediate in this oxidation. The nature of the reaction that produces cysteinesulfinic acid is not known. The subsequent metabolism of the sulfinic acid may proceed by any of three pathways. One involves further oxidation to cysteinesulfonic acid, cysteic acid.[10] The enzyme responsible has not been separated from the system responsible for the formation of cysteinesulfinic acid. Cysteinesulfonic acid is decarboxylated to taurine (I) by the decarboxylase mentioned previously (p. 284).

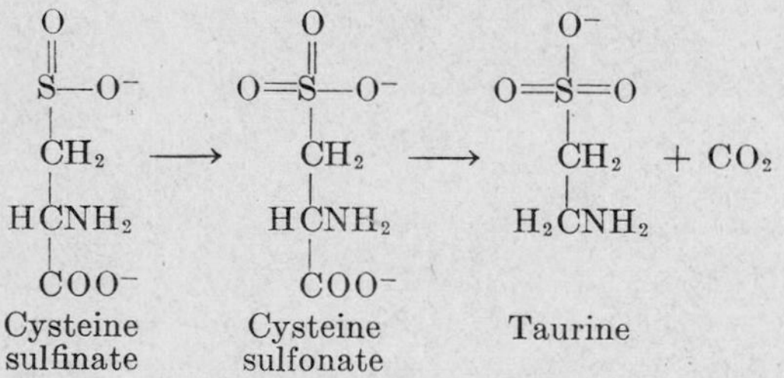

A second pathway of cysteinesulfinic acid involves decarboxylation to hypotaurine (II).[11] This reaction is followed by oxidation to taurine. This pathway appears to be more important than the cysteic acid pathway.

The major portion of cysteinesulfinic acid appears to be metabolized

[7] C. Fromageot, E. Wookey, and P. Chaix, *Enzymologia* **9**, 198 (1941); C. V. Smythe, *J. Biol. Chem.* **142**, 387 (1942); A. E. Braunstein and R. M. Azurkh, *Doklady Akad. Nauk S.S.S.R.* **71**, 93 (1950); C. Fromageot and R. Desnuelle, *Bull. soc. chim. biol.* **24**, 1269 (1942).

[8] R. E. Kallio, *J. Biol. Chem.* **192**, 371 (1951).

[9] C. Fromageot, F. Chatagner, and B. Bergeret, *Biochim. et Biophys. Acta* **2**, 294 (1948).

[10] G. Medes, *Biochem. J.* **33**, 1559 (1939).

[11] B. Bergeret and F. Chatagner, *Biochim. et Biophys. Acta* **9**, 141 (1952).

$$
\begin{array}{ccc}
O^- & & O^- \\
| & & | \\
S{=}O & & S{=}O \\
| & \longrightarrow & | \\
CH_2 & & CH_2 \\
| & & | \\
HCNH_2 & & H_2CNH_2 \\
| & & \\
COO^- & &
\end{array}
$$

Hypotaurine

(II)

through a third mechanism by both bacterial and animal systems. In this sequence the amino group is transferred to α-ketoglutarate or to oxalacetate.[12] The resulting α-keto-β-sulfinylpropionic acid (β-sulfinylpyruvate) is split in a Mn^{++} catalyzed reaction to yield sulfite and pyruvate (III). So far, no enzyme analogous to oxalacetic decarboxylase has

$$
\begin{array}{ccccccccc}
O^- & COO^- & & COO^- & O^- & & O & & O^- \\
| & | & & | & | & & \| & & | \\
S{=}O & CH_2 & \xrightarrow{\text{transaminase}} & CH_2 & S{=}O & \xrightarrow{Mn^{++}} & HO{-}S{-}O^- & \longrightarrow & O{=}S{=}O \\
| & | + | & & | & | + | & & + & & | \\
CH_2 + CH_2 & & & CH_2 + CH_2 & & & CH_3 & & O^- \\
| & | & & | & | & & | & & \\
HCNH_2 & C{=}O & & HCNH_2 & C{=}O & & C{=}O & & \\
| & | & & | & | & & | & & \\
COO^- & COO^- & & COO^- & COO^- & & COO^- & &
\end{array}
$$

(III)

been found for the desulfination reaction, but the failure to observe accumulation of the sulfinic acid in mitochondrial extracts indicates the probable existence of such an enzyme.

Beef heart preparations allow sulfite to accumulate. In bacterial and liver systems sulfate is the product.[13] β-Sulfonylpyruvate has been found to be inert in both systems, indicating that the sulfate is formed by oxidation of inorganic sulfite. A partially purified sulfite dehydrogenase has been obtained from liver. This enzyme catalyzes the reduction of methylene blue by sulfite. Similar activities have been found in some bacteria.

A variation of the sulfinylpyruvate pathway has been reported in rat liver mitochondrial acetone powders.[14] In this, DPN is used to support an oxidative deamination. This reaction also yields β-sulfinylpyruvate, which is desulfinated as described above.

Sulfur Formation. The transamination of cysteine results in the formation of β-mercaptopyruvate.[15] This product is also obtained through

[12] E. B. Kearney and T. P. Singer, *Biochim. et Biophys. Acta* **11**, 276 (1953); F. Chatagner, B. Bergeret, T. Séjourné, and C. Fromageot, *ibid.* **9**, 340 (1952).

[13] T. P. Singer and E. B. Kearney, *Biochim. et Biophys. Acta* **14**, 570 (1954); I. Fridovich and P. Handler, *Symposium on Inorg. Nitrogen Metabolism* p. 593 (1956).

[14] See p. 575 of review by Singer and Kearney (General References).

[15] P. S. Cammarata and P. P. Cohen, *J. Biol. Chem.* **187**, 439 (1950).

the action of certain amino acid oxidases.[16] A transamination reaction studied with β-mercaptopyruvate was found to yield alanine instead of the expected cysteine. This was found to result from a sequence of reactions in which pyruvate was formed initially by removal of elemental sulfur, as shown in (IV).[17] This enzyme is distinct from cysteine desulfhydrase. It is activated by sulfhydryl compounds. The sulfur-forming activity has been found in liver and bacterial preparations.

$$\begin{array}{cc}
\text{SH} & \text{CH}_3 \\
| & | \\
\text{CH}_2 & \text{C=O} \\
| & \longrightarrow \text{S} + \\
\text{C=O} & \text{COO}^- \\
| & \\
\text{COO}^- &
\end{array}$$

$$(IV)$$

Oxidation of Cystine. Cystine has been reported to initiate a series of reactions in which the S—S bond remains intact. An oxidation product is cystine disulfoxide, RSOSOR.[18] This compound can yield both cysteine and sulfate in the intact rat, but most of it is converted to taurine. Both carboxyl groups of cystine disulfoxide are removed by a decarboxylase different from cysteic decarboxylase. Oxidation and hydrolysis of the decarboxylation product would yield two equivalents of taurine, but the individual steps of this conversion have not been isolated.

Rhodanese. A product of animal metabolism is thiosulfate, $S_2O_3^=$. Two mechanisms have been shown to produce thiosulfate.[19] One is a nonspecific oxidation of sulfide catalyzed by various hemins; no protein is required. Another mechanism was found to employ liver enzymes, which form thiosulfate from β-mercaptopyruvate + sulfite. Thiosulfate is metabolized by an enzyme named rhodanese, to indicate that the enzyme synthesizes, rather than degrades, rhodanate (thiocyanate).[20] Rhodanese, which was recently crystallized by Sorbo,[21] catalyzes the reaction:

$$Na_2S_2O_3 + HCN \rightarrow HSCN + Na_2SO_3$$

The reaction is very slightly reversible, if reversible at all. An equilibrium constant of about 10^{10} was estimated, but this depended on detection of very small traces of cyanide. This reaction provides rationale for the treatment of cyanide poisoning with thiosulfate. The reaction also proceeds with various thiosulfonates at different rates.

[16] E. A. Zeller, *Advances in Enzymol.* **8,** 459 (1948).
[17] A. Meister, P. E. Fraser, and S. V. Tice, *J. Biol. Chem.* **206,** 561 (1954).
[18] G. Medes and N. Floyd, *Biochem. J.* **36,** 259 (1942).
[19] B. Sörbo, *Biochim. et Biophys. Acta* **21,** 393 (1956).
[20] K. Lang, *Biochem. Z.* **259,** 243 (1933).
[21] B. Sörbo, *Acta Chem. Scand.* **7,** 1129 (1953).

Synthesis of Sulfur Amino Acids. Of the many oxidation states of sulfur, only sulfite has been shown to be utilized by cell-free systems in the net synthesis of compounds with carbon-sulfur bonds, although mutant studies have indicated that more reduced forms can be incorporated. The formation of cysteinesulfinic acid from sulfite has been demonstrated in extracts of acetone-dried rabbit kidney;[22] it is possible that this reaction participates in the principal mechanism of sulfur incorporation. In many organisms that require preformed sulfur amino acids, cysteine may be formed from methionine. Only the sulfur of methionine is transferred to cysteine; the carbon skeleton of cysteine is derived exclusively from serine. Transsulfuration appears to require the formation of homocysteine from methionine. Homocysteine and serine condense to form a thioether, cystathionine (V).[23] Pyridoxal phosphate has been

$$
\begin{array}{ccccc}
\text{CH}_2\text{—SH} & & \text{HO—CH}_2 & & \text{CH}_2\text{—S—CH}_2 \\
| & & | & & |\qquad\quad | \\
\text{CH}_2 & + & \text{HCNH}_2 & \longrightarrow & \text{CH}_2 \quad \text{HCNH}_2 \\
| & & | & & |\qquad\quad | \\
\text{HCNH}_2 & & \text{COO}^- & & \text{HCNH}_2 \quad \text{COO}^- \\
| & & & & | \\
\text{COO}^- & & & & \text{COO}^- \\
\text{Homocysteine} & & \text{Serine} & & \text{Cystathionine}
\end{array}
$$

(V)

implicated as a cofactor in this condensation. In microorganisms the sulfur of homocysteine may be derived from cysteine also through the intermediate formation of cystathionine. Therefore a similar condensation of cysteine with homoserine or threonine may be postulated, but this reaction has not been demonstrated in an isolated system.

The enzymatic cleavage of cystathione also requires pyridoxal phosphate as a cofactor.[24] In higher animals, cysteine, α-ketobutyrate, and ammonia are the products. Crude liver extracts also contain enzymes that degrade allocystathionines, compounds in which one of the amino acid constituents has an L-configuration and the other D.[25] The products in each case include an L-amino acid containing sulfur (cysteine or homocysteine) and the α-keto acid from the other half of the substrate. D-Cystathionine, with both amino acid components in the D-configuration, is very slightly degraded. Organisms such as *Neurospora*, which accumulates cystathionine as an intermediate in methionine synthesis,[26] probably contain a cleavage enzyme that liberates homocysteine, but this enzyme has not been isolated.

[22] F. Chapeville and P. Fromageot, *Biochim. et Biophys. Acta* **14,** 415 (1954).
[23] F. Binkley, *J. Biol. Chem.* **191,** 531 (1951).
[24] F. Binkley, G. M. Christensen, and W. N. Jensen, *J. Biol. Chem.* **194,** 109 (1952).
[25] F. Binkley, W. P. Anslow, Jr., and V. du Vigneaud, *J. Biol. Chem.* **143,** 559 (1942).
[26] N. H. Horwitz, *J. Biol. Chem.* **171,** 255 (1947).

Transmethylation. Methionine is formed from homocysteine by the addition of a one-carbon moiety that becomes the methyl group of the thioether (VI). The methyl group of methionine may be derived from certain preformed methyl groups or from various one-carbon donors, such as serine and formate. In the latter case the one-carbon unit must

$$\underset{\text{Betaine}}{\underset{\substack{| \\ CH_3}}{\overset{\substack{O \\ \| }}{\underset{\substack{| \\ CH_3}}{CH_2-C-O^-}}}} \quad + \quad \underset{\substack{\text{Homocysteine} \\ \text{(VI)}}}{\overset{\substack{SH \\ | \\ CH_2 \\ | \\ CH_2 \\ | \\ HCNH_2 \\ | \\ COO^-}}{}} \quad \longrightarrow \quad \underset{\text{Dimethylglycine}}{\overset{\substack{O \\ \| \\ CH_2-C-O^- \\ | \\ N \\ \diagdown \\ CH_3 \quad CH_3}}{}} \quad + \quad \underset{\text{Methionine}}{\overset{\substack{CH_3 \\ | \\ S \\ | \\ CH_2 \\ | \\ CH_2 \\ | \\ HCNH_2 \\ | \\ COO^-}}{}} \quad + H^+$$

be "activated" through condensation with a tetrahydrofolic acid compound, but it is not clear at which oxidation level transfer to homocysteine takes place or how reduction to the level of methyl occurs. Transfer of preformed methyl groups has been known for several years. Pioneer isotope experiments by du Vigneaud and collaborators[27] showed that the carbon and all of the hydrogen of methyl groups remain together during transmethylation. Methyl donors include betaine and certain thetins.

$$\underset{\text{Dimethylthetin}}{\overset{\substack{CH_3 \\ \diagdown \\ S^+-CH_2-\overset{\overset{\text{O}}{\|}}{C}-O^- \\ \diagup \\ CH_3}}{}} + \quad \underset{\text{Homocysteine}}{\overset{\substack{SH \\ | \\ CH_2 \\ | \\ CH_2 \\ | \\ HCNH_2 \\ | \\ COO^-}}{}} \quad \longrightarrow \quad \underset{\substack{\text{Methylmercapto-} \\ \text{acetate}}}{\overset{\substack{CH_3 \\ | \\ S \\ | \\ CH_2 \\ | \\ COO^-}}{}} \quad + \quad \underset{\text{Methionine}}{\overset{\substack{CH_3 \\ | \\ S \\ | \\ CH_2 \\ | \\ CH_2 \\ | \\ HCNH_2 \\ | \\ COO^-}}{}} \quad + H^+$$

(VII)

Dubnoff and Borsook[28] have demonstrated synthesis of methionine by liver and kidney preparations, and have separated a "dimethylthetin transmethylase" from "betaine transmethylase." The dimethylthetin enzyme, renamed "dimethylthetin-homocysteine methylpherase" has been purified by Durell and Anderson,[29] who suggest that their prepara-

[27] V. du Vigneaud, *Harvey Lectures Ser.* **38**, 37 (1942).
[28] J. W. Dubnoff and H. Borsook, *J. Biol. Chem.* **176**, 789 (1948).
[29] J. Durell and D. G. Anderson, *Federation Proc.* **15**, 295 (1956).

tion may be homogeneous. It requires glutathione and is relatively specific for the two substrates in its name (VII).

Methionine shares with those compounds that contribute methyl groups to form methionine the ability to methylate *in vivo* another group of compounds that are not donors, but only acceptors, of methyl groups.

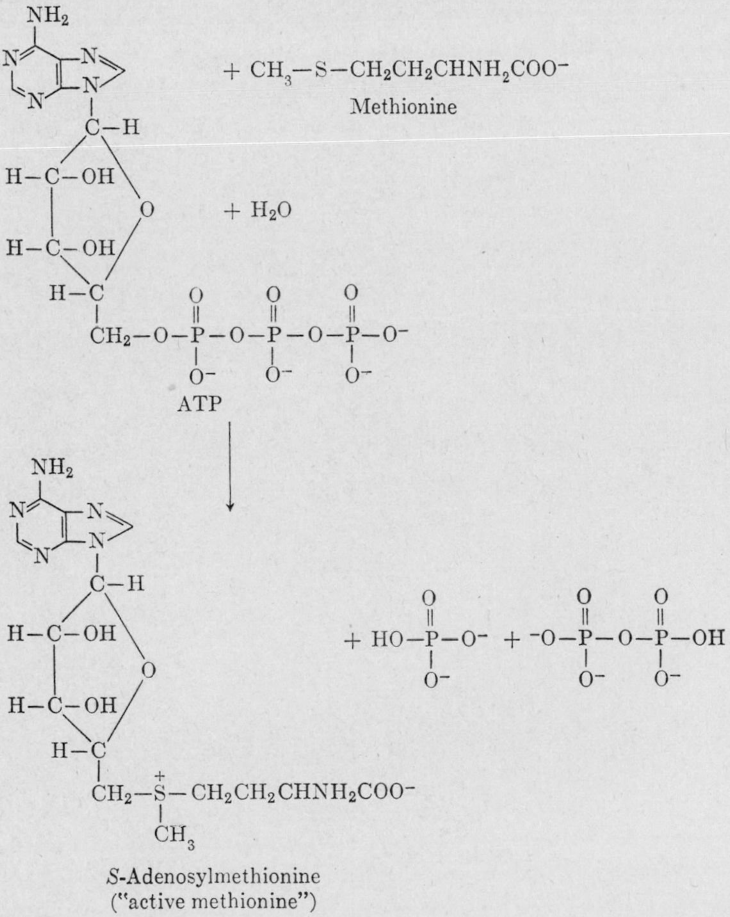

S-Adenosylmethionine
("active methionine")

(VIII)

This group includes nicotinamide, guanidoacetic acid, and ethanolamine bearing 2, 1, or no N-methyl groups. Since a large number of other compounds, including many alkaloids, occur as methyl compounds, the list of acceptor compounds will probably grow rapidly. It is probable that only methionine reacts directly in the methylation of the "acceptor" compounds, and that other donors act through methionine.

Methionine differs from the other methyl donors in being a simple thioether, whereas the others are sulfonium compounds or quarternary ammonium compounds. "Onium" compounds with an anionic group appear to be the only methyl donors; choline, an "onium" compound, is inert as a donor until it is oxidized to betaine. Methionine was found to transfer its methyl group only in the presence of ATP. Cantoni has

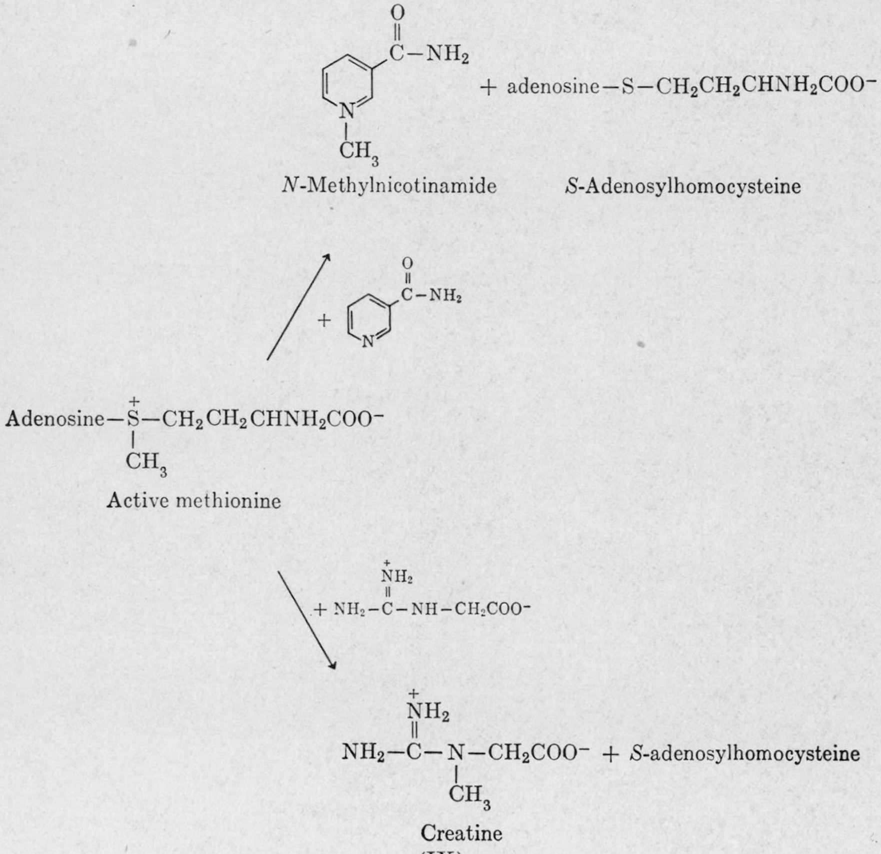

$$
\begin{array}{c}
O \\
\parallel \\
C-NH_2
\end{array}
$$

N-Methylnicotinamide   S-Adenosylhomocysteine

Adenosine—$\overset{+}{S}$—CH$_2$CH$_2$CHNH$_2$COO$^-$
          |
          CH$_3$

Active methionine

$$\overset{+}{N}H_2$$
$$+ NH_2-C-NH-CH_2COO^-$$

$$\overset{+}{N}H_2$$
$$NH_2-C-N-CH_2COO^- + S\text{-adenosylhomocysteine}$$
                |
                CH$_3$

Creatine
(IX)

purified the methionine-activating enzyme and has isolated and identified the "active methionine."[30] The activating reaction is different from all known reactions of ATP. The over-all reaction is shown in (VIII).

Extensive purification of the methionine-activating enzyme was re-

[30] G. L. Cantoni, *J. Biol. Chem.* **204,** 403 (1953); G. L. Cantoni and J. Durell, *Federation Proc.* **15,** 229 (1956).

quired to permit demonstration of pyrophosphate as a product, because of the presence of an active pyrophosphatase in cruder preparations. Experiments with P^{32}-labeled ATP have shown that the terminal phosphate of ATP becomes inorganic orthophosphate in methionine activation

$$\text{methionine} + ARP^1P^2P^3 \rightarrow \text{active methionine} + P^1P^2 + P^3$$

while the ester and middle phosphates become pyrophosphate. This reaction is obviously complex, but there is no evidence for intermediate products or a stepwise hydrolysis of ATP.

S-Adenosylmethionine is the specific methyl donor in reactions catalyzed by two partially purified enzymes; one makes N-methylnicotinamide[31] and the other creatine (IX).[32] These reactions proceed to completion, and no evidence for reversal has been found. The other reaction product in the case of creatine synthesis has been identified as adenosylhomocysteine.[33]

S-Adenosylmethionine appears to be broken down in yeast to form thiomethyladenosine. This compound was shown by Schlenk to accumulate when methionine was fed to yeast and to contain the methyl carbon and the sulfur of the methionine,[34] but it is not certain that the formation of thiomethyladenosine from active methionine is enzymatic.

Mercaptide Elimination. Kallio and Larson have described an oxidative degradation of methionine by a *Pseudomonas*.[35] A pyridoxal phosphate enzyme eliminates methyl mercaptan and ammonia, leaving α-ketobutyrate. The methyl mercaptide is oxidized to dimethyl disulfide.

Dimethylpropiothetin occurs in certain marine algae. An enzyme has been obtained from one of these that eliminates dimethyl sulfide from the thetin, leaving acrylic acid.[36]

Mercaptide Utilization. A series of reactions involving methyl mercaptan was recently found by Black and Wright.[37] A specific activating enzyme uses ATP hydrolysis to support the formation of a thioester, methylthiophosphoglycerate, from methyl mercaptan and 3-phosphoglycerate (X). This product can be hydrolyzed by a specific phosphatase to leave methyl thioglycerate. An incidental observation resulted in the finding of a kinase that forms 3-phosphoglycerate from ATP and glyceric

[31] G. L. Cantoni, *J. Biol. Chem.* **189,** 203 (1951).

[32] G. L. Cantoni and P. Vignos, Jr., *J. Biol. Chem.* **209,** 647 (1954).

[33] G. L. Cantoni and E. Scarano, *J. Am. Chem. Soc.* **76,** 4744 (1954).

[34] F. Schlenk and R. L. Smith, *J. Biol. Chem.* **204,** 27 (1953).

[35] R. E. Kallio and A. D. Larson, *Symposium on Amino Acid Metabolism, Baltimore, 1954* p. 616 (1955).

[36] G. L. Cantoni and D. G. Anderson, *J. Biol. Chem.* **222,** 171 (1956).

[37] S. Black and N. G. Wright, *J. Biol. Chem.* **221,** 171 (1956).

acid. These reactions are catalyzed by enzymes that were separated from each other from yeast. The metabolic significance of the thioesters has not been determined. It has been noted, however, that methyl mercaptan is rapidly metabolized by both microorganisms and intact animals.

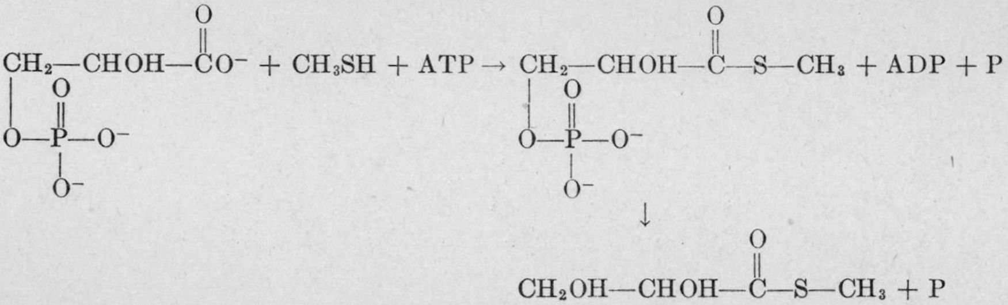

$$
\begin{array}{c}
\text{O} \\
\parallel \\
\text{CH}_2\text{—CHOH—CO}^- + \text{CH}_3\text{SH} + \text{ATP} \rightarrow \\
\mid \\
\text{O} \\
\parallel \\
\text{O—P—O}^- \\
\mid \\
\text{O}^-
\end{array}
\qquad
\begin{array}{c}
\text{O} \\
\parallel \\
\text{CH}_2\text{—CHOH—C—S—CH}_3 + \text{ADP} + \text{P} \\
\mid \\
\text{O} \\
\parallel \\
\text{O—P—O}^- \\
\mid \\
\text{O}^-
\end{array}
$$

$$\downarrow$$

$$
\begin{array}{c}
\text{O} \\
\parallel \\
\text{CH}_2\text{OH—CHOH—C—S—CH}_3 + \text{P}
\end{array}
$$

(X)

Sulfate Activation. Inorganic sulfate is used extensively for the formation of sulfate esters, which are the principal metabolic products of many cyclic compounds with hydroxyl groups. A soluble enzyme system from rat liver was found to synthesize phenyl sulfate from phenol and inor-

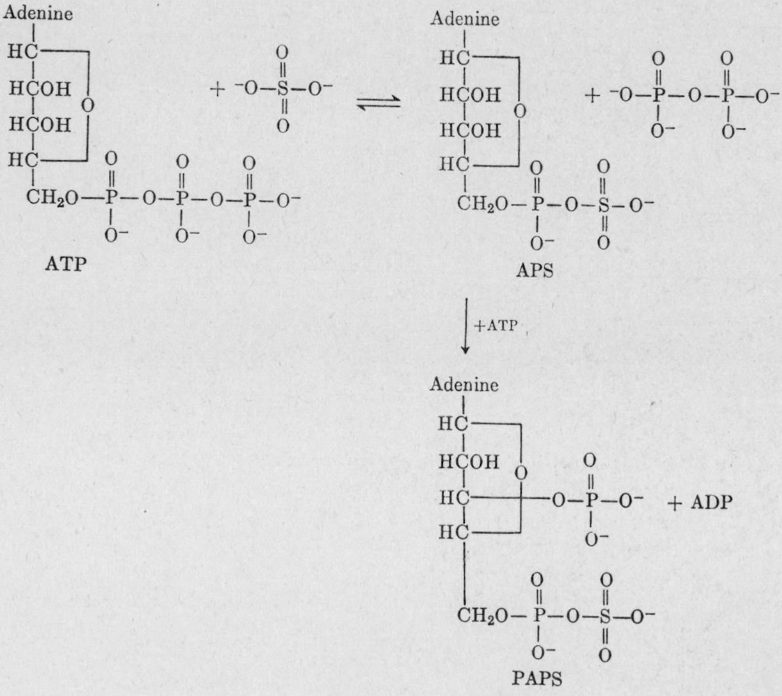

(XI)

ganic sulfate in the presence of ATP and Mg^{++}.[38] The enzymatic activation of sulfate by a liver enzyme has been shown to consist of formation of an adenosine derivative containing two phosphate groups in addition to sulfate (XI).[39a] The reaction is not a simple replacement of the terminal phosphate, however.

In a first step ATP and sulfate react to form adenosine-5'-phosphosulfate (APS), a sulfate analog of ADP, and inorganic pyrophosphate. This reversible reaction is pulled by inorganic pyrophosphatase. The sulfurylase is not specific for sulfate, but also reacts with several other anions. In a second step, APS is phosphorylated in the 3' position in an irreversible kinase reaction in which ATP is the phosphate donor. The products of the kinase reaction are ADP and 3'-phosphoadenosine-5'-phosphosulfate (PAPS).[39b]

The transfer of sulfate from "active sulfate" to acceptors such as p-nitrophenol is catalyzed by an enzyme separated from the activating enzyme.[40]

Other Sulfur-Containing Compounds. Sulfur is a component of many molecules of great biological interest, including lipoic acid, thiamine, and biotin among the vitamins whose functions have been discussed. Nothing is known about the enzymatic syntheses of these compounds. The sulfur of CoA is derived from cysteine. Many other sulfur-containing compounds occur naturally, some in high concentrations, as S-methylcysteine sulfoxide in cabbage and other plants, ergothioneine (2-thiolhistidine betaine) in molds and other organisms, and alliin (S-allylcysteine sulfoxide) in garlic. The study of the synthesis, function, and degradation of these compounds has only begun.

GENERAL REFERENCES

Challenger, F. (1951). *Advances in Enzymol.* **12**, 929 (1955); *Quart. Revs. (London)* **9**, 255.
Fromageot, C. (1951). *In* "The Enzymes" (J. B. Sumner and K. Myrbäck, eds.), Vol. II, Part 1, p. 609. Academic Press, New York.
Singer, T. P., and Kearney, E. B. (1955). "Symposium on Amino Acid Metabolism" (W. D. McElroy and B. Glass, eds.), p. 558. Johns Hopkins Press, Baltimore.
Sourkes, T. L. (1951). *In* "The Enzymes" (J. B. Sumner and K. Myrbäck, eds.), Vol. I, Part 2, p. 1068. Academic Press, New York.
Stekol, J. A. (1955). "Symposium on Amino Acid Metabolism" (W. D. McElroy and B. Glass, eds.), p. 509. Johns Hopkins Press, Baltimore.

[38] R. H. DeMeio, M. Wizerkaniuk, and E. Fabiani, *J. Biol. Chem.* **203**, 257 (1953).
[39a] P. W. Robbins and F. Lipmann, *J. Am. Chem. Soc.* **78**, 2652 (1956).
[39b] R. S. Bandurski, L. G. Wilson, and C. L. Squires, *J. Am. Chem. Soc.* **78**, 6408 (1956); P. W. Robbins and F. Lipmann, *ibid.* **78**, 6409 (1956).
[40] H. L. Segal, *J. Biol. Chem.* **213**, 161 (1955).

Histidine

Biosynthesis. Histidine is synthesized by microorganisms and by plants, but animals synthesize little if any of this amino acid. Growing mammals require histidine in their diets for growth and positive nitrogen balance.[1] Nitrogen balance studies[2] and isotope incorporation experiments[3] have indicated that histidine may be synthesized by adult mammals, but alternate explanations for these experiments have been offered,[4] and there is doubt that an enzymatic mechanism for the synthesis of histidine occurs in higher animals. Analogs of histidine, the corresponding α-hydroxy and α-keto compounds, can be converted to histidine in animals by the action of the oxidase and transaminase described earlier (pp. 292 and 286).

The early steps in the synthesis of histidine by any organism are not known. Isotope studies with microorganisms indicate that the five-carbon chain is derived from carbohydrate, not from glutamate,[3] that the γ-nitrogen comes from the amide nitrogen of glutamine,[5] and that carbon 2 of the imidazole ring is formed from formate.[6,7] The sequence of the condensations is not known, but it is apparent that a series of reactions leads directly to histidine; the imidazole of histidine is not derived from the 5-membered ring of purines, and histidine is the precursor rather than a derivative of other imidazole-containing compounds, such as histamine and ergothioneine.

The earliest established precursors of histidine are a series of imidazole-containing phosphate esters that were found to accumulate in histidine-requiring mutants of *Neurospora*.[8] Genetic evidence indicated that these compounds are formed in the sequence shown in (I).

The enzymes that catalyze these transformations have been partially purified from extracts of *Neurospora*. Imidazole glycerol phosphate is dehydrated by an enzyme named IGP dehydrase to form imidazole acetol phosphate.[9] The dehydrase requires Mn^{++} and a sulfhydryl com-

---

[1] H. Ackroyd and F. G. Hopkins, *Biochem. J.* **10**, 551 (1916); G. J. Cox and W. C. Rose, *J. Biol. Chem.* **68**, 769 (1926).

[2] W. C. Rose, W. J. Haines, D. T. Warner, and J. E. Johnson, *J. Biol. Chem.* **188**, 49 (1951).

[3] L. Levy and M. J. Coon, *Federation Proc.* **11**, 248 (1952).

[4] H. Tabor, *Pharmacol. Revs.* **6**, 299 (1954).

[5] A. Neidle and H. Waelsch, *J. Am. Chem. Soc.* **78**, 1767 (1956).

[6] L. Levy and M. J. Coon, *J. Biol. Chem.* **192**, 807 (1951).

[7] H. Tabor, A. H. Mehler, O. Hayaishi, and J. White, *J. Biol. Chem.* **196**, 121 (1952).

[8] B. N. Ames and H. K. Mitchell, *J. Biol. Chem.* **212**, 687 (1955).

[9] B. N. Ames, *Federation Proc.* **15**, 210 (1956).

pound. These requirements are similar to those of aconitase, but unlike the aconitase reaction, the dehydrase reaction can be demonstrated only in one direction.

The conversion of imidazole acetol phosphate to histidinol phosphate is catalyzed by a transaminase.[10] This pyridoxal phosphate enzyme reacts

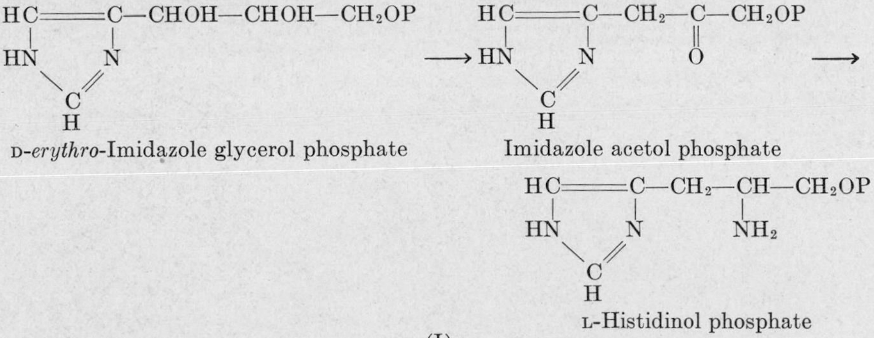

$$HC\!\!=\!\!=\!\!C\!-\!CHOH\!-\!CHOH\!-\!CH_2OP$$

HN   N

          C
          H

D-*erythro*-Imidazole glycerol phosphate

$$\longrightarrow \quad HC\!\!=\!\!=\!\!C\!-\!CH_2\!-\!C\!-\!CH_2OP$$

HN   N   O $\longrightarrow$

          C
          H

Imidazole acetol phosphate

$$HC\!\!=\!\!=\!\!C\!-\!CH_2\!-\!CH\!-\!CH_2OP$$

HN   N   NH_2

          C
          H

L-Histidinol phosphate

(I)

with glutamate and certain other amino acids. Histidine, arginine, and α-aminoadipate, as well as histidinol phosphate and glutamate, serve as amino donors in reactions with various ones of the corresponding keto acids as amino acceptors. Free histidinol and imidazole acetol do not react; only their phosphate esters are substrates. Not all of the possible

TABLE 8

RELATIVE RATES OF REACTION WITH IMIDAZOLE ACETOL
PHOSPHATE TRANSAMINASE[10]

Compound	Imidazole acetol phosphate	α-Ketoglutarate	α-Ketoadipate	α-Keto-δ-guanidovalerate
Histidinol phosphate	−	+++	++	+
Glutamate	+++	−	++	+
α-Aminoadipate	++	++	−	+
Arginine	++	+	+	−
Histidine	++	+	+	0

combinations are effective; histidine and arginine and the corresponding keto acids do not cross react (see Table 8). It appears that at least one substrate must be a dibasic acid. The reaction between glutamate and imidazole acetol phosphate is faster than reactions with other pairs of substrates. The equilibrium constant for this reaction is about 0.04. The pH optimum is near 8, but varies somewhat with the buffer used. The

[10] B. N. Ames and B. L. Horecker, *J. Biol. Chem.* **220**, 113 (1956).

transaminase is inhibited by carbonyl reagents that would be expected to react with pyridoxal phosphate and by iodoacetate.

The conversion of histidinol phosphate to histidine requires removal of the phosphate group; free histidinol was the first precursor of histidine established in genetic studies.[11] The hydrolysis of histidinol phosphate is catalyzed by a phosphatase that has recently been purified by Ames from *Neurospora* extracts. This enzyme is active with histidinol phosphate as a substrate, but has very little, if any, activity with a number of phosphate esters, including the other imidazole compounds involved in histidine biosynthesis. At least one less specific phosphatase occurs in *Neurospora;* histidinol phosphate is also hydrolyzed nonspecifically, but the activity of the nonspecific hydrolytic enzyme is slow compared with the specific histidinol phosphatase. The nonspecific reaction may account for the slow growth of mutants deficient in the specific enzyme. The specific enzyme is quite insensitive to beryllium and chelating agents, which inhibit nonspecific phosphate hydrolysis.

$$
\begin{array}{c}
\text{HC}\!=\!=\!=\!\text{C}-\text{CH}_2\text{CH}-\text{CH}_2\text{OH} \\
\mid \qquad \mid \qquad\quad \mid \\
\text{HN} \qquad \text{N} \qquad \text{NH}_2 \\
\diagdown\,\text{C}\,\diagup\diagup \\
\text{H}
\end{array}
$$

L-Histidinol

$$\underset{\text{DPNH}}{\overset{\text{DPN}}{\rightleftharpoons}}$$

$$
\begin{array}{c}
\text{HC}\!=\!=\!=\!\text{C}-\text{CH}_2\text{CH}-\text{CHO} \\
\mid \qquad \mid \qquad\quad \mid \\
\text{HN} \qquad \text{N} \qquad \text{NH}_2 \\
\diagdown\,\text{C}\,\diagup\diagup \\
\text{H}
\end{array}
$$

L-Histidinal

$$\underset{\text{DPN}}{\longrightarrow}$$

$$
\begin{array}{c}
\text{HC}\!=\!=\!=\!\text{C}-\text{CH}_2\text{CH}-\text{COOH} \\
\mid \qquad \mid \qquad\quad \mid \\
\text{HN} \qquad \text{N} \qquad \text{NH}_2 \\
\diagdown\,\text{C}\,\diagup\diagup \\
\text{H}
\end{array}
$$

L-Histidine

(II)

Histidinol is oxidized to histidine by a dehydrogenase that appears to catalyze both steps required.[12] The most definitive studies on this dehydrogenase were carried out with preparations of *Arthrobacter histidinolovorans*, a soil organism isolated by an enrichment technique using histidinol as a carbon and nitrogen source. Similar enzymes occur in *E. coli* and yeast. Two equivalents of DPNH are produced in this reaction. The reduction of the first mole of DPN by these enzymes should result in the formation of the aldehyde, histidinal. This compound, however, has not been detected in incubation mixtures, even when aldehyde binders were added. Histidinal, an unstable compound in neutral solution,[13] does serve as a substrate for the dehydrogenase, with both DPN

[11] H. J. Vogel, B. D. Davis, and E. S. Mingioli, *J. Am. Chem. Soc.* **73**, 1897 (1951).
[12] E. Adams, *J. Biol. Chem.* **217**, 325 (1955).
[13] E. Adams, *J. Biol. Chem.* **217**, 317 (1955).

and DPNH as cofactors. The reaction with DPNH and histidinal results in histidinol formation. This part of the reaction is easily reversed. The reaction with DPN leads irreversibly to histidine (II). There is no separation of these activities during purification, and mutants blocked between histidinol and histidine all lack the ability to react with either histidinol or histidinal; no extracts were found to react with one and not the other. Therefore it appears that a single enzyme catalyzes both steps, and that enzymatically formed histidinal does not accumulate, but preferentially reacts to form histidine. This was the first such enzyme to be detected, but, in the failure of an intermediate aldehyde to accumulate, UDPG dehydrogenase was subsequently found to have similar properties (p. 249).

Degradation. One pathway of degradation is started by decarboxylation to histamine by the enzyme previously discussed. This is a minor pathway in terms of quantities of histidine metabolized. The major pathway in both animals and microorganisms retains the 5-carbon chain and leads to glutamic acid.[14] Older work established that histidine degradation occurs only in the livers of animals, but this work was carried out only with crude preparations, and led to the publication and general acceptance of hypothetical schemes that have little relation to reactions found more recently.

Histidine-Glutamic Acid Pathway. Histidine is deaminated by a specific enzyme, histidase, to form urocanic acid (III).[15] This reaction was

$$
\begin{array}{ccc}
\text{HC}\!\!=\!\!\!=\!\!\text{C}\!-\!\text{CH}_2\text{CHCOOH} & \text{HC}\!\!=\!\!\!=\!\!\text{C}\!-\!\text{CH}\!\!=\!\!\text{CHCOOH} & \\
\mid \qquad \mid \qquad \mid & \mid \qquad \mid & \\
\text{HN} \qquad \text{N} \qquad \text{NH}_2 \quad \longrightarrow & \text{HN} \qquad \text{N} & +\ \text{NH}_3 \\
\diagdown \nearrow & \diagdown \nearrow & \\
\text{C} & \text{C} & \\
\text{H} & \text{H} & \\
& \text{(III)} &
\end{array}
$$

shown by isotope trapping to account for most of the histidine metabolized by liver preparations and essentially all consumed by bacterial extracts. Histidase is a rather heat-stable enzyme with a pH optimum near 9. It is inhibited by certain metal binders, and it has been claimed that a divalent cation, cadmium, mercury, or zinc, is a cofactor.[16] Unlike the analogous aspartase reaction, histidase action has not been reversed, even when high concentrations of ammonia were used; C^{14}-

[14] The isolation of glutamic acid as a product of histidine was shown by several groups of investigators (see review by Tabor in Amino Acid Metabolism). Earlier work had shown that an acid-labile derivative of glutamic acid was formed by liver preparations [S. Edlbacher, *Ergeb. Enzymforsch.* **9**, 131 (1943)].

[15] Y. Sera, *Med. J. Osaka Univ.* **4**, 1 (1951); A. H. Mehler and H. Tabor, *J. Biol. Chem.* **201**, 775 (1953).

[16] M. Suda, K. Tomihata, A. Nakaya, and A. Kato. *J. Biochem.* **40**, 257 (1953).

labeled urocanic acid did not contribute its label to histidine. The histidase reaction is conveniently assayed spectrophotometrically as urocanic acid has a large, sharp absorption maximum at 277 mμ in the pH range of enzyme activity, whereas histidine absorbs only at much lower wavelengths (see Fig. 33).

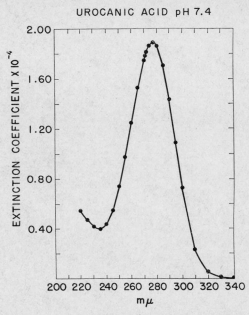

FIG. 33. The ultraviolet absorption spectrum of urocanic acid at pH 7.4.[16a]

Urocanic acid is degraded by urocanase. Although this enzyme has been purified from both animal and bacterial sources,[17] its mode of action remains obscure. The first product to accumulate is the open-chain formiminoglutamic acid.[18,19] There is no evidence for more than one enzyme participating in this reaction which involves the addition of two water molecules and opening the imidazole ring (IV). No dissociable cofactor has been detected. The enzyme that removes the 277 mμ absorption has a pH optimum near 7.

The metabolism of formiminoglutamic acid varies in different organisms. In *Pseudomonas* extracts, formiminoglutamic acid is converted to formylglutamic acid and ammonia by an enzyme that has not been purified, but is active in the presence of sulfhydryl compounds that inhibit

[16a] Taken from A. H. Mehler and H. Tabor, *J. Biol. Chem.* **201**, 775 (1953).
[17] H. Tabor and A. H. Mehler, *in* "Methods in Enzymology" (S. P. Colowick and N. O. Kaplan, eds.), Vol. II, p. 228. Academic Press, New York (1955).
[18] B. Borek and H. Waelsch, *J. Biol. Chem.* **205**, 459 (1953).
[19] H. Tabor and A. H. Mehler, *J. Biol. Chem.* **210**, 559 (1954).

the subsequent hydrolysis of formylglutamic acid.[19] The formation of formic and glutamic acids is catalyzed by glutamic formylase, which is activated by ferrous ions. In *Aerobacter aerogenes*[20] and *Clostridium cylindrosporum*[21] formylglutamic acid is not formed, but formamide is split from formiminoglutamic acid as glutamic acid accumulates.

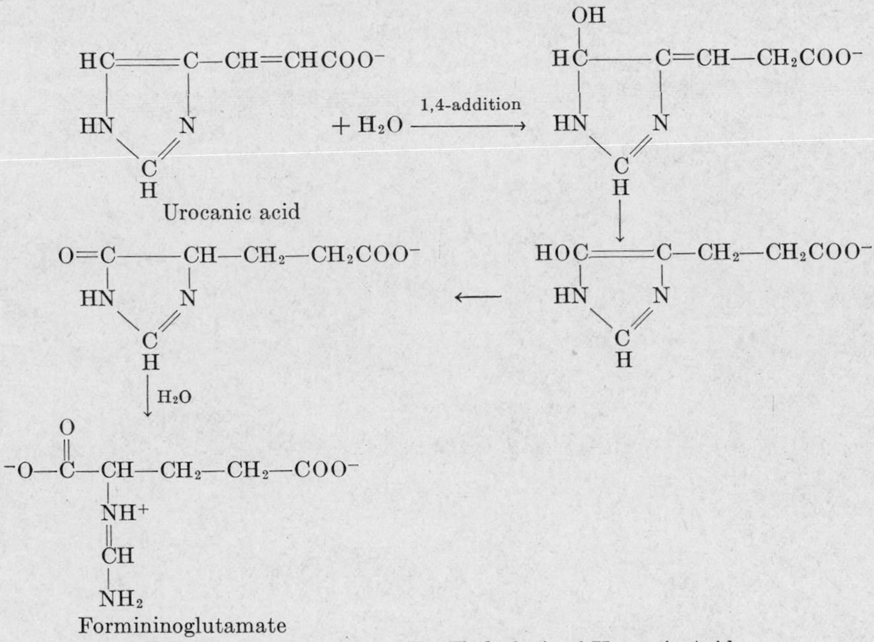

Urocanic acid

+ H₂O — 1,4-addition →

Formininoglutamate

Hypothetical Scheme for the Hydrolysis of Urocanic Acid

(IV)

In liver, formiminoglutamic acid participates in a transfer reaction with tetrahydrofolic acid[22] strictly analogous to the reaction described for formiminoglycine in Clostridial extracts (p. 272). The formimino group is transferred to position 5 of the reduced pteridine, leaving glutamic acid. The liver enzyme does not react with formiminoglycine. Subsequent reactions of formiminotetrahydrofolic acid are the same as occur in the Clostridial system: formation of cyclic anhydroformyltetrahydrofolic acid with release of ammonia, then hydrolysis to N^{10}-formyltetrahydrofolic acid, which can participate in the various transfer reactions previously described. In this way the 2-carbon of the imidazole ring of histidine is transferred to purines, serine, methionine, choline, and other compounds.

[20] B. Magasanik and H. R. Bowser, *J. Biol. Chem.* **213,** 571 (1955).
[21] J. T. Wachsman and H. A. Barker, *J. Bacteriol.* **69,** 83 (1955).
[22] H. Tabor and J. C. Rabinowitz, *J. Am. Chem. Soc.* **78,** 5705 (1956).

The histamine pathway of histidine degradation is of great interest because of the profound physiological importance of histamine. Histamine is bound in the mast cells in which it is formed.[23] Released or administered histamine is rapidly oxidized to imidazoleacetaldehyde by

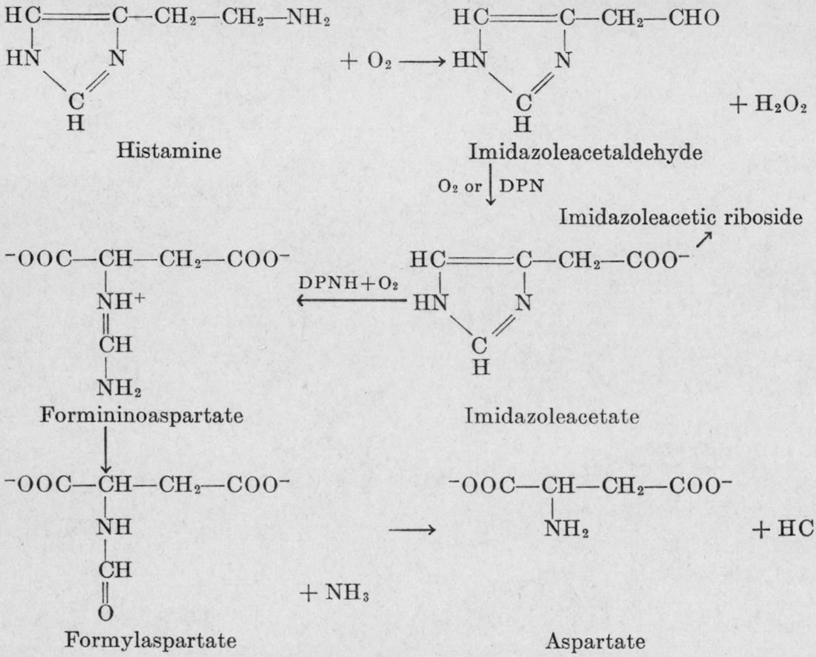

$$HC{=\!\!=}C{-}CH_2{-}CH_2{-}NH_2 \qquad\qquad HC{=\!\!=}C{-}CH_2{-}CHO$$

$$HN\diagdown\quad N \qquad + O_2 \longrightarrow HN\diagdown\quad N$$

$$\diagdown C \diagup \qquad\qquad\qquad \diagdown C \diagup \qquad + H_2O_2$$

$$\quad H \qquad\qquad\qquad\qquad\qquad H$$

Histamine   Imidazoleacetaldehyde

$$O_2 \text{ or } \big\downarrow DPN$$

Imidazoleacetic riboside

$$^{-}OOC{-}CH{-}CH_2{-}COO^{-} \qquad HC{=\!\!=}C{-}CH_2{-}COO^{-} \nearrow$$

$$\quad NH^{+} \qquad\quad \overset{DPNH+O_2}{\longleftarrow} \quad HN\diagdown\quad N$$

$$\quad \| \qquad\qquad\qquad\qquad\qquad \diagdown C \diagup$$

$$\quad CH \qquad\qquad\qquad\qquad\qquad H$$

$$\quad NH_2$$

Formininoaspartate   Imidazoleacetate

$$\big\downarrow$$

$$^{-}OOC{-}CH{-}CH_2{-}COO^{-} \qquad\qquad ^{-}OOC{-}CH{-}CH_2{-}COO^{-}$$

$$\quad NH \qquad\qquad \longrightarrow \qquad\qquad NH_2 \qquad\qquad + HCOO^{-}$$

$$\quad CH$$

$$\quad \| \qquad\qquad + NH_3$$

$$\quad O$$

Formylaspartate   Aspartate

Outline of Histamine Degradation

(V)

diamine oxidase.[24] This enzyme is widely distributed in plants and animal tissues. It attacks a number of intermediate-length dibasic compounds, forming the corresponding aldehyde, ammonia, and H_2O_2. The nature of the enzyme has not been determined. Imidazoleacetaldehyde may be oxidized by the flavoprotein aldehyde oxidases or the pyridine nucleotide aldehyde dehydrogenases to become imidazoleacetic acid. This compound is excreted when large quantities of histamine are oxidized.[25] The metabolism of imidazoleacetic acid, however, is fast enough to handle the amounts of histamine normally oxidized. In an unknown manner, imidazoleacetic acid is converted to its imidazole-β-riboside, which is ex-

[23] J. F. Riley and G. B. West, *J. Physiol.* **120,** 528 (1953); H. T. Graham, O. H. Lowry, N. Wahl, and M. K. Priebat, *J. Exptl. Med.* **102,** 307 (1955); R. W. Schayer, *Am. J. Physiol.* **186,** 199 (1956).

[24] C. H. Best, *J. Physiol.* **67,** 256 (1929); H. Tabor, *J. Biol. Chem.* **188,** 125 (1951).

[25] A. H. Mehler, H. Tabor, and H. Bauer, *J. Biol. Chem.* **197,** 475 (1952).

creted.[26] Imidazoleacetic acid riboside is hydrolyzed by the ribosidase of *Lactobacillus delbruckii.*

Imidazoleacetic acid is oxidized by extracts of *Pseudomonas.*[27] Imidazoleacetic oxidase has been purified 200-fold from extracts of the bacteria grown on imidazoleacetic acid. It behaves as a single protein and is classified as an oxygenase. The oxidation consumes a stoichiometric quantity of DPNH and oxygen in addition to imidazoleacetic acid. No partial reaction is seen when any component is omitted from incubation mixtures. The product is formiminoaspartic acid. This is converted to formylaspartic acid, which is hydrolyzed to aspartic acid by an enzyme different from that formed by cells that convert histidine to formylglutamic acid (V).[28]

Histamine is also excreted as a ring *N*-methyl derivative and as amino-*N*-acetylhistamine. The former product has not been obtained in a defined reaction; the acetylation apparently is catalyzed by the enzyme that acetylates aromatic amines with acetyl CoA, or a very similar enzyme that is present in the same liver fractions.

Histidine is converted to several other compounds of biological interest, but the enzymatic mechanisms are not known. These reactions include methylation on the ring nitrogens, condensation with β-alanine to form carnosine and formation of ergothioneine in fungi.

GENERAL REFERENCES

Ames, B. N. (1955). "Symposium on Amino Acid Metabolism" (W. D. McElroy and B. Glass, eds.), p. 357. Johns Hopkins Press, Baltimore.

Ciba Foundation Symposium on Histamine, 1956. (G. E. W. Wolstenholme and C. M. O'Conner, eds.) Churchill, London.

Tabor, H. (1954). *Pharmacol. Revs.* **6**, 299.

Tabor, H. (1955). "Symposium on Amino Acid Metabolism" (W. D. McElroy and B. Glass, eds.), p. 373. Johns Hopkins Press, Baltimore.

Aromatic Ring Biosynthesis and Metabolism of Phenylalanine and Tyrosine

Biosynthesis. The elucidation of the pathway of aromatic amino acid biosynthesis is an outstanding example of the contribution of the genetic approach to biochemical studies. A series of mutants of *E. coli* was found

[26] H. Tabor and O. Hayaishi, *J. Am. Chem. Soc.* **77**, 505 (1955); S. Karjala, *ibid.* **77**, 504 (1955).

[27] O. Hayaishi, H. Tabor, and T. Hayaishi, *J. Am. Chem. Soc.* **76**, 5570 (1954).

[28] E. Ohmura and O. Hayaishi, *Bacteriol. Proc.* p. 133 (1955).

to require aromatic compounds including phenylalanine and tyrosine.[1] A survey of potential precursors eventually revealed the ability of shikimic acid to feed these mutants.[2] Subsequent studies established shikimic acid as an intermediate in aromatic biosynthesis by showing that certain mutants accumulate this compound and that the label of synthetic shikimic acid is found in bacterial tyrosine.[3] Studies with *E. coli*, *Neurospora*,[4] and *Aerobacter*[5] established the position of certain other compounds related to shikimic acid in the biosynthetic pathway and provided the background for enzymatic studies.

Aromatic compounds are synthesized from carbohydrate precursors. These form a 7-carbon compound that cyclizes and loses water (or phosphate) and is reduced to shikimic acid.[6] Shikimic acid is converted via unknown intermediates into three groups of compounds. One pathway leads to phenylalanine and tyrosine, the second forms anthranilic acid, which is a precursor of indole, and the third produces *p*-aminobenzoic acid and related compounds (I). Many of the individual steps in these pathways have been described in the last few years.

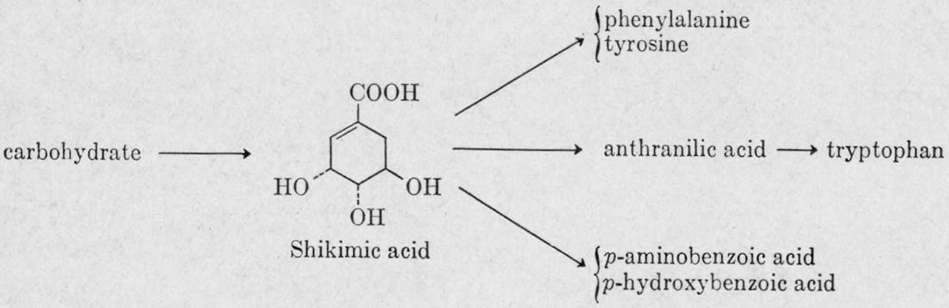

carbohydrate $\longrightarrow$ Shikimic acid $\longrightarrow$ {phenylalanine, tyrosine}; $\longrightarrow$ anthranilic acid $\longrightarrow$ tryptophan; $\longrightarrow$ {*p*-aminobenzoic acid, *p*-hydroxybenzoic acid}

(I)

The establishment of shikimic acid as an intermediate in aromatic biosynthesis gave a unique opportunity for the use of isotopes in studying the mechanism of formation of the six-carbon ring, as shikimic acid is not symmetrical. It therefore became possible to study the origin of each atom of the ring, not pairs of atoms (2 + 6, 3 + 5) as permitted by the symmetrical rings of phenylalanine and tyrosine.[7] Specifically labeled

[1] B. D. Davis, *Experientia* **6**, 41 (1950).

[2] B. D. Davis, *J. Biol. Chem.* **191**, 315 (1951).

[3] H. T. Shigeura and D. B. Sprinson, *Federation Proc.* **11**, 286 (1952).

[4] E. L. Tatum, S. R. Gross, G. Ehrensvard, and L. Garnjobst, *Proc. Natl. Acad. Sci. U.S.* **40**, 271 (1954).

[5] B. D. Davis and U. Weiss, *Arch. exptl. Pathol. Pharmakol. Naunyn-Schmiedeberg's* **220**, 1 (1953).

[6] P. R. Srinivasan and D. B. Sprinson, *Federation Proc.* **15**, 360 (1956).

[7] C. Gilvarg and K. Bloch, *J. Biol. Chem.* **199**, 689 (1952).

glucose entered shikimic acid in a pattern indicating that carbons 3, 4, 5, and 6 made up one fragment of shikimic acid, and carbons 1, 2, and 3 entered the remaining positions.[8] The availability of sedoheptulose phos-

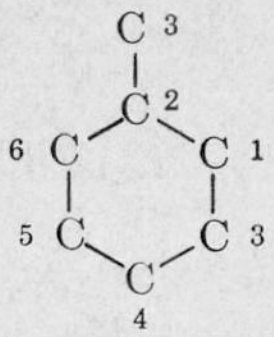

Pattern of Labeling of Shikimic acid 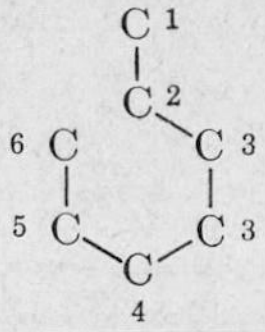   Pattern of Labeling in Sedoheptulose

(The numbers represent the carbon atoms of glucose that contribute most heavily to the positions indicated.)

phate esters at the time these results were obtained led to experiments in which sedoheptulose-1,7-diphosphate was found to be an efficient precursor of dehydroshikimic acid.[9] The isotope distribution in the carbons contributed by positions 1 and 3 of glucose were found, however, to be reversed from the positions they would have occupied if sedoheptulose were an intermediate that simply cyclized. Recently the role of sedoheptulose diphosphate has been shown to be that of donor of the two carbohydrate derivatives that enter directly into the aromatic biosynthetic pathway: erythrose-4-phosphate and phosphopyruvate.[10] Erythrose-4-phosphate is a product of cleavage of the heptose by aldolase, and phosphopyruvate is formed directly from the remaining three carbons (triose phosphate) by glycolytic enzymes.

The early steps in aromatic biosynthesis have been only partially discovered. A fraction of an extract of a *coli* mutant utilizes phosphopyruvate in the presence of D-erythrose-4-phosphate. The condensation was

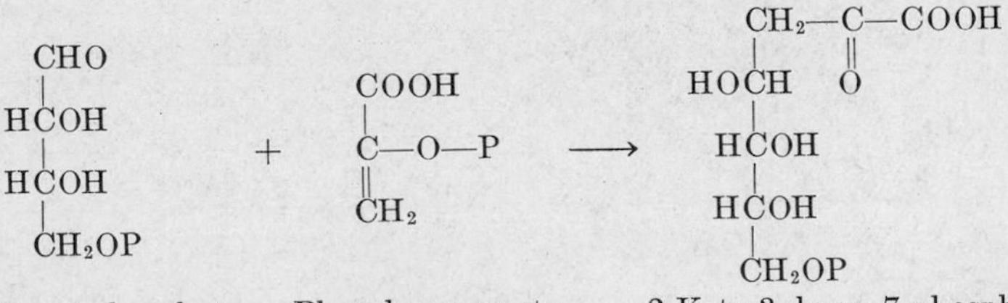

Erythrose-4-phosphate   Phosphopyruvate   2-Keto-3-deoxy-7-phospho-D-glucoheptonic acid

(II)

[8] P. R. Srinivasan, H. T. Shigeura, M. Sprecker, D. B. Sprinson, and B. D. Davis, *J. Biol. Chem.* **220**, 477 (1956).

[9] E. B. Kalan and P. R. Srinivasan, *Symposium on Amino Acid Metabolism, Baltimore, 1954* p. 826 (1955).

[10] P. R. Srinivasan, M. Katagiri, and D. B. Sprinson, *J. Am. Chem. Soc.* **77**, 4943 (1955).

presumed to proceed as shown in (II). Kinetic studies indicate that the next step involves the liberation of inorganic phosphate without the formation of a cyclized compound. The fraction that removes phosphate more slowly catalyzes the formation of 5-dehydroshikimic acid.[6] Earlier genetic evidence supported 5-dehydroquinic acid (DHQ) as a precursor of dehydroshikimic acid (DHS), and an enzyme that catalyzes the interconversion of these compounds was found in the bacteria that could carry out the reaction (III), whereas this enzyme could not be detected

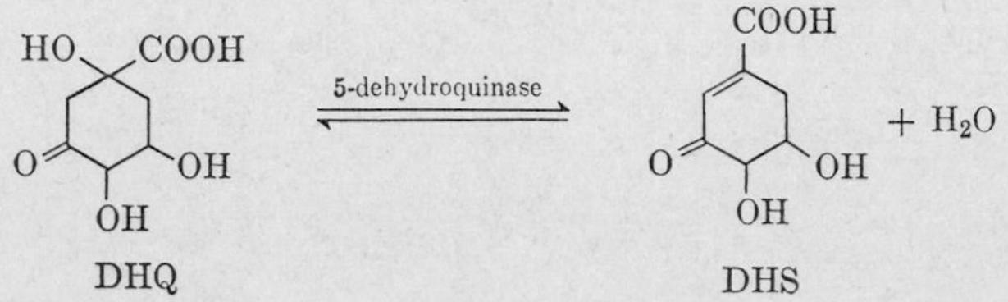

(III)

in mutants blocked at the utilization of dehydroquinic acid. Mitsuhashi and Davis purified the enzyme 10-fold.[11] The purified enzyme exhibits no cofactor requirements. The equilibrium position is reached with about 15 times as much dehydroshikimic acid as dehydroquinic. The reaction can be followed spectrophotometrically by the method used for fumarase, since the introduction of the double bond causes the appearance of an absorption peak at 234 mμ. This sensitive assay permitted the determination of the K_m for DHQ as 4.4×10^{-5}. The pH optimum for this reaction is 8.0.

The reduction of dehydroshikimic acid to shikimic acid is catalyzed by an enzyme partially purified from *E. coli* by Yaniv and Gilvarg.[12] TPNH is specifically required. The reaction is reversible and the oxidation-reduction potential of the DHS-shikimic couple at the pH optimum, 8.5, equals that of the TPN system, which is about -0.350 volts.

The next steps in aromatic biosynthesis have not been demonstrated enzymatically. A labile intermediate, prephenic acid, has been isolated;[13] the side chain is derived from glucose carbons in a pattern consistent with introduction of phosphopyruvate. This compound is converted to phenylpyruvic acid by acid and by an enzyme detected in crude extracts of *E. coli*.[14] Transamination of phenylpyruvate to give phenylalanine has already been discussed.

[11] S. Mitsuhashi and B. D. Davis, *Biochim. et Biophys. Acta* **15**, 54 (1954).
[12] H. Yaniv and C. Gilvarg, *J. Biol. Chem.* **213**, 787 (1955).
[13] S. Simmonds, *J. Biol. Chem.* **185**, 755 (1950); M. Katagiri and R. Sato, *Science* **118**, 250 (1953); B. D. Davis, *ibid.* **118**, 251 (1953).
[14] U. Weiss, C. Gilvarg, E. S. Mingioli, and B. D. Davis, *Science* **119**, 774 (1954).

An alternate reaction of prephenic acid was found in extracts of *coli* mutants blocked in the conversion to phenylpyruvate. The product of the alternate reaction is *p*-hydroxyphenyllactic acid.[15] Presumably oxidation to ketotyrosine and transamination follow the formation of *p*-hydroxyphenyllactate to yield tyrosine.

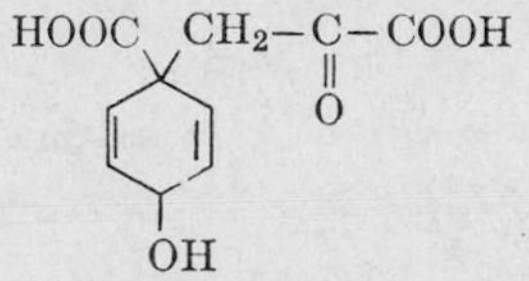

$$HOOC \quad CH_2-C-COOH$$

Prephenic acid

In animals phenylalanine is an essential amino acid, but tyrosine may be formed by hydroxylation of phenylalanine. An enzyme system that catalyzes the phenylalanine-tyrosine conversion was found in the soluble fraction of liver.[16] The requirements of the system have been established, but the mechanism of this, as of other hydroxylation reactions, remains obscure. It was originally reported that oxygen, ferrous ions, DPN, and an aldehyde were required in addition to two protein fractions.[17] The function of the aldehyde appeared to be to reduce the DPN. Recently the enzymes have been somewhat more purified, and TPNH has been found to be more effective than DPNH in supporting tyrosine formation.[18] The formation of tyrosine seems to be closely analogous to the oxygenation reactions that hydroxylate aromatic amines and steroids. The two enzyme fractions are both required for activity; no partial reactions have been found for either enzyme in the absence of the other. It is apparent that H_2O_2 is not an intermediate in this reaction, because enzymatic generation of H_2O_2 does not stimulate tyrosine production and catalase does not inhibit. The system is specific for L-phenylalanine and introduces the hydroxyl group only in the *para* position. Other phenylalanine derivatives and analogs are not attacked, and dyes have not been substituted successfully for oxygen.

During the elucidation of the aromatic biosynthetic pathway, compounds were found that appear to be made in side reactions. These include quinic acid, which is reversibly made from dehydroquinic acid through the action of a DPN-specific dehydrogenase. This enzyme is found in *Aerobacter*, but not in *E. coli*. 5-Phosphoshikimic acid is also accumulated by certain *coli* mutants, but has no known metabolic func-

[15] J. J. Ghosh, *Federation Proc.* **15**, 261 (1956).
[16] S. Udenfriend and J. R. Cooper, *J. Biol. Chem.* **194**, 503 (1952).
[17] C. Mitoma and L. C. Leeper, *Federation Proc.* **13**, 266 (1954).
[18] S. Kaufman, *Biochim. et Biophys. Acta* **23**, 445 (1957).

tions. In addition to the known products of shikimic acid metabolism, the formation of at least one more product, as yet unidentified, has been indicated by genetic studies. Six growth factors, including three amino acids (phenylalanine, tyrosine, and tryptophan) and two vitamins (p-aminobenzoic acid, a component of folic acid, and nicotinic acid, formed from tryptophan), are thus seen to be derived from shikimic acid.

Quinic acid   5-Phosphoshikimic acid

Degradation. The principal route of phenylalanine degradation by animals has been found to proceed via tyrosine. Both compounds are oxidized to acetoacetate and malate by liver slices. Isotope experiments by Lerner showed that, whereas all of the malate carbons came from the ring, acetoacetate was derived half from the side chain and half from the ring.[19] With 1, 3, 5 ring-labeled phenylalanine, Shepartz and Gurin showed that label appeared only in the terminal methyl group of acetoacetate;[20] this indicated a side-chain migration during the oxidation, as proposed earlier by Neubauer to account for homogentisic acid as an intermediate.[21] This unusual reaction has been established by the isolation of the individual enzymes concerned with the oxidation of tyrosine, although the chemistry of the migration is still conjectural.

The first step in tyrosine oxidation is a transamination to form p-hydroxyphenylpyruvic acid. Several groups of investigators independently showed a dependence of tyrosine oxidation on the presence of a keto acid.[22] Knox and LeMay-Knox showed that α-ketoglutarate is a specific partner in the transamination and that pyridoxal phosphate is a cofactor in this reaction. Partial resolution of the transaminase allowed a demonstration of parallel restoration of transaminase activity and over-all tyrosine oxidation by addition of pyridoxal phosphate.

The oxidation of p-hydroxyphenylpyruvate is a complicated reaction involving several cofactors and two proteins. This reaction results in the formation of homogentisic acid (IV). It is at this step in the degradation

[19] S. Weinhouse and R. Millington, *J. Biol. Chem.* **175,** 995 (1948); A. B. Lerner, *ibid.* **181,** 281 (1949).

[20] B. Shepartz and S. Gurin, *J. Biol. Chem.* **180,** 663 (1949).

[21] O. Neubauer, *in* "Handbuch der normalen und pathologischen Physiologie," Vol. 5, p. 671. Springer, Berlin (1928).

[22] W. E. Knox and M. LeMay-Knox, *Biochem. J.* **49,** 686 (1951); B. Shepartz, *J. Biol. Chem.* **193,** 293 (1951); B. N. LaDu, Jr. and D. M. Greenberg, *ibid.* **190,** 245 (1951).

of tyrosine that side-chain migration occurs. In a reaction that has not been dissociated from the addition of a hydroxyl group to the ring and the oxidative decarboxylation of the side chain, the side chain shifts to an adjacent carbon atom. Ascorbic acid was found to be required for the oxidation of tyrosine;[23] this compound exerts its influence in the oxidation of p-hydroxyphenylpyruvate.[22] The requirement for ascorbic acid is not specific, but can be satisfied by ascorbic acid analogs, dichlorophenolindophenol, or hydroquinone.

It is not known exactly how ascorbic acid supports the reaction. It does not act as a conventional cofactor that acts catalytically. The initial

p-Hydroxyphenylpyruvate   Homogentisate

(IV)

reaction rate is not influenced by the presence of ascorbate, but the reaction rate falls off more rapidly in the absence of one of the active reductants. The amount of ascorbate or other reductant necessary to maintain the reaction until complete substrate utilization depends on the amount of substrate to be oxidized. The requirement for a reductant is not stoichiometric with substrate; reduced dichlorophenolindophenol has been found to support the oxidation of 100 times its concentration of p-hydroxyphenylpyruvate by a partially purified enzyme. This purified preparation was found to be maintained by reduced dye, but not by ascorbic acid. It has been suggested, therefore, that ascorbic acid acts through a material removed during purification, and that the reduced dye can substitute for the natural material.[24]

Purification of p-hydroxyphenylpyruvic oxidase led to the separation of two proteins, neither of which is active alone.[25] One of these was identified as catalase. Catalase can be replaced as a component of the oxidase system by a peroxidase. These findings implicate H_2O_2 in the oxidation, and this is supported by the elimination of a lag period in the reaction by the addition of small quantities of H_2O_2. In spite of the requirement for two proteins and the effect of H_2O_2, an intermediate-level oxidant, no

[23] T. H. Lan and R. R. Sealock, *J. Biol. Chem.* **155**, 483 (1944).
[24] B. N. LaDu, Jr. and V. G. Zannoni, *J. Biol. Chem.* **217**, 777 (1955); **219**, 273 (1956).
[25] V. G. Zannoni and B. N. LaDu, Jr., *Federation Proc.* **15**, 391 (1956).

indication has been found for a partial reaction. An intriguing problem
is thus posed: to determine whether there is a sequence of reactions or
whether a peroxidase and another enzyme in some way act together to
accomplish an oxidation.

Homogentisic acid is oxidized by a ferrous ion-requiring enzyme, as
shown by Suda and Takeda,[26] and confirmed by several others. Homogentisic oxidase also requires a sulfhydryl compound. One molecule of
oxygen is consumed in the reaction, presumably in a single step. The
product has been identified as maleylacetoacetate by Knox and Edwards
(V).[27]

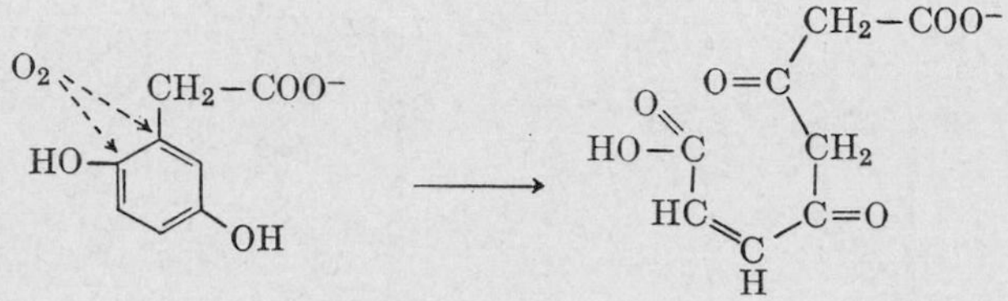

(V)

This compound differs slightly from the isomeric fumarylacetoacetate
in its absorption spectrum near 300 mμ at neutral and alkaline pH values,
but the maleyl compound has essentially no absorption at pH 1 whereas
the fumaryl compound has a strong peak at 310 mμ in acid. Fumarylacetoacetate had previously been isolated as a product of homogentisic
oxidation by Ravdin and Crandall.[28] Knox and collaborators have shown
that the fumaryl compound is formed secondarily by a *cis-trans* isomerase,
an enzyme that requires glutathione for activity and has in addition essential sulfhydryl groups on the protein.[29] The isomerization has not been
reversed.

Fumarylacetoacetate is split to fumarate and acetoacetate[28] by an
enzyme that was known previously to hydrolyze diketo acids. It has been
called acylpyruvase, triacetic acid hydrolyzing enzyme, and β-diketonase.
Since both the rate of hydrolysis of fumarylacetoacetate and its affinity
for the enzyme exceed those of other substrates there is some justification
for the name fumarylacetoacetate hydrolase. The irreversible action of
this enzyme results in the formation of products that are metabolized by
the systems previously described for fatty acid oxidation and the Krebs
cycle.

[26] M. Suda and Y. Takeda, *Med. J. Osaka Univ.* **2,** 41 (1950).
[27] W. E. Knox and S. W. Edwards, *J. Biol. Chem.* **216,** 479, 489 (1955).
[28] R. G. Ravdin and D. I. Crandall, *J. Biol. Chem.* **189,** 137 (1951).
[29] S. W. Edwards and W. E. Knox, *J. Biol. Chem.* **220,** 79 (1956).

The series of reactions outlined accounts for most of the oxidation of phenylalanine and tyrosine in animals. Conversion of phenylalanine to phenylpyruvic acid also occurs, and the keto acid is a major excretion product in the disease phenylpyruvic oligophrenia.[30] Armstrong and collaborators have shown that a very large number of phenolic compounds, presumably derived from aromatic amino acids, also occur in human urine.[31] The reactions that are responsible have not been defined.

GENERAL REFERENCES

Davis, B. D. (1955). *Advances in Enzymol.* **16,** 247.
Symposium on Amino Acid Metabolism (1955). (W. D. McElroy and B. Glass, eds.) Johns Hopkins Press, Baltimore. Various aspects of phenylalanine and tyrosine metabolism have been reviewed in this symposium in articles by B. D. Davis, p. 799; D. B. Sprinson, p. 817; and W. E. Knox, p. 836.

Tryptophan

Tryptophan Biosynthesis. Tryptophan is another of the amino acids required by animals but synthesized only by plants and microorganisms. The synthesis of tryptophan in microorganisms has been outlined in recent years, but some of the reactions are still quite obscure. The known intermediates include shikimic acid,[1] anthranilic acid,[2] and indole.[3] The mechanism of the conversion of shikimic acid to anthranilic acid has not been elucidated.

Anthranilic acid is converted by *E. coli* to indole through condensation with a derivative of ribose, probably PRPP. The carboxyl group is lost in the condensation, and carbons 1 and 2 of ribose enter the indole ring. An enzyme fraction obtained from *E. coli* carries out the reaction shown in (I). Indole-3-glycerol phosphate has been isolated and found to be a substrate for another enzyme, separated from the system that carries out the reaction above by ammonium sulfate fractionation. This enzyme catalyzes the formation of indole and triose phosphate from

---

[30] Phenylpyruvic oligophrenia has been shown to involve a specific enzyme deficiency. One of the two proteins required for the conversion of phenylalanine to tyrosine is lacking in livers of patients with the disease. The second protein, which occurs in other tissues as well as in liver, is not deficient in the disease (H. W. Wallace, K. Moldave, and A. Meister, *Proc. Soc. Exptl. Biol. Med.* **94,** 632 (1957); C. Mitoma, R. M. Auld, and S. Udenfriend, *ibid.* **94,** 634 (1957)).
[31] M. D. Armstrong, K. N. F. Shaw, and P. E. Wall, *J. Biol. Chem.* **218,** 293 (1956).

[1] B. D. Davis, *Experientia* **6,** 41 (1950).
[2] E. E. Snell, *Arch. Biochem.* **2,** 389 (1943).
[3] P. Fildes, *Brit. J. Exptl. Pathol.* **21,** 315 (1940).

indole-3-glycerol phosphate. The cleavage reaction requires the presence of the phosphate group, and appears to be reversible (II).[4]

The final step in the biosynthesis of tryptophan is the condensation of indole with serine, catalyzed by tryptophan desmolase (tryptophan synthetase).[5] This enzyme requires pyridoxal phosphate (III).

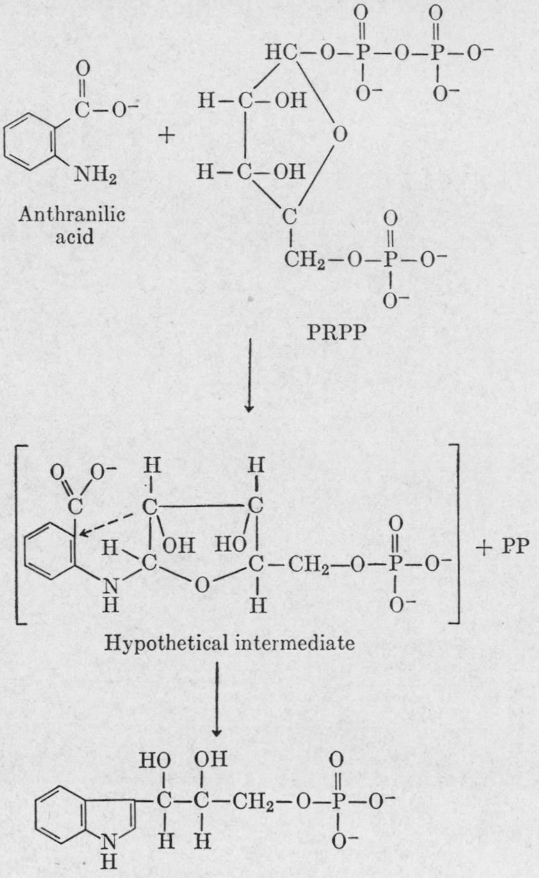

Anthranilic acid

PRPP

Hypothetical intermediate

Indole-3-glycerol phosphate

(I)

Tryptophan Degradation. Tryptophan desmolase catalyzes an essentially irreversible reaction. The formation of indole from tryptophan is catalyzed by a very similar but quite distinct enzyme, tryptophanase.[6]

[4] C. Yanofsky, *J. Biol. Chem.* **223**, 171 (1956).

[5] W. W. Umbreit, W. A. Wood, and I. C. Gunsalus, *J. Biol. Chem.* **165**, 731 (1946); W. A. Wood, I. C. Gunsalus, and W. W. Umbreit, *ibid.* **170**, 313 (1947).

[6] F. C. Happold, *Advances in Enzymol.* **10**, 51 (1950).

The tryptophanase reaction also requires pyridoxal phosphate, and differs from the synthetase reaction in that it forms pyruvate and ammonia as products, not serine (IV).

Tryptophanase has been studied as a partially purified protein,[7] and has been shown to be activated by NH_4^+, K^+, or Rb^+. The substrate requirements are quite rigid. The carboxyl group, the α-amino group and the indole —NH must all be present without substitutions. Indole-serine (β-hydroxytryptophan) is not a substrate. The only substituted tryptophan found to be attacked by tryptophanase is 5-methyltryptophan. Indole and certain 3-indolyl acids inhibit the reaction. Additional

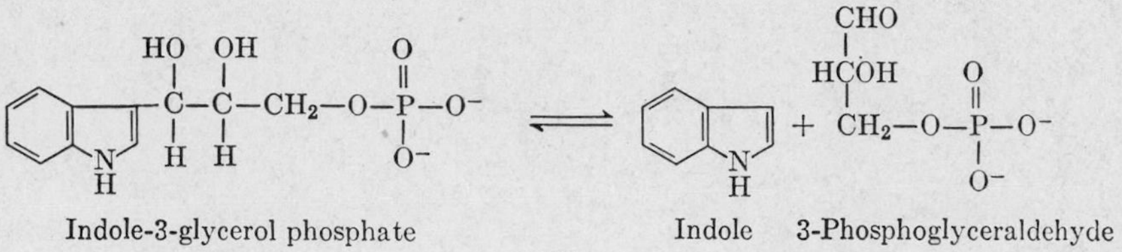

Indole-3-glycerol phosphate   Indole   3-Phosphoglyceraldehyde

(II)

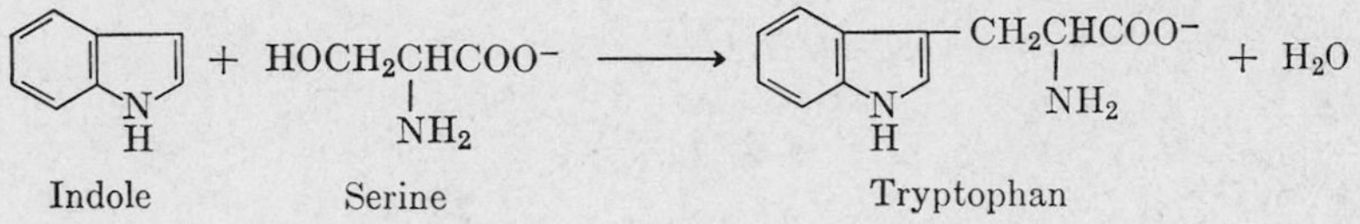

Indole   Serine   Tryptophan

(III)

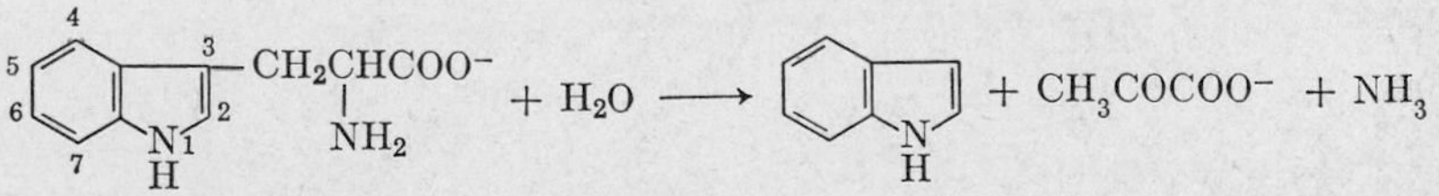

(IV)

inhibitors are cyanide, hydrazine, hydroxylamine, and other reagents that can combine with the carbonyl group of pyridoxal phosphate.

Tryptophan metabolism proceeds via several pathways that result in the production of many biologically important compounds. Among the metabolites derived from tryptophan are indoleacetic acid (a naturally-occurring auxin), 5-hydroxytryptamine (the animal hormone, serotonin), ommochrome (an insect-pigment), nicotinic acid (the vitamin, niacin), bufotenin of toad venom, a component of the mushroom toxin, phalloidin, and the plant alkaloid, gramine. The great physiological significance of these compounds emphasizes the importance of tryptophan metabolism in all forms of life.

[7] F. C. Happold and A. Struyvenberg, *Biochem. J.* **58**, 379 (1954).

Auxin Formation. Oxidation or transamination leads to the formation of indolepyruvic acid. Decarboxylation of the keto acid is presumed to result in the formation of indoleacetaldehyde, which can be oxidized by an aldehyde oxidase to indoleacetic acid (V). This compound is excreted by humans, but appears to be the natural auxin, a plant growth hormone. Degradation of auxin by plants appears to involve a peroxidase acting as an oxidase.[8] Horseradish peroxidase plus Mn^{++} carry out the same

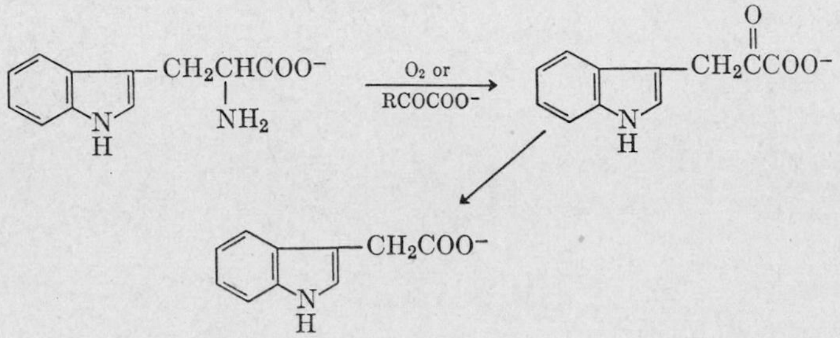

Auxin

(V)

reaction found with fractions from other plants. This reaction also is carried out with indolepropionic acid and apparently involves an attack on the indole nucleus, not on the side chain.

5-Hydroxyindole Metabolism. Another pathway of tryptophan leads to the synthesis of serotonin, a compound with great physiological and pharmacological significance in animals. This pathway is initiated by hydroxylation of tryptophan in position 5 by a liver system. The small

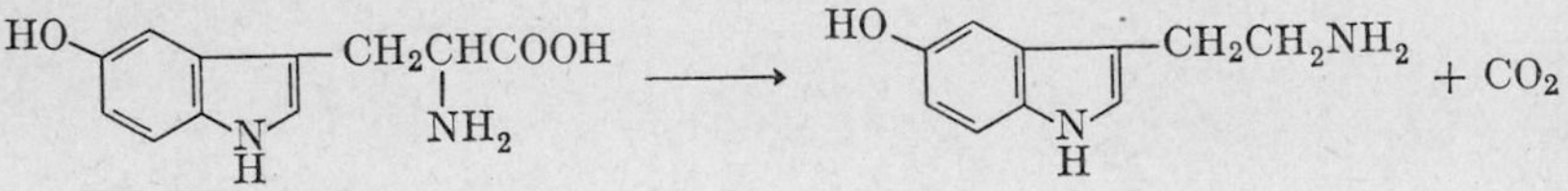

5-Hydroxytryptophan   Serotonin

(VI)

amounts of 5-hydroxytryptophan accumulated by this system are decarboxylated to serotonin by a very specific decarboxylase that appears to use pyridoxal phosphate (VI).[9] This decarboxylase does not degrade tryptophan. As yet undefined enzymes methylate the amino group of serotonin in the formation of bufotenin in toads, invertebrates, higher

[8] R. H. Kenten, *Biochem. J.* **61**, 353 (1955); P. M. Ray and K. V. Thimann, *Arch. Biochem. Biophys.* **64**, 175 (1956).

[9] C. T. Clark, H. Weissbach, and S. Udenfriend, *J. Biol. Chem.* **210**, 139 (1954).

plants, and fungi. Serotonin is presumably oxidized by an amine oxidase to the corresponding aldehyde, then to 5-hydroxyindoleacetic acid.[10]

Tryptophan Peroxidase. Most of the tryptophan degraded by animals and bacteria adapted to oxidize tryptophan is oxidized in the 5-membered ring. Tryptophan peroxidase is found in the livers of animals.[11] Its action resembles that of horseradish peroxidase in the oxidation of dihydroxymaleic acid; oxygen is required and a balance between catalase and an exogenous source of H_2O_2 must be maintained, but when catalase is removed from the enzyme, H_2O_2 producing systems are not required. The oxygen atoms introduced into the products are all derived from molecular oxygen, but it has not been determined whether hydrogen peroxide is an intermediate or whether the oxidation occurs in a single step or in sequential steps. The product of the oxidation is N'-formylkynurenine (VII). No intermediates have been detected, and two potential intermediates, α-hydroxytryptophan[12] and α,β-dihydroxytryptophan,[13] were not attacked. Tryptophan peroxidases of animal tissues and bacteria are very specific; they do not oxidize any of the analogs of tryptophan that have been tested.

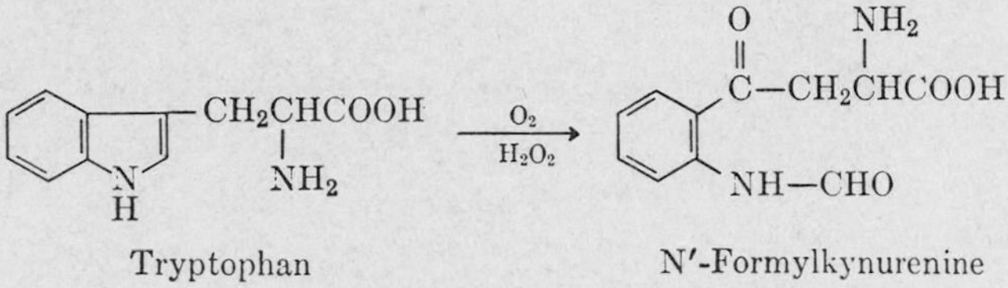

Tryptophan   N'-Formylkynurenine

(VII)

Kynurenine Formylase. Hydrolysis of formylkynurenine to kynurenine is catalyzed in liver preparations by an enzyme (formylase or kynurenine formamidase) with rather low specificity for aromatic formamides.[14] The reaction with formylkynurenine, however, is faster than that with any analogs. A very similar enzyme has been obtained from *Neurospora.*[15] In contrast, an enzyme from insects[16] does not hydrolyze formylanthranilic acid, a model substrate for formylase from other sources, or other analo-

[10] S. Udenfriend and E. Titus, *Symposium on Amino Acid Metabolism, Baltimore, 1954* p. 945 (1955).

[11] W. E. Knox and A. H. Mehler, *J. Biol. Chem.* **187**, 419 (1950).

[12] T. Sakan and O. Hayaishi, *J. Biol. Chem.* **186**, 177 (1950); C. Dalgliesh, W. E. Knox, and A. Neuberger, *Nature* **168**, 20 (1951).

[13] P. L. Julian, E. A. Dailey, H. C. Printy, H. L. Cohen, and S. Hamashige, *J. Am. Chem. Soc.* **78**, 3503 (1956).

[14] A. H. Mehler and W. E. Knox, *J. Biol. Chem.* **187**, 431 (1950).

[15] W. Jakoby, *J. Biol. Chem.* **207**, 657 (1954).

[16] E. Glassman, *Genetics* **41**, 566 (1956).

gous compounds, but hydrolyzes only formylkynurenine among the substrates tested.

Kynurenine Metabolism. Kynurenine may be metabolized in five ways: acetylation to N^α-acetylkynurenine,[17] decarboxylation to kynuramine,[18] oxidation to 3-hydroxykynurenine,[19] cyclization to a quinoline derivative,[20] and cleavage to yield anthranilic acid.[21] The oxidation, cyclization, and cleavage reactions are components of major pathways of tryptophan metabolism. Ommochrome is composed of a series of heterocyclic condensed ring systems that have been shown to be derived from tryptophan via kynurenine. The individual steps in the enzymatic formation of the pigments have not separated.[22]

Kynurenine Hydroxylase. Kynurenine is hydroxylated by an enzyme prepared from mitochondria of animal livers. The properties of this enzyme are very similar to those described from the hydroxylation of other aromatic amines, steroids, and phenylalanine; molecular oxygen is consumed and an equivalent of TPNH is oxidized simultaneously with kynurenine oxidation (VIII).

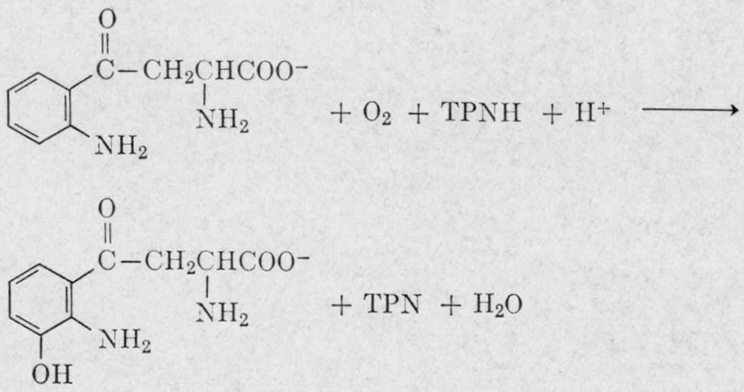

(VIII)

Kynurenine Transaminase. Both kynurenine and 3-hydroxykynurenine participate in reactions with two enzymes that use pyridoxal phosphate. One is a transaminase that presumably forms the α-keto derivative of kynurenine. This compound does not accumulate, however;

[17] C. Yanofsky and D. M. Bonner, *Proc. Natl. Acad. Sci. U.S.* **36,** 167 (1950); C. E. Dalgliesh, W. E. Knox, and A. Neuberger, *Nature* **168,** 20 (1951).

[18] A. Butenandt and U. Renner, *Z. Naturforsch.* **8b,** 454 (1953).

[19] F. T. deCastro, J. M. Price, and R. R. Brown, *J. Am. Chem. Soc.* **78,** 2904 (1956).

[20] O. Wiss, *Z. Naturforsch.* **7b,** 133 (1952); I. L. Miller, M. Tsuchida, and E. A. Adelberg, *J. Biol. Chem.* **203,** 205 (1953).

[21] W. E. Knox, *Biochem. J.* **53,** 379 (1953); W. Jakoby and D. M. Bonner, *J. Biol. Chem.* **205,** 699 (1953); O. Hayaishi and R. Y. Stanier, *ibid.* **195,** 735 (1952).

[22] A. Butenandt and R. Beckmann, *Z. physiol. Chem.* **301,** 115 (1955).

instead, a cyclic product is formed by condensation of the α-keto group with the aromatic amino group. The quinoline derivatives, kynurenic acid and xanthurenic acid, are thus formed from kynurenine and 3-hydroxykynurenine, respectively (IX). Oxidation of the α-amino acids by

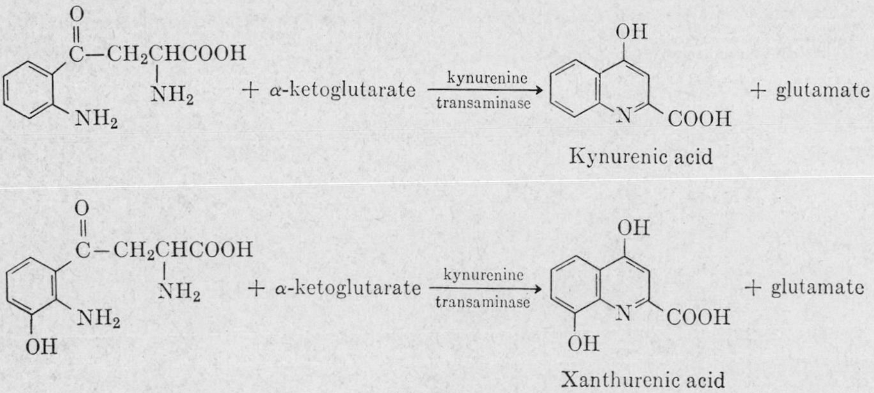

Kynurenic acid

Xanthurenic acid

(IX)

oxidases that ordinarily form α-keto acids also results in the accumulation of only the cyclic quinoline derivatives. These compounds appear to be inert end-products of animal metabolism.

Kynureninase. The cleavage of the side chains of kynurenine and 3-hydroxykynurenine is catalyzed by kynureninase. The reaction in each

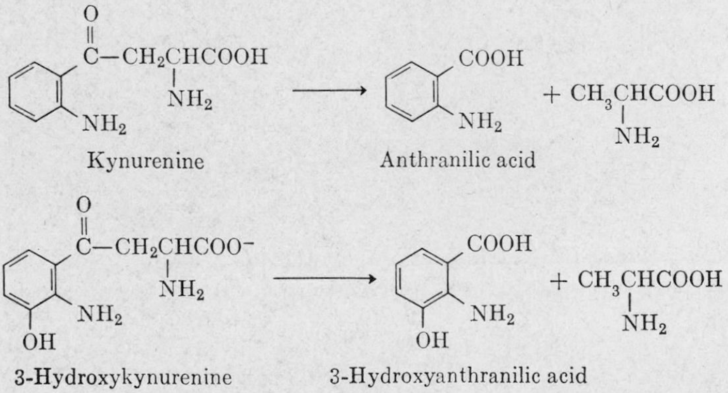

Kynurenine   Anthranilic acid

3-Hydroxykynurenine   3-Hydroxyanthranilic acid

(X)

case results in formation of alanine and a carboxylic acid, the products of hydrolysis. Kynureninase has been studied in preparations from animal liver, *Pseudomonas*, and *Neurospora*. These enzymes are all similar, but differ in substrate specificity and affinity (X).

3-Hydroxyanthranilic Oxidase. In animals an oxidation of 3-hydroxy-anthranilic acid by an oxidase from liver is known, but it is not known that other reactions do not occur. The oxidase, a soluble enzyme that survives acetone drying,[23] consumes one equivalent of O_2 in forming an unstable product believed to have the structure shown in (XI).[24,25] This

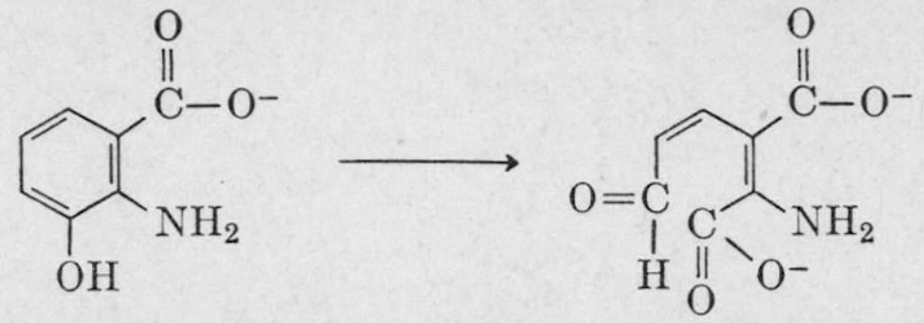

3-Hydroxyanthranilate   Proposed oxidation product

(XI)

oxidase seems to be similar to the several other oxygenases that attack aromatic rings; these use ferrous ion and glutathione, and insert an oxygen molecule directly into the substrate. 3-Hydroxyanthranilic oxidase is inhibited by α,α'-dipyridyl and CN^- and has been reported to be partially resolved and reactivated by Fe^{++}. The oxidation product includes oxygen derived from molecular oxygen, not from the medium. The oxidation product has an intense absorption band at 360 mμ, which provides the most sensitive and convenient basis for assaying the enzyme (Fig. 34). The product, however, spontaneously rearranges to form quinolinic acid. To date no enzyme has been found to catalyze quinolinic acid formation.

Picolinic Carboxylase. An enzyme in liver decarboxylates the original carboxyl group of 3-hydroxyanthranilic acid from the oxidation product.[25] The product of the decarboxylation is picolinic acid. Picolinic carboxylase has no known cofactors. The mechanism of its action is thought to involve a temporary loss of the double bond during decarboxylation. This permits rotation of the amino group into a position favoring condensation to form the pyridine ring (XII).

Picolinic Acid Metabolism. Picolinic acid is converted to its glycine conjugate when administered to mammals.[26,27] Birds use ornithine in place of glycine.[26] The mechanism of the condensation is not known, but presumably it resembles the formation of hippuric acid, in which a CoA derivative of the carboxyl group is the acylating agent. *N*-Methylpicolinic

[23] A. H. Bokman and B. S. Schweigert, *Arch. Biochem. Biophys.* **33**, 270 (1951).
[24] C. L. Long, H. N. Hill, I. M. Weinstock, and L. M. Henderson, *J. Biol. Chem.* **211**, 405 (1954).
[25] A. H. Mehler, *J. Biol. Chem.* **218**, 241 (1956).
[26] Y. Sendju, *J. Biochem.* **7**, 273 (1927).
[27] A. H. Mehler and E. L. May, *J. Biol. Chem.* **223**, 449 (1956).

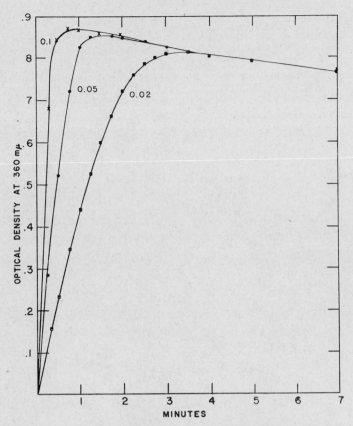

FIG. 34. Formation of oxidation product of 3-hydroxyanthranilic acid by indicated amounts of enzyme and subsequent nonenzymatic conversion to quinolinic acid.[25]

(XII)

acid, homarine,[28] occurs in many invertebrates, but the mechanism of its formation is also unknown.

Nicotinic Acid Metabolism. The sequence of reactions leading to the formation of pyridine compounds is of particular interest as a source of nicotinic acid. Nutritional, isotopic, and genetic experiments[29] have all shown that tryptophan and its metabolic derivatives including 3-hydroxyanthranilic acid are precursors of nicotinic acid in animals and in *Neurospora*. The terminal steps in this sequence are not known. Under certain physiological conditions an increase in picolinic carboxylase appears to reduce nicotinic acid synthesis.[30] This implies a common pathway as far as the oxidation of 3-hydroxyanthranilic acid. Whether quinolinic acid is a precursor of nicotinic acid is still uncertain. The enzyme that forms the amide of nicotinic acid also has not been isolated. Subsequent reactions of nicotinamide include the formation of the riboside with nucleoside phosphorylase[31] and methylation by nicotinamide methylkinase.[32] In animals *N*-methylnicotinamide is oxidized to the corresponding 6-pyridone by a liver flavoprotein.[33] Nicotinic acid also forms glycine and ornithine conjugates. Both aerobic and anaerobic bacteria have been found to oxidize nicotinic acid in the 6-position.[34]

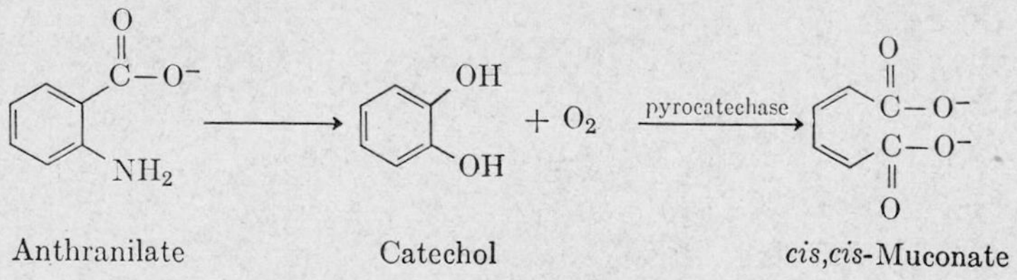

Anthranilate   Catechol   *cis,cis*-Muconate

(XIII)

Total Oxidation of Tryptophan. Certain microorganisms that use tryptophan as a primary food modify the pathway described above.[35] The scheme used by these organisms involves the formation of anthranilic acid by the reactions already described. In an unknown manner, anthranilic acid is oxidized to catechol. Catechol is oxidized to *cis,cis*-muconic acid by pyrocatechase.[36] This is an oxygenase that was shown to be

[28] F. A. Hoppe-Seyler, *Z. physiol. Chem.* **222**, 105 (1933).
[29] See review by Yanofsky and Bonner.
[30] A. H. Mehler, E. G. McDaniel, and J. M. Hundley, *Federation Proc.* **15**, 314 (1956).
[31] J. A. Rowen and A. Kornberg, *J. Biol. Chem.* **193**, 497 (1951).
[32] G. Cantoni, *J. Biol. Chem.* **189**, 203 (1951).
[33] W. E. Knox and W. I. Grossman, *J. Biol. Chem.* **168**, 363 (1947).
[34] D. E. Hughes, *Biochem. J.* **60**, 303 (1955); I. Harary, *Federation Proc.* **15**, 268 (1956).
[35] M. Suda, O. Hayaishi, and Y. Oda, *J. Biochem.* **37**, 355 (1950); R. Y. Stanier, O. Hayaishi, and M. Tsuchida, *J. Bacteriol.* **62**, 355 (1951).
[36] O. Hayaishi and K. Hashimoto, *J. Biochem.* **37**, 371 (1950).

activated by Fe^{++} and glutathione (XIII). *cis, cis*-Muconic acid is hydrated to form β-ketoadipic acid. The hydration is a complex reaction that requires at least two enzymatic steps.[37] In the first step one carboxyl group is added to a double bond, forming a lactone, $(+)\gamma$-carboxymethyl-Δ^{α}-butenolide (XIV). The enzyme also attacks *cis, trans*-muconic acid to form the $(-)$lactone. *trans-trans*-Muconic acid does not interact with the enzyme as a substrate or inhibitor. The lactonizing enzyme is activated by cations, most effectively by Mn^{++}. The lactone is irreversibly hydrolyzed by a delactonizing enzyme that attacks only one stereoisomer. The delactonizing enzyme has no known cofactors. The further metabolism of β-ketoadipate involves formation of an acyl CoA derivative by a transfer reaction in which succinyl CoA is a specific CoA donor. β-Ketoadipyl CoA is split in a thiolase reaction to acetyl CoA and succinyl CoA.

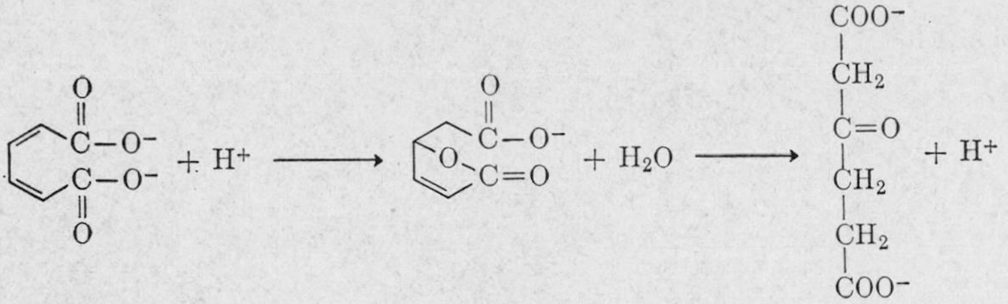

$$cis,cis\text{-Muconate} \qquad \gamma\text{-Carboxymethyl-}\Delta^{\alpha}\text{-butenolide} \qquad \beta\text{-Ketoadipate}$$
(XIV)

Succinyl CoA is thus seen to be catalytic in the production of acetyl CoA and succinate from β-ketoadipic acid (XV).[38] Succinate and acetyl CoA are, of course, oxidized via the Krebs cycle.

Pathways of metabolism have been outlined which indicate how tryptophan is a source of auxin, niacin, serotonin, ommochrome, and energy. Many other compounds are formed as side-products of these pathways and still others are formed by further reactions. Yet additional pathways remain to be elucidated to describe the formation of gramine (*N*-dimethyl-indole-3-methylamine), abrine (N^{α}-methyltryptophan), the α-hydroxy-tryptophan component of phalloidin, skatole, and possibly other complex compounds containing the indole nucleus. The reactions involved in these transformations include examples of various types of oxidation, trans-amination, cleavage, elimination, decarboxylation, and condensation as well as *cis-trans* isomerization. The wealth of biochemical variety already revealed in these studies is a stimulus toward exploration of the still unknown reactions of tryptophan metabolism.

[37] W. R. Sistrom and R. Y. Stanier, *J. Biol. Chem.* **210**, 821 (1954).
[38] M. Katagiri and O. Hayaishi, *Federation Proc.* **15**, 285 (1956).

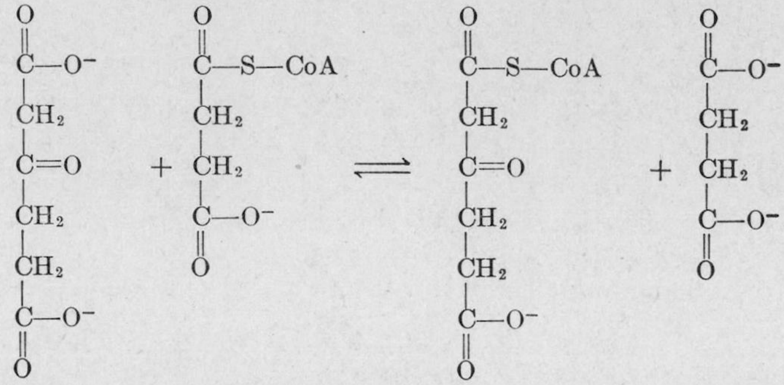

$$
\begin{array}{c}
\underset{\beta\text{-Ketoadipate}}{
\begin{array}{c}
\text{O} \\
\parallel \\
\text{C--O}^- \\
| \\
\text{CH}_2 \\
| \\
\text{C=O} \\
| \\
\text{CH}_2 \\
| \\
\text{CH}_2 \\
| \\
\text{C--O}^- \\
\parallel \\
\text{O}
\end{array}}
\; + \;
\underset{\text{Succinyl CoA}}{
\begin{array}{c}
\text{O} \\
\parallel \\
\text{C--S--CoA} \\
| \\
\text{CH}_2 \\
| \\
\text{CH}_2 \\
| \\
\text{C--O}^- \\
\parallel \\
\text{O}
\end{array}}
\;\rightleftharpoons\;
\underset{\beta\text{-Ketoadipyl CoA}}{
\begin{array}{c}
\text{O} \\
\parallel \\
\text{C--S--CoA} \\
| \\
\text{CH}_2 \\
| \\
\text{C=O} \\
| \\
\text{CH}_2 \\
| \\
\text{CH}_2 \\
| \\
\text{C--O}^- \\
\parallel \\
\text{O}
\end{array}}
\; + \;
\underset{\text{Succinate}}{
\begin{array}{c}
\text{O} \\
\parallel \\
\text{C--O}^- \\
| \\
\text{CH}_2 \\
| \\
\text{CH}_2 \\
| \\
\text{C--O}^- \\
\parallel \\
\text{O}
\end{array}}
\end{array}
$$

$$
\begin{array}{c}
\begin{array}{c}
\text{O} \\
\parallel \\
\text{C--S--CoA} \\
| \\
\text{CH}_2 \\
| \\
\text{C=O} \\
| \\
\text{CH}_2 \\
| \\
\text{CH}_2 \\
| \\
\text{C--O}^- \\
\parallel \\
\text{O}
\end{array}
\; + \text{HSCoA} \rightleftharpoons
\begin{array}{c}
\text{O} \\
\parallel \\
\text{C--S--CoA} \\
| \\
\text{CH}_3 \qquad \text{Acetyl CoA}\\
+ \\
\text{O} \\
\parallel \\
\text{C--S--CoA} \\
| \\
\text{CH}_2 \qquad \text{Succinyl CoA}\\
| \\
\text{CH}_2 \\
| \\
\text{C--O}^- \\
\parallel \\
\text{O}
\end{array}
\end{array}
$$

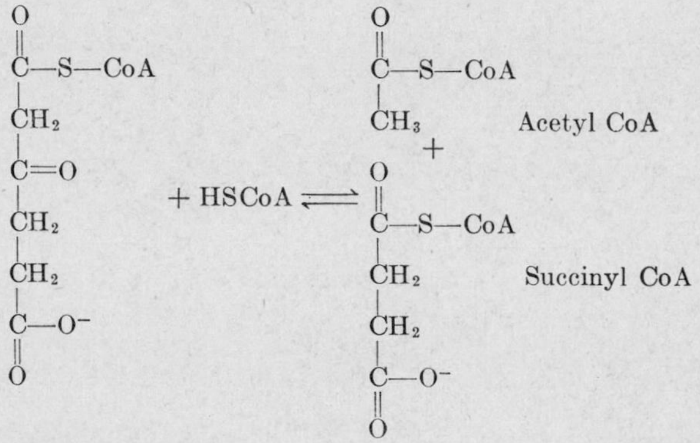

Sum: β-ketoadipate + HSCoA $\rightleftharpoons$ succinate + acetyl CoA

(XV)

GENERAL REFERENCES

Dalgleish, C. E. (1951). *Quart. Revs. (London)* **5**, 227.
Happold, F. C. (1950). *Advances in Enzymol.* **10**, 51.
Larsen, P. (1951). *Ann. Rev. Plant Physiol.* **2**, 169.
Mehler, A. H. (1955). *in* "Symposium on Amino Acid Metabolism" (W. D. McElroy and B. Glass, eds.), p. 882. Johns Hopkins Press, Baltimore.
Yanofsky, C., and Bonner, D. M. (1951). *J. Nutrition* **44**, 603.

Mechanism of Action of Pyridoxal Phosphate Enzymes

In the preceding sections pyridoxal phosphate has appeared as an essential cofactor for many apparently unrelated processes. It has been possible, however, to demonstrate common features to many different reactions of amino acids: decarboxylation, transamination, racemization, substitution, and elimination. The properties of pyridoxal are such that many of the enzymatic reactions have been duplicated nonenzymatically with pyridoxal or an analog catalyzing reactions of amino acids. These model reactions have served as the basis for a general theory of the mechanism of action of enzymes that use pyridoxal phosphate as a cofactor. Most of the experiments and interpretation discussed below were reported by Snell and his collaborators.

Transamination. Transaminations were discovered as spontaneous reactions before the enzymatic reactions were known.[1] Therefore it was first considered that transaminases carry out a direct transfer of nitrogen from amino to keto acids. The implication of pyridoxal in these reactions led to the demonstration of transfer of amino groups to pyridoxal in a reversible reaction:[2]

$$\text{amino acid} + \text{pyridoxal} \rightleftharpoons \text{keto acid} + \text{pyridoxamine}$$

This reaction is catalyzed by certain cations, Al^{+++}, Fe^{+++}, and Cu^{++}, and proceeds more rapidly at higher temperatures. Glycine alone among the amino acids reacted very slowly, but this is a consequence of the unusual equilibrium of the glycine: glyoxylate system; glyoxylate reacts rapidly with pyridoxamine. The reversible transamination of pyridoxal offers a mechanism for the catalysis of transamination between amino and keto acids. As is discussed below pyridoxal derivatives form elaborate complexes with amino acids, and studies with model reactions suggest that such complexes may facilitate transaminations as well as other reactions. The complexes that are formed involve the carboxyl group of the amino acid as well as the amino group. In the complex tautomerization can occur to equilibrate two forms, both of which can be hydrolyzed and dissociated to yield the products of transamination.

β-Hydroxy-α-amino Acids. Another reaction of glycine is catalyzed by pyridoxal. This is the reversible condensation with aldehydes.[3] Glycine and formaldehyde form serine in a model of the reaction observed in biological systems to proceed with tetrahydrofolic acid as a formaldehyde acceptor and donor. When glyoxylate reacts with pyridoxamine, the gly-

$$\text{HCHO} + \text{H}_2\text{NCH}_2\text{COOH} \rightleftharpoons \text{HOCH}_2\underset{\underset{\text{NH}_2}{|}}{\text{CH}}\text{COOH}$$

cine formed gradually condenses with glyoxylate to form β-hydroxy-aspartate. Acetaldehyde condenses with glycine in the presence of pyridoxal and metals to form a mixture of threonine and allothreonine. Propionaldehyde and pyruvate also have been shown to condense with glycine in this system.

[1] R. M. Herbst, *Advances in Enzymol.* **4**, 75 (1944).
[2] D. E. Metzler, J. Olivard, and E. E. Snell, *J. Am. Chem. Soc.* **76**, 699 (1954).
[3] D. E. Metzler, J. B. Longenecker, and E. E. Snell, *J. Am. Chem. Soc.* **76**, 639 (1954).

When glycine is heated with pyridoxal and alum between pH 4 and 6 an insoluble material is formed. The yellow material contains two equivalents of pyridoxal per aluminum, but only one of these is active as a growth factor for yeast; the other was found as a condensation product with glycine, liberated by acid from the original complex and identified as β-pyridoxylserine. This is analogous to serine formed from formaldehyde and glycine, with the aldehyde of pyridoxal becoming the β-carbon of serine. The proposed structure for the complex is shown in (I).

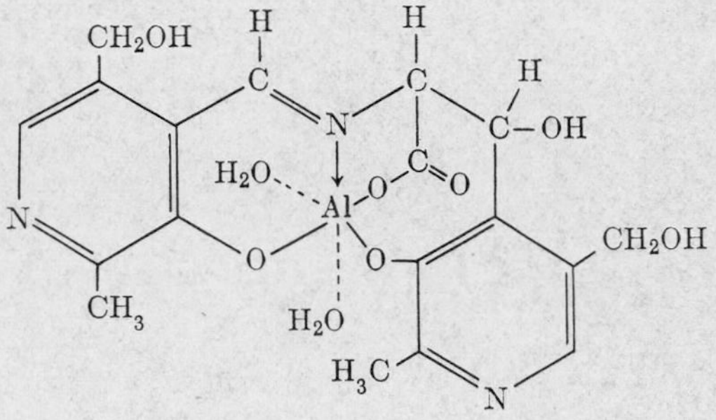

(I)

This structure illustrates the two functions of pyridoxal. The right side of the structure shows the 4-carbon condensed with glycine and the phenolic group participating in binding aluminum. The left side shows the aldehyde of a second pyridoxal condensed with the amino group of glycine in a Schiff's base. It is suggested that such Schiff's bases are intermediates in all of the reactions of pyridoxal with amino acids (and of pyridoxamine with keto acids). Transamination occurs via a tautomeric shift of the α-hydrogen, followed by hydrolysis. In enzymatic transaminations it might be considered that in some way the nitrogen is passed directly from one acid to the other, without the intermediate formation of pyridoxamine. The demonstration that pyridoxamine phosphate has the same activity as pyridoxal phosphate with a transaminase argues in favor of a mechanism involving pyridoxamine formation. Alternatively, it is possible that pyridoxamine phosphate acts as a substrate for the enzyme, and, as it is converted to pyridoxal phosphate, becomes a cofactor that combines with protein to form additional active enzyme.

Racemizations. When pyridoxal forms a Schiff's base with an amino acid in the presence of a metal ion, the structure of (II) seems likely to occur. Ionization of the α-hydrogen is favored by the formation of the quinoid structure on the right. In the racemization of amino acids, which was shown to require pyridoxal phosphate in the cases of alanine[4] and

[4] W. A. Wood and I. C. Gunsalus, *J. Biol. Chem.* **190**, 403 (1951).

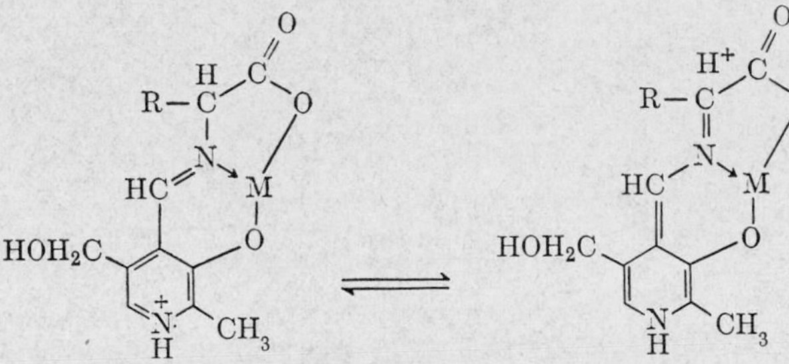

(II)

glutamic acid,[5] it is suggested that the reversal permits random addition of the hydrogen to the α-carbon.

The aromatic Schiff's base offers a variety of procedures for elimination of components by appropriate "electron pushing."[6] If the electrons

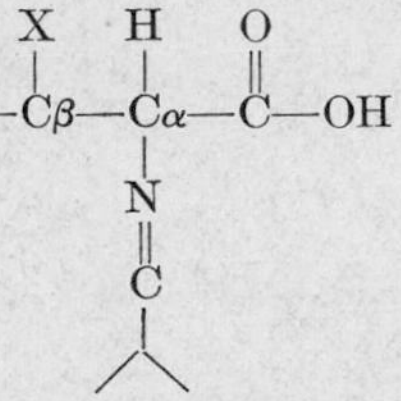

Reference Structure for Schiff's Base

of any of the bonds of the α-carbon shift to form a double bond between that carbon and the nitrogen, the remainder of the molecule may be stabilized by the formation of the quinoid structure illustrated for the case of hydrogen ionization. If the electrons were to come from the α—β bond instead of the C—H bond, and the resulting fragments were to pick up H^+ and OH^- from the medium, a mechanism of the serine and threonine cleavage reactions would result. Similarly, electrons from the carboxyl carbon-α-carbon bond would eliminate that bond and yield the CO_2 found in amino acid decarboxylations.

β-Elimination. Another type of reaction known to occur with pyridoxal both enzymatically and nonenzymatically may be described as β-elimination, the loss of X from the reference structure above. This type includes dehydration of β-hydroxy amino acids (III) and desulfhydration of cysteine. If the reaction starts with ionization of the α-hydrogen, a shift of electrons can result in the elimination of a negatively charged group from the β-position, leaving an α—β double bond. If an SH^- or OH^- group is eliminated, H_2S or H_2O would result from the sum of the

[5] P. Ayengar and E. Roberts, *J. Biol. Chem.* **197,** 453 (1952); S. A. Narrod and W. A. Wood, *Arch. Biochem. Biophys.* **35,** 462 (1952).
[6] D. E. Metzler, M. Ikawa, and E. E. Snell, *J. Am. Chem. Soc.* **76,** 648 (1954).

α-hydrogen and β-elimination. Hydrolysis of the Schiff's base would then yield the α-keto acid, ammonia, and pyridoxal. Another reaction that bears a close formal resemblance to the serine and cysteine eliminations is the tryptophanase reaction. In this case the β-indole group is eliminated, leaving an enol-pyruvate derivative. Cystathionine in similar fashion eliminates homocysteine.

$$
\begin{array}{c}
\text{R} \\
| \\
\text{HCOH} \\
| \\
\text{HC—N=R}' \\
| \\
\text{COOH}
\end{array}
\longrightarrow
\begin{array}{c}
\text{R} \\
| \\
\text{HC} \\
\| \\
\text{C—N=R}' \\
| \\
\text{COOH}
\end{array}
\xrightarrow{\text{H}_2\text{O}}
\begin{array}{c}
\text{R} \\
| \\
\text{HC} \\
\| \\
\text{C—OH} \\
| \\
\text{COOH}
\end{array}
\longrightarrow
\begin{array}{c}
\text{R} \\
| \\
\text{CH}_2 \\
| \\
\text{C=O} \\
| \\
\text{COOH}
\end{array}
$$

(III)

β-Substitution. A modification of the β-elimination reaction may be substitution of a new group. This has been proposed as the mechanism of action of tryptophan desmolase, in which indole is substituted for the OH of serine. This reaction was found to proceed to a measurable extent in model reactions, in spite of the competing β-elimination reactions of both serine and tryptophan and other side reactions of indole compounds. Additional substitution reactions of biological significance are the formation of cystathionine from homocysteine and serine[7] and the formation of S-methylcysteine from methyl mercaptan and serine.[8] These reactions are catalyzed by enzymes that require pyridoxal phosphate as a cofactor.

Miscellaneous Eliminations. Other known enzymatic reactions involving pyridoxal phosphate are more difficult to explain or to duplicate.

$$
\begin{array}{c}
\text{SH} \\
| \\
\text{CH}_2 \\
| \\
\text{CH}_2 \\
| \\
\text{HCNH}_2 \\
| \\
\text{COOH} \\
\text{Homocysteine}
\end{array}
+ \text{H}_2\text{O} \longrightarrow
\begin{array}{c}
\text{CH}_3 \\
| \\
\text{CH}_2 \\
| \\
\text{C=O} \\
| \\
\text{COOH} \\
\alpha\text{-Ketobutyric acid}
\end{array}
+ \text{NH}_3 + \text{H}_2\text{S}
$$

(IV)

These include γ-elimination, as in the cases of homocysteine[9] and homoserine[10] $\rightarrow$ α-ketobutyrate (IV). The hydrolysis of kynurenine to anthranilic acid might represent a reaction different from the others. The Schiff's base may tautomerize to form a double bond between the α-carbon and the nitrogen (V). The resulting side chain of the kynurenine deriva-

[7] F. Binkley, *J. Biol. Chem.* **191,** 531 (1951).

[8] E. Wolff, S. Black, and P. F. Downey, *J. Am. Chem. Soc.* **78,** 5958 (1956).

[9] R. E. Kallio, *J. Biol. Chem.* **192,** 371 (1951).

[10] F. Binkley and C. K. Olson, *J. Biol. Chem.* **185,** 881 (1950).

tive has the structure of an α,γ-diketone, which would have labile hydrogens on the β-carbon, and, analogous to the β-eliminations in the case of labilization of the α-hydrogen, could undergo β-elimination of a positive group, that would pick up OH^- from the medium.

Except in the case of racemases, pyridoxal phosphate enzymes are stereospecific. This has been illustrated dramatically in experiments with glutamic decarboxylase.[11] The mechanism for decarboxylation postulated earlier involves only electron shifts about the α-carbon, while the α-hydrogen remains fixed. This mechanism was supported when it was found that the α-hydrogen does not exchange with D_2O during enzymatic decarboxylation. The resulting mono-D-γ-aminobutyric acid was found to

$$
\begin{array}{c}
\overset{\displaystyle O}{\underset{\displaystyle \|\gamma}{}} \overset{\beta}{} \quad \overset{\alpha}{} \\
C-CH_2-\underset{\displaystyle \overset{\|}{N}}{C}-COOH \\
NH_2 \qquad \underset{R}{|}
\end{array}
$$

(V)

exchange the D during incubation with decarboxylase in H_2O. The isomeric mono-D-γ-aminobutyric acid, prepared by racemizing glutamic acid in D_2O, then decarboxylating the mono-deutero L-glutamic acid in H_2O, did not exchange. Thus it is seen that the enzyme reacts stereospecifically with the chemically equivalent hydrogen atoms on the γ-carbon of γ-aminobutyric acid. It might be considered that a hydrogen bond stabilizes one hydrogen, while the other is free to dissociate. The position vacated must be available to attack by CO_2, since the amino acid decarboxylases catalyze reversible reactions.

In all of these reactions mechanisms have been shown to be possible in model systems. Individual enzymes, in contrast to the model systems, react with given amino acids and carry out only one of the possible reactions. It is apparent that the function of the protein is to direct the reaction. This may be done by weakening a specific bond through attraction of a group to the protein.

Function and Structure of Pyridoxal. The mechanisms suggested for pyridoxal phosphate reactions receive strong support from experiments with analogs.[12] It was found that a nitro group *ortho* or *para* to the carbonyl of benzaldehyde serves the same function as the pyridine nitrogen. Electronically, the two are equivalent in serving as electron-attracting groups. The other essential group in analogs is the *ortho* phenolic group. The three functional groups required thus permit the formation of a

[11] S. Mandeles, R. Koppelman, and M. E. Hanke, *J. Biol. Chem.* **209**, 327 (1954).
[12] M. Ikawa and E. E. Snell, *J. Am. Chem. Soc.* **76**, 653 (1954).

Schiff's base (aldehyde), the formation of a chelate (phenol), and the shift of electrons (ring N or nitro group). Pyridoxal contains in addition a methyl group and an aliphatic hydroxyl. The hydroxyl group obviously serves as an attachment for the phosphate, which has been assumed to be essential for attachment to the enzyme. No function is known for the 2-methyl group. In some organisms the corresponding ethyl compound, ω-methylpyridoxal, serves as a growth factor.[13] Since the phenolic hydroxyl function is required in both enzymatic and nonenzymatic reactions, it has been postulated that metal chelates are involved in enzymatic reactions as well as in the model systems. An alternative suggestion was that a side chain of the protein duplicates the metal function. The recent report that resolved histidine decarboxylase requires iron in addition to pyridoxal phosphate[14] strengthens the concept of metal participation as a general property of pyridoxal phosphate reactions.

Metabolism of Amines

The oxidation of amines by mammalian liver was first demonstrated in perfusion experiments.[1] Subsequently many investigators showed the existence of enzymes that could oxidize various amines. The original substrates used contained only single amino groups, and it seemed probable that only a single enzyme, monoamine oxidase, could account for the oxidation of these substrates.[2] A separate enzyme, diamine oxidase,[3a,b] requires the presence of a second functional group in the molecule, and reacts rapidly with substrates bearing two amino groups. As has been found for many other enzyme activities, these names, monoamine and diamine oxidase, apply to families of enzymes, and the specific properties of one member are not necessarily the same as the properties of similar enzymes from other sources. Especially bacterial enzymes are known to have rigid substrate specificities not shared by other amine oxidases.

Monoamine Oxidase. Monoamine oxidase activity is found widely distributed in mammalian tissues,[4] invertebrates,[5] and plants.[6] The con-

[13] M. Ikawa and E. E. Snell, *J. Am. Chem. Soc.* **76,** 637 (1954).
[14] B. M. Guirard and E. E. Snell, *J. Am. Chem. Soc.* **76,** 4745 (1954).

[1] A. J. Ewins and P. P. Laidlaw, *J. Physiol.* **41,** 78 (1910–1911).
[2] E. A. Zeller, *Helv. Chim. Acta* **24,** 539 (1941).
[3a] C. H. Best, *J. Physiol.* **67,** 256 (1929).
[3b] E. A. Zeller, *Helv. Chim. Acta* **21,** 880 (1938).
[4] K. Bhaguat, H. Blaschko, and D. Richter, *Biochem. J.* **33,** 1338 (1939); H. Birkhauser, *Helv. Chim. Acta* **23,** 1071 (1940).
[5] H. Blaschko, D. Richter, and H. Schlossman, *J. Physiol.* **90,** 1 (1937).
[6] E. Werle and F. Roewer, *Biochem. Z.* **330,** 298 (1950).

ventional preparations of monoamine oxidase are not soluble and only slight purification has been achieved. The specificities reported must therefore be interpreted cautiously, as more than one enzyme may be participating in the reactions studied. The enzymes of animal livers have

$$RCH_2NH_2 + O_2 + H_2O \rightarrow RCHO + H_2O_2 + NH_3$$

been tested with a number of substrates. The rate of oxygen consumption depends upon chain length, as seen when aliphatic amines of one through eight carbons are substrates.[7] The very short chains are poor substrates, but very little difference is found with chains of intermediate length. A straight-chain amine with 18 carbon atoms was not oxidized.[8] The presence of a second amino group on very long chains (14 to 18 carbon atoms) instead of inhibiting the oxidation, as is found with smaller molecules, supports a reaction. Substrates with aromatic rings, including phenylethylamine, tyramine, and epinephrine, are oxidized rapidly. Secondary and tertiary amines are attacked, usually at somewhat lower rates than the corresponding primary amines.[9] Many other structural modifications, such as the presence of α- and β-substitutions and the presence of hydroxyl groups in various positions, have been found to influence the oxidation rate; this type of evidence may become useful in mapping the enzyme surface.

One monoamine oxidase has been found soluble in nature, and has been purified. This is the enzyme of blood serum, which has been purified 200-fold from steer plasma.[10] The specificity of the serum enzyme is more restricted than those of the liver enzymes; tryptamine and epinephrine are not oxidized rapidly, if at all, although phenylethylamine is a substrate. The polyamines spermine and spermidine are among the best substrates, and decamethylenediamine, but not shorter diamines, is also attacked. Aromatic substitutions on methylamine form substrates that permit spectrophotometric assays. Benzylamine and furfurylamine, for example, which do not absorb light in the region used, are converted to benzaldehyde and furfuraldehyde, respectively; these products have absorption maxima near 250 mμ and 275 mμ.

The reaction carried out in air consumes 1 mole of oxygen for each amino group oxidized, and the products are an aldehyde, H_2O_2, and ammonia (or an amine).[11] The reaction catalyzed by the liver enzyme does

$$RCH_2NR_2' + O_2 + H_2O \rightarrow RCHO + H_2O_2 + HNR_2'$$

[7] G. A. Alles and E. V. Heegard, *J. Biol. Chem.* **147,** 487 (1943).
[8] H. Blaschko and R. Duthie, *Biochem. J.* **39,** 478 (1945).
[9] H. I. Kohn, *Biochem. J.* **31,** 1693 (1937).
[10] C. W. Tabor, H. Tabor, and S. M. Rosenthal, *J. Biol. Chem.* **208,** 645 (1954).
[11] F. J. Philpot, *Biochem. J.* **31,** 856 (1937).

not require oxygen, however, and other electron acceptors have been found to support amine oxidation.

Liver monoamine oxidase is resistant to inhibition by cyanide and carbonyl reagents. The serum enzyme, on the other hand, is sensitive to these reagents.[10,12] Isonicotinoyl hydrazide and some related compounds inhibit amine oxidases.[13]

Diamine Oxidase. The original description of the enzyme that oxidizes diamines was based on the oxidation of histamine, and the enzyme was named histaminase.[3a] The name diamine oxidase was applied when evidence was obtained that the same enzyme also attacks cadaverine, putrescine, and agmatine.[3b] This activity has also been found widely distributed among animals,[14] plants,[15] and microorganisms.[16]

Diamine oxidase has been purified extensively from acetone-dried kidney.[17] The purified enzyme attacks many dibasic compounds, but does not oxidize spermine. More recent studies[18] have employed 62 diamines. Of these, the compounds with one primary amino group and no α-substitution were oxidized. Substitutions on the second nitrogen or the carbon which bears it affect the rate, but do not prevent the reaction. These experiments are interpreted as indicating that the oxidation takes place at the primary amine and is inhibited by adjacent groups, while attachment to the enzyme also occurs through a second group whose structure may be varied. Thus, in histamine the imidazole group serves as a site

$$\underset{HN\diagdown N}{\boxed{}}{-CH_2CH_2NH_2} + O_2 + H_2O \longrightarrow \underset{HN\diagdown N}{\boxed{}}{-CH_2CHO} + H_2O_2 + NH_3$$

(I)

of attachment, but oxidation occurs only at the primary amine (I). Hydroxyl groups and other electronegative groups can substitute for the second amine in providing a means for interacting with the protein.[19] The position of the second group can vary widely, but very short chains are not good substrates; ethylenediamine and imidazolylmethylamine are not oxidized.[20]

[12] J. G. Hirsch, *J. Exptl. Med.* **97**, 345 (1953).
[13] E. A. Zeller, J. Barsky, J. R. Fouts, W. F. Kirchheimer, and L. S. Van Orden, *Experientia* **8**, 349 (1952).
[14] C. H. Best and E. W. McHenry, *J. Physiol.* **70**, 349 (1930).
[15] E. Werle and A. Zabel, *Biochem. Z.* **318**, 554 (1948).
[16] E. Werle, *Biochem. Z.* **309**, 61 (1941).
[17] H. Tabor, *J. Biol. Chem.* **188**, 125 (1951).
[18] E. A. Zeller, J. R. Fouts, J. A. Carbon, J. C. Lazanas, and W. Voegtle, *Helv. Chim. Acta* **39**, 1632 (1956).
[19] J. R. Fouts, *Federation Proc.* **13**, 217 (1954).
[20] See reviews by Tabor and Zeller.

Although purified animal enzymes appear to have broad substrate specificity, extremely specific oxidases have been found in bacteria.[21] *Achromonas* sp. was found to produce oxidases when incubated with histamine or putrescine. The enzymes induced by one amine were incapable of reacting with the other. These enzymes are able to use indophenols as well as oxygen as electron acceptors, but they are inhibited by methylene blue.

The animal diamine oxidase is inhibited by cyanide[22] and other carbonyl reagents.[23] In contrast to liver monoamine oxidase, diamine oxidase is not sensitive to *p*-chloromercuribenzoate or other SH reagents.[24] Isonicotinoyl hydrazide is a potent inhibitor of diamine oxidase.[13] The chemical basis for the inhibitions are not known. It has been suggested that the bacterial enzymes contain flavins,[21] but there is no evidence for any cofactors in the animal enzymes. Again in contrast to monoamine oxidase, diamine oxidase is inhibited by excess substrate.[22] This is interpreted as showing combination of the enzyme with separate substrate molecules at the two adsorbing sites; in this situation there is no effective reaction, as the catalysis appears to require combination of the substrate at two points.

Diamine oxidase is influenced profoundly by physiological states. In pregnancy in humans there is a dramatic increase in this enzyme in the blood;[25] the circulating enzyme appears to have its origin in the placenta.[26] In cats the oxidase is concentrated in the intestinal mucosa; adrenalectomy has been reported to cause the enzyme to leave the intestine and appear in the lymph. Cortisone causes the leakage of enzyme to stop and permits normal levels to accumulate in the intestine.[27]

Several amines are known to have specific biological functions. Some obvious examples are histamine and serotonin (5-hydroxytryptamine), which are produced by decarboxylation of the corresponding amino acids. The secondary amine, epinephrine, is formed in an incompletely elucidated pathway from phenylalanine.[28] In addition to these compounds with potent pharmacological activity, other amines seem to play vital

[21] K. Satake and H. Fujita, *J. Biochem.* **40,** 547 (1953).

[22] E. A. Zeller, *Helv. Chim. Acta* **23,** 1418 (1940).

[23] E. Werle, *Biochem. Z.* **304,** 201 (1940); E. A. Zeller, *Advances in Enzymol.* **2,** 93 (1942).

[24] T. P. Singer and E. S. G. Barron, *J. Biol. Chem.* **157,** 241 (1945); E. A. Zeller, *Helv. Chim. Acta* **21,** 1645 (1938).

[25] A. Ahlmark, *Acta Physiol. Scand.* **9,** Suppl. 28 (1944).

[26] G. V. Anrep, G. S. Barsoum, and A. Ibrahim, *J. Physiol.* **106,** 379 (1947).

[27] G. Kahlson, S. E. Lindell, and H. Westling, *Acta Physiol. Scand.* **30,** Suppl. 111, 192 (1953).

[28] L. C. Leeper and S. Udenfriend, *Federation Proc.* **15,** 298 (1956).

but less obvious functions. The polyamines spermine and spermidine occur in high concentrations in many mammalian organs, but no function has been determined for them.[29] There is reason to suspect a function, however, since microorganisms have been found to have nutritional requirements that are met by putrescine or the polyamines.[30] When putrescine is incubated with *Aspergillus* or *E. coli*, it is rapidly and efficiently incorporated as a unit into spermidine.[31a] The formation of spermidine involves the utilization of a three-carbon fragment derived from methioneine.[31b] Active methionine appears to be decarboxylated to the corresponding amine, which then transfers the propylamine group to putrescine in a reaction analogous to the methylation reactions of active methionine. The products of the transfer are spermidine and thiomethyladenosine.[31c] The oxidation of diamines, such as cadaverine, results in the formation of ω-amino aldehydes, which have been shown to cyclize spontaneously through Schiff's base formation, yielding heterocyclic rings.[32] These can carry out further nonenzymatic condensations, and it is possible that these reactive molecules may be intermediates in the biosynthesis of heterocyclic rings (II). The formation of heterocyclic rings has been demonstrated with amine oxidase preparations from plants[33] and animals.[17]

$$
\begin{array}{ccc}
\text{NH}_2 & & \text{O} \\
| & & \| \\
\text{CH}_2 & & \text{CH} \\
| & & | \\
\text{CH}_2 & & \text{CH}_2 \\
| & \xrightarrow{\text{diamine oxidase}} & | \\
\text{CH}_2 + \text{O}_2 + \text{H}_2\text{O} & & \text{CH}_2 \longrightarrow \\
| & & | \\
\text{CH}_2 & & \text{CH}_2 \\
| & & | \\
\text{CH}_2 & & \text{CH}_2 \\
| & & | \\
\text{NH}_2 & & \text{NH}_2
\end{array}
$$

(II)

GENERAL REFERENCES

Ciba Foundation Symposium on Histamine, 1956. (G. E. W. Wolstenholme and C. M. O'Conner, eds.) Churchill, London.

Tabor, H. (1954). *Pharmacol. Revs.* **6,** 299.

Zeller, E. A. (1951). *In* "The Enzymes" (J. B. Sumner and K. Myrbäck, eds.), Vol. II, Part 1, p. 536. Academic Press, New York.

[29] S. M. Rosenthal and C. W. Tabor, *J. Pharmacol. Exptl. Therap.* **116,** 131 (1956).

[30] E. J. Herbst and E. E. Snell, *J. Biol. Chem.* **176,** 989 (1948); P. H. A. Sneath, *Nature* **175,** 818 (1955).

[31a] H. Tabor, S. M. Rosenthal, and C. W. Tabor, *Federation Proc.* **15,** 367 (1956).

[31b] R. C. Greene, *Federation Proc.* **16,** 189 (1957).

[31c] H. Tabor, S. M. Rosenthal, and C. W. Tabor, *J. Am. Chem. Soc.* **79,** 2978 (1957).

[32] C. Schöpf, A. Komzak, F. Braun, and E. Jacobi, *Ann.* **559,** 1 (1948).

[33] P. J. G. Mann and W. R. Smithies, *Biochem. J.* **61,** 89, 101 (1955).

Chapter VIII

ACIDS AND ACID DERIVATIVES

Esters

Esters of carboxylic acids are prominent among the chemical structures of biological systems. The biosyntheses of only a few of the naturally occurring esters have been studied enzymatically. Among the known reactions are the syntheses of acetylcholine and of phosphatidic acids. Acetylcholine is formed by choline acetylase, which has been purified from nervous tissue. Nachmansohn and associates have prepared choline acetylase from rat brain, the electric organ of the electric eel, squid ganglia, and other tissues. It has been detected in muscle also, but not in liver or kidney.[1] The reaction is very similar to the acetylation of amines.

$$
CH_3\overset{\overset{O}{\parallel}}{C}-SCoA + HO-CH_2CH_2-\overset{+}{N}(CH_3)_3 \longrightarrow
$$

$$
CH_3\overset{\overset{O}{\parallel}}{C}-O-CH_2CH_2-\overset{+}{N}(CH_3)_3 + CoASH
$$

Choline acetylase and the amine-acetylating enzyme are distinct, and do not cross react with the substrates of the other enzyme.

The synthesis of phosphatidic acids is a very similar reaction catalyzed by insoluble preparations of liver. Acyl CoA derivatives (formed as described for fatty acid oxidation, p. 141) are acyl donors and L-α-glycerophosphate is the acyl acceptor. Two fatty acid groups may be added to one α-glycerophosphate (I).[2] This reaction requires fatty acids with 12 or

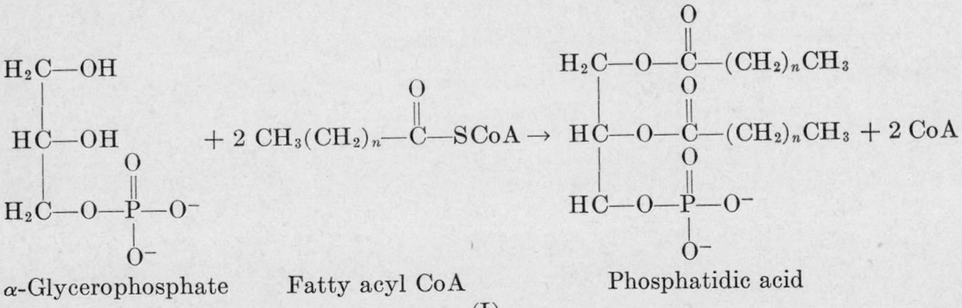

α-Glycerophosphate   Fatty acyl CoA   Phosphatidic acid
(I)

[1] See reuiew by Nachmansohn and Wilson.
[2] A. Kornberg and W. E. Pricer, Jr., *J. Biol. Chem.* **204,** 345 (1953).

more carbons in the chain, and reacts more rapidly with longer chains. Unsaturated fatty acids also are esterified.

It was thought that phosphatidic acid formation might lead to the formation of the phospholipids, lecithin and cephalin. At this time direct incorporation of phosphatidic acids into the cationic phospholipids has not been demonstrated, but it is possible that the known precursors of lecithin, α,β-diglycerides,[3] are formed by hydrolysis of the phosphate esters of phosphatidic acids. The formation of the cationic lipids requires the following steps.

Choline is converted to phosphorylcholine by an enzyme found in yeast and several animal tissues. This enzyme, choline phosphokinase, catalyzes the reaction:

$$\text{HOCH}_2\text{CH}_2\overset{+}{\text{N}}(\text{CH}_3)_3 + \text{ATP} \longrightarrow {}^-\text{O}-\overset{\overset{\text{O}}{\|}}{\underset{\underset{\text{O}^-}{|}}{\text{P}}}-\text{O}-\text{CH}_2\text{CH}_2-\overset{+}{\text{N}}(\text{CH}_3)_3 + \text{ADP}$$

Choline   Phosphorylcholine

Ethanolamine and its other N-methyl derivatives also serve as substrates.[4]

Both the phosphorus and carbon of phosphorylcholine are introduced into lecithin together.[5] An intermediate activation is required in which

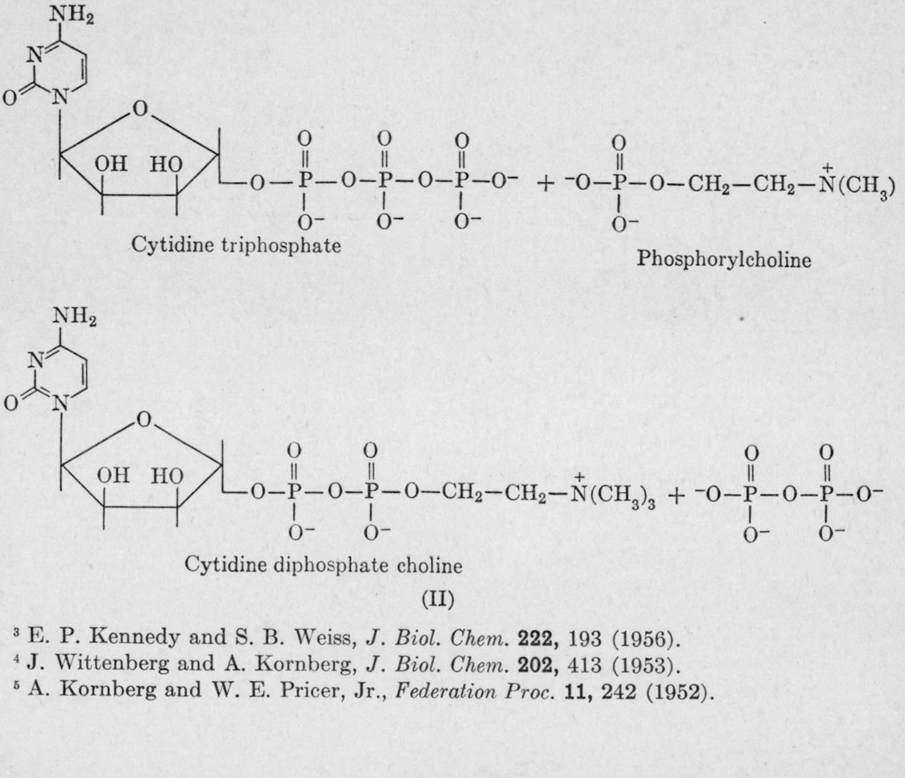

Cytidine triphosphate   Phosphorylcholine

Cytidine diphosphate choline

(II)

[3] E. P. Kennedy and S. B. Weiss, *J. Biol. Chem.* **222**, 193 (1956).
[4] J. Wittenberg and A. Kornberg, *J. Biol. Chem.* **202**, 413 (1953).
[5] A. Kornberg and W. E. Pricer, Jr., *Federation Proc.* **11**, 242 (1952).

cytidine nucleotides specifically participate. An enzyme studied in rat liver particles carries out the pyrophosphorolysis shown in (II).[3] Uridine, guanosine, and adenosine nucleotides cannot be used in place of cytidine. Cytidine diphosphate choline (CDP-choline) is a donor of phosphoryl-choline to α,β-diglycerides to form lecithin (III).

$$
\begin{array}{l}
\text{H}_2\text{C}-\text{O}-\overset{\overset{\text{O}}{\|}}{\text{C}}-\text{R} \\
\text{H}\text{C}-\text{O}-\overset{\overset{\text{O}}{\|}}{\text{C}}-\text{R}' \quad + \quad \text{CDP-choline} \longrightarrow \\
\text{H}_2\text{C}-\text{OH}
\end{array}
\qquad
\begin{array}{l}
\text{H}_2\text{C}-\text{O}-\overset{\overset{\text{O}}{\|}}{\text{C}}-\text{R} \\
\text{H}\text{C}-\text{O}-\overset{\overset{\text{O}}{\|}}{\text{C}}-\text{R}' \\
\text{H}_2\text{C}-\text{O}-\overset{\overset{\text{O}}{\|}}{\underset{\underset{\text{O}^-}{|}}{\text{P}}}-\text{O}-\text{CH}_2-\text{CH}_2-\overset{+}{\text{N}}(\text{CH}_3)_3
\end{array}
$$

α,β-Diglyceride   Lecithin

(III)

Hydrolysis of Esters

Enzymes that hydrolyze a bewildering variety of esters have been studied extensively for over half a century. An elaborate literature describes many different enzyme preparations, but since little of the early work included effective attempts to purify individual enzymes, the number of discreet enzymes is unknown. There obviously are many. Among the substrates studied are simple esters, fats, cholesterol esters, esters of unsaturated acids, and chlorophyll.

Simple esters are split by extracts of all tissues. The number of different esterases is not known. The digestive esterases produced by the pancreas are called lipases because they hydrolyze the triglycerides that are the most prominent lipids. They also hydrolyze simple esters. A special group of enzymes hydrolyze phospholipids. So-called lecithinase A removes one acyl group from lecithin to form lysolecithin, which causes hemolysis of erythrocytes. This enzyme has been crystallized from snake venom.[6] Other animal toxins and bacteria also form lysolecithin by hydrolysis of lecithin. The removal of the second acyl group is catalyzed by phospholipase B, which has been studied in plant and animal extracts and also occurs in bacteria.[7] Other enzymes, phospholipase C and D, specifically remove phosphorylcholine[8] and choline,[9] respectively, from lecithin.

[6] K. H. Slotta and H. L. Fraenkel-Conrat, *Ber.* **71**, 1076 (1938); S. S. De, *Ann. Biochem. and Exptl. Med.* (*Calcutta*) **4**, 45 (1944).
[7] B. Shapiro, *Biochem. J.* **53**, 663 (1953); O. Hayaishi and A. Kornberg, *J. Biol. Chem.* **206**, 647 (1954).
[8] M. G. MacFarlane and B. C. J. G. Knight, *Biochem. J.* **35**, 884 (1941).
[9] D. J. Hanahan and I. L. Chaikoff, *J. Biol. Chem.* **172**, 191 (1948).

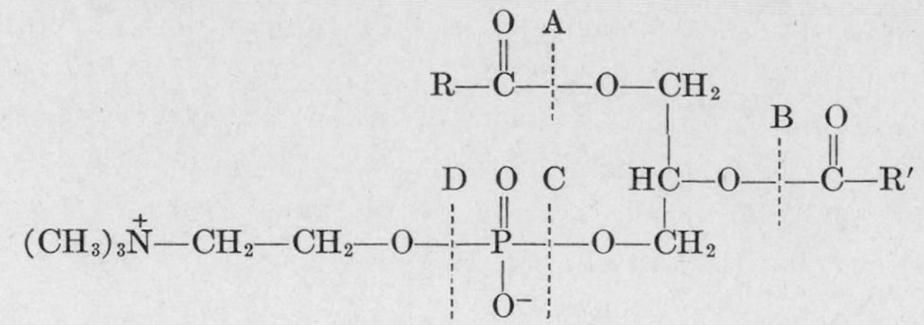

$$\text{R—C—}\overset{O}{\overset{\|}{}}\overset{A}{\vdots}\text{—O—CH}_2$$

Lecithin, Showing Points of Cleavage by Lecithinases

Chlorophyllase appears to accompany chlorophyll in green plants. This enzyme is responsible for much of the difficulty in isolating chlorophyll. It hydrolyzes the phytyl ester of the propionic acid side chain.[10] The reaction is a transfer, in which water is but one possible acceptor. The enzyme is extremely tough; it is active at high temperatures (75°C.) and in the presence of organic solvents (70 per cent or more acetone or ethanol). In the presence of high concentrations of alcohols, the transfer reaction yields predominantly the corresponding esters, e.g., ethylchlorophillide.

Acetylcholine is an ester that occurs in animal tissues and is of great physiological importance. Its precise role in nerve and muscle function has not been defined to the satisfaction of all workers in this field, but the high concentrations of enzymes that synthesize and degrade this ester attest to its metabolic importance.

Two types of enzyme that hydrolyze acetylcholine rapidly have been identified. One is represented by an enzyme present in the serum of many animals. This esterase attacks many other esters at faster rates than acetylcholine, and has been called nonspecific or pseudocholinesterase, in distinction to the so-called specific or true cholinesterase found in erythrocytes, nervous and electrical tissue.[11] The latter hydrolyzes acetylcholine more rapidly than other choline esters. Extensive kinetic studies have been made with both types of enzyme. Studies with highly purified true cholinesterase have been particularly revealing in explorations of the chemical nature of enzymatic action.

A model of cholinesterase has been proposed that is consistent with all of the information available (Fig. 35).[12] In the model there are two sites of attachment, an anionic site that binds the cationic quarternary nitrogen group and an esteratic site that reacts with the carbon of the carbonyl moiety of the ester. When acetylcholine is adsorbed to the enzyme, a transfer of the carbonyl group to the enzyme takes place, forming

[10] R. Willstatter and A. Stoll, "Untersuchungen uber Chlorophyll." Springer, Berlin (1913).

[11] B. Mendel and H. Rudney, *Biochem. J.* **37**, 59 (1943).

[12] See review by Wilson.

an acyl enzyme. The cationic group then dissociates freely, and the acyl group is removed by hydrolysis.

The properties ascribed to the enzyme were derived from studies with many different substrates and inhibitors. Two types of ionic effects can be distinguished. When acetylcholine is the substrate, there is no change in the acidic dissociation of the substrate over the pH range of enzyme

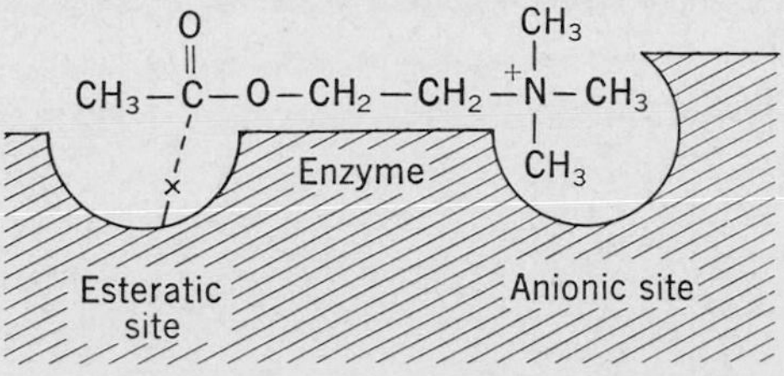

FIG. 35. Hypothetical adsorption of acetylcholine to cholinesterase. From Wilson.[12]

activity. Therefore, all effects of pH change must be associated with dissociable groups of the enzyme. Since the enzyme activity falls off on both sides of the pH optimum of 8.3, dissociable groups with pK values above and below 8.3 are assumed to be essential for enzyme activity. pK values of 7.2 and 9.3 have been calculated for these groups;[13] these may be associated with imidazole and ammonium or phenolic groups in the protein.

Two inhibitors of cholinesterase are classical pharmacological reagents; prostigmine and eserine. As inhibitors of cholinesterase, these

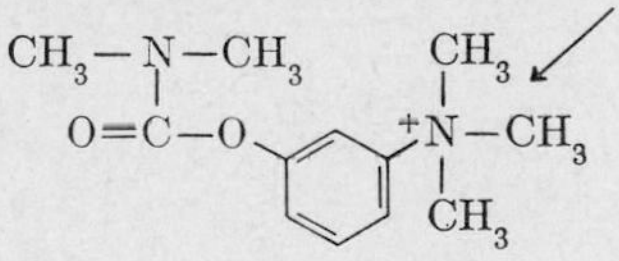

Prostigmine 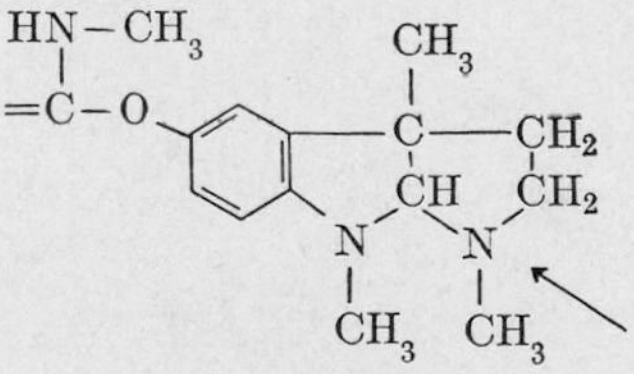   Eserine

compounds differ principally in the basicity of the nitrogen atoms indicated by arrows. Prostigmine contains a quarternary ammonium group, which is charged throughout the pH range of enzyme activity, whereas eserine contains a basic group with a pK near 8.2. When fixed concentrations of these inhibitors are studied at different pH values, the effect of prostigmine remains constant, whereas the inhibition caused by eserine is great on the acid side and low on the alkaline side of its dissociation constant (Fig. 36). Therefore it is concluded that a cationic group in the inhibitor is essential for interacting ionically with the enzyme.[14]

[13] I. B. Wilson and F. Bergmann, *J. Biol. Chem.* **186,** 683 (1950).
[14] I. B. Wilson and F. Bergmann, *J. Biol. Chem.* **185,** 479 (1950).

Experiments with model substrates support the conclusions derived from inhibitor studies. For example, dimethylaminoethyl acetate has been compared with acetylcholine as a substrate for cholinesterase.[14] The essential difference between these substrates is the strength of the

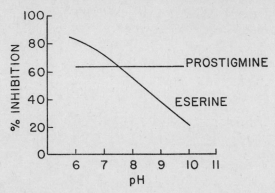

FIG. 36. Inhibition of cholinesterase by eserine and prostigmine.[14]

basic group; dimethylaminoethyl acetate has a pK of 8.3. The curve of relative activity versus pH shows clearly that the cationic form is the susceptible one (Fig. 37).

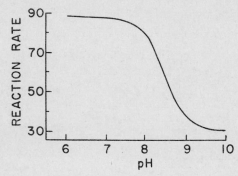

FIG. 37. The relative rate is the rate of hydrolysis of dimethylaminoethyl acetate compared with the rate of hydrolysis of acetylcholine at the same pH. From Wilson.[12]

Simple ammonium compounds are competitive inhibitors of cholinesterase.[15] The binding of these inhibitors to the enzyme depends upon their charge. At pH 7, where all are cations, a series of inhibitors $R—N(CH_3)_xH_y$ illustrates the influence of methyl substituents on the affinity. When x is increased from 0 to 1 to 2, the affinity increases by amounts indicative of forces of 1.2 kcal. per mole per methyl group. The addition of a third methyl group, however, has no influence on the

[15] I. B. Wilson, *J. Biol. Chem.* **197**, 214 (1952).

binding. This observation indicates that the anionic site on the enzyme is flanked by groups that interact with methyl groups bound to the tetrahedral nitrogen, but that the fourth position of the tetrahedron is directed away from the enzyme surface, and a methyl group in this position does not interact with the enzyme.

An interaction of the esteratic site with the carbonyl carbon is supported by results of experiments with both substrates and inhibitors. A series of derivatives of nicotinic acid was found to inhibit cholinesterase.[16] The inhibitory strength was correlated with the electrophilic character of the carbon in the order shown in (IV). The postulation of an acyl enzyme

$$\underset{\displaystyle N}{\overset{\displaystyle \overset{O}{\underset{\|}{C}}-O^-}{\bigcirc}} \quad < \quad \overset{O}{\underset{\|}{-C}}-NH_2 \quad < \quad \overset{O}{\underset{\|}{-C}}-N(C_2H_5)_2 \quad < \quad \overset{O}{\underset{\|}{-C}}-CH_3 \quad < \quad \overset{O}{\underset{\|}{-C}}-O-C_2H_5$$

(IV)

intermediate receives support from the ability of numerous nucleophilic reagents to act as acyl acceptors. These include hydroxylamine, alcohol, and choline as well as water.[13] Acyl donors are not necessarily choline esters; simple esters and the acetic anhydride also react with the enzyme. As was shown for chymotrypsin (p. 29), cholinesterase also reacts with free acids. Thus, acethydroxamic acid formation from acetate and hydroxylamine has been shown to be catalyzed by cholinesterase,[17] although at only 1 per cent of the rate found when an ester of acetic acid is the substrate. Similarly the formation of acetylcholine has been demonstrated. A striking example of the formation of acyl derivatives is the liberation of H_2S from thioacetic acid, CH_3COSH.[18]

A topical relation between the esteratic and anionic sites is implicit in any consideration of enzyme action in which the two sites participate. It seems reasonable to assume that the two sites are spaced to accommodate choline esters. The proximity of the sites is indicated by the ability of prostigmine to inhibit the reaction with thioacetic acid, which does not involve interaction with the anionic site. The necessity for binding acetylcholine at both sites for efficient catalysis has been used to explain inhibition by excess substrate.[19] When high concentrations of acetylcholine are present, it is possible for one molecule to interact with the anionic site of the enzyme, while a second molecule associates with the

[16] F. Bergmann, I. B. Wilson, and D. Nachmansohn, *Biochim. et Biophys. Acta* **6**, 217 (1950).

[17] S. Hestrin, *Biochim. et Biophys. Acta* **4**, 310 (1950).

[18] I. B. Wilson, *Biochim. et Biophys. Acta* **7**, 520 (1951).

[19] E. A. Zeller and A. Bissegger, *Helv. Chim. Acta* **26**, 1619 (1943).

esteratic site. The two bound substrate molecules interfere with each other, and neither forms the doubly bound catalytically active complex with the enzyme. Inhibition by excess substrate is characteristic of true cholinesterase, but not of the nonspecific pseudocholinesterase.

A more subtle analysis of the active surface of cholinesterase is being approached through the use of inhibitors. An illustration of this approach is found in the study of hydroxyproline derivatives.[20] The N-dimethyl derivatives of hydroxyproline and allohydroxyproline, the betaines betonicine and turacine, illustrate the effect of steric hindrance on the combination of inhibitors with the enzyme. The quarternary nitrogen groups interact with the anionic site on the enzyme and the hydroxyl group is attracted to the esteratic site. Betonicine is an effective inhibitor, with a K_I of 8.5×10^{-5}; turacine at concentrations as high as 6×10^{-2} M does not inhibit the hydrolysis of acetylcholine. In betonicine the carboxyl group is on the side of the ring opposite to the hydroxyl group, and does not interfere with the approach of the hydroxyl group to the enzyme. The *cis* arrangement of the carboxyl and hydroxyl groups in turacine places the carboxyl group in a position that prevents the ring from approaching close enough to the enzyme to permit attachment at more than one point at a time.

Cholinesterase shares with many other hydrolytic enzymes a sensitivity to substituted phosphoric anhydrides, typified by diisopropyl fluorophosphate, DFP.[21] DFP has been shown to be a substrate for this type of enzyme. The initial reaction, formation of a diisopropyl phospho-enzyme, occurs rapidly, but this intermediate is hydrolyzed very slowly. Thus a stoichiometric reaction effectively inhibits the enzyme. The subsequent hydrolysis has been demonstrated, however, and acceptors other than water have been found to remove phosphorus from the inhibited enzyme at sufficient rates to diminish the toxicity of DFP *in vivo*. The reaction of the phosphorylated enzyme with hydroxylamine is much faster than the reaction with water, but high concentrations of hydroxylamine are required for efficient enzyme reactivation. When the reactive group is combined with a cationic group at the appropriate distance, as in pyridine-2-aldoxime methiodide, the specific binding of the reagent to the site of choline binding permits low concentrations of the reagent to remove the inhibitor very rapidly.[22] Pyridine-2-aldoxime methiodide is relatively specific as a reactivator for cholinesterase, since other esterases that are inhibited by DFP do not bind the cationic group as cholinesterase does. The specific reactivation of DFP-inhibited cholinesterase has been

[20] S. L. Friess, A. A. Patchett, and B. Witkop, *J. Am. Chem. Soc.* **79**, 459 (1957).
[21] A. Mazur and O. Bodansky, *J. Biol. Chem.* **163**, 261 (1946).
[22] H. Kewitz, I. B. Wilson, and D. Nachmansohn, *Arch. Biochem. Biophys.* **64**, 456 (1956).

demonstrated *in vivo*, and the reagent pyridine-2-aldoxime methiodide has been found to be an effective antidote for DFP poisoning.

GENERAL REFERENCES

Ammon, R., and Jaarma, M. (1950). *In* "The Enzymes" (J. B. Sumner and K. Myrbäck, eds.), Vol. I, Part 1, p. 390. Academic Press, New York.

Augustinsson, K. B. (1950). *In* "The Enzymes" (J. B. Sumner and K. Myrbäck, eds.), Vol. I, Part 1, p. 443. Academic Press, New York.

Augustinsson, K. B. (1952). *In* "The Enzymes" (J. B. Sumner and K. Myrbäck, eds.), Vol. II, Part 2, p. 906. Academic Press, New York.

Nachmansohn, D., and Wilson, I. B. (1951). *Advances in Enzymol.* **12**, 259.

Wilson, I. B. (1954). "Symposium on Mechanism of Enzyme Action" (W. D. McElroy and B. Glass, eds.), p. 642. Johns Hopkins Press, Baltimore.

Zeller, E. A. (1951). *In* "The Enzymes" (J. B. Sumner and K. Myrbäck, eds.), Vol. I, Part 2, p. 986. Academic Press, New York.

Carbonic Anhydrase

Carbonic anhydrase catalyzes the equilibration of CO_2 with carbonic acid.

$$CO_2 + H_2O \rightleftharpoons H_2CO_3$$

This reaction occurs spontaneously and is accelerated by anions and by certain nitrogenous bases.[1] Nevertheless, the nonenzymatic hydration of CO_2 and the dehydration of carbonic acid are very slow compared with the reactions that require or produce one of these forms, and carbonic anhydrase is found in many tissues, where it performs essential physiological functions. Its presence in red cells[2] is associated with the necessity of transferring CO_2 efficiently both in the removal of CO_2 from body tissues and in the elimination of CO_2 in lungs or gills. This activity has also been found in secretory cells and in green plants, but its function in these places is not established. The formation of carbamic acid from CO_2 and ammonia and the subsequent formation of carbamyl phosphate by the kinase are greatly accelerated by the presence of carbonic anhydrase.

Carbonic anhydrase has been purified from erythrocytes.[3] It is purified by conventional precipitation and adsorption techniques after separation from hemoglobin by the Tsuchihashi procedure of precipitating denatured hemoglobin by shaking with alcohol–chloroform.[4] The purified enzyme has a molecular weight of about 30,000[5] and contains 1 zinc atom per molecule.[2]

[1] F. J. W. Roughton and V. H. Booth, *Biochem. J.* **32**, 2049 (1938).
[2] D. Keilin and T. Mann, *Biochem. J.* **34**, 1163 (1940).
[3] D. A. Scott and A. M. Fisher, *J. Biol. Chem.* **144**, 371 (1942).
[4] M. Tsuchihashi, *Biochem. Z.* **140**, 63 (1923).
[5] M. L. Petermann and N. V. Hakala, *J. Biol. Chem.* **145**, 701 (1942).

Very extensive kinetic studies have been made with carbonic anhydrase. The activity increases linearly with pH from 6 to 10.[6] A K_m of 9×10^{-3} M was found for CO_2, independent of pH. This value was obtained at zero degrees, where blank reactions are minimized. Even at zero degrees, the enzyme is very unstable, and special apparatus has been devised to permit extremely rapid assays, both manometric[7] and acidimetric.[8]

Carbonic anhydrase is a useful reagent in analyzing the reaction mechanisms of CO_2 production. When active enzymes (such as urease and carboxylase) act on limiting amounts of substrate, manometric observations sometimes show an "overshoot"; that is, the pressure increases initially to a larger value, then falls to the expected value. This phenomenon was interpreted as indicating the production of CO_2, not HCO_3^- or H_2CO_3, which escapes into the gas phase, then slowly dissolves as it is hydrated. This explanation was substantiated when it was found that carbonic anhydrase decreased the initial rate of CO_2 evolution and abolished the overshoot; that is, the rapid hydration of CO_2 prevented its escape from solution except in the quantities permitted by the equilibrium.[9]

A number of substances are inhibitory to carbonic anhydrase, including heavy metals[7] and oxidizing agents,[10] which indicate essential sulfhydryl groups. Acids or anions that can complex with zinc also inhibit. A specific type of inhibition is found with sulfonamides with unsubstituted —SO_2NH_2 groups.[11] These inhibitors have found applications in clinical and laboratory studies concerned with secretions. The rather specific effects of the sulfonamides which inhibit carbonic anhydrase[12] in causing diuresis[13] and in reducing the intraocular pressure of glaucoma[14] support the concept that carbonic anhydrase participates in the secretory processes.

GENERAL REFERENCE

Roughton, F. J. W., and Clark, A. M. (1951). In "The Enzymes" (J. B. Sumner and K. Myrbäck, eds.), Vol. I, Part 2, p. 1250. Academic Press, New York.

[6] F. J. W. Roughton and V. H. Booth, Biochem. J. 40, 309, 319 (1946).
[7] N. V. Meldrum and F. J. W. Roughton, J. Physiol. 80, 143 (1933).
[8] K. Wilbur and N. G. Anderson, J. Biol. Chem. 176, 147 (1948).
[9] H. A. Krebs and F. J. W. Roughton, Biochem. J. 43, 550 (1948).
[10] M. Kiese and A. B. Hastings, J. Biol. Chem. 132, 281 (1940).
[11] D. Keilin and T. Mann, Nature 146, 164 (1940).
[12] W. H. Miller, A. M. Dessert, and R. O. Roblin, Jr., J. Am. Chem. Soc. 72, 4893 (1950).
[13] C. K. Friedberg, M. Halpern, and R. Taymor, J. Clin. Invest. 31, 1074 (1952).
[14] B. Becker, Am. J. Ophthalmol. 37, 13 (1954).

Chapter IX

ORGANIZATION OF STRUCTURE AND FUNCTION

Integration of Biochemical Reactions

Multi-enzyme Systems. The rates of reactions of enzymes that participate in metabolic processes are controlled by many factors. Some of these factors are variable, and result from the actions of other enzymes or from changes in the environment of the cell. These include temperature, pH, ion concentration, cofactor concentration, oxidation-reduction potential, concentration of substrates, and concentration of products. Another type of factor that influences rate of reaction is the localization of enzymes in cell structures. The organization of cells associates certain enzymes physically, so that the product of one enzyme system is rapidly and efficiently utilized by the next. On the other hand, cellular organization separates some enzymes, so that reactions that proceed well in broken-cell preparations occur only to a limited degree or not at all in the intact cell. The complex series of reactions described in previous chapters require the participation of many enzymes whose activities are interrelated and coordinated. Groups of enzymes that act together to accomplish a chemical conversion form a multi-enzyme system. Since very many reactions may influence the concentrations of the reagents, cofactors, etc., and even of the enzymes themselves, entire cells or organisms are multi-enzyme systems, but the term may be profitably applied to groups of two or more enzymes artificially isolated *in vitro* or arbitrarily considered *in vivo*.

The conversion of glucose to its fermentation products furnishes a convenient example of a multi-enzyme system. In such a system the production of lactic acid requires the sequential action of at least 10 enzymes. In intact organisms this process is controlled so that ATP is generated to meet the requirements of synthetic and kinetic reactions, but the available substrate is not squandered. There are several possible mechanisms that may control the rate of a multi-enzyme system. One would be to have the rate-limiting reaction determined by the concentration of an enzyme that occurs in fixed amounts per cell. In such a case the rate of the over-all process would be constant for a given amount of cell material. Measurements of the activities of individual enzymes extracted from cells have shown that each of the glycolytic enzymes is capable of acting at a different maximum rate. The enzyme with the slowest V_{max} has been considered to be rate-limiting. However, a kinetic study of the glycolysis of leucocytes has shown that the over-all rate is slower than any of the individual reactions,[1] which indicates that the rate is determined by an interaction between enzymes, not by the activity of a single enzyme.

Substrate-Limited Reactions. As individual steps in a sequence approach their equilibria, the rate of reaction decreases. The fermentation of glucose catalyzed by yeast extracts illustrates the effect of product accumulation. Normally ATP is considered a desirable product, and, as it plays a catalytic role in glycolysis, is not considered in estimating the progress of the reaction. In most crude extracts active phosphatases hydrolyze ATP and prevent its accumulation. Glycolyzing yeast extracts, however, were found to be stimulated by the addition of apyrase, which formed ADP from ATP.[2] Thus glycolysis was shown to be limited at the phosphokinase steps by limiting concentrations of ADP, the phosphate acceptor, or by excessive concentrations of ATP that favored the back reactions.

Kinetic Limitations. The accumulation of end-products, such as ATP, does not seem to explain the limitation of the rate of glycolysis of leucocytes. Instead, a kinetic explanation seems to apply. In a sequence of reactions

$$A \underset{k_2}{\overset{k_1}{\rightleftharpoons}} B \underset{k_4}{\overset{k_3}{\rightleftharpoons}} C$$

the rate of formation of C is limited by the concentration of B as well as by the rate constant, k_3. The steady-state concentration of B may be limited either by equilibration with A or by the ratio of the actual rates of formation and utilization. In the case that the steady-state concentration of B is less than the amount required to saturate the enzyme catalyzing the conversion to C, the over-all formation of C from A will be less than the maximum rate of either step.[3]

Enzyme-Limited Processes. Even though the over-all rate of a biochemical process may not be a simple function of any single enzyme, in many cases changes in the concentration of a single enzyme alter the over-all rate. The fate of glucose phosphate and glycogen in liver has been shown to be controlled by the activity of phosphorylase.[4] Phosphorylase activity in liver slices limits the breakdown of glycogen, since it can be shown that liver phosphoglucomutase and glucose-6-phosphatase produce free glucose from glucose-1-phosphate much faster than the rate of glucose

[1] W. S. Beck, *J. Biol. Chem.* **216**, 333 (1955).
[2] O. Meyerhof, *J. Biol. Chem.* **157**, 105 (1945).
[3] Some properties of steady-state systems have been described by A. C. Burton, *J. Cellular Comp. Physiol.* **9**, 1 (1936) and by C. N. Hinshelwood, "The Chemical Kinetics of the Bacterial Cell." Oxford Univ. Press London and New York (1947).
[4] E. W. Sutherland, *Phosphorus Metabolism Symposium* **1**, 53 (1951).

production from glycogen. When glycogen must be mobilized to supply material for increased blood sugar requirements, as in the presence of epinephrine, the demand is met by the formation of active phosphorylase from dephosphophosphorylase (p. 234). The increase in phosphorylase can be measured both in terms of glycogenolysis and of glycogen synthesis in the presence of glucose-1-phosphate and fluoride (to inhibit the mutase reaction). The two enzymes that degrade and resynthesize phosphorylase, the phosphatase and the kinase, thus control carbohydrate metabolism by determining the effective concentration of liver phosphorylase. The factors that influence the phosphatase and kinase are currently being investigated.

The Pasteur Effect. Pasteur noticed that the production of fermentation products could be greatly reduced by the presence of oxygen. The inhibition of fermentation by oxygen has been named the Pasteur effect, and results in a net decrease in substrate utilization. Many explanations have been suggested to account for this phenomenon.[5] These include the postulation of a hypothetical "Pasteur enzyme" that diverts material from an anaerobic to an aerobic pathway. Another explanation proposes that the presence of oxygen causes inactivation of glycolytic enzymes by some device such as oxidation of essential SH groups. Simpler explanations are currently more usually advanced. Oxidation alters the degree of phosphorylation of adenosine nucleotides and the ratio of oxidized to reduced pyridine nucleotides. Since oxidation is associated with the esterification of several times as much phosphate as is esterified by the fermentation of a given amount of carbohydrate, the adenosine nucleotides tend to accumulate as ATP. This might result in a limitation of fermentation in the manner described for yeast extracts. Similarly, in the presence of air, DPNH does not accumulate, but is oxidized by the electron-transport system. Each equivalent of DPNH removed prevents the formation of an equivalent of ultimate fermentation product. The accumulation of intermediates may then inhibit the operation of the early steps of fermentation. It has also been proposed that inorganic phosphate may become limiting for glycolysis when oxidative phosphorylation occurs. One or more of the proposed explanations may eventually be determined to apply to specific cells.

Rate-Limiting Reactions in vivo. The determination of the rate-limiting step in an intact cell is very difficult because of the possible influences on enzyme activity both inside and outside the cell: inhibitors, activators, cofactor concentrations, and other environmental factors in

---

[5] The current status of this phenomenon has been described by Dickens in his review that acknowledges the many explanations offered during the past 30 years. (See General References.)

the cell cannot be evaluated. Alteration in activity when enzymes are liberated from cells may be caused by many factors in addition to those mentioned as normal control mechanisms. These include dilution of competing systems, liberation of autolytic enzymes, and physical changes caused by surface phenomena, osmotic effects, or specific ion effects. Indeed, the alteration of enzyme activity may be so great that caution must be exercised in concluding that reactions found *in vitro* occur at all in the intact cell. The physical separation of enzymes from their substrates within cells has been postulated to explain the failure of certain enzymes *in vivo* to carry out reactions that occur rapidly when the cells are disrupted. The oxidation of phenolic compounds by tyrosinase illustrates this situation. It is possible that enzymes so separated from apparent substrates serve other functions; that is, the reaction studied *in vitro* may not represent the physiological function of an enzyme. In the case of glyoxalase, there is no known function for this enzyme since methylglyoxal has been shown to be an artifact, not an intermediate of glycolysis. The question of the normal substrate for glyoxalase continues to serve as a subject for philosophical inquiry.

Some information may be obtained about the rate-limiting steps by means of specific inhibitors, but the interpretation of effects of poisons on whole cells must be interpreted cautiously. Poisons are often found to be less specific than originally thought, and may exert inhibitory effects on many reactions. For example, iodoacetate, once considered as a specific inhibitor for triose phosphate dehydrogenase, has been found to inactivate a large number of sulfhydryl enzymes. The failure of an inhibitor to influence a reaction sequence also cannot be interpreted simply. Unless the enzyme involved is already rate-limiting, or can be reduced in activity until it is rate-limiting, an amount of inhibitor that has an effect on the isolated reaction may fail to influence the over-all rate of a series of reactions including the sensitive enzyme. Other possibilities, such as destruction of the inhibitor by other enzymes, may also intervene to complicate the interpretation of experiments with inhibitors *in vivo*.

Enzymes and Cell Structure. Cytologically many cell structures have been identified. One approach to the localization of enzymes within cells involves incubation of substrates with sections fixed to microscope slides. Reactions that yield insoluble products are coupled with the enzyme reaction if the enzyme product is not already insoluble, and the sections are studied microscopically. There are serious limitations to this technique. Since it involves fixing sections to slides, it may (and has been shown to) involve inactivation of part or all of a given activity. More important is the possible translocation of the reaction products. These do not necessarily precipitate where they are formed, but may be adsorbed on cell structures far from the enzyme being studied.[6] In spite of the technical difficulites, the histochemical technique has been applied successfully in many instances. The use of frozen sections and of lyophilized tissues has permitted the assay of many enzymes that are destroyed by older methods of preparing slides.

Another approach to the elucidation of enzymatic organization of cells depends upon separation of cell components from disrupted cells. It is obvious that methods for breaking cells should be selected to avoid fragmenting organelles. The method most widely used today employs the Potter-Elvehjem homogenizer, in which a fitted cylindrical pestle is rotated within a tube.[7] Other methods of fragmenting cells include grinding with abrasives, rapid shaking with glass beads, exposure to sonic or ultrasonic vibrations, freezing and thawing and rapid extrusion through small orifices.[8] The suspending medium is selected to maintain the integrity of cell components; the ideal medium is still being sought. Isotonic or hypertonic sucrose solutions are most commonly used.[9] These may be fortified with specific salts. Polyvinylpyrrolidone has been used, as has Carbowax, and other water-soluble high polymers, to increase tonicity with inert large molecules that do not enter cells or cell fragments.

Dispersed cells are fractionated by differential centrifugation.[10] Various techniques are used to permit separation without alteration of individual cell structures and isolation of single components with minimum contamination by other parts of the cell. Workers in this field have emphasized the difficulty in obtaining homogeneous preparations, and have noted the adsorption on one cell component of enzymes extracted from another fraction. Another obvious limitation in the cell fractionation technique is the necessity of working with large populations of cells. Not only are tissues composed of various cell types, but differences may occur between cells of a given type as a result of differences in location in an organ or age of cells. Nevertheless, this technique has yielded many results that seem to locate certain enzymes unambiguously.

Nuclei. Nuclei are the largest and densest structures isolated from animal cells. Considerable variation in methods for isolating nuclei has

[6] These problems have been explored by J. F. Danielli, *J. Exptl. Biol.* **22**, 110 (1946), who notes them to be obstacles that hinder but do not prevent the satisfactory use of histochemical methods.

[7] V. R. Potter and C. A. Elvehjem, *J. Biol. Chem.* **114**, 495 (1936).

[8] These and other methods are described in the first section of "Methods in Enzymology" (S. P. Colowick and N. O. Kaplan, eds.), Vol. I. Academic Press, New York (1955).

[9] G. H. Hogeboom, W. C. Schneider, and G. E. Palade, *Proc. Soc. Exptl. Biol. Med.* **65**, 320 (1947).

[10] A. Claude, *Harvey Lectures Ser.* **43**, 121 (1947–48).

characterized recent work. Many of the enzymes claimed by some workers to be associated with nuclei were shown by others to be present in contaminants. DPN pyrophosphorylase was found to be present only in nuclei of liver cells, and the fact that it was liberated by water extraction after damage to the nuclear membrane has been used as evidence that the aqueous medium used for fractionation does not extract water-soluble enzymes from nuclei.[11] Other enzymes, including some involved in the metabolism of certain uridine nucleotides[12] and polynucleotides,[13] are conveniently prepared from nuclei, but further studies will be required to ascertain which enzymes are present in nuclei *in vivo*.

Mitochondria. The best studied cell particles are the mitochondria, which sediment at intermediate gravitational fields. These are delicate structures that, according to electron microscopy, are rods 0.3–0.7 μ by 1–4 μ, surrounded by a membrane and containing a matrix divided by many lamellae, the cristae mitochondriales.[14] Many functions have been ascribed to mitochondria. The histochemical identification of mitochondria by means of Janus green staining[15] depends upon the concentration of oxidative enzymes in these structures.[16] Current investigations are still occupied with determining whether certain enzymes are found exclusively in mitochondria or whether they are distributed through various cell components.

Mitochondria are capable of carrying out all of the reactions of the Krebs cycle and of fatty acid oxidation and consume oxygen. Since all of the integrated cytochrome system of cells is found in mitochondria, it is not surprising that no other cell components consume oxygen at significant rates. The rate of oxygen consumption of mitochondria alone is less than the rate found with the original homogenate and the full rate is obtained only on recombination of cell fragments. Aconitase and isocitric dehydrogenase were found concentrated in the soluble portion of the homogenate. These results have been interpreted as indicating that mitochondria do not act alone in carrying out the Krebs cycle reactions.[17] The demonstration that aconitase can be liberated very rapidly from some cell particles forces a consideration of the possibility that in the intact cell all of the Krebs cycle enzymes are localized in mitochondria, and that they are liberated differentially during cell fragmentation.

[11] G. H. Hogeboom and W. C. Schneider, *J. Biol. Chem.* **197**, 611 (1952).
[12] G. T. Mills, R. Ondarza, and E. E. B. Smith, *Biochim. et Biophys. Acta* **17**, 159 (1954).
[13] L. A. Heppel, P. Ortiz, and S. Ochoa, *Science* **123**, 415 (1956).
[14] G. E. Palade, *J. Histochem. and Cytochem.* **1**, 188 (1953).
[15] R. R. Bensley, *Trans. Chicago Pathol. Soc.* **8**, 78 (1910).
[16] A. Lazarow and S. J. Cooperstein, *J. Histochem. and Cytochem.* **1**, 234 (1953).
[17] See reviews by Schneider and by Hogeboom. (General References.)

Several enzymes are not detected when fresh mitochondria are assayed, but are detected when the preparations are aged, warmed, frozen, or otherwise maltreated. These treatments, that cause loss of some activities, permit measurement of "latent" enzymes, including ATPase, glutamic dehydrogenase, and DPN cytochrome reductase. Some of the latent enzymes are readily solubilized by disruption of the mitochondria, while others remain associated with large sedimentable fragments.[17]

Oxidative Phosphorylation. The value to a cell of the oxidative reactions carried out by the electron-transport mechanism of mitochondria lies primarily in the esterification of inorganic orthophosphate that is coupled with the oxidations. Phosphorylation coupled with electron transport is designated *oxidative phosphorylation* in distinction from substrate-level phosphorylation, such as accompanies the oxidation of phosphoglyceraldehyde and α-ketoglutarate. Most of the ATP found in aerobic cells is produced by oxidative phosphorylation, but despite many years of intensive investigation the reactions that form this ATP remain unknown. The nature of oxidative phosphorylation has been investigated by many techniques, and a large amount of information about the properties of the responsible systems has been obtained. It is hoped that the indirect evidence accumulating will lead to an understanding of the fundamental process of utilization of the energy of biological oxidations.

Independent investigations of Belitzer and Tsibakowa[18] and of Ochoa[19] demonstrated that more than 1 atom of phosphorus could be esterified per atom of oxygen consumed by tissue dispersions. Ochoa calculated that electron transport could support the coupled esterification of 3 atoms of P per atom of O consumed. For many years controversy has raged about the maximum P:O ratio, with arguments marshalled for values of 2, 3, and 4; in general it has been assumed that fractional values indicate that the correct value is the next higher whole number, but schemes have been devised that can accomodate values of $\frac{1}{2}$. While there is still no unanimity about the true value of the P:O ratio under ideal conditions, there seems to be a tendency for workers in the field to accept a value of 3 for oxidations involving pyridine nucleotides. When α-ketoglutarate is the substrate, the P:O ratio is probably 4, since there is a substrate-level phosphorylation in addition to oxidative phosphorylation. Succinate oxidation does not involve pyridine nucleotides, and appears to have a P:O ratio of 2. The definitive assignment of a value for the P:O ratio is of more significance than a philosophical evaluation of the efficiency of utilization of the energy of respiration. Since a coupled phosphorylation must involve a stoichiometric reaction

[18] V. A. Belitzer and E. T. Tsibakowa, *Biokhimiya* **4**, 516 (1939).
[19] S. Ochoa, *J. Biol. Chem.* **138**, 751 (1941).

with an intermediate compound of the electron-transport system, the P:O ratio enumerates the number of such reactions along the oxidative chain.

A tremendous obstacle in the study of oxidative phosphorylation is the lability of the phosphorylating system. Careful preparation of mitochondria results in particles whose ability to respire is closely associated with their ability to esterify phosphate.[20] These "tightly coupled" particles do not consume oxygen in the absence of a phosphate acceptor, and in this respect are believed to represent the physiological state of mitochondria in intact cells. Exposure of mitochondria to slightly unfavorable environments for even short periods of time results in a loosening of the relationship between oxidation and phosphorylation. Some preparations respire in the absence of phosphate acceptors, but fix phosphate when systems such as hexokinase, ADP, and glucose are added. Other preparations have the ability to respire when substrate is present, but have lost the ability to esterify phosphate. The maximum rates of oxygen consumption of the various preparations are similar, which leaves an open question as to whether the same intermediate catalysts transfer electrons in the coupled and uncoupled systems.

Several compounds have been found to uncouple phosphorylation from oxidation. The best known are 2,4-dinitrophenol (DNP)[21] and some of its analogs. At low concentrations these compounds either have no effect on or stimulate oxygen consumption, while completely eliminating oxidative phosphorylation. The site of action of these uncouplers is not known, but it has been established that they do not interfere with the transfer of phosphate in known kinase reactions and they do not affect substrate-level phosphorylations. Other uncouplers include azide, dicoumarol, and thyroxine. The similarity between the effects of thyroxine and DNP both *in vivo* and on mitochondrial phosphorylation has led to speculation that the hormone might act by selective uncoupling.[22] The nature of the uncoupling is not necessarily the same for both reagents; preparations have been made that are insensitive to thyroxine but sensitive to DNP.[23]

The localization of the steps that couple with phosphorylation has been explored with the use of specific electron donors and acceptors. Friedkin and Lehninger succeeded in preparing particles that esterified phosphate when exogenous DPNH was oxidized,[24] but greater efficiency

[20] H. A. Lardy and H. Wellman, *J. Biol. Chem.* **195**, 215 (1952).
[21] W. F. Loomis and F. Lipmann, *J. Biol. Chem.* **173**, 807 (1948).
[22] C. Martins and B. Hess, *Arch. Biochem. Biophys.* **33**, 486 (1951); G. F. Maley and H. A. Lardy, *J. Biol. Chem.* **204**, 435 (1953).
[23] D. F. Tapley and C. Cooper, *J. Biol. Chem.* **222**, 341 (1956).
[24] M. Friedkin and A. L. Lehninger, *J. Biol. Chem.* **178**, 611 (1949).

was observed when β-hydroxybutyrate was used as the primary electron donor.[25] The treatment of the mitochondria to render them "permeable" to DPNH is critical in these experiments. α-Ketoglutarate, succinate, and other citric acid cycle intermediates have also been used as substrates for oxidative phosphorylation. Cytochrome c and ferricyanide have been found to serve as electron acceptors, and have permitted the demonstration of oxidative phosphorylation in systems restricted to the first steps in electron transport (excluding cytochrome oxidase and oxygen). Similarly, it has been shown that reduced cytochrome c can be oxidized by mitochondria in a reaction coupled with phosphorylation. The results from several laboratories indicate that two steps in the oxidation of DPNH by cytochrome c are coupled with phosphorylation, and a third step esterifies phosphate in the oxidation of reduced cytochome c.[26]

Fragments of mitochondria have been obtained by Cooper et al., who treated the mitochondria with digitonin.[27] These esterify phosphate in reactions associated with the oxidation of β-hydroxybutyrate, and can be uncoupled by DNP but not by thyroxine.[23] These preparations also oxidize exogenous DPNH, but the coupled phosphorylation is less efficient than when β-hydroxybutyrate is the substrate. The phosphorylation associated with DPNH oxidation appears to be associated only with the electron transport between cytochrome c and oxygen. Another particle has been obtained by Green and collaborators, who name their material PETP, phosphorylating electron-transferring particle.[28] PETP differs from the preparation of Cooper et al. in that it oxidizes DPNH but does not carry out a coupled phosphorylation; when a substrate known to reduce the endogenous DPN is used, however, these particles do esterify phosphate.

Another approach to the nature of the phosphate esterification reaction has been made with the use of isotope exchange methods. Boyer et al. and Swanson have shown that an exchange reaction between ATP and inorganic P is catalyzed by anaerobic mitochondria, and suggest that the exchange represents a reversible step in the process of oxidative phosphorylation.[29] Similar conclusions were reached by Cohn and Drysdale,[30] who found in experiments with mitochondria that the oxygen of

[25] J. H. Copenhaver and H. A. Lardy, J. Biol. Chem. 195, 225 (1952).

[26] T. M. Devlin and A. L. Lehninger, J. Biol. Chem. 219, 507 (1956); C. Cooper and A. L. Lehninger, ibid. 219, 519 (1956).

[27] C. Cooper, T. M. Devlin, and A. L. Lehninger, Biochim. et Biophys. Acta 18, 159 (1955).

[28] D. Ziegler, R. Lester, and D. E. Green, Biochim. et Biophys. Acta 21, 80 (1956).

[29] P. D. Boyer, W. W. Luchsinger, and A. B. Falcone, J. Biol. Chem. 223, 405 (1956); M. A. Swanson, Biochim. et Biophys. Acta 20, 85 (1956).

[30] M. Cohn and G. R. Drysdale, J. Biol. Chem. 216, 831 (1955).

both inorganic P and adenine phosphate exchanged with water much more rapidly than was expected from the net rate of oxidation and phosphorylation. The oxygen exchange reaction could not be separated from the oxidizing system, and so far no purification of the oxygen exchange catalyst has been possible.

Oxidative phosphorylation has been observed in particles derived from both bacteria and plants in addition to those derived from animal cells. A special type of oxidative phosphorylation has been found in photosynthetic organisms.[31] Particles that contain the photochemical apparatus also appear to contain a series of enzymes that can recombine the products of photolysis and couple this process with the esterification of phosphate. The photochemical system is distinct from more conventional oxidative enzymes that use molecular oxygen as an electron acceptor.

Cytochromes in Mitochondria. The majority of cytochromes occur entirely in mitochondria, where they are associated with large amounts of lipid. An intimate association exists between the cytochromes and succinic dehydrogenase to form succinoxidase, which retains the ability to oxidize succinate after all other oxidative reactions have been washed out of the particles. The activity of the succinoxidase system is sensitive to many environmental factors, which apparently influence the physical state of the particles. When mitochondria are exposed to distilled water, for example, cytochrome c is lost. Exogenous cytochrome c is used much less efficiently than particle-bound cytochrome. It is concluded from observations of this sort that the individual components, succinic dehydrogenase, cytochrome c, cytochrome a, and cytochrome oxidase, together with any other factors that may be involved, are fixed in position to react rapidly with each other, and do not depend on diffusion.

Heterogeneity of Mitochondria. Mitochondria contain many enzymes in addition to those of oxidative pathways. These include oxidases, transaminases, phosphatases, condensing enzymes, peptidases, and many others. Centrifugal techniques have been used to sub-fractionate mitochondrial preparations.[32] Fractions differing in color were obtained and assayed for various enzymes.[33] Uricase, DNAase, and succinic dehydrogenase were found to have different relative distributions. It has been suggested that true mitochondria, containing the succinoxidase system, are homogenous, but that other particles contain cytochrome reductases and other enzymes. The name leposomes was applied to particles that con-

[31] A. Frenkel, *J. Am. Chem. Soc.* **76**, 5568 (1954); F. R. Whatley, M. B. Allen, and D. I. Arnon, *Biochim. et Biophys. Acta* **16**, 605 (1955).

[32] A. B. Novikoff, E. Podber, J. Ryan, and E. Noe, *J. Histochem. and Cytochem.* **1**, 27 (1953).

[33] E. L. Kuff and W. C. Schneider, *J. Biol. Chem.* **206**, 677 (1954).

tain high concentration of hydrolytic enzymes: ribonuclease, deoxyribo-nuclease, "cathepsin," β-glucuronidase, and acid phosphatase.[34] Particles from melanoma have been fractionated by chromatography on "inert" columns, and have also been found to be heterogeneous.[35]

Microsomes. Microsomes are much smaller particles than mitochondria and sediment in higher gravitational fields.[36] Several enzymes have been associated almost exclusively with microsomes. Early work indicated similarities between microsomes and mitochondria, but it is now felt that fragments of mitochondria contaminated the microsome preparations. Microsomes are believed to contain the glucose-6-phosphatase of liver cells.[37] Studies on incorporation of labeled amino acids into proteins associate microsomes with protein synthesis.[38] The incompletely understood reactions that introduce hydroxyl groups into certain aromatic rings are catalyzed by microsomes, using TPNH and molecular oxygen.[39] The phenylalanine–tyrosine conversion does not use microsomal enzymes, however, and the ferrous iron oxidases are also not found in these particles.

Other Cell Structures. Recently other cell structures have been separated: Golgi substance, nucleolus, chromosomes, mitotic apparatus, and secretory granules. The identification of some of these fractions is not certain, and therefore the association of specific enzymes with these structures cannot be made at this time.

Soluble Enzymes. Very many enzymes appear to be present in solution in cytoplasm. The glycolytic enzymes are all found in the soluble fraction of homogenates. The report that glycolysis is stimulated by the addition of other components cannot be evaluated now; it may be that glycolysis does involve additional elements of the cell, or it may be that some enzymes are adsorbed to particles. The enzymes of many ATP-requiring reactions are found in the soluble fraction. Early work on many of these systems, such as those that form citrulline and arginine, showed a requirement for mitochondria, but this requirement was eventually found to be related only to the ability of mitochondria to form ATP.

The enzymes of fatty acid oxidation are known to catalyze reversible reactions, and therefore were assumed to be responsible for fatty acid synthesis. Synthesis of fatty acids, however, does not occur with mito-

[34] C. deDuve, B. C. Pressman, R. Gianetto, R. Wattiaux, and F. Appelmans, *Biochem. J.* **60**, 604 (1955).

[35] V. T. Riley, M. L. Hesselbach, S. Fiala, and M. W. Woods, *Science* **104**, 361 (1949).

[36] A. Claude, *Science* **97**, 451 (1943).

[37] H. G. Hers and C. deDuve, *Bull. soc. chim. biol.* **32**, 20 (1950).

[38] E. B. Keller and P. C. Zamecnik, *J. Biol. Chem.* **221**, 45 (1956).

[39] B. B. Brodie, J. Axelrod, J. R. Cooper, L. Gaudette, B. N. LaDu, C. Mitoma, and S. Udenfriend, *Science* **121**, 603 (1955).

chondria, which contain the enzymes of fatty acid oxidation. Soluble enzymes of mammary gland and liver, on the other hand, have been found to synthesize long-chain fatty acids from acetate.[40]

Teleological Considerations of Cell Structure. It is apparent that the organized structures of the cell contain specific enzymes and carry out defined biochemical functions. Most of the conclusions discussed above were derived from studies with liver preparations. In general the same properties are found when fractions of other cells are studied, but there is not sufficient information to conclude that the enzyme distribution of cells necessarily follows a given pattern.

One suggestion to explain the existence of a given activity in different cell fractions is the existence of different enzymes with similar functions. Differences in pH optima and apparent K_m values support this suggestion, but it must be seriously considered that enzymes bound to or in particles may exhibit properties characteristic of the particles instead of the enzyme. Especially in the case of mitochondria, permeability of substrate and products may determine the ability of the enzyme to act.

The elaborate structure of cells serves not only to permit association of functions that require integration but also to prevent destructive reactions from acting without restraint. Thus, several enzymes have been found to be separated physically from the bulk of their substrates. Much has been learned about the relation of enzymes to cell structure, but as yet only a small fraction of the known enzymes have been assigned to definite parts of the cell. Many questions of fact and interpretation remain to be answered by future work in this area.

GENERAL REFERENCES

Chance, B., and Williams, G. N. (1956). *Advances in Enzymol.* **17**, 65.

Dickens, F. (1951). *In* "The Enzymes" (J. B. Sumner and K. Myrbäck, eds.), Vol. II, Part 1, p. 624. Academic Press, New York.

Dixon, M. (1949). "Multi-enzyme Systems." Cambridge Univ. Press, London and New York.

Dounce, A. L. (1950). *In* "The Enzymes" (J. B. Sumner and K. Myrbäck, eds.), Vol. I, Part 1, p. 187. Academic Press, New York.

Hogeboom, G. H., and Schneider, W. C. (1955). *In* "The Nucleic Acids" (E. Chargaff and J. N. Davidson, eds.), Vol. II, p. 199. Academic Press, New York.

Holter, H. (1952). *Advances in Enzymol.* **13**, 1.

Lardy, H. A. (1956). *Proc. 3rd Intern. Congr. Biochem., Brussels, 1955*, p. 287.

Martius, C. (1956). *Proc. 3rd Intern. Congr. Biochem., Brussels, 1955*, p. 1.

Schneider, W. C. (1953). *J. Histochem. and Cytochem.* **1**, 212.

Schneider, W. C. (1956). *Proc. 3rd Intern. Congr. Biochem., Brussels, 1955*, p. 364.

Slater, E. C. (1956). *Proc. 3rd Intern. Congr. Biochem., Brussels, 1955*, p. 264.

[40] G. Popjak and A. Tietz, *Biochem. J.* **60**, 147 (1955); A. J. Wakil, J. W. Porter, and A. Tietz, *Federation Proc.* **15**, 377 (1956).

Adaptive Enzymes and the Synthesis of Proteins

The biosynthesis of proteins is another of the major problems of biochemistry which has been investigated for years, about which many intriguing suggestions have been made and ingenious experiments performed, but which remains a mystery since the nature of the reactions involved is completely unknown. The properties of protein-synthesizing systems have been studied by measuring the incorporation of labeled amino acids into well-characterized proteins, by determining the factors that influence the formation of increased quantities of enzymes (adaptive enzymes) in microorganisms, by finding conditions that permit enzyme synthesis in cell fragments, and by relating enzyme synthesis to genetic changes. Another approach that is directed toward an understanding of protein synthesis is the analysis of the syntheses of small peptides and the activation of amino acids.

Studies with Labeled Amino Acids. It is generally accepted that proteins are synthesized rather directly from free amino acids. Two general pictures of protein synthesis have developed from studies with labeled amino acids. When labeled substrates were administered to animals and individual proteins from milk or hemoglobin were isolated, no evidence was found for differential labeling of any portion of the molecules; that is, the specific activity of a given amino acid in one fragment does not differ from the specific activity of that amino acid in another fragment.[1] These experiments are compatible with models in which all of the amino acids are incorporated at essentially the same time. Another model has received support from similar experiments *in vitro*.[2] Ovalbumin, insulin, and ribonuclease have been labeled by incubating labeled amino acids with appropriate organs. The proteins have been crystallized in pure form and degraded by enzymatic and chemical means that allow the separation of characterized peptides. For example, an N-terminal hexapeptide is split from ovalbumin by the protease, subtilisin. The specific activities of one aspartic acid and the alanine residues differed markedly from others in the plakalbumin produced by removal of the hexapeptide. Peptic digestion of ovalbumin also yielded separable fragments containing alanine with specific activities differing by factors of as much as 4. Similar results were obtained with ribonuclease and insulin. These experiments indicate that different portions of the proteins are assembled at different times, as the specific activity of the amino acid pool changes.

[1] B. A. Askonas, P. N. Campbell, C. Godin, and T. S. Work, *Biochem. J.* **61**, 105 (1955); H. M. Muir, A. Neuberger, and J. C. Perrone, *ibid.* **52**, 87 (1952).
[2] D. Steinberg, M. Vaughn, and C. B. Anfinsen, *Science* **124**, 389 (1956).

The incorporation of amino acids into the enzymes of rabbit muscle has been studied by Velick *et al.*[3] Aldolase, triose phosphate dehydrogenase, and phosphorylase have been isolated following the administration of groups of labeled amino acids. Although the individual enzymes acquire and lose the labeled compounds at rates characteristic of the specific protein, the ratios of specific activities of the different amino acids in a given protein were found to be the same as the corresponding ratios in the other proteins. These experiments were interpreted as favoring simultaneous incorporation of all the amino acids of the protein. This interpretation has been questioned, however, since only average specific activities of whole proteins were measured, and these reflect over-all rate of synthesis, and may obscure differences in various portions of the molecule. The incorporation of labeled glycine into silk fibroin has been found to give a product with extremely nonuniform labeling.[4] It remains to be seen whether further experiments will reconcile the apparent differences between the various experiments or whether some proteins are formed within shorter times and from more homogeneous pools than others.

Adaptive Enzymes. Karstrom coined the terms constitutive and adaptive enzymes to describe those enzymes present in fixed and variable concentrations, respectively.[5] An extensive list of adaptive enzymes has accumulated as many microorganisms have been found to produce large amounts of specific enzymes in response to the presence of specific inducers in their media. Not all enzyme inductions, however, are caused by substrates, or even by analogs of substrates. Careful measurements of enzyme levels have indicated that constitutive and adaptive are only relative terms, since the concentrations of all enzymes appear to vary to some extent. Nevertheless, some organisms contain extremely small or undetectable amounts of certain enzymes before exposure to specific inducers, and contain large amounts of these enzymes after exposure; these are still called adaptive or induced enzymes, in contrast to the less variable cell constituents.

A very intensive study of the adaptive enzyme β-galactosidase of *E. coli* has been made by Monod and Cohn and their associates.[6] Their studies have been carried out with growing cells. In these organisms free amino acids are rapidly incorporated into proteins, and there is no evidence of preformed protein precursors. β-Galactosidase formation is

---

[3] M. V. Simpson and S. F. Velick, *J. Biol. Chem.* **208**, 61 (1954); M. Heimberg and S. F. Velick, *ibid.* **208**, 725 (1954).

[4] K. Shimura, H. Fukai, J. Sato, and R. Sacki, *J. Biochem.* **43**, 101 (1956).

[5] H. Karstrom, *Ergeb. Enzymforsch.* **7**, 350 (1938).

[6] See reviews by J. Monod and M. Cohn, *in* "Enzymes: Units of Biological Structure and Function" (O. H. Gaebler, ed.), pp. 7, 41. Academic Press, New York.

induced by a large variety of β-galactosides and by certain analogs. There is no correlation between sensitivity as a substrate or affinity for the enzyme and ability to induce enzyme formation. For example, phenyl-β-galactoside is a substrate but not an inducer, whereas methyl-β-thio-galactoside is an inducer but not a substrate.

Induced cells contain a mechanism for concentrating inducers within the cells, and this mechanism appears to play a part in the induction process. The formation of enzyme depends on the continued presence of inducer. The rate of enzyme synthesis with adequate amounts of inducer is proportional to the growth of the bacteria. When the inducer is removed (by suspending the centrifuged bacteria in fresh medium), enzyme synthesis stops abruptly. The enzyme already formed, however, is stable, and persists unchanged for many generations. Sulfur-labeled amino acids have been used to demonstrate that the induced enzyme is formed directly from free amino acids, and that proteins already in the bacteria do not contribute amino acids to the new enzyme. In the absence of the inducer, the adaptive enzyme retains its label. Some properties of inducers were found in a study of penicillinase production by *Bacillus cereus*.[7] With this system it was shown that in a brief exposure a small amount of penicillin is specifically bound within the cells, and is not hydrolyzed, but stimulates the production of several equivalents of penicillinase.

Sequential Adaptation. Adaptive enzyme formation has been demonstrated with many microorganisms growing in the presence of specific substrates. In several cases it has been found that the adaptation is not restricted to a single enzyme, but involves a series of enzymes that act on the series of intermediates formed from the original substrate. The formation of enzymes that specifically metabolize intermediates has allowed this technique to be used to describe a metabolic pathway; suspected intermediates that are attacked by adapted cells or enzymes obtained from them are assigned a metabolic role, while those compounds that are not metabolized by the adapted system are presumed to be foreign to the pathway. Two groups working with *Pseudomonas* independently and simultaneously developed this method;[8] coincidentally, both studied the oxidative degradation of tryptophan. Adaptive enzymes have been useful in the preparation of unusual molecules as well as in the analysis of biochemical pathways.

Glucose Inhibition of Enzyme Induction. The formation of adaptive enzymes usually requires measurable time, and lag periods during which

[7] M. R. Pollack and A. M. Torriani, *Compt. rend.* **237**, 276 (1953).

[8] R. Y. Stanier and M. Tsuchida, *J. Bacteriol.* **58**, 45 (1949); R. Y. Stanier, *ibid.* **54**, 339 (1947); M. Suda, O. Hayaishi, and Y. Oda, *Med. J. Osaka Univ.* **2**, 21 (1950).

no reaction is seen may extend from a few minutes to an hour or more. The presence of an inducer does not necessarily evoke an adaptive enzyme. In general, the presence of glucose inhibits the formation of adaptive enzymes. When glucose and another oxidizable substrate are incubated together with an organism, a curve of oxygen consumption or growth versus time is obtained that shows three portions: an initial rapid rate, a lag, and a second rapid rate. This so-called diauxic curve indicates glucose oxidation followed by a period of adaptation to the second substrate, then oxidation of the second substrate (Fig. 38).[9]

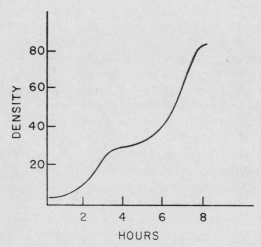

FIG. 38. Diauxie, illustrated by the growth of *E. coli* on a mixture of glucose and dulcitol. From J. Monod "La croissance des cultures bacteriennes," Hermann et Cie., Paris (1942).

The nature of the inhibition of adaptive enzyme formation by glucose is not clear. It has been suggested that glucose may interfere with penetration of inducer into the cell, but while this may be true for some systems, such as the β-galactosidase induction of *E. coli*, it is not true for various inductions in *A. aerogenes*. In this organism Neidhart and Magasanik[10] have shown that histidine, galactose, and other substrates enter the cells in both the presence and absence of glucose, but that adaptive enzymes are formed only in its absence. Other explanations, such as competition for the "building blocks" of enzyme synthesis and the effect of rapid growth supported by glucose, also fail to explain the inhibition. Neidhart and Magasanick have shown that inhibition by glucose does not occur when the adaptive enzyme is needed to supply a requirement

[9] J. Monod, *Growth* 11, 223 (1947).
[10] F. C. Neidhart and B. Magasanik, *Biochim. et Biophys. Acta* 21, 324 (1956).

for growth. Thus histidase and succeeding enzymes are formed when histidine is the only source of nitrogen or when a glutamic acid-requiring mutant is grown with histidine in the absence of glutamate. A teleological explanation of glucose inhibition based on feed-back mechanisms has been suggested; that is, the materials the cell would derive from the induced enzyme system are formed more efficiently from glucose, and the accumulations of these products inhibit enzyme formation. It remains to be seen whether this suggestion will apply in general and whether a chemical mechanism will be found to implement it.

Protein Turnover. The failure of induced enzymes to acquire label from bacterial proteins or to liberate their own amino acids has led to the conclusion that there is no turnover of proteins in growing cells. This is in contradiction to the generally held interpretation of the Schoenheimer concept of dynamic equilibrium.[11] From observations on the incorporation of labeled amino acids, fats, and carbohydrates into various parts of animals, Schoenheimer and his collaborators postulated that there are essentially no inert biochemical structures, but that apparently stable structures are continually broken down and renewed.

The Schoenheimer concept has received support from the experiments of Velick *et al.*[3] who found that the enzymes of rabbit muscle "turned over" at different rates. Since these are presumably found within the same cells, which are in a steady state, lysis of old cells and regeneration cannot account for the different rates. Additional support comes from the finding of adaptive enzymes in animal tissues. Many animal enzymes have been found to change with nutritional and physiological states. Tryptophan peroxidase is increased by the action of adrenal cortical hormones on the liver, but, even in the absence of adrenals, large doses of tryptophan evoke an adaptive response.[12] Within a few hours the level of enzyme increases 10 to 20-fold, and remains at high levels for a few more hours. The levels return to normal values within 18–24 hours. These changes appear to be changes in enzyme protein, and, since the rates of the changes far exceed the rates of changes in the cell population, it seems likely that liver proteins are actively turned over in response to physiological requirements. The degradation of liver proteins does not seem to be a spontaneous reaction catalyzed by cathepsins. Conditions that limit available energy, such as anaerobiosis or presence of DNP, minimize protein degradation.[13a] If the release of amino acids were correlated with cell death and lysis, these conditions should favor protein

[11] R. Schoenheimer, "The Dynamic State of Body Constituents." Harvard Univ. Press, Cambridge, Massachusetts (1942).

[12] W. E. Knox and A. H. Mehler, *Science* **113**, 237 (1951).

[13a] D. Steinberg and M. Vaughn, *Biochim. et Biophys. Acta* **19**, 589 (1956).

breakdown. The inhibition of protein breakdown by amino acid analogs that are known to inhibit protein synthesis indicates that specific mechanisms are used for protein turnover.

Hormonal Influences on Enzyme Levels. Many reagents influence the biochemical activity of animal tissues, and, in general, the effects must occur at the enzyme level. The hormones that regulate animal metabolism are among the reagents known to influence enzyme activity. The changes described for tryptophan peroxidase are representative of similar changes in the effective levels of many animal enzymes. In some cases there appears to be only simple response to substrate, as in the case of bacterial enzyme adaptation. These changes are probably measures of net synthesis of enzyme protein, and require several hours to effect. The effects of cortisone and many other hormones are often produced even more slowly. For example, glucose-6-phosphatase of rat liver microsomes increases two-to-threefold gradually for one to two days after cortisone administration,[13b] in contrast to the few hours required for the ten- or more-fold increase in tryptophan peroxidase. The mechanism of cortisone action is not known, but appears to be very complex; several enzymes in different cell components change at individual rates.

Epinephrine is a hormone that causes extremely rapid responses. One of the responses to this hormone is an increased liberation of liver glycogen as glucose. This reaction has been found to be limited by the enzyme phosphorylase. Liver phosphorylase exists as the active enzyme and as the inactive, dephosphophosphorylase. Epinephrine, and also the hyperglycemic factor, glucagon, have been found to stimulate the formation of active phosphorylase. This stimulation is mediated by an adenine nucleotide liberated from cell particles by an action of the hormones. The liberation of the nucleotide has been demonstrated in cell-free systems. The nucleotide in an unexplained manner, perhaps indirectly or after metabolic conversion to another form, alters the activity of either the kinase or the phosphatase so that the steady-state concentration of active phosphorylase is increased.[13c]

In addition to the examples cited, many other enzymes have been found in altered amounts following changes in the levels of hormones of the thyroid, parathyroid, and pituitary glands, pancreas, and sex organs. The accompanying physiological changes are so profound, however, that it has not been possible to determine the immediate causes of the changes in enzyme levels.

[13b] J. Ashmore, A. B. Hastings, F. B. Nesbett, and A. E. Renold, *J. Biol. Chem.* **218**, 77 (1956).
[13c] T. W. Rall, E. W. Sutherland, and W. D. Wosilait, *J. Biol. Chem.* **218**, 483 (1956); T. W. Rall, E. W. Sutherland, and J. Berthet, *J. Biol. Chem.* **224**, 463 (1957).

Chemical Mechanisms in Enzyme Synthesis. Enzyme synthesis has been demonstrated in bacterial protoplasts, which are cells without cell walls, and in cell fragments obtained by supersonic vibrations.[14] The advantages of these preparations are found in their permeability to both enzymes and substrates. With these preparations both Gale and Spiegelman and their collaborators have demonstrated relationships between nucleic acids and enzyme synthesis. Such a relationship was indicated by the ultraviolet microscope studies of Caspersson.[15] The results obtained with various systems are not entirely consistent with each other, and the roles of nucleic acids cannot be determined with certainty. In general, there is an impression that DNA does not participate directly in enzyme synthesis, but exerts an influence in determining which enzymes may be formed. RNA seems to be more closely associated with protein synthesis. In Gale's broken cells, RNA synthesis from nucleotides seems to proceed together with protein synthesis. All of these systems require a complete complement of amino acids and a source of energy (usually fructose diphosphate and ATP).

A preliminary report has indicated that soluble preparations from acetone-dried pancreas are capable of synthesizing amylase.[16] This system requires supplementation with amino acids and an energy source, and presumably is catalyzing a net synthesis of enzyme protein. The reaction is inhibited by the omission of any of the amino acids or by the presence of amino acid analogs.

Many investigators have studied the incorporation of labeled amino acids into proteins. The system involved in mammalian cells seems to be associated with microsomes, but requires factors from the soluble portion of the cell. Both ATP and GTP are required.[17] The function of the ATP may be to form activated amino acids. Several systems have been found with relative specificity for various amino acids that catalyze the exchange of pyrophosphate with ATP in the presence of the amino acid, and presumably form adenyl-amino acid compounds.[18] The activated amino acids react with hydroxylamine to form hydroxamic acids. Several reactions have been considered as models for peptide bond formation. These include the formation of hippuric acid, in which benzoyl CoA condenses with glycine in a reaction similar to the acetylation of

[14] E. F. Gale, *in* "Enzymes: Units of Biological Structure and Function" (O. H. Gaebler, ed.), p. 49. Academic Press, New York (1956); S. Spiegelman, *ibid.* p. 67.

[15] T. Caspersson, *Naturwissenschaften* **29**, 33 (1941).

[16] F. B. Straub, A. Ullman, and G. Acs, *Biochim. et Biophys. Acta* **18**, 439 (1955).

[17] E. B. Keller and P. Zamecnik, *J. Biol. Chem.* **221**, 45 (1956).

[18] M. Hoagland, *Biochim. et Biophys. Acta* **16**, 288 (1955); J. A. DeMoss and G. D. Novelli, *ibid.* **22**, 49 (1956); P. Berg, *J. Biol. Chem.* **222**, 1025 (1956).

aromatic amines.[19] Another is the synthesis of glutathione.[20] This tripeptide is formed in two dissociable steps; one enzyme splits one ATP in catalyzing the condensation of the γ-carboxyl group of glutamic acid with cysteine. The resulting dipeptide condenses with glycine in a similar reaction, catalyzed by a second enzyme. These reactions may be related to protein synthesis, but, as discussed below, it is unlikely that proteins are assembled amino acid by amino acid with specific enzymes.

Genetic Control of Enzyme Production. The development of biochemical genetics has coincided with other developments in the genetics of microorganisms, and is part of the most exciting development in biological science today.[21] This area of investigation is concerned with the elucidation of the chemistry of genetic material and the mechanisms by which this material determines the structures and functions of cells. The basic observations in genetics deal with detectable changes in properties of organisms that are inherited by progeny. These anatomical or physiological changes are the results of alterations (mutations) in the genes, the functional elements of chromosomes. The measurable gentic change is mediated through biochemical reactions that are determined by the genes. The analysis of a large number of mutants of various organisms has given great strength to the concept of a "one gene–one enzyme" relationship. This indicates that a single minute genetic area (500–1000 nucleotides of DNA in a macromolecule) are required for the production of a given enzyme. This does not imply that no other genetic sites are also required. From the evidence that RNA is more closely associated with protein synthesis than DNA is, the function of the gene may be to form the appropriate RNA. In a completely unknown manner the nucleic acids must form a template for protein synthesis. The chemical nature of the template and the manner of its replication are completely unknown. There is reason to consider mechanisms that use pools of activated molecules that are oriented in some fashion by nucleic acids. If specific enzymes were required to fashion each protein, additional enzymes in a never-ending chain would be required to synthesize each other; at a minimum, a second (synthesizing) enzyme would be required for each metabolic enzyme in addition to a group of enzymes to make the enzyme formers. The forma-

[19] H. Chantrenne, *J. Biol. Chem.* **189**, 227 (1951).

[20] J. E. Snoke and K. Block, *J. Biol. Chem.* **199**, 407 (1952).

[21] It is beyond the scope of a text on enzyme chemistry to discuss adequately an independent discipline such as biochemical genetics. The subject must be mentioned, however, for it has a common border with enzyme chemistry in the area of protein synthesis. For an introduction to biochemical genetics, see the textbook of R. P. Wagner and H. K. Mitchell, "Genetics and Metabolism." Wiley, New York (1955) and for recent developments see "Chemical Basis for Heredity" (W. P. McElroy and B. Glass, eds.). Johns Hopkins Press, Baltimore, Maryland (1956).

tion of complex, specific proteins seems to call for a template mechanism, in contrast to the stepwise enzymatic mechanisms of the less specific synthesis of polysaccharides.

Several types of genetic studies lend support to the conclusion that DNA, localized in chromosomes, transmits inherited properties. Strains of organisms that either possess or lack a given characteristic can transfer the characteristic to other organisms. Two phenomena have been found to explain the transfer of small bits of genetic material from one bacterial cell to another. One of these, transduction, is a process in which a bacteriophage carries a piece of DNA into a new host. The fragment so transferred contains only a few genes, and therefore usually is detected as a transfer of a single characteristic. Two or more genetic characteristics can be transferred together by transduction in cases of altered genes in adjacent locations. The other phenomenon, transformation, justifies the continual reference to DNA as genetic material. DNA isolated from some strains of bacteria can be incorporated into other strains, thereby introducing characteristics of the donor strain, such as drug resistance, immunological specificity, and the ability to synthesize a specific enzyme.

Genetic loci are sufficiently large for several independent alterations to cause the same loss of function. It might be speculated that some changes would not result in an all-or-none change but would result in the formation of a modified enzyme. In some cases proteins related immunochemically to the enzyme are found in increased quantities when the enzyme is not synthesized.

Three cases of genetic alterations causing the formation of active, but altered, enzyme have been described. The enzyme of *E. coli* that condenses pantoic acid with β-alanine to form pantothenic acid has been found to be more labile to heat when extracted from a mutant than when obtained from the parent strain.[22] A similar case of heat-lability in the case of *Neurospora* tyrosinase has been found to be attributable to a change in the locus known to control production of the enzyme.[23] An altered glutamic dehydrogenase has also been obtained from *Neurospora*;[24] this enzyme reversibly loses activity at 20° and regains it at 35–50°. The altered glutamic dehydrogenase also is activated by high concentration of substrates. In spite of being activated by warming, it is less stable than the normal enzyme at both high and low temperatures. The altered glutamic dehydrogenase was obtained from a back-mutation of an organism unable to synthesize this enzyme.

This brief mention of approaches to the study of enzyme formation

[22] W. K. Maas and B. D. Davis, *Proc. Natl. Acad. Sci. U. S.* **38**, 785 (1952).
[23] N. H. Horowitz and M. Fling, *Genetics* **38**, 360 (1953).
[24] J. R. S. Fincham, *Biochem. J.* **65**, 721 (1957).

only defines certain properties of enzyme-forming systems. The status of nucleic acid synthesis, which also must yield products of subtle specificity, is similarly in a primitive state. These two interrelated systems represent the area of greatest biochemical challenge today.

GENERAL REFERENCES

"Enzymes: Units of Biological Structure and Function" (O. H. Gaebler, ed.). Academic Press, New York (1956).

Gale, E. F. (1943). *Bacteriol. Revs.* **7**, 139.

Gale, E. F. (1955). "Symposium on Amino Acid Metabolism" (W. D. McElroy and B. Glass, eds.), p. 171. Johns Hopkins Press, Baltimore.

Knox, W. E., Auerbach, V. H., and Lin, E. C. C. (1956). *Physiol. Revs.* **36**, 164.

Monod, J., and Cohn, M. (1952). *Advances in Enzymol.* **13**, 67.

Spiegelman, S. (1950). *In* "The Enzymes" (J. B. Sumner and K. Myrbäck, eds.), Vol. I, Part 1, p. 267. Academic Press, New York.

Spiegelman, S. (1955). "Symposium on Amino Acid Metabolism" (W. D. McElroy and B. Glass, eds.), p. 124. Johns Hopkins Press, Baltimore.

ABBREVIATIONS

It has become acceptable to substitute initials and other abbreviations for the many lengthy and unwieldy terms in the nomenclature of biochemistry. Many of the abbreviations apply to materials that figure prominently in current literature, where the complete names are rarely met. The following list is intended to aid in the recognition of some of the more common abbreviations. The extent of the practice of abbreviation would require a complete glossary of biochemistry to list all of the symbols employed. No enzymes are included in this list, although there is a growing tendency to abbreviate enzyme names also; e.g. ADH represents alcohol dehydrogenase, PPase represents inorganic pyrophosphatase, etc.

ATP, ADP, AMP   adenosine triphosphate, diphosphate, and monophosphate
BAL   British anti-lewisite; 2,3 dimercaptopropanol
CoA   Coenzyme A
CoI; CoII   DPN; TPN
CTP, CDP, CMP   cytidine triphosphate, diphosphate, and monophosphate
DFP   diisopropyl fluorophosphate
DNA   deoxyribonucleic acid
DNFB   dinitrofluorobenzene
DNP   dinitrophenol
DPN, DPNH   diphosphopyridine nucleotide, reduced diphosphopyridine nucleotide
EDTA   ethylenediamine tetraacetic acid
FAD   flavin adenine dinucleotide
FMN   flavin mononucleotide
F-1-P, F-6-P, F-1,6-P   fructose-1-phosphate, -6-phosphate, and -1,6-diphosphate
GSH; GSSG   glutathione; oxidized glutathione
G-1-P, G-6-P, G-1,6-P   glucose-1-phosphate, -6-phosphate, and -1,6-diphosphate
GTP, GDP, GMP   guanosine triphosphate, diphosphate and monophosphate
HDP   hexose (fructose) diphosphate

401

IAA   iodoacetic acid; iodoacetamide; indoleacetic acid

ITP, ADP, IMP   inosine triphosphate, diphosphate, and monophosphate

MB   methylene blue

NMN   nicotinamide mononucleotide

P   inorganic ortho phosphate

PALPO; PAMPO   pyridoxal phosphate; pyridoxamine phosphate

PCMB   p-chloromercuribenzoate

PEP   phosphoenolpyruvate

PGA   phosphoglyceric acid; pteroylglutamic acid

PNA   pentose nucleic acid

PP   pyrophosphate

PPA   phosphopyruvic acid

PRPP   5-phosphoribosyl-1-pyrophosphate

R-1-P   ribose-1-phosphate

RNA   ribonucleic acid

TCA   trichloracetic acid; tricarboxylic acid

Tris   Tris (hydroxymethyl) aminomethane

TPN, TPNH   triphosphopyridine nucleotide; reduced triphosphopyridine nucleotide

TPP, ThPP   thiamine pyrophosphate

UDPG; UDPGal   uridine diphosphate glucose; uridine diphosphate galactose

UTP, UDP, UMP   uridine triphosphate, diphosphate, and monophosphate

SUBJECT INDEX

A

Abrine, 357

Acetaldehyde, and alcohol dehydrogenase, 45, 67, 68, 157, 160, 163, 164,
166, 167
and aldehyde dehydrogenase, 76, 77
and aldolase, 53
and carboxylase, 45, 64–67
and threonine aldolase, 317, 359
and triose phosphate dehydrogenase, 58
in deuterium labeling experiments, 160
oxidation-reduction couples, 166

Acetanilide, 217

Acetate, active, 70, 85, 138
activation, 86–88
from acetaldehyde, 76, 77
from acetyl phosphate, 85
from citrate, 133, 134
from glucose, 38
from lactate, 64
from lysine, 306
incorporation into fatty acids, 145, 146
in Thunberg condensation, 146, 147
in tricarboxylic acid cycle, 113, 114, 116
oxidation, 137
oxidation-reduction couple with
acetaldehyde, 166
reaction with cholinesterase, 375

Acetate-ATP reaction, 86

Acetate replacing factor, 73

Acethydroxamic acid, 87, 144, 375

Acetic acid (see Acetate)

Acetic thiokinase, 86, 140, 141

Acetoacetate
from acetoacetyl CoA, 85, 140
from aromatic amino acids, 344, 346
from fatty acids, 137–139, 151, 152
metabolism, 145
-succinyl CoA transferase, 88, 111

Acetoacetyl CoA, 85, 90, 140, 144

Acetoacetyl-succinic thiophorase, 140,
145

Acetoin, 38, 65, 66, 67

Acetokinase, 87, 88, 91

Acetolactate, 67

Acetol phosphate, 53

Acetone, 38, 137, 145

Acetylation
of amines, 70, 88–90
of choline, 70, 370
of enzyme, 9
of histamine, 339

Acetyl AMP, 87

Acetylcholine, 369, 370, 372, 373, 375

Acetyl CoA
and CoA transferase, 88
and thiolase, 144
and transacetylase, 85, 86
determination, 88
formation, 70, 73–77, 87, 139–141, 151,
152, 277, 357, 358
free energy of hydrolysis, 85
hydrolysis, 110
utilization, 70, 88–93, 108, 145, 147,
250, 298, 339, 369

Acetyl coenzyme A (see Acetyl CoA)

N-Acetylglucosamine, 50, 239, 240

N-Acetylglucosamine-1-phosphate, 249,
250

N-Acetylglucosamine-6-phosphate, 250

N-Acetylglutamic acid, 298

N-Acetylglutamic semialdehyde, 298

N-Acetylhyalobiuronic acid, 239, 240

ϵ-N-Acetyl-α-hydroxycaproic acid, 306

N^{α}-Acetylkynurenine, 352

S-Acetyllipoic acid, 74, 75

Acetylornithinase, 299

α-N-Acetylornithine, 298, 299

Acetylornithine δ-transaminase, 298

Acetyl phosphate and acetokinase, 87
and transacetylase, 86, 89
and triose phosphate dehydrogenase, 58
determination of, 87, 88
free energy of hydrolysis, 85
from citrate, 91
from pentose phosphate, 124, 125

403